HAZARDOUS MATERIALS

Regulations, Response, and Site Operations

Second Edition

Paul Gantt

Australia • Brazil • Japan • Korea • Mexico • Singapore • Spain • United Kingdom • United States

Hazardous Materials: Regulations, Response, and Site Operations, Second Edition
Paul Gantt

Vice President, Technology and Trades ABU: David Garza

Director of Learning Solutions: Sandy Clark

Managing Editor: Larry Main

Product Development Manager: Janet Maker

Senior Product Manager: Jennifer A. Starr

Marketing Director: Deborah S. Yarnell

Senior Marketing Manager: Erin Coffin

Marketing Coordinator: Shanna Gibbs

Director of Production: Wendy Troeger

Production Manager: Mark Bernard

Senior Content Project Manager: Jennifer Hanley

Art Director: Benjamin Gleeksman

Technology Director: Joe Pliss

Editorial Assistant: Maria Conto

Library of Congress Control Number: 2008930635

ISBN-13: 978-1-4180-4992-8

ISBN-10: 1-4180-4992-1

Delmar
5 Maxwell Drive
Clifton Park, NY 12065-2919
USA

Cengage Learning is a leading provider of customized learning solutions with office locations around the globe, including Singapore, the United Kingdom, Australia, Mexico, Brazil, and Japan. Locate your local office at: **international.cengage.com/region**

Cengage Learning products are represented in Canada by Nelson Education, Ltd.

For your lifelong learning solutions, visit **delmar.cengage.com**

Visit our corporate website at **www.cengage.com**

Printed in the United States of America
2 3 4 5 XX 11 10 09

CONTENTS

PREFACE

INTENT OF THIS BOOK

This book is specifically designed for HAZWOPER courses, but anyone whose work activities place him in potential contact with hazardous substances will benefit from reading this book. The book is written by a team of working professionals in both the public and private sectors. Through this experience, we have recognized that many of the concepts of safe handling, personal protection, and the implementation of safety systems are universal whether the work involves handling materials at a site, cleaning up the materials if they are released, or in bringing control to a chaotic situation where an emergency is created by the accidental or intentional release of a hazardous material. The book provides a broad range of easy-to-follow concepts for use in such areas.

WHY WE WROTE THIS BOOK

The world of dealing with site operations and emergency response activities where hazardous materials are found is quite complex and continually changing. Unfortunately, many of the textbooks and materials that are used to help train personnel who work in these fields are themselves complicated and confusing. As a result of this, personnel are often confused or may even have conflicting or incorrect information.

Because of this, our team began the process of providing a complete update to the first edition of the textbook that was published a few years ago. Since then, much has changed in the world of hazardous materials response and site operations. The dramatic events of September 11, 2001, and the anthrax scare that followed, showed that our systems of response must be ready for not just accidental releases of hazardous substances, but also for those that are intentional. Add to this the issues identified in the post–Hurricane Katrina responses in August 2005. Cleanup of sites where hazardous materials were released has now taken on an order of magnitude never recognized before.

During all this time our team has been developing response systems, implementing safety programs, investigating accidents, and presenting training programs to a host of groups on these topics. As we presented those systems and programs, we felt the need to keep things simple, for in simplicity comes a greater assurance that the response and activities would be conducted in the right way. We found that often more is not better, for more protection does not necessarily ensure the right amount of protection. So the text is written in such a way that it is easy to understand and the concepts are more easily implemented.

HOW THIS BOOK IS ORGANIZED

This book progresses logically through the various topics required for a HAZWOPER course, and focuses on simplifying the complex nature of this technical subject. Each chapter is presented in a straightforward manner, which is easy to understand.

- **Chapter 1** begins with the basics and highlights the regulations that guide the response to hazardous materials spills.
- **Chapter 2** documents the importance of hazard control in order to help reduce risk. It details the elements of each control and how to use this information to conduct self-analysis and provide protection from hazards.
- **Chapter 3** dives into toxicology and reveals the impact of various chemicals and other hazards on those who work and are involved in response to these materials.
- **Chapter 4** identifies the different hazardous materials classes and the physical hazards associated with each class.
- **Chapter 5** explains the various identification systems and how they can be an essential aid for workers and responders when dealing with hazardous materials.
- **Chapter 6** covers respiratory protection and the advantages and disadvantages of different classes in specific situations.
- **Chapter 7** outlines personal protective equipment—listing the various types and identifying the conditions that require certain levels of protection.
- **Chapter 8** steps the reader through decontamination. It covers the procedures and processes of this undertaking, and the factors that determine the extent to which decontamination is required for various situations.
- **Chapter 9** covers the associated hazards of working in confined spaces and the necessary steps to provide protection to those who work in these environments.
- **Chapter 10** explains the all-important aspect of air monitoring and how proper use of air monitoring equipment can help to save lives.
- **Chapter 11** provides an overview of site control, supervision, and incident management. Essential to the successful outcome of an incident, this chapter describes the function and necessity of the procedures in leading a response team.
- **Chapter 12** pulls everything learned in the previous chapter together by applying the various concepts to different scenarios. Included in this chapter are a step-by-step explanations for a clean-up scenario and emergency response to an unknown substance.

FEATURES OF THIS BOOK

Each chapter combines technical content along with learning features to help aid in the understanding of various concepts.

- **Case studies** depict actual events in the United States which required hazardous materials response. These cases tie into the concepts presented in the chapters, and explain the importance of learning and understanding these concepts for protection against hazards and the successful outcomes to incidents.
- **Key terms** define and explain the concepts for a more thorough understanding of the subject.
- **Notes** highlight important information that readers need to learn. These notes also serve a helpful tool for review.
- **Safety notes** outline information essential to the health and safety of workers and responders dealing with hazardous materials. These notes are aimed at protecting and preventing on-the-job injuries and deaths.
- **Bulleted summaries** outline important concepts learned in the chapters and serve as a helpful tool for review.

- **Review questions** tie into the learning objectives and pinpoint important concepts in the chapters. The questions are helpful for evaluating knowledge of the content in the chapter.
- **Activities** at the end of each chapter present various scenarios that require the application of content learned in the chapters. Presenting realistic response situations, these activities are essential for evaluating the skills required for response.

NEW TO THIS EDITION

While our first edition was successful in its approach to presenting a comprehensive overview of the topics, the second edition benefits from a lot more years of experience in watching those concepts at work. Like students, good teachers also learn as they gain experience and further expertise in their respective fields. During the time that the first edition was used, the team involved in the development of the second edition was also taking courses, learning from the best of the best. Our certifications and qualifications have expanded to include certification by the Board of Certified Safety Professionals, Office of the State Fire Marshal, and as EPA Certified Registered Environmental Assessors. Team members have also served as expert witnesses and have provided testimony in a large number of cases where accidents have resulted. When such accidents occur, every detail of the operation or response activity is subject to extreme scrutiny and significant learning can result.

This learning is also passed on in the development of the book. The extent of revision for this new edition was determined by these collaborative experiences, which resulted in some key additions:

- **New chapter on Hazard and Risk Assessment**, an important topic in HazMat response, focuses on reducing risk to protect workers and responders—both in public and private settings. This chapter also highlights the overlap between HAZWOPER response personnel and those responsible for homeland security.
- **Up-to-date regulations and standards**: this book serves as a guide to the policies that determine the actions of the hazardous materials responders. The book opens with an overview of these regulatory documents and articulates between the various levels of HazMat operations.
- **New technology** is introduced in this book, including the latest in air monitoring, to keep responders apprised of the most current equipment in the industry.
- **Current cases** and scenarios integrated into the chapters provide a realistic framework for the job of a hazardous materials responder—including hazardous materials, terrorism, and natural disaster response.

SUPPLEMENT TO THIS BOOK

We have created an instructor's **e.resource CD-ROM** that contains many helpful tools to prepare for classroom presentations and student evaluation:

- **Lesson plans** outline important concepts in each chapter and correlate to the accompanying *PowerPoint presentations*, creating a seamless set of plans for classroom instruction.
- **PowerPoint presentations** correlated to the accompanying *Lesson plans* combine chapter outlines with graphics and photos from the book to enhance classroom presentations. PowerPoint presentations are editable so that instructors may incorporate their own notes if they choose.

- **Quizzes** in Word format allow instructors to evaluate student knowledge of the concepts presented in each chapter. Quizzes are editable so that instructors may edit or add questions based on the needs of the class.
- The **Image Library** is available as an option for instructors. Organized by chapter and figure number, graphics and photos from the book may be printed out or used to supplement the *PowerPoint presentations* for the classroom.

Order#: 1-4180-4993-X

ABOUT THE AUTHOR

Paul W. Gantt is the current President of Safety Compliance Management, Inc. (SCM), and is one of the founding partners of the corporation. Paul has over 30 years of experience working in the emergency management field, serving in a variety of positions. His career includes over 15 years of working in the Fire Service, where he held a variety of positions from Firefighter through Fire Chief. He has also been actively involved in providing training for over 25 years to a large number of varied audiences.

While in the Fire Service, Paul worked as a Firefighter/Paramedic, Paramedic Operations Officer, Training Officer, Fire Captain, Battalion Chief, Fire Marshal, Deputy Fire Chief, Radiological Officer, and Fire Chief. He held these positions through his career, which included service in four different fire agencies. In addition, Paul was elected to serve on the Board of Directors for the California Fire Training Officers' Association and also served as its President. He is a Life Member in the Fire Training Officers' Association.

In addition to his experience in the Fire Service, Paul is a Certified Safety Professional (CSP) and Registered Environmental Assessor with California EPA, and works in the field of health and safety services, where he develops and delivers training programs, provides consultation to clients on areas involving regulatory compliance, and serves as an expert witness. He is in wide demand as a public speaker and is regularly invited to be a guest lecturer for a number of programs and audiences. As a trainer, he has taught at a variety of colleges and universities throughout California and has served as an instructor for the University of California Berkeley Extension programs. He is certified by the State of California Office of the State Fire Marshal, Governor's Office of Emergency Services, Federal Emergency Management Agency, California EPA, and the University of California to provide a wide array of training programs in areas that include hazardous materials handling and response, emergency preparedness and response, occupational health and safety, and regulatory compliance.

ACKNOWLEDGMENTS

When it comes to dealing with hazardous materials, whether as a site worker or emergency responder, there are no Lone Rangers. As we know, these materials can present a complex range of hazards to us, and no one person can effectively deal with these by himself. We are taught to use the buddy system as we enter areas where hazardous materials are present, to have a number of personnel serve as part of a backup team in our response efforts, and to involve others who have specific expertise in dealing with these materials when we handle them.

So too is the process of writing a comprehensive textbook covering site and response operations; even this is a team effort. No one person can possibly know all of what is required in such a broad and complex field. So in writing this book, it is appropriate to acknowledge the efforts of a team of people who have contributed to the second edition.

These people include a large number of the students to whom we have trained over the many years of teaching various hazardous materials topics. These students share their individual knowledge in class, knowledge that we then use in future classes. They also ask the relevant questions in these courses and identify topics that we included in this second edition. These additional topics help to complete the list of materials that make this edition even more complete than the first.

There is the team at our firm, Safety Compliance Management (SCM). The SCM team of instructors and trainers include those with expertise in site operations and who have worked in industries around the world. Rob Williamson is one of those whose broad knowledge of these types of operations has added to this edition. Another team member is Ron Gantt, whose expertise in regulations related to handling and transporting hazardous wastes has contributed to a more detailed discussion of these topics. Both of these professionals were very helpful in identifying the types of topics that should be explained as well as providing significant insight on how best to explain some of the more complicated items.

Also, there was the direct help of the team who worked directly to pull all of the pieces together. The graphics and pictures were largely the result of considerable effort of Melody Benedict, whose talent and experience in these areas provided the clarity of the artwork for this edition. And the head of the team was my wife Laura, whose tireless efforts led to the, albeit late, submissions of the materials to the team at Cengage Learning. Laura was the one who herded the cats to edit the manuscripts, coordinated the submittals, and packaged the materials together so that the experts at Cengage Learning could work their magic.

And then there are those who offered up their help to make available facilities and equipment. These include members of the San Ramon Valley Fire Protection District, Contra Costa County Health Services, and a number of clients of our firm. Without their help, the text would not have been completed.

The authors and publisher also wish to thank the reviewers who participated in reviewing the manuscript and offered recommendations for the new edition:

Richard Bahena
Instructor/Coordinator
El Paso Community College Fire Technology Academy
El Paso, TX

Corey Molinelli
President
Molinelli Consulting Services
Deltona, FL

Harold Richardson
Chief of Training/Safety Officer
Yarmouth Fire Department
Yarmouth, Nova Scotia

Tracy Rickman
Fire Coordinator
Rio Hondo College Fire Academy
Santa Fe Springs, CA

Jim Thomas
Corporate Safety Specialist
Leprino Foods
Denver, CO

HAZARDOUS MATERIALS REGULATORY OVERVIEW

Learning Objectives

Upon completion of this chapter, you should be able to:

- Identify the goal of the Occupational Safety and Health Administration (OSHA) and list examples of how it works to reach its goal.
- Identify the relationship between the following laws and the HAZWOPER regulation:
  - Resource Conservation and Recovery Act of 1976 (RCRA).
  - Comprehensive Environmental Response, Compensation, and Liability Act (CERCLA).
  - Superfund Amendment and Reauthorization Act (SARA).
- Identify the types of issues that led to the promulgation of the HAZWOPER regulation.
- Identify the major components of 29 CFR, Part 1910.120—Hazardous Waste Operations and Emergency Response (HAZWOPER).
- List the three major operations covered by the HAZWOPER regulation.
- Identify the requirements prescribed for each of the major operations covered by the HAZWOPER regulation.

CASE STUDY

In August 2005, several major storms struck the Gulf Coast of the United States, causing extensive damage due to winds and flooding. That area is filled with many of the nation's petrochemical facilities as well as a large number of other major industries where a range of hazardous substances are used. The storms, including Hurricane Katrina, wreaked havoc and resulted in the release of many hazardous substances into the waterways, local communities, and surrounding environment. Employees involved in the cleanup operations were faced with conditions that placed them at risk if they were not adequately protected. The risks included potential exposure to hazardous materials released by the disaster, reactions of materials brought together by flooding and storms, mold exposure due to the weather and environmental extremes, and a host of other conditions not normally encountered by workers.

The attacks on the World Trade Center building on September 11, 2001, created conditions in the surrounding area that had not been previously experienced. Harmful levels of dust, including various contaminants such as asbestos, silica, and lead, were present during the rescue, response, and cleanup efforts that continued for months. Workers at the site reported various symptoms months later, with many suffering from disease and related effects years after the work was completed. In late 2006, there were reports that nearly 70% of the workers involved in the rescue and recovery efforts were experiencing respiratory symptoms.

The aftermath of these recent disasters, one natural and the other man-made, revealed the need for regulations and procedures to prevent workers' exposure to hazardous materials and other harmful conditions. Unfortunate events such as these often become catalysts that are used by regulatory bodies such as the Occupational Safety and Health Administration (OSHA) and the Environmental Protection Agency (EPA) to help develop and promulgate regulations that protect workers, the public, and the environment.

While not perfect, the HAZWOPER regulation, which is the focus of this chapter, was also developed following a series of incidents involving the cleanup of contaminated waste sites around the United States. The application of the regulatory requirements from the HAZWOPER regulation has been shown to promote a much higher degree of safety in activities involving hazardous materials, whether the cause is natural or man-made. Even today, the application of this regulation extends to those operations described in the preceding paragraphs. The requirements and procedures prescribed in the regulation will promote a higher degree of safety and help reduce the serious consequences that can occur at the next disaster.

INTRODUCTION

HAZWOPER
the term used to describe the Hazardous Waste Operations and Emergency Response Regulation

The study that you are about to undertake is centered on the specific issues involved with the proper handling, operations, and emergency response activities associated with hazardous materials and wastes. At the core of this study is a lengthy regulation that was first introduced in the United States in March, 1989. It is called the **HAZWOPER** regulation (HAZardous Waste OPerations and

Emergency Response). This regulation significantly changed the way businesses and organizations conducted themselves in the handling of and in response activities associated with hazardous materials and wastes. It has had, and continues to have, a major impact on the manner in which we must deal with operations that fall within the scope of the regulation. This regulation, which we will use as the basis for this study, is found in **Title 29 of the Code of Federal Regulations** and is codified in Part 1910.120. A copy of the regulation at the time of publication of this text is contained in Appendix A. While the HAZWOPER regulation has not changed significantly since its inception, it would be wise to review any potential changes by reviewing the current version of the regulation on the federal Web site, http://www.OSHA.gov/.

Title 29 or 29 CFR
the section of regulation where federal OSHA Safety Regulations are found

Background on OSHA: The Occupational Safety and Health Administration

Before we study the topics associated with hazardous waste programs, we must understand the history that led to the regulation mandating specific rules and training requirements for hazardous waste handling and response programs. So, as we begin to study the major regulation that deals with the training requirements mandated for those who work in the worlds of hazardous materials response and site operations, we should first establish some of the background that will help us better understand the rules that we will be following in these fields. When we understand the rules and how they apply to the world of hazardous materials training regulations, we will be better able to select the appropriate information required by the **Occupational Safety and Health Administration (OSHA)**.

Occupational Safety and Health Administration (OSHA)
the branch of the federal government charged with developing regulations to promote health and safety in the workplace

OSHA was formed by an Act of Congress in 1970. The Act, called the Williams-Steiger Occupational Safety and Health Act, is more commonly known as the OSH Act. The Act established two new branches in the federal government for the purpose of providing a higher level of protection for the nation's workforce. Prior to the establishing of OSHA, the American workforce was composed of approximately 68 million workers. In a campaign speech in 1969, President Lyndon Johnson stated that it was a shame nearly 14,000 American workers died each year as a result of occupational accidents. Since that time, the number of workers has risen significantly, yet the number of occupational deaths has declined. As of 2004, the number of workers is estimated to be 115 million and the annual death rate has been reduced to approximately 5703. Quite a drop when we consider the significant changes that also have occurred over that period—changes which in some cases have involved the introduction of new materials and significant new hazards.

> **Note**
> OSHA was developed by an Act of Congress in 1970 to help reduce the death and injury rates in the nation's workforce.

National Institute of Occupational Safety and Health (NIOSH)
the agency in the federal system that is used to conduct research and make recommendations to OSHA for the development of regulations to protect the nation's workers

How did OSHA make such a decline in the death rates possible? Simply put, the OSH Act established a research arm in the federal government, the **National Institute of Occupational Safety and Health (NIOSH)**, whose mission was to assist OSHA in working to reduce death and injury rates. Each of these two federal agencies was placed in a different cabinet department in the federal government. The idea was that NIOSH would conduct research and make recommendations, and OSHA would take those recommendations and make regulations or rules, which OSHA would then enforce. The goal of the program was to reduce the occupational death rates, and clearly the statistics show that this program has been largely successful. However, not all of the recommendations made by NIOSH became regulations, since there must be a public hearing process to involve those who may be impacted by the new regulations. During the public hearing process, some recommendations are disregarded entirely or are subject to compromises to get them enacted.

State Plan State
a state or U.S. territory that has its own OSHA programs

Another area important in the understanding of OSHA and how it works is the **State Plan State**. Simply put, the OSH Act allows any of the individual states and territories in the United States to administer their own programs to ensure that every workplace is "free from recognized hazards that are causing or are likely to cause death or serious harm to employees" (29 USC 654, Section 5). Essentially, each jurisdiction has the option of either allowing federal OSHA programs to be implemented in the states and enforced by federal personnel, or for the individual state to develop its own *plan* to provide protection equal to or greater than what is required on the federal level. Many states and territories have done this. A list of the states and territories that had approved State-Plan programs in 2006 is contained in Appendix B. Updates to the information can be found on the federal OSHA Web site.

Because the focus of this text will be on one particular regulation created and enforced by OSHA, it would be wise to discuss this regulation and how it fits into what we know about OSHA and its various State Plan States. While a number of states have their own OSHA programs, it is also true that most states have adopted the federal HAZWOPER regulation in a form that is very close to the original. This is partly a tribute to the considerable effort that went into the development of the original regulations. Federal HAZWOPER regulations are the minimum for any particular state. State Plan States are allowed to adopt stricter regulations for application in their particular state. As you review the complexity of the regulation later in our study, you will see that it is a very expansive document with numerous areas of coverage. Even in California, which is a State Plan State with a very high level of environmental consciousness, the HAZWOPER regulation is similar to the federal version. Because of this, it is safe to use the federal version of the regulations as the basis of our study.

In understanding the state and federal role of OSHA, we must recognize that the primary purpose of any OSHA regulation is to provide a higher degree of workplace safety and to ensure that employees are adequately protected from workplace hazards. To meet its General Duty Clause goal that "each employer shall furnish to each of his employees employment and a place of employment which are free from recognized hazards that are causing or are likely to cause death or serious physical harm to his employees," OSHA has developed and implemented a number of rules and regulations which will overlap into our study of hazardous materials handling, operations, and response activities. These include the following program requirements that will be incorporated in our study of the various HAZWOPER program and training requirements:

■ Note
The primary purpose of OSHA, whether state or federal, is to develop a safety system that will provide a higher degree of safety in the workplace and to ensure that employees are adequately protected from the hazards associated with their workplace and job responsibilities.

- Provide competent inspection of each work site.
- Prohibit the use of unsafe equipment by employees.
- Require trained, experienced equipment operators to use various types of equipment.

- Instruct employees in hazard recognition and avoidance.
- Instruct employees in pertinent regulations and procedures.

As we conclude this overview of how the OSHA regulations apply to operations involving hazardous materials and wastes, we should understand that even OSHA has limitations. Perhaps the most important of these is that in selected cases, some groups of employers and their employees are not covered by the OSHA regulations, whether federal or state. This is mostly the case for those employees who are in public employment and who are paid by a governmental agency. Even in non-State Plan States, federal OSHA regulations are not always applicable to employees of the federal or state governments. Because this particular regulation was considered to be so important, the reach of it was extended by the federal Environmental Protection Agency (EPA), which mandated its use in one of its own regulations, which we will review a bit later in our study. This law is called the Superfund Amendment and Reauthorization Act (SARA). Because the EPA has jurisdiction independent of OSHA, the SARA law helped to extend the coverage of the HAZWOPER regulation to include other groups of employees who might not have been included given the limitation of the OSHA regulations to specific employee groups employed in the public sector. In effect, this regulation has one of the broadest reaches of any OSHA regulation because its reach has been extended to cover public employees who might otherwise not be covered by the OSHA regulation.

■ Note

The SARA law helped to extend the coverage of the HAZWOPER regulation to include other groups of employees who might not have been included because of the limitation of the OSHA regulations to specific employee groups who are employed in the public sector.

REGULATORY DRIVERS

In addition to the introduction of OSHA, several other key pieces of legislation were implemented in the 1970s. Many of these dealt with the issues of hazardous waste disposal that had long been practiced by industries in the United States. The list of these legislation is far reaching. While the following is not a complete list, it shows some of the major regulations that impacted the handling of hazardous materials and wastes and helped to set up the introduction of the HAZWOPER regulation that we will study. These regulations helped to drive the final rules OSHA adopted in the HAZWOPER regulation that became law in 1990. To help us better understand what OSHA was trying to do, let us look at the following list and examine a few of those regulations that more directly led to the HAZWOPER regulation, which is the basis of our study. Also note the significant increase in the promulgation of these types of laws or acts in the 1970s and 1980s.

1938—federal Food, Drug and Cosmetic Act

1948—federal Water Pollution Control Act

1953—Flammable Fabrics Act

1970—Occupational Safety and Health Act

1970—Poison Prevention and Packaging Act

1970—Clean Air Act

1972—Consumer Product Safety Act

1972—federal Insecticide, Fungicide and Rodenticide Act

1972—Clean Water Act

1975—Hazardous Materials Transportation Act

1975—Toxic Substance Control Act

1976—Resource Conservation and Recovery Act of 1976

1980—Comprehensive Environmental Response, Compensation, and Liability Act

1986—Superfund Amendment and Reauthorization Act

1989—HAZWOPER regulation

Note
Numerous environmental regulations helped establish the need for OSHA to develop regulations specifically designed to protect workers involved in cleaning up contaminated sites.

HAZARD COMMUNICATION STANDARD: 29 CFR, PART 1910.1200

Prior to the enactment of the HAZWOPER regulation, there were few regulations in effect that provided only limited protection to hazardous materials and waste handling workers. As we will see in our discussion of some of the regulations listed in the previous section, the U.S. government was actively participating in the oversight of many toxic waste sites cleanups. These were sites where hazardous materials and wastes had been mishandled and were causing significant environmental and human health concerns.

One such site was infamous Love Canal in upstate New York. Many people in the country had heard about Love Canal as significantly toxic. In fact, the result of the contamination was so extreme that the residents of the area were forced to move out so that trained workers could come in and try to clean up the toxic soup left behind from the mishandling of hazardous wastes at a chemical facility in the area. What was even more startling was that workers who undertook that project had little in the way of regulations to protect them. At this time, OSHA had not dealt with problems associated with site cleanup activities at places like Love Canal and several hundred others that had been identified. While the EPA had identified sites, mandated their cleanup, and even funded the cleanups with governmental money, workers were only protected with a minimal regulation, the **Hazard Communication** or Employee Right-to-Know regulation. Before we get into the details of the HAZWOPER requirements, let us quickly review the Hazard Communication regulation that was in effect.

Hazard Communication the OSHA regulation sometimes called the Employee Right-to-Know rule, which requires employers to make information available to employees on all of the hazardous substances in the workplace to which they may be exposed

The Hazard Communication regulation required employers to inform workers of the hazards of the materials in the workplace as well as detailed information on the safe handling of those materials. In most workplaces, this was a relatively simple job since a complete list of the chemicals used was easily obtained. However, the Hazard Communication regulation fell far short of protecting those engaged in cleanup activities at sites like Love Canal where contamination had occurred over a prolonged period and where the materials may have mixed together forming toxic soups.

Essentially, the Hazard Communication regulation has a number of requirements that we will see are not overly applicable to waste sites where the materials and their hazards are unknown. In examining the regulation, we will see the need for more protective requirements that OSHA ultimately put in the HAZWOPER regulation in its final form. Some of the major components of the OSHA Hazard Communication regulation, which is found in 29 CFR, Part 1910.1200, include:

- *Identify and list hazardous chemicals used in the workplace.* Before any definitive programs can be enacted in the workplace to protect the workers, it is essential that the hazards be first identified. This process must take place both at the onset of the program and on a regular basis to ensure that the materials identified as part of the program are maintained up to date with those used in the facility.
- *Obtain a **Material Safety Data Sheet (MSDS)** for each hazardous substance or material found in the workplace.* The law requires that each manufacturer or importer of the chemical provide a copy of the MSDS to those who use the product. The information contained on an MSDS is standardized, although the format varies significantly between manufacturers. A portion of a sample MSDS is found in Figure 1-1. Not all MSDSs look the same, and while the minimum information contained on them is standard, their formats are not. You may want to take some time and review several different MSDSs to fully understand the differences between them.

Material Safety Data Sheet (MSDS)
the written information on a specific hazardous substance or material that includes information on the health and physical hazards, signs and symptoms of exposure, proper handling procedures, and personal protective equipment necessary for the safe handling of the material

- *Ensure all hazardous chemicals are labeled properly.* Proper labeling of the material is critical if the employee is to recognize the material as hazardous and take the appropriate action. The labeling requirements include a variety of warnings and systems such as that used by the Department of Transportation (DOT), which we will review later.
- *Develop and implement a written Hazard Communication program.* Written programs are the foundation of many OSHA regulations. Written programs force each employer to commit to various actions required by OSHA and have the advantage of providing a mechanism whereby individual employees or labor organizations can monitor whether the activities listed are being carried out. After the items have been identified and an MSDS has been obtained for each item, the employer is required to develop a written plan to disseminate the information and provide for the maintenance of the program.
- *Train workers to recognize the hazards associated with chemicals used in the workplace.* The foundation of the training program is that workers not only have a right to know, but also have a right to understand. This understanding is best conveyed through a comprehensive employee training program about the information found on the MSDS, where the MSDS file is maintained, and selected terms and information found on the MSDS. While everyone agrees on the importance of such training programs, there is little or no agreement on the amount or type of training that is necessary to comply. This item is often cited as one of the main problems with the Hazard Communication regulation in that, while it specifies that workers should be trained, like most other OSHA regulations, it does little to specify the amount of time or the specific topics to be included in the training.

■ Note

While the Hazard Communication regulation requires that employees be trained on the hazards of substances in the workplace, it does not require a specific number of training hours.

Material Safety Data Sheet

Version 1.5
Revision Date 07/20/2004

MSDS Number 300000000004
Print Date 08/12/2007

1. PRODUCT AND COMPANY IDENTIFICATION

Product name : Argon

Chemical formula : Ar

Synonyms : Argon, Argon gas, Gaseous Argon, GAR

Product Use Description : General Industrial

Company : Air Products and Chemicals,Inc
7201 Hamilton Blvd.
Allentown, PA 18195-1501

Telephone : 800-345-3148

Emergency telephone number : 800-523-9374 USA
01-610-481-7711 International

2. COMPOSITION/INFORMATION ON INGREDIENTS

Components	CAS Number	Concentration (Volume)
Argon	7440-37-1	100 %

Concentration is nominal. For the exact product composition, please refer to Air Products technical specifications.

3. HAZARDS IDENTIFICATION

Emergency Overview

High pressure gas.
Can cause rapid suffocation.
Self contained breathing apparatus (SCBA) may be required.

Potential Health Effects

Inhalation : In high concentrations may cause asphyxiation. Asphyxiation may bring about unconsciousness without warning and so rapidly that victim may be unable to protect themselves.

Eye contact : No adverse effect.

Skin contact : No adverse effect.

Ingestion : Ingestion is not considered a potential route of exposure.

Chronic Health Hazard : Not applicable.

Exposure Guidelines

1/7
Air Products and Chemicals,Inc
Argon

Figure 1-1 *An example of a Material Safety Data Sheet. (Courtesy of Air Products and Chemical, Inc.)*

Material Safety Data Sheet

Version 1.5
Revision Date 07/20/2004

MSDS Number 300000000004
Print Date 08/12/2007

Primary Routes of Entry : Inhalation

Target Organs : None known.

Symptoms : Exposure to oxygen deficient atmosphere may cause the following symptoms: Dizziness. Salivation. Nausea. Vomiting. Loss of mobility/consciousness.

Aggravated Medical Condition

None.

Environmental Effects

Not harmful.

4. FIRST AID MEASURES

General advice : Remove victim to uncontaminated area wearing self contained breathing apparatus. Keep victim warm and rested. Call a doctor. Apply artificial respiration if breathing stopped.

Eye contact : Not applicable.

Skin contact : Not applicable.

Ingestion : Ingestion is not considered a potential route of exposure.

Inhalation : Remove to fresh air. If breathing has stopped or is labored, give assisted respirations. Supplemental oxygen may be indicated. If the heart has stopped, trained personnel should begin cardiopulmonary resuscitation immediately. In case of shortness of breath, give oxygen.

5. FIRE-FIGHTING MEASURES

Suitable extinguishing media : All known extinguishing media can be used.

Specific hazards : Upon exposure to intense heat or flame, cylinder will vent rapidly and or rupture violently. Product is nonflammable and does not support combustion. Move away from container and cool with water from a protected position. Keep containers and surroundings cool with water spray.

Special protective equipment for fire-fighters : Wear self contained breathing apparatus for fire fighting if necessary.

6. ACCIDENTAL RELEASE MEASURES

Personal precautions : Gas/vapor heavier than air. May accumulate in confined spaces, particularly at or below ground level. Evacuate personnel to safe areas. Wear self-contained breathing apparatus when entering area unless atmosphere is proved to be safe. Monitor oxygen level. Ventilate the area.

Environmental precautions : Do not discharge into any place where its accumulation could be dangerous.

2/7

Air Products and Chemicals,Inc Argon

Figure 1-1 (*Continued*)

Material Safety Data Sheet

Version 1.5
Revision Date 07/20/2004

MSDS Number 300000000004
Print Date 08/12/2007

Prevent further leakage or spillage if safe to do so.

Methods for cleaning up	:	Ventilate the area.
Additional advice	:	If possible, stop flow of product. Increase ventilation to the release area and monitor oxygen level. If leak is from cylinder or cylinder valve, call the Air Products emergency telephone number. If the leak is in the user's system, close the cylinder valve, safely vent the pressure, and purge with an inert gas before attempting repairs.

7. HANDLING AND STORAGE

Handling

Protect cylinders from physical damage; do not drag, roll, slide or drop. Do not allow storage area temperature to exceed 50°C (122°F). Only experienced and properly instructed persons should handle compressed gases. Before using the product, determine its identity by reading the label. Know and understand the properties and hazards of the product before use. When doubt exists as to the correct handling procedure for a particular gas, contact the supplier. Do not remove or deface labels provided by the supplier for the identification of the cylinder contents. When moving cylinders, even for short distances, use a cart (trolley, hand truck, etc.) designed to transport cylinders. Leave valve protection caps in place until the container has been secured against either a wall or bench or placed in a container stand and is ready for use. Use an adjustable strap wrench to remove over-tight or rusted caps. Before connecting the container, check the complete gas system for suitability, particularly for pressure rating and materials. Before connecting the container for use, ensure that back feed from the system into the container is prevented. Ensure the complete gas system is compatible for pressure rating and materials of construction. Ensure the complete gas system has been checked for leaks before use. Employ suitable pressure regulating devices on all containers when the gas is being emitted to systems with lower pressure rating than that of the container. Never insert an object (e.g. wrench, screwdriver, pry bar, etc.) into valve cap openings. Doing so may damage valve, causing a leak to occur. Open valve slowly. If user experiences any difficulty operating cylinder valve discontinue use and contact supplier. Close container valve after each use and when empty, even if still connected to equipment. Never attempt to repair or modify container valves or safety relief devices. Damaged valves should be reported immediately to the supplier. Close valve after each use and when empty. Replace outlet caps or plugs and container caps as soon as container is disconnected from equipment. Do not subject containers to abnormal mechanical shocks which may cause damage to their valve or safety devices. Never attempt to lift a cylinder by its valve protection cap or guard. Do not use containers as rollers or supports or for any other purpose than to contain the gas as supplied. Never strike an arc on a compressed gas cylinder or make a cylinder a part of an electrical circuit. Do not smoke while handling product or cylinders. Never re-compress a gas or a gas mixture without first consulting the supplier. Never attempt to transfer gases from one cylinder/container to another. Always use backflow protective device in piping. When returning cylinder install valve outlet cap or plug leak tight. Never use direct flame or electrical heating devices to raise the pressure of a container. Containers should not be subjected to temperatures above 50°C (122°F). Prolonged periods of cold temperature below -30°C (-20°F) should be avoided.

Storage

Full containers should be stored so that oldest stock is used first. Containers should be stored in a purpose build compound which should be well ventilated, preferably in the open air. Stored containers should be periodically checked for general condition and leakage. Observe all regulations and local requirements regarding storage of containers. Protect containers stored in the open against rusting and extremes of weather. Containers should not be stored in conditions likely to encourage corrosion. Containers should be stored in the vertical position and properly secured to prevent toppling. The container valves should be tightly closed and where appropriate valve outlets should be capped or plugged. Container valve guards or caps should be in place. Keep containers tightly closed in a cool, well-ventilated place. Store containers in location free from fire risk and away from sources of heat and ignition. Full and empty cylinders should be segregated. Do not allow storage temperature to exceed

3/7
Air Products and Chemicals,Inc
Argon

Figure 1-1 *(Continued)*

Material Safety Data Sheet

Version 1.5
Revision Date 07/20/2004

MSDS Number 300000000004
Print Date 08/12/2007

50°C (122°F). Return empty containers in a timely manner.

Technical measures/Precautions

Containers should be segregated in the storage area according to the various categories (e.g. flammable, toxic, etc.) and in accordance with local regulations. Keep away from combustible material.

8. EXPOSURE CONTROLS / PERSONAL PROTECTION

Engineering measures

Provide natural or mechanical ventilation to prevent oxygen deficient atmospheres below 19.5% oxygen.

Personal protective equipment

Respiratory protection	:	Self contained breathing apparatus (SCBA) or positive pressure airline with mask are to be used in oxygen-deficient atmosphere. Air purifying respirators will not provide protection. Users of breathing apparatus must be trained.
Hand protection	:	Sturdy work gloves are recommended for handling cylinders. The breakthrough time of the selected glove(s) must be greater than the intended use period.
Eye protection	:	Safety glasses recommended when handling cylinders.
Skin and body protection	:	Safety shoes are recommended when handling cylinders.
Special instructions for protection and hygiene	:	Ensure adequate ventilation, especially in confined areas.
Remarks	:	Simple asphyxiant.

9. PHYSICAL AND CHEMICAL PROPERTIES

Form	:	Compressed gas.
Color	:	Colorless gas
Odor	:	No odor warning properties.
Molecular Weight	:	39.95 g/mol
Relative vapor density	:	1.379 (air = 1)
Density	:	0.106 lb/ft3 (0.0017 g/cm3) at 70 °F (21 °C) Note: (as vapor)
Specific Volume	:	9.68 ft3/lb (0.6043 m3/kg) at 70 °F (21 °C)
Boiling point/range	:	-302 °F (-185.8 °C)
Critical temperature	:	-188 °F (-122.4 °C)

4/7

Air Products and Chemicals,Inc

Argon

Figure 1-1 *(Continued)*

Melting point/range	:	-309 °F (-189.3 °C)
Water solubility	:	0.061 g/l

10. STABILITY AND REACTIVITY

Stability	:	Stable under normal conditions.
Hazardous decomposition products	:	None.

11. TOXICOLOGICAL INFORMATION

Acute Health Hazard

Ingestion	:	No data is available on the product itself.
Inhalation	:	No data is available on the product itself.
Skin.	:	No data is available on the product itself.

12. ECOLOGICAL INFORMATION

Ecotoxicity effects

Aquatic toxicity	:	No data is available on the product itself.
Toxicity to other organisms	:	No data available.

Persistence and degradability

Mobility	:	No data available.
Bioaccumulation	:	No data is available on the product itself.

Further information

This product has no known eco-toxicological effects.

13. DISPOSAL CONSIDERATIONS

Waste from residues / unused products	:	Contact supplier if guidance is required. Return unused product in orginal cylinder to supplier.
Contaminated packaging	:	Return cylinder to supplier.

14. TRANSPORT INFORMATION

CFR

Proper shipping name	:	Argon, compressed

Figure 1-1 *(Continued)*

Material Safety Data Sheet

Version 1.5
Revision Date 07/20/2004

MSDS Number 300000000004
Print Date 08/12/2007

Class : 2.2
UN/ID No. : UN1006

IATA

Proper shipping name : Argon, compressed
Class : 2.2
UN/ID No. : UN1006

IMDG

Proper shipping name : ARGON, COMPRESSED
Class : 2.2
UN/ID No. : UN1006

CTC

Proper shipping name : ARGON, COMPRESSED
Class : 2.2
UN/ID No. : UN1006

Further Information
Avoid transport on vehicles where the load space is not separated from the driver's compartment. Ensure vehicle driver is aware of the potential hazards of the load and knows what to do in the event of an accident or an emergency.

15. REGULATORY INFORMATION

OSHA Hazard Communication Standard (29 CFR 1910.1200) Hazard Class(es)
Compressed Gas.

Country	Regulatory list	Notification
USA	TSCA	Included on Inventory.
EU	EINECS	Included on Inventory.
Canada	DSL	Included on Inventory.
Australia	AICS	Included on Inventory.
Japan	ENCS	Included on Inventory.
South Korea	ECL	Included on Inventory.
China	SEPA	Included on Inventory.
Philippines	PICCS	Included on Inventory.

EPA SARA Title III Section 312 (40 CFR 370) Hazard Classification:
Sudden Release of Pressure Hazard.

US. California Safe Drinking Water & Toxic Enforcement Act (Proposition 65)
This product does not contain any chemicals known to State of California to cause cancer, birth defects or any other harm.

16. OTHER INFORMATION

6/7

Air Products and Chemicals,Inc Argon

Figure 1-1 *(Continued)*

Material Safety Data Sheet

Version 1.5
Revision Date 07/20/2004

MSDS Number 300000000004
Print Date 08/12/2007

NFPA Rating

Health	:	0
Fire	:	0
Instability	:	0
Special	:	SA

HMIS Rating

Health	:	0
Flammability	:	0
Physical hazard	:	3

Prepared by : Air Products and Chemicals, Inc. Global EH&S Product Safety Department

For additional information, please visit our Product Stewardship web site at http//www.airproducts.com/productstewardship/

7/7

Air Products and Chemicals,Inc

Argon

Figure 1-1 (*Continued*)

Given the differences between a standard workplace where these provisions can be carried out and a waste site where the hazards could be hidden or where little is known, we can see why OSHA needed to enact the very restrictive HAZWOPER regulation. While it is true that the Hazard Communication regulation was in effect prior to the enacting of the HAZWOPER regulation, almost everyone involved in the hazardous waste industry worked under the belief that the basic provision of the Hazard Communication regulations did not provide adequate protection for those involved in hazardous waste operations. The rationale in this belief was the fact that there were no MSDSs for the materials involved in many of the cleanup operations because the materials were either unknown or a combination of two or more substances that had mixed at the site.

ENVIRONMENTAL REGULATIONS

The number of laws and regulations enacted to protect the environment from the effects of hazardous chemicals rapidly expanded beginning in 1970. As the list in the section Regulatory Drivers shows, regulations dealing with a variety of topics to protect people and the environment were developed at a fever pitch. While these regulations did a good job of increasing public awareness, it was not until most of the environmental laws were well in effect that OSHA was able to develop and implement its HAZWOPER regulation. Before we study that, it is important that we understand some of the other major laws and regulations that led to the promulgation of the HAZWOPER regulation.

Resource Conservation and Recovery Act of 1976 (RCRA)

Recognition of the hazardous waste issues in the United States and the need to clean up contaminated sites became very obvious in the late 1970s and early 1980s. While many regulations were developed to deal with various aspects of the problems, one of the earliest ones that had far-reaching effects is the **Resource Conservation and Recovery Act of 1976 (RCRA)**. RCRA (pronounced *rick-rah* or *reck-rah*) gave the EPA the authority to control hazardous wastes from "cradle to grave" by mandating that all organizations follow specific rules for the handling and disposal of all materials classified as hazardous wastes. The cradle-to-grave concept established that the operations that generate the hazardous wastes must be responsible for those materials through their entire lifespan. The rules define what is a hazardous waste; identify the programs required for the generation and handling of these materials at all sites; define transportation requirements for the wastes to be transported to designated sites; define the treatment, storage, and disposal of the hazardous wastes once they have reached the designated waste disposal sites. Essentially, the Act stopped the further generation and illegal disposal of wastes and helped control the environmental and human damage caused by improper handling and disposal of hazardous wastes.

Resource Conservation and Recovery Act of 1976 (RCRA)
the law that began the process of regulating the handling of hazardous waste; it defines what is a hazardous waste and outlines the programs that must be followed to properly handle, store, and dispose of these materials

In 1986, some amendments were made to RCRA that enabled EPA to address environmental problems resulting from leaks in underground tanks storing petroleum and other hazardous substances. Despite its reach, RCRA focused only on active and future facilities and did not address abandoned or historical sites; these sites were covered by later laws such as CERCLA and SARA.

Comprehensive Environmental Response, Compensation, and Liability Act (CERCLA)
an early law that authorized the collection of taxes to pay for the cleanup of abandoned hazardous waste sites; sites covered by this law were called Superfund sites

Comprehensive Environmental Response, Compensation, and Liability Act (CERCLA)

The **Comprehensive Environmental Response, Compensation, and Liability Act (CERCLA)**, more commonly called the Superfund Act, was enacted by Congress on December 11, 1980. This law taxed the chemical and petroleum industries and provided broad federal authority to respond directly to releases or threatened releases of hazardous substances that might endanger public health or the environment. Over 5 years, $1.6 billion was collected and went to a trust fund for cleaning up abandoned or uncontrolled hazardous waste sites.

Additionally, CERCLA established prohibitions and requirements concerning closed and abandoned hazardous waste sites, provided for liability of persons responsible for releases of hazardous waste at these sites and established a trust fund to provide for cleanup when no responsible party could be identified. The trust fund became known as the Superfund because it created a fund to be used to clean up specific waste sites.

The law authorizes two kinds of response actions:

1. Short-term removals, where actions may be taken to address releases or threatened releases requiring prompt response.
2. Long-term remedial response actions, which permanently and significantly reduce the dangers associated with releases or threats of releases of hazardous substances that are serious, but not immediately life-threatening. These actions can be conducted only at sites listed on EPA's National Priority List (NPL) of Superfund Sites.

CERCLA also enabled the revision of the National Contingency Plan (NCP). The NCP provided the guidelines and procedures needed to respond to releases and threatened releases of hazardous substances, pollutants, or contaminants. The NCP also established the NPL.

> **Note**
> The NPL is a list of contaminated sites that come under the Superfund cleanup program.

Superfund Amendment and Reauthorization Act (SARA)
a law enacted to add money to the Superfund cleanup program and to expand its scope in other key areas of hazardous materials safety

Superfund Amendment and Reauthorization Act (SARA)

CERCLA was amended by the **Superfund Amendment and Reauthorization Act (SARA)** on October 17, 1986. Its scope was broadly expanded and funding was increased. SARA helped to expand what previously had been an environmental cleanup focus to also include planning for handling emergency releases of hazardous materials and wastes. As part of this, the Act required reporting of hazardous materials releases, emergency planning by state and local agencies, establishment of State Emergency Response Commissions (SERCs), and the formation of Local Emergency Planning Committees (LEPCs). Additionally, the Act led to the formation of what is now termed the Emergency Planning and Community Right-to-Know Act.

Finally, SARA provided considerable guidance to governmental agencies by establishing specific requirements and programs. Some of these led to the development of the HAZWOPER regulation within the OSHA system. Other mandates include the following:

- Stressed the importance of permanent remedies and innovative treatment technologies in cleaning up hazardous waste sites
- Required Superfund actions to consider the standards and requirements found in other state and federal environmental laws and regulations
- Provided new enforcement authorities and settlement tools
- Increased state involvement in every phase of the Superfund program
- Increased the focus on human health problems posed by hazardous waste sites

- Encouraged greater citizen participation in deciding on how sites should be cleaned up
- Increased the size of the trust fund to $8.5 billion

THE HAZWOPER REGULATION

Considering the background on the specific issues faced by those involved in this broadening field, it became clear that OSHA needed to step in and develop programs to keep up with the rapid changes taking place. Because the Hazard Communication requirements fell short of protecting workers, OSHA developed what we now know to be the pinnacle of all regulations for workers employed in some facet of the hazardous waste industry, the HAZWOPER regulation. Now we will shift our attention from the general issues to the specifics of the HAZWOPER regulation. It is very complex, as we will see, and because of this took considerable time to develop.

The process started in the early 1980s. The Interim Final Rule, which contained most of the provisions now in effect, was issued on December 19, 1986. After some discussion, the Proposed Final Rule was issued on August 10, 1987. The final version of the HAZWOPER regulation went into effect on March 6, 1990.

> **■ Note**
> HAZWOPER was developed during the 1980s and went into effect in its final form in March, 1990.

In our study of the specific points contained in the regulation, it is probably best to look at it in its entirety, section by section. OSHA refers to these sections as "paragraphs," so we will do the same as we discuss them. To do this, we must understand the layout that is used in the codification process. Unlike the format used by most of us if we were to develop an outline, the federal regulations use a system of hierarchy that is as follows:

- Lowercase letters in parenthesis such as (a), (b), (c) are used to denote the main divisions or paragraphs.
- Subdivisions of the paragraphs or main divisions are denoted in the following order:
  - Arabic numbers such as (1), (2), (3) are used to denote the next major divisions.
  - Lowercase Roman numerals such as (i), (ii), (iii), (iv) are used to denote the next major subdivisions.
  - Uppercase letters such as (A), (B), (C) are used to denote the next major subdivisions.

As we begin to look at the regulation, we will explore the 17 major sections or paragraphs as well as some of the subsections contained in it. Because the major paragraphs are identified by lowercase letters, these 17 paragraphs are denoted by the letters (a) through (q).

> **■ Note**
> It is important to recognize that the format for the regulation is different from other standard formats. The 17 sections of the regulation are identified by lowercase letters from (a) to (q).

Paragraph (a): Scope, Application, and Definitions

Some of the 17 sections of the regulation contain multiple parts, such as the first. Paragraph (a) is one of those that is complex and covers three important areas that need to be understood prior to beginning to ascertain if the regulatory requirements contained in the HAZWOPER regulation apply to the operations in question.

Paragraph (a)(1): Scope One of the most important, and probably misunderstood, issues involving the HAZWOPER regulation is which types of operations this regulation is intended to cover. Imagine driving down the freeway and seeing a speed limit sign that applies to trucks or cars with trailers. Does that mean all vehicles must drive at that particular limit, or would the scope of that requirement not apply to all? Of course, if you are driving a car without a trailer, you would not need to comply with that requirement and could drive at a different speed since the posted speed limit does not apply to you. Conversely, truck drivers and the cars with trailers would need to comply or risk receiving a citation. Simply put, we need to know *if* the specific regulation applies to us and also *which* sections apply, because many times OSHA separates its requirements in subgroups, so not all groups are covered by the general regulation. Seems complex, doesn't it? But we will see it really isn't, as we consider the sections and explore them in detail. In doing this, we will discover what OSHA really had in mind for our operations. The following points from section (a)(1) detail the five groups of operations covered by the HAZWOPER regulation:

1. Cleanup operations required by a governmental body, whether federal, state, local or other involving hazardous substances that are conducted at uncontrolled hazardous waste sites (including, but not limited to, the EPA's National Priority List (NPL), state priority site lists, sites recommended for the EPA's NPL, and initial investigations of government identified sites which are conducted before the presence or absence of hazardous substances has been ascertained);
2. Corrective actions involving cleanup operations at sites covered by the Resource Conservation and Recovery Act of 1976 (RCRA) as amended;
3. Voluntary cleanup operations at sites recognized by federal, state, local or other government bodies as uncontrolled hazardous waste sites;
4. Operations involving hazardous wastes that are conducted at treatment, storage and disposal (TSD) facilities regulated by 40 CFR parts 264 and 265 pursuant to RCRA or by agencies under agreement with the U.S. EPA to implement RCRA regulations;
5. Emergency response operations for releases of, or substantial threats of releases of, hazardous substances without regard to the location of the hazard.

OSHA is stating that five types of operations are covered by the HAZWOPER regulation. If you are involved in any one of them, you will need to comply with the parts of the regulation that apply to those particular activities. If your activities do not involve one of the five listed in the HAZWOPER regulation, there is no legal need to comply with the HAZWOPER regulation. It is interesting to see how many people are confused by this. In fact, many regulatory bodies including OSHA inspectors and other regulators may not fully understand this and could give out the wrong information about the scope of the regulation. But the language is clear; later in our study, we will explore those groups to ensure that we fully understand who is covered, since this is one of the most important parts to the regulation.

As we look into the five groups listed, we find a similarity among the first three groups. These three groups are all engaged in some form of hazardous waste cleanup activities. These may be done at a Superfund site, at a RCRA site, or at a local site as a voluntary cleanup operation. Because of the similarity, OSHA groups these three types of cleanup workers into a single group and mandates programs for all of them equally. In the last two groups listed, TSD operations and emergency response activities are each separate groups and will have different rules that apply to each of them.

■ Note
The HAZWOPER regulation applies to five types of operations that involve three distinct groups.

Another important consideration is that while not specifically required, cleanup operations can still be mandated to other groups by regulatory bodies or industry practices. Many specific industries have adopted the training requirements of the HAZWOPER regulations as the minimum requirement for those who work in their specific industry. This is done despite the fact that training is not mandated, since that particular operation is not part of the scope of the regulation.

Take, for example, the case of the petrochemical industry. At a typical operating petroleum refinery, while it could be argued that there would be some hazardous waste generated and handled at the site, the site does not really fall into any of the listed categories noted in the scope of the regulation. It is certainly not a Superfund cleanup site, nor is it a site where there is an ongoing uncontrolled release of hazardous wastes, yet many petroleum refineries mandate that workers and contractors certify to the 40-hour HAZWOPER level as a **General Site Worker**. A General Site Worker is an employee who works at a cleanup site covered by the HAZWOPER regulation and whose work exposes or potentially exposes him to high levels of hazardous substances.

General Site Worker
an employee who works at a cleanup site regulated by the HAZWOPER regulation and whose work exposes or potentially exposes him to high levels of hazardous substances

The reason why a particular industry that uses significantly hazardous materials in its operations mandates HAZWOPER training is relatively easy to understand. If we reflect back on the original concerns that led to the promulgation of the regulation in the first place, we find that few other types of training provide information to workers on the hazards of the materials. Recall that the Hazard Communication regulation, which was the regulation that provided protection for workers when HAZWOPER was developed, is still one of the few regulations in effect today that mandates workers be trained on the hazards of materials that they might come in contact with. But remember too that the Hazard Communication regulation is very nonspecific in its language regarding training. It is vague at best in its training requirements, does not identify specific topics to be covered, does not list competencies to be demonstrated, and gives no mandated number of hours to ensure that the topic is adequately covered. In fact, some people *satisfy* the training requirement for this program by having their employees watch a video on the topic. Certainly this minimal training may not work for many people. Given the absence of consistent and in-depth OSHA-mandated training programs, many refinery workers and contractors are required to take the non-mandated HAZWOPER courses as a means of training to a recognized standard relative to the safe handling of hazardous materials and wastes. So the scope is far greater than the actual regulation that mandates it.

■ Note
HAZWOPER is a key industry standard for many types of operations, including those not covered by its scope.

Paragraph (a)(2): Applications The second subdivision in section (a), denoted as (2), defines the application of the regulation. In this case, the term *application* is used to further describe the specific areas of the regulation relative to whom they are intended to apply. Remember that there are 17 sections to the regulation and that the scope lists five groups of operations covered by the regulation. Earlier we identified that the five groups involve three distinct operations: cleanup; treatment, storage and disposal facility operations; and emergency response activities. So not all of what is contained in the regulation will apply to all groups equally.

In reading this portion of the regulation, we find that sections (a) through (o) apply to operations specified in (a)(1)(i) through (a)(1)(iii). That is, the majority of the regulation applies to those operations that are hazardous waste cleanup operations as noted above. As we look at those three groups, we find that they share the common theme of cleaning up waste sites as we discussed earlier. Each specifies a type of waste site where the work is conducted, but each is also involved in cleaning them up.

> **■ Note**
> The majority of the HAZWOPER regulations apply to cleanup operations involving hazardous materials or wastes.

Recognizing that sections (a) through (o) account for 15 of the 17 paragraphs of the HAZWOPER, it is clear that waste cleanup activities were the primary focus of the HAZWOPER regulation. As we know from our review of some of the environmental activities and laws that were being passed at the time of the development of the HAZWOPER regulation, it would make sense that this group was the primary focus for some time.

OSHA did include information and regulations related to other groups of personnel. If we continue to read the regulation, we will find that section (p) contains specific regulations that apply to the operations listed in section (a)(1)(iv). Recall that this section covers the operations conducted at licensed **treatment, storage and disposal facilities (TSDFs)**. OSHA correctly realized that once the materials were cleaned up at the waste site where the contamination and original disposal occurred, they would need to be transported to another site for further treatment, storage, or disposal by some method since that was now mandated by RCRA and other regulations. Remember that our earlier discussion of the RCRA regulation requires that wastes be handled from cradle to grave, and the grave in this case is the TSDF where the materials ultimately end up. Obviously, the workers at those sites would also potentially be exposed to the same hazardous substances which were covered by the early part of the regulation. So it makes sense that OSHA would cover these personnel with the regulation since they too handle these materials.

treatment, storage and disposal facilities (TSDFs)
regulated sites where hazardous wastes are taken for final disposal or treatment

> **■ Note**
> Workers at TSDF operations are exposed to hazards similar to those involved in cleaning up the sites, so OSHA extended the scope of the HAZWOPER regulation to include these workers.

The last paragraph of the HAZWOPER regulation, paragraph (q), is the section applicable to the operations discussed under section (a)(1)(v), workers engaged in emergency response activities. Therefore, if you are involved in emergency response operations, paragraph (q) is the section that applies to you, and most of the other parts of the HAZWOPER regulation do not apply unless specifically referenced in paragraph (q). As we will see later in our study, paragraph (q) references a number of other paragraphs and subsections such as medical surveillance. However, it is clear that the majority of the HAZWOPER regulation is not intended for those engaged in emergency response activities, and the requirements not found in or referenced by paragraph (q) do not apply to emergency response activities. This is perhaps one of the most misunderstood concepts of the regulation, but the facts are clear if you take the time to read and understand the regulation's content in terms of applicability. To help us better understand the intent of OSHA in terms of applying this emergency response section to real-world issues, federal OSHA issued a very comprehensive directive in 2005 that provides considerable guidance on applying the HAZWOPER emergency response section (q) to all phases of hazardous waste activities. A copy of this directive can be found on the OSHA Web site.

Note
OSHA requires that those engaged in emergency response activities be trained in accordance with section (q) of the HAZWOPER regulation.

One last item to review, as we discuss application, is that a number of overlapping OSHA regulations have some applicability to certain operations. The HAZWOPER regulation references many other OSHA regulations in its 17 sections. In all cases, it is important to recognize that if more than one OSHA regulation applies to a specific operation, OSHA requires that the more-restrictive procedures in the more-restrictive regulation apply. For this reason, you need a good understanding of not just HAZWOPER as codified in 29 CFR, Part 1910.120, but also other regulations that are referenced in this regulation. Examples of this could include operations involving work in trenches or excavations in the ground that are contaminated with hazardous materials. In such cases, not only the requirements of the HAZWOPER regulation would come into play, but also the protective systems required by the OSHA regulations concerning trench safety.

Safety
It is important to recognize that if more than one OSHA regulation applies to a specific operation, OSHA requires that the more-restrictive procedures in the more-restrictive regulation apply. For this reason, you need to have a good understanding of the various OSHA regulations that apply in each operation.

Paragraph (a)(3): Definitions One of the most useful things that OSHA does in this regulation is to use part of the first paragraph to define certain terms that are used in it. These are contained in paragraph (a), subsection (3). Unlike other sections, this subsection does not contain any further subdivisions. Instead, the terms are listed alphabetically. It is in this section that we will find the OSHA definitions used for terms such as *uncontrolled hazardous waste site* and even *emergency response*.

Note
Definitions of key terms are found in the HAZWOPER regulation.

Paragraph (b): Safety and Health Program

Paragraph (b) contains some important subdivisions and describes a very important aspect of hazardous waste-site cleanup operations. This paragraph mandates that all employers to which this applies must develop and implement a written safety and health program for their employees who are involved in hazardous waste operations. Remember that this section only applies to those operations covered under the scope as defined in (a)(1)(i), (ii), and (iii). These are hazardous waste cleanup operations that are the major focus of the HAZWOPER regulation and not TSDFs or emergency response operations.

In summarizing this section, we find it requires that a written health and safety program be implemented. The program must have the following *minimum* components:

- An outline of the organizational structure
- A comprehensive work plan for activities
- An outline of the training program
- An outline of the medical surveillance program
- A list of standard operating procedures
- A listing of the necessary interface between general program and site-specific activities
- A program to inform contractors and subcontractors of the hazards
- A list of the requirements for a pre-entry briefing
- Review procedures for the health and safety program
- The requirements for a site-specific Health and Safety Plan (often referred to as a **HASP**) that contains the following:
  - A hazard and risk analysis of the site
  - Specific training requirements for workers at the site
  - **Personal protective equipment (PPE)** requirements for operations at the site
  - Medical surveillance requirements for those operating at the site
  - Air monitoring requirements for the site
  - A site-control plan
  - Decontamination procedures
  - An Emergency Response Plan to be used in the event of an emergency
  - Confined-space entry procedures
  - A spill containment program and procedures

HASP
a Health and Safety Plan that is written for a specific cleanup activity required under the HAZWOPER regulation; it is broad in its scope and very detailed, covering the specific operations that must be conducted

personal protective equipment (PPE)
items worn by the individual to provide protection from a hazard in the workplace; examples include chemical resistive gloves, safety glasses, or some type of respirator

Incident Action Plan
a specific Emergency Response Plan that is used to identify the specific hazards and operations that must be conducted in the event of an emergency hazardous materials incident; it is different from the Site Safety Plan in that it is less detailed and deals only with the emergency phase of the incident

Again, the requirements of this section apply to operations that fall under the cleanup portion of the HAZWOPER regulation. Obviously, in an emergency situation, personnel would not take time to develop such a plan before they initiated actions to abate the emergency situation. Certainly, they will develop their individual **Incident Action Plan**, which is the specific Emergency Response Plan used to outline the activities to be taken in the event of an emergency response to a hazardous material. This topic will be presented in more detail in Chapter 10. The development of such a comprehensive HASP would bring all emergency activities to a halt for a considerable period.

Paragraph (c): Site Characterization and Analysis

This paragraph mandates something that should already be known to anyone working at a hazardous waste cleanup site. This is the need to evaluate the hazards present prior to any significant activity taking place. In the Paragraph, we find that minimum information regarding the site is required to be obtained and added into the site safety plans.

Among other things, the Paragraph requires that a preliminary evaluation of the site's characteristics be made prior to entry by a trained person to identify any potential site hazards and to aid in the selection of appropriate employee protection methods. Included in this evaluation would be the identification of all suspected conditions that OSHA terms **immediately dangerous to life or health (IDLH)**. The term IDLH will be discussed in a later chapter because it represents a level of atmospheric contamination in the workplace where only properly trained and protected workers can operate. The IDLH level of contamination is based on an amount that will harm workers who are exposed to it without the appropriate protective systems in place. So if a site contained a hazardous material at an IDLH level, the rules for working in those areas will by necessity be very strict to provide the high level of protection that is required.

immediately dangerous to life or health (IDLH) the level of exposure that would pose a danger to the life of the person exposed, would result in significant and irreversible health effects, or would render the person exposed unable to escape without assistance

This paragraph requires that personnel who are part of the initial evaluation of the site hazards be provided with appropriate levels of PPE. OSHA and EPA have adopted standard levels of protection based on the hazards present. These levels will be discussed in Chapter 7 when we discuss PPE requirement. The levels start with Level D, the least protective, and work up to Level A, the level that provides the highest amount of protection. At a minimum, the personnel who are part of the initial entry are required to have respiratory protection if a respiratory hazard is suspected, and that the level of protection provided must be at least to Level B.

Safety

Sites must be evaluated prior to the implementation of cleanup activities in accordance with paragraph (c) of the HAZWOPER regulation.

Paragraph (d): Site Control

Another key safety issue relating to activities at hazardous waste cleanup sites is that of access to the site, or more correctly, restricting access to the site. The regulation mandates that prior to any activity taking place, the site must be fully mapped out to determine the extent of the hazards, work zones must be established, and workers must be instructed in the use of the "buddy system" for activities taking place at the site. The term *buddy system* is one of those we find defined in paragraph (a) under the definitions section. It simply refers to the system of organizing personnel into work groups so that appropriate oversight and assistance can be obtained in the event of an emergency situation.

Note

Access to hazardous waste sites must be strictly controlled to avoid untrained personnel becoming exposed.

Paragraph (e): Training

One of the most important sections of the HAZWOPER regulation is paragraph (e), which covers the minimum training requirements for those engaged in cleanup activities. Because this is the section that causes most of us to undertake this course of study in the first place, it is important that we review in detail the requirements outlined in it. Then we will know what we can and cannot do with the training programs that are prescribed in this section of the regulation. Keep in mind that this is one of three areas where training is discussed in the HAZWOPER regulation. This paragraph covers cleanup activities, as we know from our earlier review. Paragraph (p), which covers TSDF operations, has its own training programs, and paragraph (q) has the training requirements that apply to emergency response personnel. In total, nine initial training levels are mandated by the HAZWOPER regulation covering three groups. Of those, eight have mandated refresher courses. Of the 17 courses, paragraph (e) contains the specific training mandates for five training programs including three initial training certifications and two refresher training programs.

> **Note**
> Training mandated by paragraph (e) covers those involved in cleanup operations and not other activities such as emergency response.

As we have previously discussed, prior to the issuance of the HAZWOPER regulations, there were few or no requirements specific to worker training for those persons who worked at hazardous waste sites. This resulted in a high potential for injury or exposure for those mostly untrained employees who worked at hazardous waste cleanup sites. Because Superfund money was used in many cases to fund the cleanup of such sites, the federal government felt compelled to develop minimum guidelines for activities at these waste cleanup sites. An overview of the specific training requirements for those persons who work at hazardous waste sites is given in the following section. What we see as we review is that there are three training levels specified: General Site Worker, Occasional Site Worker, and Supervisor.

General and Occasional Site Workers Training The General Site Worker classification is one of two classifications of workers who work at a waste site and for whom specific safety training programs are mandated. The regulation lists examples of those who qualify as General Site Workers and includes persons "such as equipment operators, general laborers, and supervisory personnel." Clearly, these are not the only ones who work at waste cleanup sites and might need to be trained in safety practices at such sites, but it provides an example of the types of workers who should be trained to this level. This level of training is the default level for those workers at a hazardous waste site unless the workers meet the criteria for the other level of training that is specified in this paragraph. That is the Occasional Site Worker level, which we will discuss later.

The term *General Site Worker* is descriptive in that it denotes those who are generally at the site where the cleanup operations take place and whose job responsibilities possibly place them in contact with the hazardous substances above the **Permissible Exposure Limit (PEL)** established by OSHA. The term PEL, which will be expanded on in Chapter 3, essentially represents the maximum safe level of exposure that a worker can have to a hazardous substance. Above this level of exposure, workers would be required to wear a specific PPE, including the use of some type of respirator. With this level of training, workers are authorized to use the various levels of PPE that provide protection while engaged in activities that are considered to be hazardous and could expose them to hazardous levels if they are not protected.

Permissible Exposure Limit (PEL)
the level of exposure established by OSHA that an employee can be exposed to on an average basis over an 8-hour work period; exposure above this level would require the use of some type of respiratory protection

Because of this potential exposure, training for certification as a General Site Worker is the highest level mandated in the entire HAZWOPER Regulation. It is required that the training be at least 40 hours with an additional 3 days of on-the-job supervised field training/experience. The need for the on-the-job experience is clear when one reviews the list of items that must be covered during the training. The list of required items for these training programs is as follows:

- Names of personnel and alternates responsible for site safety and health
- Safety, health, and other hazards present on the site
- Use of PPE
- Work practices by which the employee can minimize risks from hazards
- Safe use of engineering controls and equipment on the site
- Medical surveillance requirements
- Recognition of signs and symptoms that might indicate exposure to hazards
- Elements of the Site Health and Safety Plan, including:
    - Decontamination procedures
    - Site Emergency Response Plan
    - Confined-space entry procedures
    - Site-specific spill containment procedures

> **Note**
> Specific training topics are required for cleanup workers and are found in paragraph (e).

Topics such as "names of personnel and alternates responsible for site safety and health" or the "safety, health, and other hazards present on the site" could only be addressed during the site-specific training programs because that information would be specific to the site and perhaps even not known during the initial 40 hours of training. In fact, as personnel move to other sites, the original 40 hours of training is transferable, but the required 3 days of on-the-job training and monitoring must be repeated given that the site-specific hazards and information would be different.

> **Note**
> Site-specific training is mandated to ensure that workers are made aware of site-specific hazards and programs.

Occasional Site Workers those cleanup workers at a HAZWOPER-regulated site whose exposure to hazardous materials is below the established PEL for the material

Occasional Site Workers are identified in the regulation as employees who meet one of two criteria, each having to do with their possible exposure to hazardous substances that might be present. The first is the group who are at the site on only an occasional basis and whose potential exposure is below the established PEL. Such workers might include well drillers or engineers who only occasionally visit the site for a specific purpose and whose exposure levels are very low given their work responsibilities. It is from the designation as *only occasionally* on site that we derive the term Occasional Site Worker.

The second group of workers who fall in the Occasional Site Worker classification are those who work at a hazardous waste site on a regular or continuous basis, but whose exposure is not above the PEL. The regulations mandate that the sites be "fully characterized" and determined not to present any potential for exposure to the workers since the workers will be there on a continuous basis. At this level, the workers would not be required to wear respiratory protection, since the need for respiratory protection implies that an exposure level above the PEL is present. So Occasional Site Workers are only allowed to work in areas where the hazards are known and below the PEL for the materials.

> **Safety**
> Occasional Site Workers cannot work in areas where the level of exposure is above the PEL or where the use of a respirator is required.

Based on the lesser potential for exposure, the Occasional Site Worker training programs are less involved in both the initial training requirements as well as the on-the-job training. The programs are required to be at least 24 hours with an additional 1 day of on-the-job, supervised field training/experience. This is well below the level required for the General Site Worker, as we can see. While their potential exposure to hazardous substances varies significantly, both classifications of site workers are mandated to have a training program that contains the same basic elements. Because the General Site Workers course contains an extra 16 hours of instruction plus two additional days of on-the-job supervised field training/experience, the topics are covered in greater depth and focus on protecting these workers from the hazardous materials present. Topics such as the use of PPE and decontamination practices are emphasized because General Site Workers may be involved in such activities.

Once the workers are trained and certified, the regulations mandate that the training program be regularly refreshed and updated on an annual basis. The refresher time lines are true for either the General or Occasional Site Worker level in that each must receive an annual refresher training program of at least 8 hours per year. Once this refresher training program is complete, the certification is valid for another year. During this time the employee is required to receive another 8-hour refresher class in order to remain certified.

> **Note**
> HAZWOPER training is required to be refreshed every year.

Supervisor Training As we have seen, the training for both the General and Occasional Site Workers requires that each receive a certain amount of "supervised field experience" under the supervision of a trained and experienced Supervisor. The **Waste Site Supervisor** is another one of the levels of training mandated by the HAZWOPER regulation.

Waste Site Supervisor an individual trained and certified under the HAZWOPER regulation whose job includes oversight of other certified HAZWOPER waste-site cleanup personnel

In this case, Supervisors are required to undergo the same initial training program mandated for the workers they will supervise. Those Supervisors who will be engaged in supervising General Site Workers (employees who need the 40-hour course plus 3 days of field experience) would need to have completed the same 40-hour training and supervised field experience as those they are supervising. Supervisors of Occasional Site Workers would take only the 24-hour program and 1 day of field training/supervision initially, because they are supervising people whose exposure is expected to be minimal. As with the other site workers, Supervisors are required to have the supervised field experience prior to being certified and also are required to take the annual refresher training to maintain their certification as a site worker. However, no additional refresher training is necessary once certified as a Supervisor other than to the requirement to maintain their original General or Occasional Site Worker refresher programs.

■ Note
Waste Site Supervisors must maintain certification at either a General or Occasional Cleanup Worker level.

The actual training needed to qualify as a Supervisor following completion of the General or Occasional Site Worker program is an additional 8-hour program. The regulation detailing the training for Supervisors is not as specific as we saw in the training requirements for waste-site cleanup personnel. In this case, the regulation states that the Supervisor training program should cover:

> *such topics as . . . the employer's safety and health program and the associated employee training program, PPE program, spill containment program, and health hazard monitoring procedures and techniques.*

There is considerable latitude in this and because of that we find Supervisor programs that range from being very good to those that are far less effective. Figure 1-2 depicts students receiving instruction on the use of PPE required by the HAZWOPER regulation.

■ Note
Supervisor training is required to be at least 8 hours and to cover a range of topics.

One last reminder regarding the training requirements as outlined in paragraph (e) is that while the regulation is specific as to the types of personnel who are mandated to take the site worker training programs, it does not preclude other types of employees from taking this training program. Recall our earlier discussion that in many facilities, the 40-hour training class has become the standard for training employees who work in that industry. Such is the case in the petrochemical industry, the high-technology industry, and others, where we find that many firms require parts of their workforce to receive the 40-hour training. The students receive certification as General Site Workers even though they are not involved in activities at hazardous waste cleanup sites.

Figure 1-2 *Students receiving HAZWOPER PPE training.*

The rationale for this is that these workers are exposed to similar hazards on the job at the refinery or high-technology firm as found in waste cleanup sites. Many other industries have followed suit and require 40-hour certification for employees and contractors whose job responsibilities place them in close contact with hazardous materials and hazardous wastes.

Paragraph (f): Medical Surveillance

This paragraph outlines the requirements for medical examinations and medical surveillance programs for certain groups of workers covered by the regulations. In summary, paragraph (f) mandates that the following workers undergo annual medical examinations:

- Those who are exposed to levels of the chemicals above the PELs for more than 30 days per year
- Those who wear respirators for more than 30 days per year
- Members of **HazMat teams**: workers defined in paragraph (a)(3) as those who are part of an organized response team such as outlined in paragraph (q), which include the Hazardous Materials Technician and Specialist levels. In fact, paragraph (q) references this section as part of its requirements
- Those who are exposed to a hazardous substance and who exhibit signs and symptoms of exposure

HazMat team
those employees who are trained to respond to an emergency response involving hazardous substances and who will assume an aggressive role in stopping the release

Certain other groups of employees working in the hazardous waste field are also required to have medical examinations provided to them on an as-needed basis. This includes workers who are injured as a result of exposure to chemicals. In such cases, they would be seen by a physician who would determine the frequency of follow-up examinations, if any.

A final note on this section is that while medical examinations are required, there is no mandated type of examination that must take place. The regulation states that the examination be done on an annual basis and include the patient's medical and work history. This allows for a wide range of physical examinations based on the judgment of the physician. I am sure that we can all recall doctors who are very thorough, and others who do a cursory examination and take little time or have little real understanding of the issues involved. To provide some guidance, the paragraph does reference a number of standards that can be consulted, but in no way are any of these required.

■ Note

Medical evaluations are required for specific types of workers covered under the HAZWOPER regulation.

Paragraph (g): Engineering Controls, Work Practices, and PPE for Employee Protection

This paragraph outlines the manner in which OSHA mandates that employers control the hazards present in any workplace. In many of its regulations, OSHA requires that hazards be controlled in a particular order. Following is a summary of these "controls," as OSHA refers to them. We will discuss these items again as we introduce other hazards covered by the regulation.

- *Engineering controls.* This form of control is the most effective and involves the elimination of the hazard through the use of items such as ventilation systems to reduce airborne hazards, guardrails to reduce the possibility of falling, design of the workplace to eliminate ergonomic concerns, and so on.

- *Administrative controls.* This second form of controlling hazards involves using rules and work practices that reduce the potential exposure. Examples of this could include staying out of an area where loud equipment is in operation or rotating personnel in high-heat areas to reduce the potential of heat stress.
- *Personal protective equipment.* Recognizing that using PPE is the least effective means of protecting workers given its wide range of variables, OSHA requires that this level of control be used as a last resort when the other two means of controlling hazards are not fully effective in protecting workers.

As noted, a more complete discussion of these concepts will be given when we discuss hazard and risk assessment in later chapters.

Note

OSHA mandates that employers provide systems to eliminate hazards and/or control activities as part of the HAZWOPER regulation.

Paragraph (h): Monitoring

This paragraph covering monitoring addresses two major issues. These include establishing a monitoring plan for the environment or site and a plan to monitor the exposure of the personnel working at the site. Some of the items contained in these monitoring programs include requirements for determining the presence of a flammable atmosphere and whether the levels of an airborne contaminant reach the IDLH levels.

Safety

Monitoring the workplace and personnel exposure is a required element of a HAZWOPER program.

Paragraph (i): Informational Program

Another of the basic sections, and frankly one that seems to go without saying, is this paragraph's requirement for employers to provide employees with information regarding the hazards at the site. Much like the expanded Hazard Communication program that was discussed earlier, paragraph (i) states that the program must be in writing and that it must define the actual level of hazards present at the site and to which the employees may be exposed.

Paragraph (j): Handling Drums and Containers

Because many of the activities related to work with hazardous waste cleanup involves the use of drums and containers, the HAZWOPER regulations make specific requirements for the handling, movement, and transportation of such items. In this paragraph, we find requirements such as "unlabeled drums shall be considered to contain hazardous materials and handled accordingly" and "drums and containers that cannot be moved without rupture, leakage, or spillage shall be emptied" prior to moving. Figure 1-3 shows students practicing product transfer from one container to another.

Figure 1-3 *Students learning the proper method of transferring flammable liquids.*

> ■ **Note**
> Drum handling practices are outlined in section (j) of the HAZWOPER regulation.

Paragraph (k): Decontamination

decontamination the systematic process of removing hazardous materials from personnel and their equipment; it is necessary to reduce the potential for exposure to personnel and to minimize the spread of contamination from the site

The issues of **decontamination** are paramount in activities involving hazardous waste cleanup sites, TSDFs, and emergency response activities. In fact, this paragraph is referenced by both paragraph (p), which covers TSDF operations, and paragraph (q), which covers emergency response activities, making it applicable to all three types of activities covered by the HAZWOPER regulation. As we will study in Chapter 8, the process of decontamination is necessary in order to limit the exposure of those working at the site to the hazards present on the site, and to reduce the spread of the materials beyond the boundaries of the site. Implementation of a decontamination procedure is required before any employee or equipment leaves an area of potential hazardous exposure. Further, paragraph (k) requires that standard operating procedures be established to minimize exposure through contact with exposed equipment, other employees, or used clothing; and that showers and changing rooms be provided where needed. Figure 1-4 shows students training on the proper procedures involved in decontamination activities.

> ■ **Note**
> Decontamination is required to be performed as part of a HAZWOPER program.

Paragraph (l): Emergency Response at Uncontrolled Hazardous Waste Sites

Paragraph (l) is designated to cover the topic of an emergency occurring at a hazardous waste cleanup site. In this paragraph, we find the requirements to

Figure 1-4 *Students practicing decontamination activities.*

Emergency Response Plan
a plan that is developed in advance of an emergency situation that identifies the actions to be taken by all employees at the site in the event of an emergency

develop a written **Emergency Response Plan** for the handling of emergencies that could occur during hazardous waste operations. Such plans must address the following topics:

- Personnel roles
- Lines of authority
- Training and communications
- Emergency recognition and prevention
- Safe places of refuge
- Site security
- Evacuation routes and procedures
- Emergency medical treatment
- Emergency alerting

Keep in mind that these are the basic requirements for a waste site and that they are not in conflict with the requirements of paragraph (q), which describes the specific requirements for emergency response activities. In clarifying how emergency response is to take place, OSHA has provided guidance documents that state that all offensive emergency response activities regardless of location must be done in accordance with the requirements of paragraph (q).

> **■ Note**
> Basic duties to be performed in the event of an emergency by employees at a waste site is a component of the HAZWOPER regulation.

Paragraph (m): Illumination

Because not all cleanup sites are in areas with fixed electrical service and/or the work is done at night when lighting may be limited, this paragraph outlines the basic amounts of lighting required at hazardous waste sites. These requirements are consistent with other OSHA standards related to other types of work, including those done at construction sites.

Paragraph (n): Sanitation for Temporary Workplace

In this paragraph, we find the requirements regarding toilets, sleeping areas, and washing facilities for sites that are remote and that might not have access to such facilities given their location. If we assume that some of the sites that are contaminated and covered by the HAZWOPER regulation could be remote, these requirements would be very appropriate. As with the lighting requirements found in paragraph (m), the requirements for number of toilets and washing facilities are consistent with other OSHA regulations for other industries.

On reviewing the requirements in this section, we also find some that seem to be common sense, although OSHA makes a point of listing them. One example of this is found in paragraph (n)(3)(v), which lists the requirement that "doors entering toilet facilities shall be provided with entrance locks controlled from inside the facility" and that "employers shall assure that employees shower at the end of their work shift and when leaving the hazardous waste site."

> **■ Note**
> OSHA mandates that specific sanitation practices be used during hazardous waste operations.

Paragraph (o): New Technology Programs

Another of the small paragraphs covered by the regulation is the one involving new technologies. Here we find the requirement for employers to look for new ways to accomplish hazardous tasks in a less hazardous manner. As new technologies, concepts, or equipment become available, employers covered under the HAZWOPER regulation are required to review them for possible inclusion in site work plans.

Paragraph (p): Certain Operations Conducted Under the Resource Conservation and Recovery Act of 1976 (RCRA)

In the discussion regarding the scope of the regulation, we identified that the majority of the HAZWOPER regulation, 17 paragraphs or sections, is related to activities at hazardous waste cleanup sites. Paragraph (p) then is a transition from the requirements for waste site cleanup operations to those involving TSDF activities. It contains all of the requirements related to these operations in the single paragraph and references to some of the previous paragraphs. Paragraph (p) also makes those referenced items part of the requirements for the TSDF sites.

> **■ Note**
> Paragraph (p) covers the requirements for those who are working at a TSDF operation.

In some cases, paragraph (p) is a mini version of previously discussed items, such as training requirements for workers who work at TSDF sites, requirements for written programs and procedures, decontamination programs, new technology programs, and medical surveillance requirements that are related to activities at TSDF sites. In other cases, these requirements are significantly different from those for the waste cleanup workers.

An area where this is clearly the case is the identification of training requirements for employees at TSDFs. When we look carefully at these requirements, we find that they are similar to those for Occasional Site Workers, but are not actually the same. The training requirements for workers at TSDF sites are found in paragraph (p)(8)(iii). This states that the minimum training requirement for these workers is 24 hours. However, unlike the Occasional Site Worker requirement, there is no mention of a mandated number of hours of on-site training. Additionally, the type of training topics is not specifically addressed in as much detail as in paragraph (e) for the waste site cleanup workers.

Like the site cleanup workers, employees at TSDFs are also required to have an annual 8-hour refresher class. A commonly asked question in this area is whether the 8-hour refresher class for the waste site cleanup workers would satisfy the requirements for certification as a TSDF worker. In some cases, the training could be the same. However, depending on the specific operations, the refresher training may need to be different.

> **■ Note**
> Initial training for workers at a TSDF is required to be at least 24 hours.

Paragraph (q): Emergency Response to Hazardous Substance Releases

As with paragraph (p), paragraph (q) introduces a new group of workers to the HAZWOPER requirements. This section covers those workers who are engaged in emergency response activities to hazardous materials and wastes. Paragraph (q) addresses a number of areas, including a detailed description of how to handle emergency responses, the requirement to prepare and implement an Emergency Response Plan, the need to use the **Incident Command System (ICS)**, training requirements for the five levels of response personnel identified in the paragraph, and procedures to clean up spilled materials following the completion of emergency response activities. Regardless of whether the response activities are conducted by a governmental organization or a private company, the requirements apply to all workers who have a role in the emergency response program.

Incident Command System (ICS) a standardized tool to help manage an emergency response; OSHA mandates the use of the ICS in handling emergency responses to hazardous materials incidents

> **■ Note**
> Paragraph (q) provides the requirements of employees engaged in emergency response activities.

In Chapter 11, we will discuss the details of the ICS and some of the other systems that are often used to manage emergency response operations. After the original development of the OSHA regulation and its requirement to use the standardized ICS, a variation of the system titled the National Incident Management System (NIMS), has now been developed for use throughout the United States. As we will learn, this system, and others that are used at various state levels, are based on the concepts created in the original ICS, making the use of these systems in compliance with the OSHA requirement to use the ICS for emergency response operations.

Because of its reach, anyone engaged in emergency response activities should take time to thoroughly review this section. It begins with a lengthy list of things that must be done in any emergency activity. A list of some of the specific requirements includes the following:

- *The senior response official must be in charge of a site-specific Incident Command System (ICS) and all emergency responders and their communications shall be coordinated and controlled through the individual in charge of the ICS assisted by the senior official present for each employer.*

- *The individual in charge of the ICS shall identify, to the extent possible, all hazardous substances or conditions present and shall address as appropriate site analysis, use of engineering controls, maximum exposure limits, hazardous substance handling procedures, and use of any new technologies.*
- *Based on the hazardous substances and/or conditions present, the individual in charge of the ICS shall implement appropriate emergency operations and assure that the personal protective equipment worn is appropriate for the hazards to be encountered.*
- *Employees engaged in emergency response and exposed to hazardous substances presenting an inhalation hazard or potential inhalation hazard shall wear positive pressure self-contained breathing apparatus while engaged in emergency response, until such time that the individual in charge of the ICS determines through the use of air monitoring that a decreased level of respiratory protection will not result in hazardous exposures to employees.*
- *The individual in charge of the ICS shall limit the number of emergency response personnel at the emergency site, in those areas of potential or actual exposure to incident or site hazards, to those who are actively performing emergency operations. However, operations in hazardous areas shall be performed using the buddy system in groups of two or more.*
- *Back-up personnel shall stand by with equipment ready to provide assistance or rescue. Qualified basic life support personnel [properly trained, advanced first-aid support personnel or those with a higher level, such as an Emergency Medical Technician or Paramedic], as a minimum, shall also stand by with medical equipment and transportation capability.*
- *The individual in charge of the ICS shall designate a safety official, who is knowledgeable in the operations implemented at the emergency response site, with specific responsibility to identify and evaluate hazards and to provide direction with respect to the safety of operations for the emergency at hand.*
- *When activities are judged by the safety official to be an IDLH condition and/or to involve an imminent danger condition, the safety official shall have the authority to alter, suspend, or terminate those activities. The safety official shall immediately inform the individual in charge of the ICS of any actions needed to be taken to correct these hazards at the emergency scene.*
- *After emergency operations have terminated, the individual in charge of the ICS shall implement appropriate decontamination procedures.*

HAZWOPER regulation, 29 CFR 1910.120, paragraph (q)(3)(i)

■ Note

Specific emergency response practices and training requirements are mandated in section (q).

Emergency Responder Training Another of the more important areas of paragraph (q) relates to the training requirements that emergency response personnel must receive before they are allowed to engage in emergency response activities. When we review this topic, we find detailed information regarding the levels of training and what is required at each level, but none regarding which level applies to which specific groups of workers. Specifically, the regulations cite that the training "shall be based on the duties and function to be performed by each responder of an emergency response organization." This leaves the responsibility for determining the level of training that will be provided to the specific workers up to the employer. The employer is charged with the responsibility of determining what action they want their employees to undertake in the event of a release of hazardous materials and then to train and certify them to that level. This makes sense since the regulations state that it is the employer who is certifying its employees to perform the tasks. While there is a responsibility for the trainers to provide the training, the regulation is clear that it is the employer who *so certifies* the employee to respond at a particular level.

> **Note**
> Training for emergency response personnel is based on the duties that employees are expected to perform.

Despite this being a little unclear as to who gets what level of training, there are some general standards that are commonly followed throughout the United States. During the discussion of the specific training levels, common organizations are listed that generally certify their employees to the specific levels.

Once it is clear what the employees are expected to do in the event of an emergency, the regulations mandate that training be provided to one of five responder levels:

Level 1	First Responder Awareness (FRA)
Level 2	First Responder Operations (FRO)
Level 3	Hazardous Materials Technician
Level 4	Hazardous Materials Specialist
Level 5	On-scene Incident Commander

In looking at the specific requirements for each of these levels of training, we find that Levels 1 through 4 are placed in order of progressively increasing knowledge and job responsibilities. For example, to be certified and to function at Level 3 requires that you have been trained to and have the knowledge of the previous two levels. In other words, the levels build as you go up. The exception to this rule comes into effect at Level 5. Personnel certified to Level 5 do not have to have completed all of the levels below them. In this case, they are mandated to be trained up through Level 2 and then receive additional training to certify them to Level 5.

To help us understand the basic levels of responders, what they can do, and what training is required, we must take time to review the regulation in some depth. It is extremely important that responders understand which of the levels they have been trained and certified to, since this will dictate the level of involvement in a hazardous materials release.

First Responder Awareness (FRA) level the first of the five levels of responder found in the HAZWOPER regulation; employees certified to this level are trained to recognize hazardous materials emergency incidents and to activate the Emergency Response Plan for the site

First Responder Awareness Level The first level of emergency response training is the **First Responder Awareness (FRA) level**. While termed a responder, they are not really *first responders* as the term implies because their primary role is not to respond to an emergency scene, but rather to identify the incident as an emergency and leave the area. Their actions do not involve a response in the traditional sense, but rather a reaction to the incident.

The regulation specifies that this level is designed for those employees who are:

> *likely to witness or discover a hazardous substance release and who have been trained to initiate the emergency response sequence by notifying the proper authorities of the release.*

They would take no further action beyond notifying the authorities of the release. As we can see, this level of certification is severely limited in that the individuals at this level are no more than someone who is *aware* of the spill. What do they do with this awareness? They simply make others with more training aware so that more highly trained personnel can conduct an emergency response.

In studying the regulation for this group, we find that there is no required number of hours for the training of those eligible for certification to this level. If you recall, this is consistent with the manner in which OSHA deals with almost all of its regulations such as the Hazard Communication regulation that we discussed earlier. In almost every other type of training, OSHA does not list minimum hours of training required for most of the programs that they regulate. This is the case with this first level of responder as well. Instead, OSHA lists a number of areas to which the student must *objectively demonstrate competency*. In the case of the FRA level, these include the following:

- *An understanding of what hazardous substances are and the risks associated with them in an incident.*
- *An understanding of the potential outcomes associated with an emergency created when a hazardous material is released.*
- *The ability to recognize the presence of a hazardous substance in an emergency.*
- *The ability to identify the hazardous substances, if possible.*
- *An understanding of the role of the First Responder Awareness individual in the employer's Emergency Response Plan, including site security and control, and the U.S. Department of Transportation's Emergency Response Guidebook.*
- *The ability to realize the need for additional resources and to make appropriate notifications to the communications center.*

HAZWOPER regulation, 29 CFR 1910.120, paragraph (q)(6)(i)

If we look at these, we find that this is a significant list of items to cover in a training program. While it is true that there is no set number of hours listed for certification to this level, most training programs designed to provide this certification range from 4 to 8 hours in their attempt to meet the listed competencies. Remember, however, that because there are no listed hours for this level of certification, you might find some employers who certify their staff on the basis of a very minimal program composed mostly of watching videotapes.

The next question regarding this level of training is just who might benefit from this level of certification. Some of the examples are obvious and include employees who work in areas where hazardous materials are used, stored, or transported. This could include forklift drivers in a warehouse, shipping and receiving personnel at a site where chemicals are used, or almost any worker in a typical facility where chemicals are routinely handled. Another group of personnel who often receive this training are some of the more traditional emergency response personnel including police officers, Emergency Medical Services (EMS) personnel (paramedics and emergency medical technicians), and Public

Works or other governmental employees whose job places them in areas where chemical spills are possible.

> **Safety**
> FRA-trained personnel are only trained to recognize the emergency and call for more highly trained personnel to respond.

> **Note**
> FRA training is often provided to police and EMS personnel.

First Responder Operations (FRO) level
the second level of responder under the HAZWOPER regulation; it defines the training requirements for those personnel who respond to a release of hazardous substances and whose job involves the defensive containment of those materials; no contact with the hazardous substance is allowed with this level of training

defensive
actions conducted from outside the hazardous areas and are limited to protecting nearby areas from the effects of a release of a hazardous substance

First Responder Operations Level The second level of emergency response training is called the **First Responder Operations (FRO) level**. The FRO level of training is designed for those workers who respond to an incident from an area outside of the spill. Once at the scene, personnel at this level are expected to initiate some type of action to minimize the effects of a spill or release of a hazardous material. As the regulations state, these workers:

> *respond to releases or potential releases of hazardous substances as part of the initial response to the site for the purpose of protecting nearby persons, property, or the environment from the effects of the release. They are trained to respond in a* **defensive** *fashion without trying to stop the release. Their function is to contain the release from a safe distance, keep it from spreading and prevent exposures.*

So the key term in all of that is the word *defensive*. The regulation gives considerable guidance on this by specifying that they work at a safe distance and not approach the point of release, as other more highly trained personnel might.

If we think about some examples of the types of actions that someone might initiate which are defensive, we could list things such as turning off the flow of a gas or liquid from a safe area, evacuating an area that is potentially threatened by a chemical release, or putting down materials to keep a spill of material from entering the environment. In the course of their activities, they may have to use some types of PPE; however, this is an area where they get close to overstepping their boundaries since they are not expected to ever enter an area where they might contact the spilled material. In this case, the PPE is only to serve as a backup in the event that the material does go beyond where it is expected to be. The FRO level personnel are to use distance as their primary source of protection. PPE is secondary protection for them. It could be argued that if they are relying on the PPE for protection, they are too close and might be operating outside of the regulations since they might be in an area where there is the potential for an exposure.

Unlike the Awareness level, the regulations do specify a minimum number of hours for certification to this level. The regulations specify that the training programs be at least 8 hours and that the students "objectively demonstrate competency" in the following areas in addition to those listed for the Awareness level:

> **Note**
> FRO training is mandated to be at least 8 hours and covers specific competencies.

- *Knowledge of the basic hazard and risk assessment techniques.*
- *Know how to select and use proper personal protective equipment provided to the First Responder Operations level.*

- *An understanding of basic hazardous materials terms.*
- *Know how to perform basic control, containment, and/or confinement operations within the capabilities of the resources and PPE available with their unit*
- *Know how to implement basic decontamination procedures.*
- *An understanding of the relevant standard operating procedures and termination procedures.*

HAZWOPER regulation, 29 CFR 1910.120, paragraph (q)(6)(ii)

Although the regulations specify that this level of training must be at least 8 hours, this may not be enough time to demonstrate competency in the areas listed, as is required by the regulations for all groups who are trained to this level. For this reason, many programs that provide this training require more than the minimum number of hours. It is not unusual to find programs up to 24 hours in length for training to the FRO level.

What groups of employees are typically trained to the FRO level? Considering that this is for those who respond to the release and initiate defensive actions, we might conclude that this group could include Fire Department personnel who respond from outside the site, or Emergency Response Team members who are part of the site emergency response organization at the facility.

The key point to remember for personnel certified to respond at this level is that their response efforts are composed of defensive actions that will reduce the impact of the spill on people or limit the spread of the material in the environment. Their actions should not involve product contact.

Hazardous Materials Technician
an individual trained to the third level of HAZWOPER emergency response training and whose job function involves an aggressive/offensive response to a release of a hazardous material; personnel at this level are trained to select and use appropriate chemical protective equipment that will allow them to approach a release for the purpose of stopping the release

> **■ Note**
> The FRO level training is the level regularly provided to Fire Department personnel

Hazardous Materials Technician Level The **Hazardous Materials Technician** level is designed for employees whose job is to respond to a release of a material and do what is necessary to correct the problems encountered. They are trained in dealing with spills at almost the highest level and are protected with the appropriate level of personal protection that would be necessary to handle the types of emergency situations encountered by the release. The regulations state that they are individuals who:

> *respond to releases or potential releases of hazardous substances for the purpose of stopping the release. They assume a more aggressive role than a First Responder Operations level in that they approach the point of release to plug, patch, or otherwise stop the release of a hazardous substance.*

offensive
actions allowed to be performed by Hazardous Materials Technicians include entering the hazardous areas with appropriate levels of protection, rescuing exposed personnel, and stopping the release of the hazardous substances

Offensive actions are allowed to be conducted by personnel with this level of training and include entering an area where the material is present and leaking; identifying the hazards; taking samples for later analysis; performing field analysis; stopping the flow of material using various plugging, patching, or containment techniques; and even cleaning up or neutralizing the material to make it safe. Certainly, entering an extremely hazardous area to rescue someone would best be done by personnel having this level of training and certification.

Unlike training at the previous two levels, there are no restrictions relative to actions at hazardous materials releases for personnel certified to this level. They can wear the full range of chemical protective equipment and perform all types of offensive activities in accordance with the training program that they received.

As with the FRO level, there is a minimum number of hours of training required for this level of certification. The regulations specify that personnel who are certified at this level receive at least 24 hours of training, of which 8 hours shall be equivalent to the FRO level, and that additionally they have competency in the following areas:

> **■ Note**
> Technician training is mandated to be at least 24 hours and includes specific competencies that must be demonstrated.

- *Know how to implement the employer's emergency response plan.*
- *Know the classification, identification, and verification of known and unknown materials by using field survey instruments and equipment.*
- *Be able to function within an assigned role in the ICS (Incident Command System).*
- *Know how to select and use proper specialized chemical personal protective equipment provided to the Hazardous Materials Technician.*
- *Understand hazard and risk assessment techniques.*
- *Be able to perform advanced control, containment, and/or confinement operations and rescue injured or contaminated persons within the resources and personal protective equipment available with their unit.*
- *Understand and implement decontamination procedures.*
- *Understand termination procedures.*
- *Understand basic chemical and toxicological terminology and behavior.*

HAZWOPER regulation, 29 CFR 1910.120, paragraph (q)(6)(iii)

Again, while there is a minimum number of hours specified for certification to the Technician level (24 hours), meeting the competencies listed for this level might require a much more extensive training program. Unless the training program is designed for some site-specific hazards of a given industry, and only covers dealing with a specific range of hazardous materials (such as response to ammonia in an ammonia refrigeration system), it would be reasonable to see that far more than the minimum 24 hours of training could be needed. In fact, for many hazardous materials response team members whose job requires them to respond to a range of emergency situations involving both known and unknown materials, it is easy to see that a program of only 24 hours would be insufficient. For this reason, programs that certify personnel to this level generally range from 40 hours to as many as 240 hours.

> **■ Note**
> Many Technician level training programs exceed the minimum number of hours based on the types of duties and materials expected to be encountered.

Again, it is not OSHA that mandates the type of employees who receive this level of certification. What we find when we look at groups with this certification is that they are employees who are part of a HazMat team. In some locations, this involves members of the site Emergency Response Team for a particular business or operation. In other cases, public safety agencies such as fire or police personnel receive this training.

Hazardous Materials Specialist
the fourth level of emergency response personnel in the HAZWOPER regulation; the Hazardous Materials Specialist's actions are similar to those of the Technician level, but may also include training on the handling of specific materials and interaction with outside agencies

Hazardous Materials Specialist Level The fourth level of response training is the **Hazardous Materials Specialist**. While this level is listed separately in the regulations, in practical terms there is little difference in actual practice between this and the Hazardous Materials Technician level. In many cases, Specialists and Technicians function interchangeably in hazardous materials emergencies. Many agencies often do not have anyone certified at the Specialist level and choose to certify their personnel only to the Technician level since there is little that the Specialist can do that the Hazardous Materials Technician cannot. However, for organizations whose employees respond to a limited number of materials (such as a worker in an ammonia refrigeration facility), this level of training is ideal in that it allows employees to specialize in response activities to the specific materials present at their sites.

The regulation states that the Specialist level is for:

> *individuals who respond with and provide support to Hazardous Materials Technicians. Their duties parallel those of the Hazardous Materials Technician. However, those duties require a more directed and specific knowledge of the various substances that they are called upon to contain. The Hazardous Materials Specialist also acts as the site liaison with federal, State, local and other governmental authorities in regards to site activities.*

■ Note
Duties of the Hazardous Materials Specialist are similar to those of the Technician. They can include a more focused training on specific materials.

As with the previous two levels, there is a minimum number of hours required for certification to this level and the participants must have competency in a number of areas as outlined. The regulations specify that the training program shall be at least 24 hours, equal to the Technician level, and have the following competencies in addition:

- *Know how to implement the local emergency response plan.*
- *Understand classification, identification, and verification of known and unknown materials by using advanced survey instruments and equipment.*
- *Know of the state emergency response plan.*
- *Be able to select and use proper specialized chemical personal protective equipment provided to the Hazardous Materials Specialist.*
- *Understand in-depth hazard and risk techniques.*
- *Be able to perform specialized control, containment, and/or confinement operations within the capabilities of the resources and personal protective equipment available.*
- *Be able to determine and implement decontamination procedures.*
- *Have the ability to develop a site safety and health control plan.*

- *Understand chemical, radiological, and toxicological terminology and behavior.*

HAZWOPER regulation, 29 CFR 1910.120, paragraph (q)(6)(iv)

Once again, we face the problem of minimum hours specified versus the actual time it takes to meet the competencies outlined for the Hazardous Materials Specialist level training. As with the Technician, if the persons being certified are working with a limited number of known materials, the programs can come close to the minimum hours listed by the regulations. On the other hand, if the persons receiving the training are to be part of a regional HazMat team with multiple responses to unknown types of spills, the training programs can and should be much longer.

It is a fact that many of the programs that certify personnel to this level range from 100 to 400 hours, depending on the complexity of the duties that Specialists will perform and on the range of chemicals that Specialists will be expected to handle.

Because the job responsibilities are similar, the types of personnel who receive this level of certification are similar to those listed for the Technician. In this group, we find members of HazMat teams from both the private and public sectors.

On-scene Incident Commander
the fifth level of emergency responder under the HAZWOPER regulation; these individuals direct the activities of all emergency response personnel at the scene

On-scene Incident Commander The final level of training specified for emergency response is the **On-scene Incident Commander** level. This level is for those who "assume control of the incident scene beyond the FRA level." In many cases, this is the role of the local law enforcement or fire safety agency that is responsible for public health and safety in a given jurisdiction. In the case of those working at a particular site, it is generally management personnel or key Emergency Response Team personnel.

Safety
An On-scene Commander is required when response activities are conducted above the FRA level.

The training required for certifying personnel to this level is to be at least 24 hours long, be equal to the FRO level, and in addition have demonstrated competency in the following areas:

- *Know how to implement the employer's incident command system.*
- *Know how to implement the employer's emergency response plan.*
- *Know and understand the hazards and risks associated with employees working in chemical protective clothing.*
- *Know how to implement the local emergency response plan.*
- *Know the State emergency response plan and the federal Regional Response Team.*
- *Know and understand the importance of decontamination procedures.*

HAZWOPER regulation, 29 CFR 1910.120, paragraph (q)(6)(v)

It should be noted that this last level of certification does not require completion of all of the levels below this. Command level personnel are only required to receive training to the FRO level and may never have been exposed to any type of experience involving chemical protective clothing. They are the managers of the incident, and not the technicians. Because of this, they must often rely on personnel trained to the Technician or Specialist level to help them implement their command and control system.

> **Note**
> On-scene Command training is required to be at least 24 hours and includes training to the FRO level plus demonstration of specific competencies.

Refresher Training: Emergency Responders One of the major problems that creates confusion with the training programs for emergency response personnel is the lack of specified refresher training requirements. Unlike the specific requirements for site worker training programs and TSDF personnel, the emergency response certification levels do not mandate any number of refresher training hours. Essentially, it is up to the employer to determine the type and duration of response-oriented refresher training. The regulations specify for all five emergency response levels that personnel shall receive annual refresher training of "sufficient content and duration to maintain their competencies, or shall demonstrate competency in those areas at least annually." How this is accomplished is entirely up to the employer, and the range is quite broad.

In some cases, employers may choose to enroll their employees in a formal recertification class. Because the regulations do not mandate any specific number of hours for each of the levels, these recertification classes vary significantly depending on who is providing the programs. In other cases, formal drills and exercises may be a part of the recertification process, or the employer might simply administer a written examination or give credit to the employees for the responses that they were involved with in the previous year. There is considerable leeway in the regulation, and each employer must make their own individual determination based on their own situation. Generally, programs ranging from 8 to 24 hours are given to the various responders above the FRO level.

Regardless of the manner selected to certify or recertify personnel, the importance of documentation cannot be overstressed. Because such a range of topics and methods can be included in a recertification program, it is vital that the employer make careful notes as to the method and manner chosen. The regulation specifies that employees may be recertified based on training or a demonstration of competency. It requires that:

> *a statement shall be made of the training or competency, and if a statement of competency is made, the employer shall keep a record of the methodology used to demonstrate competency.*

> **Note**
> Refresher training requirements for emergency response personnel are more flexible than those for the other two major types of HAZWOPER training.

Determining Emergency Response Levels In deciding which of the levels to use in a particular situation, it is often wise to employ a series of questions and answers to help determine the appropriate training levels needed. Following is a short summary of the regulation that helps to establish some parameters as to the level of training required. The response dictated by the following questions will help establish the level of training required by the HAZWOPER regulation.

- When a material is released, I expect my workers to immediately leave the area and to not take any action to limit the spill. They should simply leave the area, initiate an evacuation of the area, and notify the appropriate authorities.

  - If this is the level of response expected, the personnel should be trained and certified to the FRA level.

- When a material is released, I expect my workers to initiate actions to limit the spill of the material without becoming contaminated by it. They should not expose themselves to any of the material at any time. If possible, I would like them to try to minimize the spill by diking the material and keeping it from getting into storm drains, by turning off valves to stop the flow of materials if it is safe to do so, and to initiate evacuation efforts.
  - If this is the level of response expected, the personnel should be trained and certified to the FRO level.
- When a material is released, I expect my workers to initiate whatever actions are needed to limit the size of the spill, reduce the consequences of the spill, and clean the spill up when appropriate. These workers will be provided with the required PPE (respirators and personal protective clothing) to safely work with the materials, and to approach the point of release to control it without becoming contaminated by it.
  - If this is the level of response expected, the personnel should be trained and certified to either the Hazardous Materials Technician or Specialist levels.
- When I have a spill of material in my facility, I expect someone from my organization to oversee the operation and the handling of that spill through its cleanup. This person would be someone on my staff who would know how to establish a system to safely manage such an incident and provide a command and control system for the safe handling of the incident.
  - If this is the level of response expected, the personnel should be trained and certified to the Command level.

> **■ Note**
> The employer must determine what they want their employees to do and provide the appropriate level of training and certification.

SUMMARY

- During the 1970s and 1980s, a number of laws came into effect that involved the proper handling, disposal, and treatment of hazardous wastes. Three of the most notable were the Resource Conservation and Recovery Act of 1976 (RCRA); the Comprehensive Environmental Response, Compensation, and Liability Act (CERCLA), and the Superfund Amendment and Reauthorization Act (SARA).
- The federal Occupational Safety and Health Administration (OSHA), which came into effect in 1970, has primary jurisdiction for employee safety and health throughout the United States. The OSH Act, which established OSHA, allows states to have their own version of OSHA in the state. Twenty-six states and territories that have their own versions of OSHA are called State Plan States.
- The Hazard Communication regulation, enacted in the 1970s, outlines basic worker safety training for workers who are exposed to potentially hazardous materials in the workplace. It is limited in its ability to provide protection during waste site cleanup activities and does not specify any number of hours of training needed.

- Because of the limited effectiveness of the Hazard Communication regulation, the Hazardous Waste Operations and Emergency Response (HAZWOPER) regulation was enacted in March, 1990.
- HAZWOPER is, without a doubt, one of the most far-reaching regulations ever developed by OSHA. It covers a wide range of activities related to hazardous materials and waste handling, and even when not required, it provides a recognized standard of training and has become a standard in many industries.
- The HAZWOPER regulation covers three distinct areas. These include cleanup work at uncontrolled hazardous waste sites; operations at a treatment, storage and disposal facility (TSDF); and emergency response to hazardous materials releases.
- Training for each of the covered operations is contained in the HAZWOPER regulation. There are a total of nine different levels and types of training required. Eight of the nine have mandated refresher training requirements.
- Hazardous waste cleanup site workers receive training in accordance with their expected job functions. This includes General Site Workers, who receive 40 hours of training plus 3 days of on-the-job training; Occasional Site Workers, who receive 24 hours of training plus 1 day of on-the-job training; and Supervisors, who receive site worker training and an additional 8 hours of training.
- TSDF workers are required to receive 24 hours of initial training and have an 8-hour refresher training class each year.
- Emergency response personnel receive training in accordance with their expected job functions. There are five levels: the First Responder Awareness level, First Responder Operations level, Hazardous Materials Technician level, Hazardous Materials Specialist level, and On-scene Incident Commander level.
- Training programs for emergency response personnel range from a mandated 8 to 24 hours. In practice, many programs can be over 240 hours.
- All workers covered by the HAZWOPER regulations are required to have annual refresher training. Refresher training programs for hazardous waste-site cleanup workers and those involved in TSDF operations are specified to be at least 8 hours. Emergency responder refresher requirements do not specify any number of hours for recertification or refresher programs.
- In addition to defining the levels of training, the 17 sections of the HAZWOPER regulation cover a range of topics such as requirements for Health and Safety Programs, decontamination practices, sanitation and lighting requirements, and medical surveillance requirements for groups covered under the HAZWOPER regulation.

REVIEW QUESTIONS

1. Briefly describe the primary mission of each of the following regulations.
   a. RCRA
   b. CERCLA
   c. SARA
2. What is the mission of OSHA?
3. List three components of a Hazard Communication program.
4. List the three major types of operations covered by the HAZWOPER regulation.
5. List the five levels of responders covered under HAZWOPER and identify their functions.
6. What are the differences between a General and an Occasional Site Worker?
7. List four sections of the HAZWOPER regulation and describe two requirements in each.
8. What is a State Plan State?
9. List the three types of controls mandated by OSHA.
10. Explain why the HAZWOPER regulation is often used as an industry standard.

ACTIVITY

HAZWOPER REGULATORY REVIEW

To help in the application of your knowledge of the regulation, review the following questions and use the regulation that is given in Appendix A to answer them.

You are the Safety Manager of a large fixed facility. (Use your own or one that you are familiar with if it will help.) Members of upper management are aware of your recent HAZWOPER training and have asked you to address the following questions. They expect that you will present the information at one of the upcoming staff meetings.

Read each question and refer to the HAZWOPER regulation. The regulation can be reviewed either through your online link or in Appendix A. Answer the questions and note the section where you found the answer. For example, the answer to question 1 can be found in Section (e)(5).

1. If our firm wanted to train and certify our people, could we do so? What are the qualifications of those who instruct the General and Occasional Site Worker HAZWOPER courses?
2. We have several persons who have worked with our waste products for a number of years. Can they be "grandfathered" into compliance with the training requirements? If so, what is the process?
3. We heard some rumors that a site supervisor is required at our hazardous waste site. Do we really need one, and if so, what, if any, additional training is required?
4. If we work with hazardous waste as part of a required cleanup, will we need to develop a site health and safety plan and what should be included in it?
5. If we were determined to be a hazardous waste cleanup site, would any of our employees be required to have physical examinations by a doctor? How often would they need this and what is involved in these physicals?
6. We understand that one of our employees has already had the 24-hour Occasional Site Worker course. If we need the 40-hour certification, does this person need to take the whole 40-hour class? What do we need to do to get him certified?
7. What level of training is required for those workers who respond to emergencies and who are allowed to approach the point of release in an effort to stop the release?

HAZARD AND RISK ASSESSMENT

Learning Objectives

Upon completion of this chapter, you should be able to:

- List the two types of hazards posed by hazardous substances.
- Identify the components of a hazard assessment.
- Identify the elements of a risk assessment.
- List the three levels of controlling hazards that are required to be used by the Occupational Safety and Health Administration (OSHA).
- List examples of each type of control and the limitations presented.
- Describe how the controls can be used as a system to provide protection.
- List the steps required in completing a job safety analysis.

CASE STUDY

A New England fire department was called to respond to an explosion at a facility in their jurisdiction. The facility where the emergency occurred used metallic sodium to manufacture components for high-technology equipment. Sodium metal is a highly hazardous material that is classified as a Flammable Solid. Once ignited, it presents a serious hazard in that it cannot be extinguished with water, results in extremely high temperatures, and requires special materials to extinguish it. The explosion resulted from sodium contacting water during the course of firefighting operations. As a result of the accident, 11 firefighters were injured and unable to return to duty as a result of the burns, hearing damage, and other medical conditions.

This accident illustrates many of the concepts of hazard and risk assessment that are key to the safety of those who work with or respond to incidents involving hazardous materials. Without a complete understanding of the hazards, personnel are often placed into conditions where the risks exceed acceptable levels and where unfavorable consequences such as those in the incident described in the previous paragraph occur. In some cases, the risks that are taken exceed the ability of the protective clothing to provide an acceptable level of safety. Such was the case in which the hazards of burning sodium made the risk associated with the response effort extreme. For this reason, OSHA requires that all emergency response operations include a thorough and accurate hazard and risk assessment before any activities are allowed to take place during emergency response efforts involving hazardous materials.

INTRODUCTION

Chemicals and other hazardous substances have been used for a variety of reasons for all of recorded history. In ancient Egypt, such substances were used to help mummify dead bodies, and information on the proper use of these materials was recorded and left in some of the tombs.

Nature itself manufactures a range of materials that find their way into plants, the air, and other parts of the environment. Some of these materials are beneficial and lead, for example, to the pleasant odors emitted from a rose as it opens and blossoms. Others are pesticides manufactured by plants to kill or hold off the range of predators that would otherwise attack and harm the plant. It is estimated that thousands of different natural pesticides are produced by plants themselves, a number that dwarfs the number of synthetic or man-made pesticides that often cause us concern. Today, in a growing industrialized world, there are tens of thousands of materials, both natural and synthetic, in use for a range of purposes. The proper handling, use, storage, and transportation of these materials warrants considerable study if we are to protect ourselves, others, and the environment from the effects of these materials.

> **■ Note**
> There are thousands of natural and man-made materials that can be considered hazardous.

As someone who will work with hazardous materials or respond to emergency situations created when there is a release or potential release of them, you must understand that the hazards fall into two distinct categories. The first category includes the ability of hazardous materials to elicit a health effect as a result of their toxicity if they are allowed to enter or contact our bodies. The second category is their ability to present a physical concern such as a fire or an explosion, as in the Case Study. It is only with a proper understanding of these concepts that we can do our jobs in a safe manner. Without this information, we place ourselves, others, and the environment at risk.

toxicology
the study of the adverse effects chemicals and other hazardous substances have on the human body

carcinogenic
carcinogenic materials are capable of causing cancer

Potential health effects resulting from exposure to hazardous substances can include a whole range of problems depending on the materials involved and a number of other factors that will be studied in Chapter 3, where the topic of **toxicology** will be explored in detail. Toxicology is simply the study of the adverse or harmful effects that chemicals and other hazardous substances have on the human body. These effects can include chemical burns to the skin or eyes, irritation or damage to the lungs causing severe coughing, contracting cancer from exposure to a **carcinogenic** substance, organ damage as a result of prolonged exposure to certain materials, depression of the central nervous system caused by breathing solvent vapors, and a host of other conditions. Often, the effects on the health of the worker are not recognized immediately as they build inside the body as a result of repeated exposure. A good illustration of this concept is the example of a frog placed in water that is slowly heated. The story explains that as the water temperature gradually increases, the frog is unaware of the growing hazard. It is not until the temperature is too high and damage has already been done that the frog is able to recognize the danger—only to find out it is too late. In many instances, exposure to hazardous substances can be like the frog in the heating water, where the results of the exposure are not noticed because they build up inside the body incrementally.

■ Note
Many health effects from chemical exposure build up slowly.

Physical hazards are often a lot more noticeable to those who work with hazardous materials. They can include fires that result from the ignition of flammable vapors, corrosion of equipment as a result of contact with a corrosive material, and explosions where significant physical damage occurs. In the case of physical hazards, the effect of the material is less directly with our bodies, and the damage is more the result of circumstances created by the materials than by contact with the material. Physical hazards will be studied in greater depth in Chapter 4. In some cases, the material can present both types of hazard. This is the case with many flammable gases that, in addition to presenting a fire hazard, also can harm us through their presence in the air that we breathe, resulting in suffocation as a result of the low oxygen levels present. Without an understanding of the hazards of a substance, we run the risk of placing ourselves in conditions that are too hazardous and that could result in exposure, injury, or even death. Figure 2-1 shows a response to a release of highly hazardous material.

As we know from the discussion of the intent of the OSHA regulations in Chapter 1, the topic of safety is one that must be integrated into all places of employment throughout the United States. In fact, many other countries adopt rules and regulations that help to provide a safe workplace for the employees. Many of these rules are patterned after the regulations developed in the United States by OSHA, or the OSHA regulations are taken to other countries by multinational corporations that do business in the United States and other countries. While

Figure 2-1
Personnel working at an emergency response situation involving high risk. (Courtesy of Los Angeles County Fire Department.)

it is impossible to write rules for every type of hazardous condition that could ever be encountered, there are some basic concepts of safety that permeate into the way we conduct business. It is through an understanding and application of these concepts that work involving the most hazardous of substances can safely be performed. In this chapter, we will identify the concepts and help in the application to specific types of operations involving hazardous materials.

Safety
Safety concepts must be applied regardless of specific rules mandated by OSHA.

In addition to the hazards posed by the materials themselves, the dawning of the 21st century ushered in additional concerns and issues that must be addressed by those who work with the range of hazardous materials in use today. The events of September 11, 2001, raised concerns not only about the effects of these materials as they are accidentally released, but also about the potential for an intentional release of various materials. Part of the thought process of all of those involved in industries where hazardous materials are used, stored, or transported must now include the security of these materials from potential use by others in harmful ways directed against people, operations, and/or the environment. Given the range of hazards presented by many types of commonly used materials, it is easy to see that such materials have the potential to cause significant harm. While some groups of materials such as radioactive, explosive, or biohazardous substances present greater concern than others, even commonly used fertilizers can become bombs in the hands of those who have a desire and determination to create terror or cause harm.

■ Note
Hazardous materials can be released intentionally in criminal acts or acts of terrorism.

Even when used and stored appropriately, natural disasters can also create significant catastrophic events when areas with major operations involving hazardous materials are impacted. Consider the storms that occurred on the Gulf Coast of the United States in the summer of 2005. In addition to the storm damage that commonly accompanies major hurricanes, the aftermath of the storms on the area caused significant problems due to the release of a range of hazardous substances found in the region. Such events are not isolated to any one area of the country, as we know, and events such as storms or earthquakes can create serious consequences because they may involve areas where hazardous materials are found.

> **■ Note**
> Natural disasters also present challenges when hazardous materials are released.

Because of this, it is imperative that those who work in the hazardous materials or hazardous waste industries have a thorough and complete understanding of the real dangers associated with the materials that are present. No longer can we simply assume that because things have not happened before, they cannot happen in the future. Planning for disasters, both natural and man-made, is now part of the process that all users of hazardous substances must do. Workers who handle with these materials will also be called on to think outside of the box about the consequences in such events because the potential for harm is great.

> **■ Note**
> Current safety programs must include discussion of the effects of both natural and man-made disasters on the hazardous materials present at the site.

However, while this may sound like an overwhelming task, consider that the materials and their properties do not change regardless of the circumstances. Chlorine gas has the same basic properties and hazards in every situation where it is found. Whether released as a result of an intentional act or as a consequence of a failure of a valve in a facility, the hazards that the material presents are the same. Given this, a thorough understanding of those hazards and properties is the focus of the next few chapters. To help you understand how hazards can affect workers, this chapter will introduce some extremely important safe handling and response concepts.

> **■ Note**
> Understanding the hazards of the materials present is the key to safety.

HAZARD AND RISK ASSESSMENT

Regardless of the type of material present, the hazards of that material, or the circumstances involved, a key component in any safety system is the hazard and risk assessment. The concept is so fundamental that it is found in almost every one of the required competencies for each of the nine levels of the Hazardous Waste Operations and Emergency Response (HAZWOPER) training. From workers involved in cleaning up sites contaminated by dumping hazardous wastes, to the processing of those materials at a treatment site, and through all aspects of emergency response, all who work with hazardous substances must be able to recognize and understand the hazards associated with their activities and take appropriate steps to ensure their safety. The process of assessing the hazards and taking the appropriate steps to protect employees is simply known as a **hazard and risk assessment**. This process incorporates the specific hazards of a hazardous material into the tasks that are required.

hazard and risk assessment
the process of determining the hazards of a particular material and deciding action that will be taken

Information on the hazards of the material is part of the site Hazard Communication program where the material is commonly found. The Material Safety Data Sheet (MSDS) for the material will provide the employee with information on the hazards, both toxicological and physical, and help to ensure that the necessary information for the safe use, handling, or transportation of the material is present. Determining the hazards is one of the first steps in the hazard and risk assessment process that involves potential exposure to hazardous materials.

> ■ **Note**
> Hazard and risk assessment is a process required by OSHA whenever hazardous materials are present.

Next, anyone who works with those materials must then consider how the hazards will present themselves in relation to his job functions or tasks when he is working with those materials. If the hazards are high, and there is considerable risk of exposure to the materials, that person will need to take the necessary steps to limit or minimize the harmful effects in order to avoid injury. The steps can include a range of protective actions, which will be discussed later in this chapter, and will likely be more complex as the hazards increase and the risk of exposure expands. This process in effect forms the basis for what OSHA terms a hazard and risk assessment. Without going through this process, workers who are involved in the operations run a higher risk of exposure than is necessary for themselves and others who might be present. So what are the steps required to conduct a hazard and risk assessment? The answer is actually quite easy to understand.

> ■ **Note**
> A Hazard Communication program is designed to both identify the hazardous materials present at the site and ensure that the information concerning the hazards is communicated to those potentially exposed.

hazard assessment
a process that identifies the potential harm a material presents

As we noted, the **hazard assessment** part of the process simply determines the types of hazard presented by the hazardous material(s) that are encountered. Take, for example, a material such as diesel fuel. The hazards of this material are very easily obtained from a range of reference sources including the MSDS. From reading the MSDS, we find that diesel fuel presents a very minimal health concern and has some limited potential for ignition and fire. The hazard assessment part of this process is relatively easy because the material is known, adequate reference data is available, and the data shows that the hazards present are low. The first step in the process is to identify the material present and then research all of the hazards that they present. Depending on a number of factors, some situations involving hazardous materials are more complicated than others.

For example, a spill in a facility where the materials present are well known makes the process of identification easy. In these cases, an MSDS for all substances used at the site is immediately available, helping to identify not only the material but also the actual hazards that are present. This is far different from an accident in which, for example, an overturned tank vehicle is leaking its cargo. In this case, there may be limited information on what material is leaking. Properly identifying what is present in some cases requires that we have considerable knowledge of identification systems. This is often done as the appropriate label or other identification system. Identification systems will be studied in Chapter 5. Without understanding the material present, the hazard assessment process becomes complicated, potentially resulting in need to follow a worst-case scenario protocol.

■ Note
Hazard and risk assessment begins with proper identification of the specific materials present.

Spills or releases of unknown materials where proper identification of the material is not possible severely limits our ability to design a program for working safely with the material and increases our risk when handling these materials. No one type of chemical protective clothing provides protection from all hazards, and the term *chemical-proof* relating to suits and clothing is inaccurate because such materials do not exist. Handling or responding to situations with unknown hazards requires the utmost care and limitations on the activities that are allowed. In such cases, all activities should be done in the most conservative manner and should be geared toward gaining information about the hazardous materials or properties involved.

! Safety
Without a complete identification of the material, the risks associated with handling and response increase significantly.

However, once the material is identified or known, a number of excellent reference sources can be used to describe the types of hazards present by the substance involved. The most complete source of information on most types of materials continues to be the MSDS. MSDS files can be either paper or electronic, and the information is provided in a relatively consistent format. Essentially, most of what is needed to assess the hazards of a particular substance is found on the MSDS.

Other useful sources of information on the material include standard reference books that often are designed for a particular group. The *Emergency Response Guidebook*, for example, is designed for use in transportation emergencies. The use and limitations of this book will be covered in Chapter 5. Other helpful materials come from the National Fire Protection Association (NFPA), which has several standards related to identifying chemical hazards. This includes NFPA 704 and other, older standards such as NFPA 49 and NFPA 325, standards that still contain some useful information even though they are not updated on a regular basis.

■ Note
Reference materials are an important part of the hazard assessment process in that they identify the specific hazards that materials pose.

Regardless of the reference materials used, specific information regarding the hazards and properties of the substances must be known before any action is taken. Will the material release flammable vapors that could be ignited by static electricity or through the use of tools such as electrical devices and have electrical sparks, or by a tool creating a spark as it strikes other metallic substances? Could it explode, and under what conditions? Will it cause health effects? Is it hazardous if inhaled or accidentally touched with our bodies? These are all questions that should be asked and answered as part of the hazard assessment. Additionally, the properties of the material must be understood as part of the hazard assessment. Will the vapors rise or sink once released into the atmosphere? Is the material able to mix with water if water is present at the site? Does it react with water or other materials that could be present? Again, the more you know of the material, its direct hazards, and its chemical properties, the better you will be identifying the risk that will be assumed as you do your job. A large portion of this book will deal with these issues and discuss the hazardous conditions presented by specific types of materials, and also help you gain a working knowledge of the properties and determine how they can impact the situation.

> ■ **Note**
> Part of the hazard assessment process is identifying the properties of the material present.

risk assessment
the process used to determine how the hazards will potentially harm those involved in the operation

Risk assessment is far less objective and involves some degree of judgment on the part of those involved. Unlike the hazard assessment part of the process, no reference books are available to consult to fully understand the risks associated when working with a particular hazardous substance. Often, one of the few sources of information available may be the worker's or expert's experience. And that experience may not always be the best information available because of the range of circumstances that could be present. In some cases, personal experience may be limited and may not be accurate, making the situation even more complex. So how then does one assess the risk once the hazards are known? The process often goes like this.

> ■ **Note**
> Risk assessment is the more subjective part of the process and involves a review of what actions or operations are expected to be conducted.

Once the material is properly identified, the hazard assessment is conducted using all of the available reference materials. Based on the hazard assessment, we have a lot of information available to us. If the hazards of the material are low, such as with diesel fuel, oils, or chalk, the risks associated with most activities involving these materials will be low, with a reduced potential for harm. Generally, when the hazards are low, the associated risk of working with them is low enough already and minimal protection would be required. We may simply ensure that we have good ventilation, appropriate protective equipment, and standard procedures for working with these materials because the potential harm is very low.

> ■ **Note**
> Materials that have a low potential for harm generally expose the worker to less risk.

On the other hand, if the hazard assessment reveals a list of significant hazards for the material, our risk will be high if we do not take serious steps to minimize it. Consider a release of chlorine gas, a material that can cause immediate harm to anyone exposed to it. In this case, because the hazards are significantly high, any activity that we conduct in the area of the release will expose us to a much higher risk, given the properties of the material involved. In such cases, the risk assessment looks at the activities that will be conducted, and ensures that the steps taken will not overly expose us to the hazardous conditions present. Knowing the type of hazard—health or physical—is an important step because there are systems that we can use to control the effects and minimize the risk of hazardous materials on workers. These systems are prescribed by OSHA for use in controlling the risk of working with highly hazardous materials. They help control the risks associated with specific materials or operations, and include a three-step process of implementing what OSHA terms **controls**. The controls are ways to eliminate the hazards present or control the potential for exposure and limit it to an acceptable level. It is through an implementation of one or more of the controls that we can minimize our risk when working with hazardous substances in any capacity.

controls
methods required by OSHA to eliminate the hazards or reduce the risk when working with hazardous materials

HAZARD CONTROL: REDUCING THE RISK

Whenever a hazard is present in the workplace, and employees could be exposed to the hazard, OSHA mandates that the employer responsible for the workplace eliminate or control the hazard through the use of one or more of the OSHA-prescribed controls. Essentially, the controls form the basis for the protection systems that are used to ensure employees are protected while performing various tasks in the workplace. In other words, these controls reduce the risk of harm that could be present if the hazards were not controlled. Because the use of hazardous substances presents a significant challenge in terms of protection, it is important that any study of hazards include a discussion of these controls.

In one sense, the term *safety* can actually be equated to the term *control*. Anyone who has ever felt a loss of control can certainly relate to that feeling of impending hazard or the lack of safety caused by that loss of control. Take, for example, the situation where someone walks across a slippery floor. Despite being careful, at some point the person may sense his traction is compromised and that he is starting to slip and will likely fall. This loss of control evokes a feeling of panic for just a second as he recognizes that he is no longer in control, and the effects of gravity will soon take over. As he recognizes the situation, he might try in vain to catch himself, and then realize that the fall is now inevitable. What takes place in only seconds plays out in the mind in a protracted way as he now looks for ways to limit the damage. He may reach out with his hands in an attempt to cushion the landing, but it is clear that control is lost and his safety is now at considerable risk. Others may see this and the embarrassment of the fall is the least of his worries. In some cases, injuries could result from the fall, and damage to equipment or property is possible. All of this is due to that momentary loss of control.

> **Safety**
> Loss of control often translates to higher hazards and less safety for the job being done.

Countless other examples could be cited to illustrate the point that loss of control compromises the safety of the people involved. Certainly, the ability to effect a level of control when working with hazardous materials is critical to reducing the risk associated with working with those substances. So in its efforts to provide protection for employees as they engage in operations in the workplace, OSHA mandates that the three-step process be undertaken to eliminate the hazards, reduce exposure, and provide barriers to protect the exposed personnel from the hazards. This is required whether the circumstance involves simply walking across a floor or responding to a hazardous substance release. The consequences of the loss of control may vary, but our need to provide protection remains the same. It is the application of the controls to each type of situation that will be presented as we begin the study of providing safety when working with or responding to incidents involving hazardous substances.

The three controls mandated by OSHA are engineering controls, administrative controls, and the use of personal protective equipment (PPE). On reviewing the three controls, we find that they are not necessarily designed to work independently. They work best when they are part of a *well-thought-out system* to minimize or eliminate the risk associated with specific operations. If we think of the controls as a system of shields, or layers of protection, we will better understand how they work to ensure our safety when working with hazardous substances or in any situation where there is a potential for harm. In all cases, the more layers we place between us and the hazards we face, the more protection we have as we encounter those hazards and the less risk we assume. More protection translates into better control of the situation and circumstances, versus the other option, where we become the victims of the hazards present. If we rely on only one layer of protection, a failure of that one system could result in exposure, injury, or even death. This is particularly true when that one system is PPE, because PPE has the most variables and is more likely to fail, as we will see in Chapter 7. To see how these systems work together, review Figure 2-2.

engineering controls
the first level of controlling hazards in the workplace; includes the use of built-in safety systems such as guardrails, ventilation systems, and machine-guarding devices

You might also recall from the discussion of the HAZWOPER regulation in Chapter 1 that an entire section of the HAZWOPER regulation, paragraph (g), requires that we use "engineering controls, work practices, and PPE." The term **engineering controls** is listed first because these controls are the required initial step in the protection of employees under most circumstances. This is because the purpose of engineering controls is to eliminate or significantly reduce the hazard associated with a particular operation or condition. Common examples of engineering controls include the use of a guardrail around elevated locations

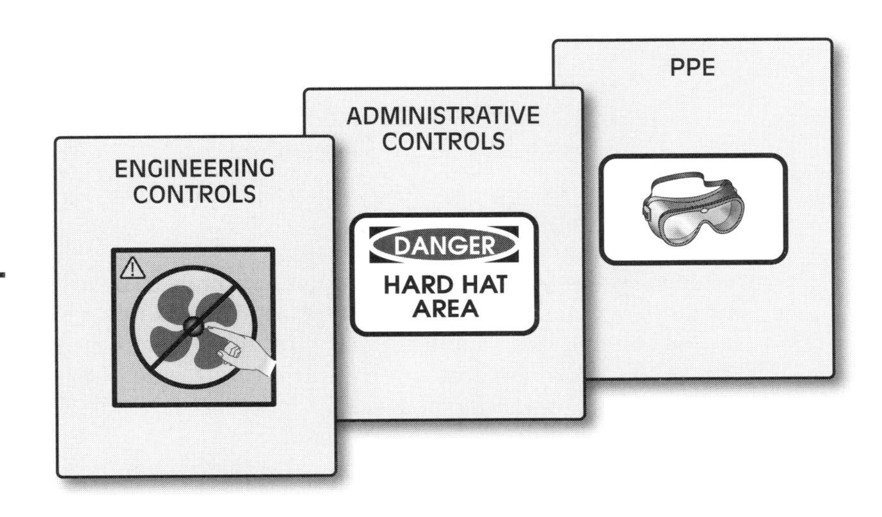

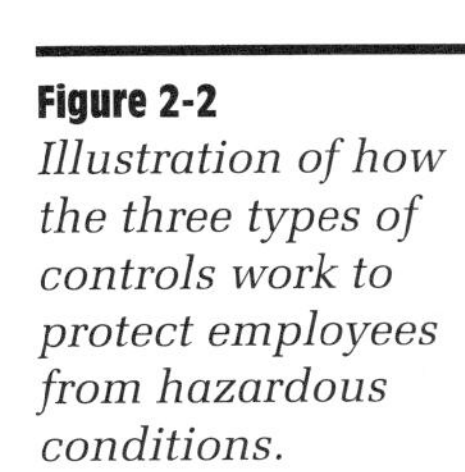

Figure 2-2
Illustration of how the three types of controls work to protect employees from hazardous conditions.

to prevent employees from falling, guards on saws to prevent injuries due to contact with the saw blade, ventilation systems to dilute the concentration of hazardous vapors that might be present in a given area, and mufflers on loud equipment to reduce the noise levels and the subsequent damage to ears. If these types of controls are in place, the risk associated with operations involving the hazards can often be well controlled or minimized because the goal of the engineering control is to help control the hazardous situations through eliminating the hazards. Consider how ventilation systems can dilute the concentration of hazardous substances present in the air to a level that is safe, thereby completely eliminating the hazards present. Additionally, the use of guards on machinery keeps personnel from exposing their bodies to the hazards associated with the equipment. Engineering controls are the most effective in controlling hazards due to their ability to eliminate the hazard completely or minimize the risk of persons potentially exposed to the hazards present. The use of engineering controls is constant and generally does not require that the employees take action on their own. The guardrail around an elevated location protects the employee from falling without the employee having to do anything. An example of this can also be found in automobiles where the air bag forms the basis of the engineering controls and works independently of the actions of the occupant.

> **■ Note**
> Engineering controls work independently of the employee and most often provide protection independent of employee action.

Examples of commonly used engineering controls and their protective mechanisms are as follows:

- Ventilation systems that provide adequate volumes of air in a space to dilute the concentration of vapors to a safe level
- Ventilation systems in an area that provide air movement that assists in ensuring that the environmental temperature remains at a safe level
- Design of a workstation that provides protection from bending or repetitive motion injuries
- Noise-dampening enclosures inside operations to reduce noise levels
- Photoelectric light beams that stop the rotation of equipment if employees get too close to the area of hazard
- Antilock braking systems in vehicles to provide additional control and potentially reduce stopping distances
- Air monitoring systems that detect the presence of hazardous materials, and that shut down the system and provide additional ventilation in the area if unsafe levels are detected

Engineering controls are not always perfect and the protection that they provide can be compromised in some situations. Consider the example of a guard on an electric saw. The guard is designed to move as materials are passed along the edge of the blade. However, some employees have been known to disable saw guards, compromising the protection it would provide. Numerous other examples can be given, including both intentional and unintentional actions that limit or minimize the protection afforded by engineering controls. In some cases, intentionally eliminating engineering controls is done to further the impact of a release, as in a criminal or terrorist act. It is for this reason that we do not rely on the use of this level of control alone and that we employ other types of controls in combination with engineering controls to provide the required level of safety for workers who are potentially exposed to the hazards present.

> **■ Note**
> Engineering controls provide a high level of protection for workers, but can be disabled or rendered ineffective.

> **■ Note**
> Additional types of controls can be used with engineering controls to provide a higher degree of safety.

administrative controls include work rules, schedules, standard procedures, and other methods to limit the potential exposure of the worker

The second type of control, called **administrative controls**, uses a combination of work rules, standard operating procedures, signage, and other methods to alert the employee to the hazards, in an effort to control the risk associated with the hazard. Common examples of administrative controls include the following:

- Signs that prohibit entry into a confined space that has a hazardous condition inside
- Work rules that require working in teams in hazardous areas using the buddy system to ensure that any exposure is immediately noted by others
- Rotation of personnel into areas where hazards are present to maintain safe exposure levels
- Requirements for specific types of air monitoring to ensure that safe levels of exposure are maintained
- Requirements for equipment inspections prior to their use, such as daily forklift inspection
- Mandated standard procedures to be followed by anyone conducting hazardous tasks

While effective in reducing exposure, administrative controls are generally not as effective as engineering controls for a number of reasons. First, their ability to control the hazard is limited. Signs do not control the hazard. It is only through application of the controls that the hazard is reduced. Because of this, administrative controls often require that the worker pay attention and follow the rules, since they are not automatic. Figure 2-3 shows an example of the signs on a gasoline pump at a typical gas station. While the signs may warn of the hazards, the signs alone do not eliminate the hazards nor do they reduce the exposure potential. The signs that are the basis for administrative controls require that the worker read them and follow the directions regarding what is required. Unlike engineering controls that eliminate or reduce hazardous conditions independent of action by the worker, administrative controls are effective only when workers are aware of the conditions in which they work, pay attention to warning signage, understand what is indicated by the signage or the rules applicable to the situation, and take the appropriate action. Many of us have seen people violating the rules at the gas station by leaving their car running while fueling or by smoking cigarettes; both are violations of the rules shown on signs at the gas pump.

The third level of control is the use of PPE. Like administrative controls, the third control requires that the worker take action. As with most of the

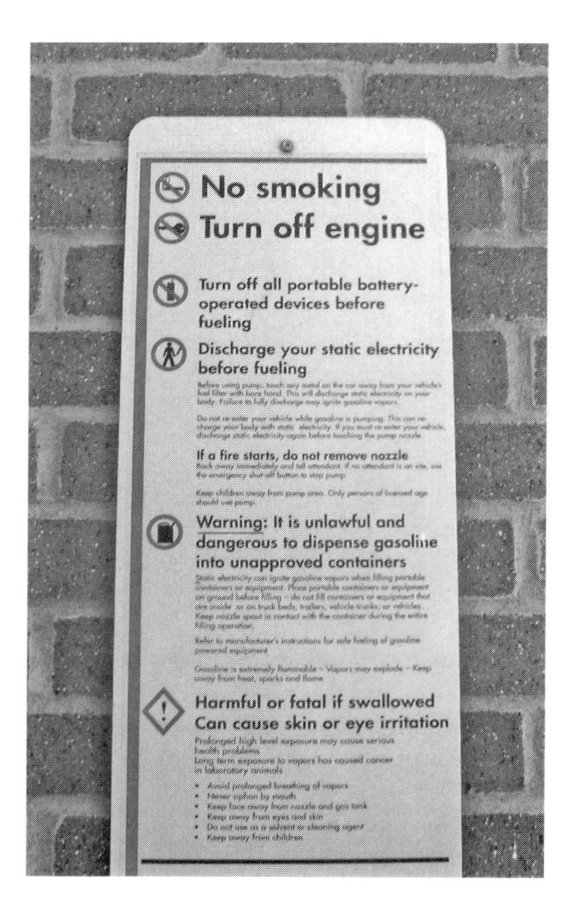

Figure 2-3
Example of signs that are used to administratively control the pumping of gas.

administrative controls, the protection afforded by this level of control is not automatic and requires both action and some degree of skill on the part of the worker. Additionally, because of a range of variables, the use of this level of control requires considerable expertise, training, and additional equipment. Because of this, the topic will be discussed in depth when we review the use of respiratory protection equipment in Chapter 6 and other types of personnel protective clothing in Chapter 7. However, it is important to keep in mind that although the use of PPE is a necessary component of our overall protection, because of its complexity it is not the best way to provide protection. If we rely on the use of this level of protection as our sole method of controlling our risk of exposure, we may find that we are not adequately protected because the use of the protective equipment has the highest potential for failure among all of the methods of controlling hazards. Not only does the hazard need to be recognized, but also the range and limitations of the equipment that can be used, and the limitations of that equipment must be fully understood. Again, the best method for controlling the risk presented by the hazards is to use a combination of all three controls to maintain an adequate degree of safety.

Finally, to further illustrate how each of the controls works to reduce exposure, it is helpful to point out how each of the topics presented in this book relates to the three controls.

1. *Engineering controls.* These controls are a required element of all safety programs. Their use will be highlighted in Chapter 3, where controlling health hazards is presented; in Chapter 5, where the control of physical hazards is discussed; and in Chapter 11, where site control programs are

discussed and the use of engineering controls for specific situations is expanded on. Additionally, Chapter 10 has a discussion on other dangerous situations presented by hazardous energy and how engineering controls can be employed to eliminate these hazards.

2. *Administrative controls.* These controls are discussed in depth in Chapter 5, where identification systems are described. Chapters 4 and 5 discuss how to limit exposure to health and physical hazards. Chapter 8 covers decontamination, which is a work practice that limits the potential of exposure by requiring specific work practices. Chapter 10 provides a discussion on the use of administrative controls to provide protection to those working with hazardous energy and other physical conditions. Chapter 11 discusses administrative controls in emergency response programs, such as the buddy system and backup teams. Finally, Chapter 12 identifies a number of standard practices that are used in the reduction of risk in specific hazardous operations.
3. *The use of PPE.* This topic is covered throughout the book. Two chapters are devoted exclusively to PPE and provide in-depth details of how to properly select and use the equipment. The limitations of this level of control are also discussed throughout the book.

DOCUMENTING HAZARD AND RISK ASSESSMENTS

OSHA mandates that hazard and risk assessments be conducted for various types of activities, but it does not specify the exact format for documenting these assessments. Many variations of plans and programs have been used to help with this process. As we know from Chapter 1, OSHA mandates that written plans and programs be developed to identify the hazards present and the work rules, including the specific types of protection required. This provides us with some of the documentation regarding the hazards and risk, but again the specific information may be buried in a lengthy document of plan.

To assist with the process of identification of the hazards and risks associated with various jobs, many safety experts rely on standardized programs that can be used for a variety of operations. These programs are called by different names depending on the organization involved and include the terms *job hazard analysis (JHA)*, *job safety analysis (JSA)*, or even *critical task analysis (CTA)*. While the programs are not limited solely to chemical exposure, their use in this area is well suited. Typically, these programs are used when the operations involve any of the following:

- Jobs with historical high injury/incident rates
- Jobs with the potential to cause severe or disabling injuries or illness, even if there is no history of previous accidents/incidents. This can include jobs that involve the use or presence of most types of hazardous materials.
- Jobs in which one simple human error or loss of attention could lead to a severe accident or injury
- Jobs that are new to your operation or that have undergone changes in processes, procedures, or equipment
- Jobs that are complex enough to require written instructions

Given this, standardized programs are well suited for jobs involving hazardous materials. The process focuses on the job tasks that are to be performed in a way that identifies the associated risks before they occur. It focuses on the relationship between the worker, the task, the tools, and the environment that in many cases can contain hazardous materials. Once the hazards and steps are identified, the JHA will identify the steps to eliminate or reduce the hazards to an acceptable level of risk. These will be in the form of one or more of the controls that we introduced earlier in this chapter.

While there is no one standard format that is mandated for use, each of the programs has some common components. Following is an example of a JHA program that can be adapted to many types of hazardous operations. It is a variation of a JHA program that is available on the OSHA Web site and that is well suited for setting up a hazard and risk assessment for specific operations that involve hazardous materials. It involves the following steps:

1. *Involve employees.* This is a critical step in the process because employee involvement helps to ensure that the employee has bought into the program. Employees have a unique and often more complete understanding of the job and the tasks necessary to complete it. Involving them will help minimize oversights, ensure that a more quality analysis is done, and allow them to share ownership of the process.
2. *Review the accident history for the job.* It is always a good idea to determine how the tasks have been done previously. This is done by a quick review of the history to learn from what has happened. History is a good teacher if we listen to the lessons that are there and learn from them. Repeating a prior mistake when dealing with hazardous materials is too costly a mistake to make. If your organization does a good job of incident and **near miss** reporting, the information that you get at this step will be critical in your overall results. A near miss is simply documenting an event that could have resulted in an accident.

near miss
an incident that could have resulted in an exposure or accident

3. *List, rank, and set priorities for hazardous jobs.* List the jobs that present unacceptable risks based on the hazards most likely to occur and with the most severe consequences. These jobs should be your first priority for analysis. Many times the hazard presented by the various materials associated with the job will help in this process. Jobs with potential exposure to higher hazardous substances often pose the most risk.
4. *Outline the steps or tasks needed to complete the job.* Nearly every job can be broken down into steps or tasks. When beginning a JHA, you may need to watch employees perform the job and list each step required to complete the task. Be sure to record enough information to describe the job action performed. Avoid breaking down the steps into too many components, because each step will need to be analyzed. After you do this, review the steps with the employees who do the job to ensure that the steps you listed are the ones performed when the job is done. You may need to involve other employees who do the same job to ensure that the method you describe is used consistently. A JHA that lists steps not done by others doing the same job is of little value. Make sure you tell employees that you are not grading them on how they do the job, but simply identifying the steps. Photos or videotapes are often helpful in your documentation process.
5. *Complete the worksheet for each task.* This is the detective work part of your job. You are now on a hazard hunt. You should think about things such as:
   - What hazardous materials are present?
   - What are their properties?
   - Are there ergonomic issues present?

- ○ Are there potential energy sources that need to be controlled?
- ○ What can go wrong?
- ○ What are the consequences?
- ○ How could it happen?
- ○ What factors contribute to a hazard?
- ○ How likely is it that a hazardous situation will occur?

Then, using the worksheet given in Figure 2-4, document the answers to these questions in a consistent manner. Describing a hazard in this way will help you eliminate or control it. There is an old adage that says you can't fix what you don't know is broken. And it is true with JHAs.

When all of this is captured, the final document worksheet will describe the following:

- Where is it happening? (the environment, including the hazards present)
- Who is involved? (such as what group of employees)
- What could happen? (the exposure)
- What precipitates the hazard? (the trigger)

Job Title:	*Job Location:*	*Analyst:*	*Date:*
Task #:	*Task Description:*		
Hazard Type:	*Hazard Description:*		
Consequence:	*Hazard Controls:*		
Rationale or Comment:			

Figure 2-4
Sample job hazard analysis worksheet.

- What are the outcomes of the hazardous situation? (the consequence)
- Are there any other contributing factors?

To understand how this process works, let us use an example of a simple task, an operation where an employee is going to transfer a hazardous material from a large drum to a smaller, portable container. All that is required is to move the material with a pump from the drum to the smaller container. We will assume that the employees involved in the operation are assisting with the completion of the worksheet, that there is no prior history of incidents involving the process, and that the only task required is the actual transfer. The material the employees are transferring is gasoline. Its hazards are relatively well known, including its ability to catch fire, emit vapors that can present health concerns in high concentrations, and damage exposed skin. We will identify this as Task #1 on the worksheet.

The task description will be: Transfer 3 gallons of gasoline from Drum 1 to Container 2 using the fixed hose and manual pump.

The steps in the process have been identified as follows:

1. Obtain a portable container from the storage area.
2. Place the container inside the containment berm near the storage drum.
3. Open the top of the portable container.
4. Remove the hose and nozzle assembly and insert it into the portable container.
5. Manually pump the liquid into the container using four strokes for each gallon to be transferred.
6. Replace the hose and nozzle when done.
7. Place the top on the portable container and remove the container from area.

Hazard description:

1. Fire due to possible release of flammable vapors
2. Breathing toxic vapors that are evolved from the process
3. Skin contact hazard should liquids be released

Hazard control:

1. When entering the area, ensure that the ventilation system for the room is active and remains on during the transfer operation. (This is one form of engineering control that will dilute vapors and reduce the potential for fire or exposure.)
2. Ensure that all other equipment and sources of ignition are not present in the area at the time of transfer. (This is an administrative control to help ensure that no ignition sources are present in case vapors are released.)
3. Confirm that the fire extinguisher is present just outside the area. (This is another administrative control to provide a degree of rapid response to an emergency situation should it be needed.)
4. Ensure that the bonding strap is in place from the drum to the portable container. (This is an administrative control to help reduce the potential for static electricity to develop during the transfer.)
5. Obtain and wear appropriate gloves, safety glasses or chemical goggles, and a face shield. (The use of PPE will help prevent accidental contact with the material.)
6. Ensure that the pump is designed for the safe transfer of flammable materials.

7. Manually pump four times, stopping after each four to confirm the level of liquid in the smaller container. (This is the administrative control that ensures no overflow issues develop.)

Additional information can be put into the comment area such as the information contained in the parentheses in the list. This field is to be used for additional information that will highlight areas of concern or explain the necessity of specific steps.

From this we can see the benefit of evaluating the job that needs to be done, as it relates to the hazards present. If the same job involved a less hazardcus material, such as diesel, the hazard assessment would conclude that flammability was less of a concern given that diesel is less likely to create ignitable vapors at normal temperatures. In this case, the steps to reduce the risk can be shortened, given that the primary hazard is the direct contact with the liquid itself.

Later in our discussion of emergency response activities, we will see that OSHA mandates that emergency response personnel conduct a hazard and risk assessment for emergency response activities. The value of this process is clear when we consider all of the potential for harm in emergency response operations, although a standard worksheet is not always completed. Generally, the amount of information available to those who must determine the appropriate response actions is less than would be found at a typical job site where hazardous materials are used. Additionally, the need to rapidly intervene in an emergency response situation places additional pressure on those who must determine specific tactics to be employed. Despite the relative lack of information available and the need for rapid action, the actual mental evaluation should take place and the process of conducting the assessment of the hazards and the risks is performed.

The steps in the process for an emergency response operation would be as follows:

1. Identify the specific materials present.
2. Determine the properties of the material as they relate to the ability of the material to produce harm by themselves or in combination with other materials that could be present.
3. Identify the action plan for containing or controlling the hazards present. Detail the steps to be taken by those who will be potentially exposed to the hazard.
4. Assess the risk of conducting those activities as they relate to the hazard present.
5. If the risk is significant, review alternative methods to accomplish the tasks.
6. Implement the actions, making sure to continually review them if the hazards are high.
7. Evaluate and make adjustments as needed.

Again, the key to ensuring that work is conducted in a safe manner is to properly identify the material and determine the hazards and properties of the material. This is always one of the first steps in the hazard and risk assessment process. If this cannot be determined, the hazard assessment will work under the premise of an unknown material being present, and some degree of judgment will need to be made by the person in charge of the operations. However, the conclusion does not always need to be reached that the material is the most toxic, most flammable, or most reactive material possible. It must take into account the circumstances where the materials are found. For example, fluid leaking from an automobile engine compartment could be gasoline, diesel, radiator fluid, or some type of oil. The hazard assessment for an unknown liquid leaking from a vehicle would use the worst of the hazardous substances that could likely be present. While there may always be second-guessing, a rational approach to a leak involving a vehicle would normally not require the use of specialized chemical resistive suits, as seen in Figure 2-5. In this case, the most appropriate response may be to use the chemical resistive clothing appropriate to the hazards present.

As we conclude our discussion of this topic, it is important to remember that it is always best to overrespond to an incident than to underrespond. By this we mean that when there are serious doubts, and the possibility exists for a serious hazard if the exact material is not known, you should be conservative in both the use of the protective clothing worn and the actions taken. In such cases, it is always prudent to wear higher levels of protective clothing and equipment and to be cautious in your approach. Use the initial activities at the scene to confirm the materials present or to do the most immediate actions, such as rescue. Delay subsequent and nonessential activities until a confirmation can be made of what material is present and in what form. But also keep in mind that working with hazardous materials is a thinking game. To default to the highest form of protection in all cases is simply not the best course of action. As we will learn, higher levels of protection come at a cost and can expose those who use them to additional and unnecessary hazards. The use of the most appropriate type of protective equipment and taking the most appropriate actions does not always require that the highest protection and most conservative actions be selected. This is the purpose of the next few chapters in the book, chapters that will help us make appropriate determinations about addressing the hazards. It is through a complete understanding of the information contained in this study that the process of a hazard and risk assessment will be most successful and result in the most appropriate decision.

Figure 2-5
Example of an emergency response activity using specialized chemical-resistant PPE. (Courtesy of Los Angeles County Fire Department.)

SUMMARY

- Current safety programs must include discussion of and planning for the effects of both natural and man-made disasters on the hazardous materials at the site.
- A key component in any safety system is the hazard and risk assessment. The concept is so fundamental that it is found in almost every one of the required competencies for each of the nine levels of HAZWOPER training.
- Hazard and risk assessment is the process of assessing particular hazards and taking appropriate steps to protect employees. It incorporates the specific hazards of a hazardous material into the required tasks to be performed.
- Risk assessment is the more subjective part of the process and involves reviewing what actions or operations are expected.
- The three controls mandated by OSHA are engineering controls, administrative controls, and the use of PPE. They work best when they are part of a *well-thought-out system* to minimize or eliminate the risk associated with the specific operations.
- The purpose of the engineering controls is to eliminate or significantly reduce the hazard associated with a particular operation or condition.
- Administrative controls are the combination of work rules, standard operating procedures, signage, and other methods to alert the employee to the hazards in an effort to help control the risk associated with the hazards.
- The third level of control is the use of PPE. This is the weakest of the levels of control and contains the most variables for use.
- Safety experts rely on programs that help to identify the specific hazards associated with performing operations. These programs are called by different names depending on the organization involved and include job hazard analysis (JHA), job safety analysis (JSA), or even critical task analysis (CTA).
- In a JHA, the hazards and steps for the job to be performed must be identified. Then, the JHA can identify the steps to eliminate or reduce the hazards to an acceptable level of risk. These steps will be in the form of one or more of the controls.
- The questions to answer in a JHA include:
  - What hazardous materials are present?
  - What are their properties?
  - What can go wrong?
  - What are the consequences?
  - How could it happen?
  - What factors contribute to the hazard?
  - How likely is it that a hazard will occur?

REVIEW QUESTIONS

1. Describe the differences between health hazards and physical hazards. Give an example of each.
2. Cite the similarities and differences between an accidental release of a hazardous material and an intentional release of that material.
3. List the three levels of control in order of use as mandated by OSHA. Give two examples of each type of control.
4. Which level of control provides the highest form of protection in most cases? Explain why this is the case.

5. Why is it better to use more than one level of control for a given situation?
6. Describe the types of tasks that benefit from a hazard and risk assessment.
7. List the primary steps in conducting a JHA.
8. Explain the differences between emergency and nonemergency hazard and risk assessment techniques.

ACTIVITY

JOB HAZARD ANALYSIS

1. Using the form provided in this chapter, complete a JHA on a job of your choice. Be sure to consider all aspects of the job and the hazards presented by the materials.
2. Conduct a hazard and risk assessment for an emergency response operation of your choice.

PRINCIPLES OF TOXICOLOGY

Learning Objectives

Upon completion of this chapter, you should be able to:

- Define toxicology.
- Describe the seven branches of toxicology.
- Describe the difference between a *toxic* and a *poison.*
- Identify the two categories of chemical effects—local versus systemic.
- Describe the differences between acute and chronic exposure.
- List the routes of entry for materials to enter the human body.
- Identify the route of entry that produces the highest potential hazard to most workers.
- List the three time frames of toxic exposure.
- Define LD50, LC50, LDlo, LClo, LDhi, LChi, odor threshold, PEL, TLV-TWA, STEL, and IDLH.
- Define asphyxiant, carcinogen, and Teratogen.
- Describe additive, potentiation, antagonistic, and synergistic chemical interactions.

CASE STUDY

Many people appreciate the taste of buttered popcorn. The flavor is so popular that even jelly beans are made with artificial flavoring to simulate the taste. But while the flavor is appreciated by many, for those who work with them, the flavoring materials can be deadly. What is not widely known by consumers is that the buttered popcorn flavoring is artificially created by using a series of chemicals. Like many other materials, when the chemicals reach their final form, they are safe. But getting them there is a process that can be highly hazardous.

Another example of safe chemical compounds that are dangerous in the making would be the various types of plastic pipe used in our homes. When formed into the black plastic that makes up most drainpipes, the materials used to make the pipe are safe. But the materials by themselves are highly hazardous and potentially deadly. In the case of the pipe, the materials include acrylonitrile, butadiene, and styrene. Each of these materials is significantly hazardous in and of itself. Workers exposed to these materials run added risks, including higher incidences of cancer and other health concerns.

Such is the case with the materials that make up the buttered popcorn flavor. Reports have emerged from several sources noting that workers who produce the artificial flavoring have been diagnosed with a number of rare and serious lung diseases caused by exposure to the heating of the chemical Diacetyl. The hazards associated with this material are so widespread that a new term, "popcorn worker's lung," has been coined to identify the condition whose only treatment in some cases is a complete lung transplant. Lawsuits related to the exposure have been filed, and in a 2005 case a worker was awarded $2.7 million for his claim of lung damage caused by his exposure to the material. Because of the potential to cause serious health effects, several organizations including the Occupational Safety and Health Administration (OSHA) are looking into ways to protect workers who are exposed to these types of materials. In some instances, recommendations for an outright ban on the use of these materials are being considered.

It is clear from this example that common materials, which are often taken for granted by those who develop or use the products, can present clear and present dangers. Workers exposed to hazardous substances in the workplace must understand that their exposure may cause harm to their health if they allow the materials to enter their bodies in a high enough level. As with other materials that have been identified as causing harm to people, workers must be aware of the conditions present in their workplaces and take the necessary precautions that are recommended or required. Failure to do so could result in significant health damage or death.

INTRODUCTION TO TOXICOLOGY

For many of us, the thought of a chemical exposure may create a sense of panic. This could stem from the fact that chemicals are often associated with terrible health effects. This belief is compounded by reports in the media, including the Internet, which we often find ourselves turning to for information on what takes place in our world. In those reports, we are bombarded with terms such as *toxic, carcinogenic, combustible,* and *dangerous.* Rarely are any positive effects of hazardous substances reported, as stories are related of the damage and injury that result from these hazardous materials.

Is what we hear from these sources accurate, or does it simply play into our preconceived notions of these materials? When we look at the facts, we often find that reports of hazardous materials in the media are filled with gross generalizations that help to sell the news. But do they actually convey the facts? As someone who works with these materials, or who responds to emergencies involving released substances, we need to know these facts in order to be safe and to truly understand the hazards of the substances. Relying on partial truths, inaccuracies, or emotional accounting will not make us safe and in fact may result in the opposite. So we need to get to the truth, as it relates to the real hazards of the substances we work with.

> **■ Note**
> Many sources of information on the subject of toxicology do not always provide accurate and complete information.

This chapter provides a discussion on the health effects of hazardous substances commonly encountered in the workplace and in our society. This discussion will help identify the real dangers of these materials because many of them are potentially harmful or deadly, especially if we do not have accurate information about their properties. But in doing this, it will not overstate the subject area and will help us decide where to find the most accurate and most complete information on the materials and their properties that may affect our health. The chapter is not intended to replace a more extensive discussion of toxicology, which would be found in a course specifically on the subject, since the topic is quite broad and is growing all the time. It is intended to give us the information we need from a health perspective, as responders to hazardous materials or those who work with chemicals to allow us to work safely.

> **■ Note**
> An accurate understanding of the health effects of a particular substance is critical to working safely with the material.

BACKGROUND

Studies of the harmful effects of chemicals have been going on for a long time. Scientists began experimenting with chemicals to find medicines and other materials that could be used to cure diseases. As they began working with various materials, they found that many were as capable of causing harm as they were of healing. As this type of study expanded, an entire new science evolved. Those involved in this new field called themselves toxicologists and began to investigate the harmful health effects that chemicals could have rather than just the beneficial effects. It is from this new field that we now have a much clearer understanding of the potential and reality of our exposure to various types of hazardous materials.

toxicology
the study of the adverse or harmful effects of materials on living organisms

Toxicology, for the purpose of our discussion, is the study of the adverse or harmful effects of materials on living organisms. It is far-reaching in its scope when we consider that "adverse effects" can mean a number of things, from simply causing a minor rash to causing death in a short time or producing cancer years after exposure.

It is important to recognize that living organisms involve a whole host of subjects from single cell amoebas to human beings. In fact, if we look at the field of toxicology as a whole, we find that there are a number of disciplines involving many types of toxicologists. For example, the field of toxicology has several major branches, including the following:

- *Clinical toxicology.* The branch of toxicology that is concerned with studying diseases caused by, or uniquely associated with, various toxic substances. This area of toxicology is often involved in the study of drug overdose treatments.
- *Descriptive toxicology.* The branch of toxicology that is directly concerned with toxicity testing, such as determining the harmful effects that a substance may have on an organism.
- *Mechanistic or biochemical toxicology.* The branch of toxicology that is concerned with understanding the mechanism by which chemicals and materials exert their toxic effects on living organisms. The branch of toxicology that tries to determine why something causes harm to an organism rather than the effects of the exposure.
- *Forensic toxicology.* The branch of toxicology that is concerned with medical and legal aspects of the effects of chemicals and substances on humans and animals. It is concerned with using techniques of analytical chemistry to answer medical questions about harmful effects of chemicals.
- *Environmental toxicology.* The branch of toxicology that is dedicated to developing an understanding of chemicals in the environment and of their effects on humans and other organisms. This branch is broad in its scope because the environment covers a variety of media, including air, groundwater, surface water, and soil. Additionally, this branch considers a whole host of life forms in the environment, ranging from the very lowest form of animal life through humans. Ultimately, this branch is involved in determining the environmental fate of the materials as they are found in the natural environment.
- *Regulatory toxicology.* The branch of toxicology that has responsibility for deciding what safe levels of chemicals may be allowed in the various environmental media, including the workplace. The safe levels are then refined and adopted as standards, guidelines, or policies, and are developed on the basis of the data from studies provided by the other branches of toxicology.
- *Industrial toxicology.* A relatively new branch of toxicology that deals with the disorders produced in individuals who have been exposed to harmful materials during the course of their employment. This branch includes the study of materials most relevant to workers in the field, whose exposure is caused by their work. This population is not the most sensitive to exposure because it involves people who are physically fit, so the data gathered by this study differ slightly from that gathered by other branches, such as environmental toxicology, which takes into account the most sensitive populations.

■ Note

There are many branches of toxicology that study the various aspects of the adverse effects of hazardous materials and chemicals.

Because we now have much more information on the topic than we had in previous years, we have a much clearer understanding of the truth. As those who work with the materials, we should be encouraged that entire branches of toxicology are focused on this aspect and are working with the other branches to ensure that we are not exposed to unsafe levels of materials in the course of our employment. Additionally, as a responder to releases of hazardous substances, we can be encouraged that our exposure has been reviewed in a similar manner and that the fate of the material in the environment is also a subject of study as we engage in protecting not only the population exposed to the hazard, but the environment as well.

EXPOSURE MECHANISMS

As we consider what is involved in chemical exposure, we first learn that the actual process of exposure is quite complex. Common sense will tell us that simply working with or around a toxic material does not mean that we have been exposed to its harmful effects. There is a whole range of factors that come into play, as we may know. A number of factors in the mechanisms of exposure can be identified to help us understand our actual risk or the potential hazards we face in any situation where an exposure could occur. These factors can be summarized by the following equation:

HAZARD or RISK = TOXICITY FACTORS + EXPOSURE FACTORS + YOU FACTORS

■ Note

The toxicological risk or hazards associated with working with hazardous substances is a function of three factors—toxicity, exposure, and you. We must consider each of them in order to work safely with hazardous materials.

Toxicity Factors

The easiest of these factors to quantify, or know specifically, is the toxicity of the substance. Recall that there is an entire branch of toxicology that studies this and nothing else. The mechanistic or biochemical branch reviews the material as a whole and quantifies the amount and type of harm that can occur as a result of an exposure to the material.

Once the exact hazards or toxicity of the material is determined, the information is required by the Hazard Communication regulation to be provided to employees exposed to the material via the Material Safety Data Sheet (MSDS). The MSDS is required to be available for employee review at all places where a potential exposure could occur. While it is more difficult for those engaged in emergency response activities to obtain this information, once a positive identification of the substance is made, there are a range of databases and other sources for the information to be available at the scene. Once that information is accessed, the specific toxicology of the material is provided to emergency responders to use in reducing their own potential exposure and ensuring that others are not harmed by the materials involved in the incident.

To understand the degree of toxicity that materials present, there are a number of terms that define or explain the levels of toxicity of a particular material. Following is a discussion of some of the terms that will help us fully identify the actual risk from the material.

LD50 and LC50
Lethal Dose 50% and Lethal Concentration 50%; these terms are used to identify the level of exposure to the material where death will occur in 50% of the test population

lethality
the ability of the material to kill the exposed subject

Toxic materials
materials that have an LD50 less than 500 mg/kg or an LC50 below 2000 ppm.

Highly Toxic
material that has an LD50 of less than 50 mg/kg or an LC50 below 200 ppm

LDlo
the level at which the animal in a given population with the lowest tolerance dies when exposed to a particular dose of the material

LDhi
the dose level that would kill the strongest test animal in a given population

LClo
airborne concentration that would kill the weakest test animal when it is exposed to a particular concentration in the air

LChi
the highest concentration of material in the air that would kill the strongest test animal in a given population

LD50/LC50: In determining the specific effects of a material on a living organism, toxicologists expose test animals to specific levels of the material in a controlled environment. Depending on the material being studied, test animals such as rats, rabbits, guinea pigs, and others are used to identify whether the material will have harmful effects and the levels at which those effects will occur. This is the pure science of toxicology. The data is quantifiable and available for our use or use by others who can then relate the data to specific situations such as workplace exposure.

The "LD" and "LC" portion of the terms stand for Lethal Dose and Lethal Concentration. The Lethal Dose is determined by measuring the amount of the material inside the test animal in milligrams, relative to the body weight of the animal. It is most often expressed as mg/kg. Lethal Dose is most often used to express the amount taken into the animal by ingestion, by absorption through the skin, or by injection. The term *Lethal Concentration* is used to describe the amount of material present in the air that causes death. It is most often noted by describing how much in parts per million (ppm) of contaminant is present in the air.

Perhaps the most basic testing that is done on potentially hazardous substances to determine their actual potential for harm is in the study of the lethality of the material. **Lethality** is the term used to describe the ability of the material to kill the exposed subject. While other effects of exposure to materials are also studied, such as their ability to produce cancer, cause sickness, or create birth defects, the ability to kill and the level at which this occurs are critical in our overall understanding of the material. Once it is determined that the material is capable of causing death, we then need to know at what level this occurs. Once that level is determined, additional testing and research are done to determine other issues related to the specific toxicity of the material.

As an example, a material such as diesel fuel, which can kill people in a high-enough concentration or dose, is far less hazardous than a material such as chlorine, which is deadly in very low amounts. To determine the levels at which the deadly effect occurs, toxicologists have developed terms to show the relative lethality of the material—in other words, to quantify just how dangerous the material is relative to all other materials. Next, the materials can be broken into classes such as **Toxic** and **Highly Toxic**. OSHA and other regulatory agencies use these classifications to help ensure that exposures remain below safe limits. While other regulatory agencies list different definitions for these terms, the OSHA definition of a toxic material is one that has an LD50 less than 500 mg/kg or an LC50 that is below 2000 ppm. To be classified as a Highly Toxic material, a Toxic material must have an LD50 that is less than 50 mg/kg or an LC50 below 200 ppm. Materials that are classified as Toxic or Highly toxic demand our respect, and considerable care should be exercised whenever these materials are present.

Other terms are also used to describe other levels where the test population suffers death, but these levels are not part of determining the classification of the material. These include the terms **LDlo**, **LDhi**, **LClo**, and **LChi**. These stand for Lethal Dose Low, Lethal Dose High, Lethal Concentration Low, and Lethal Concentration High. Specifically, these terms describe the levels where the test animal with a range of tolerance to a material dies when exposed to a particular dose of the material. LDlo is the level where the animal in a given population with the lowest tolerance dies when exposed to a particular dose of the material. This dose is given to animals through a variety of mechanisms such as oral intake, contact with the skin, or injection into their bodies. LClo is the dose level capable of killing the weakest test animal when it is exposed to a particular concentration in the air. LDhi is the term that describes the dose level that would kill the strongest test animal in a given population. LChi is the highest concentration of material in the air that kills the strongest test animal in a given population. Of these values, the 50% level is chosen for most studies because it represents the average member of the particular population group and is more indicative of the hazards of the material on that member.

In the testing, the animals are exposed to various levels of a study material. The animals are monitored to determine the dose taken into their bodies. When the first test animal dies, the dose of the material that caused the death is determined and the value is set as the LDlo. The dose continues to be raised and eventually more members of the population die. The point where the animal that represents one-half of the population, or 50% dies, is determined, and that level becomes the LD50 for that particular material. As the dose is increased, it eventually kills off the entire population of the study group, and that value is determined to show the LDhi for the material.

Once the level or dose of the material is reached where death occurs, the dose level is quantified as to how much of the material is required to cause the effect versus the weight of the test animal, allowing the data to be somewhat relevant to other animals that may be larger. So an LD50 value of 1 mg/kg would indicate that it takes 1 mg of the material for each kilogram of body weight of the test animal to result in death. Nicotine has such a value, and if a test animal weighed approximately 176 pounds, the animal would likely die if exposed to only 80 mg of the material.

A similar process is used to determine the Lethal Concentration of the material. In this case, the exposure is an airborne material that must be inhaled by the test population. Unlike the Lethal Dose, which is quantified in mg/kg, the Lethal Concentration shows the level of contamination in an air sample.

Because the 50% level is most indicative of knowing the effects of the material on an average member of a population, the LD50 and LC50 values are most useful in toxicology studies and in extrapolation into other areas. One of the areas of concern is occupational exposure to materials that are used or found in the workplace. It is because studies are conducted on hazardous substances that we can establish safe limits of exposure for a responder, a worker who handles hazardous substances, or a member of a population exposed to the material in the event of an unplanned release or accident.

Exposure Limits For those who work with, or who are exposed to, hazardous substances, it should be reassuring that a number of research groups such as OSHA and the National Institute for Occupational Safety and Health (NIOSH) are involved in the process of identifying safe chemical exposure levels. Using all of the data available from various toxicological studies on these materials along with information from actual levels found in the workplace, a safe exposure amount is identified. These **exposure limits** are levels established by studies that determine and define the safe levels to enable workers to maintain a margin of safety when functioning in contaminated atmospheres. They are designed to provide the requirements for employers and employees alike in the areas of chemical safety. The working supposition is that if workers stay below these predetermined chemical exposure levels, they will not be exposed to harmful effects and will be able to work comfortably and safely with the chemicals. These limits are bolstered by volumes of laboratory research and are established to provide safe working conditions.

exposure limits
the levels of exposure established by studies that determine the safe levels to enable workers to maintain a margin of safety when functioning in contaminated atmospheres

A word of caution should be expressed regarding any published exposure limit. Keep in mind that there are a number of individual factors to be considered when discussing chemical exposure. Some of these factors will be identified later in our study. However, recall from our earlier discussion, many animals die at lower levels before the testing for lethality reaches the 50% value most often used to determine the hazards of the material. As NIOSH and OSHA determine these limits, which tend to be quite conservative, these organizations cannot possibly take into account all of the individual differences that come into play. They are assumed to be applicable to the average worker, who can be from 18 to 65 years old, male or female, of any ethnic background, and with a whole range of health conditions. In no way can the values take all of these variables into account, nor can they be based on the most cautious approach.

> **■ Note**
> Occupational exposure limits represent safe limits for the average employee.

Another point to consider in determining the levels of occupational exposure is that they are often difficult to determine scientifically. Unlike lethality studies, where the test animals may be used to determine toxicity, it is relatively difficult to get rats to work in a factory for 8 hours a day, 40 hours a week, for 30 or more years to determine if the material will hurt them at a particular level or keep them from doing the work. Partly because of this, the levels are often the educated "best guess" of those who have considerable knowledge of the materials in question, taking into account the scientific data that is available. As more data and information about the materials becomes available, the levels are adjusted accordingly, so these levels often change. Additionally, some of the State Plan States have developed their own levels, which may be considerably stricter than the federal versions or other state levels.

> **■ Note**
> Exposure limits are not easily determined by specific testing and are often estimates based on available information and experience. These levels may vary between the federally established ones and those in the various State Plan States.

A last important point relative to exposure limits is that they are primarily intended to protect us from airborne exposure to the materials listed. As we will study later, the inhalation route should be of the utmost concern, and the formation of airborne exposure limits reinforces this point. Therefore the levels that we will identify in the study of occupational exposure limits are expressed in ppm or mg/m^3 of a substance in air because these are the ways to quantify the levels of contamination in the air.

> **■ Note**
> Exposure limits are primarily intended to protect us from airborne exposure to the materials listed. The values expressed in exposure limits are expressed in ppm or mg/m^3.

To understand the exposure limits, it is necessary to understand that the levels of our exposure to airborne materials vary significantly depending on a number of factors. The most basic of these is that the exposure to the material varies throughout a given period. As we move closer to the material, our exposure increases. As we move around, it changes. As we leave the area, it goes down. Essentially, it varies based on a number of factors, including the amount of time that we spend in the area.

Time Weighted Average (TWA)
the average amount of the material in the air over a given time period; many TWA studies represent an 8-hour work period

PEL
the Permissible Exposure Limit represents the 8-hour Time Weighted Average exposure limit for the material and is set by OSHA

STEL
the Short Term Exposure Limit; the 15-minute TWA safe exposure limit set by OSHA for a particular material

For this reason, it is important to understand that most of the established exposure limits are not absolute values such as those found with the study of Lethal Doses. The exposure levels for the most part represent the average amount of the material in the air over a given period. We call these values a **Time Weighted Average (TWA)** because they represent the average of the amount of material in the air over a given time period. This is done because the exposure levels of a given individual will go up and down throughout the period that they are in the area. To determine the specific TWA, we measure the total amount of exposure and average it out over the work period. Many of the TWAs that we will study are based on an average 8-hour workday. Hence, a typical TWA study may reflect the exposure amounts averaged over that 8-hour period, as can be seen in Figure 3-1.

The **PEL** is the Permissible Exposure Limit for a substance. This value is established by OSHA, either state or federal, and usually identifies the 8-hour TWA exposure for a 40-hour workweek over a working career of exposure to the material. It is expected that exposures at this level would not result in any adverse health effects to the average worker.

The **STEL** is the Short Term Exposure Limit for a particular material. Because it is possible that the levels of exposure to a particular material could vary considerably and during any given 8-hour period there could be high levels of exposure that could be harmful even for that short period, OSHA limits the exposure in any 15-minute period during a workday to a second exposure limit. In so doing, OSHA acknowledges that there could be times during the workday when the amount spikes up to an unsafe level, creating a hazard, and which would still be below the PEL when averaged over the 8-hour work period. Like the PEL, the STEL is a value that reflects the average exposure, or TWA, for that 15-minute period because the concentration may vary over this period.

Because the exposure to the STEL level is relatively brief, and because it is relatively low, workers are allowed to reenter an area where the level of contamination is at the STEL more than once in a given workday, assuming that they do not stay there more than 15 minutes at a time. Workers are allowed to work at the STEL level up to four times per day, assuming they have at least a 60-minute break between these exposures and that their overall 8-hour level is below the established PEL. This allows for short and managed entries into an area that has a higher than PEL level of material in the air.

> **Note**
> Exposures to the STEL level are allowed for 15-minute intervals during a workday, assuming that there is a 1-hour period between exposures.

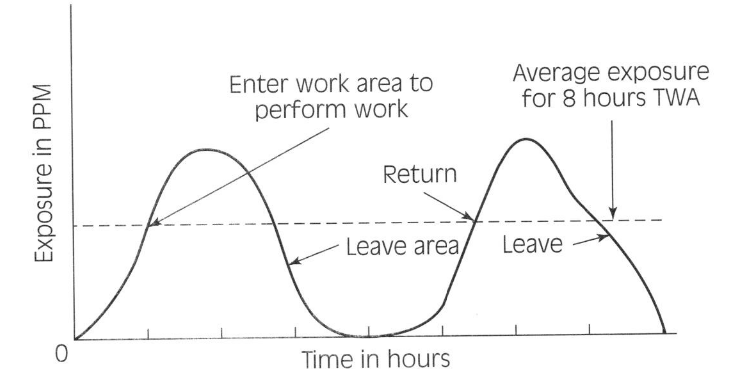

Figure 3-1
Graph depicting a typical workplace exposure based on a TWA of the period of exposure.

TLV/C
the Threshold Limit Value–Ceiling level; this is the maximum level of exposure that must never be exceeded at any point in the workday

TLV–TWA
the Threshold Limit Value–Time Weighted Average; this level is similar to the PEL in that it represents an 8-hour TWA exposure but is different in that the values are not regulations but standards set by the ACGIH

TLV/C stands for the Threshold Limit Value–Ceiling level. As the name implies, this is the upper limit of exposure, or ceiling, for anyone in the workplace. It is a level that could produce harm or at least enough irritation that workers might suffer some type of health effect or would not be able to function in the working environment, even if the exposure were only for an instant. TLV/C levels are never to be exceeded in any workday, not even for an instant. These levels have not been set for all materials, and we rely on the PEL and STEL averages to provide the safe limits required for the particular material.

The **TLV–TWA** is the Threshold Limit Value–Time Weighted Average. While the TLV–TWA is similar to the PEL, it is only a recommendation developed by the American Conference of Governmental Industrial Hygienists (ACGIH). The ACGIH is a private, nonprofit organization that studies occupational exposure and establishes guidelines or standards for various exposures. Because the levels are recommendations published by a nongovernmental agency, they are not considered to be regulations that must be followed. They can be used to set limits that must be followed only in a few cases where OSHA may reference or refer to them. However, these levels represent current studies and recommendations and are often below the established PEL values.

While not an occupational limit, the term *Immediately Dangerous to Life or Health (IDLH)* is very important to understand. IDLH is a level that identifies a value where serious harm to the exposed individual could occur. The IDLH levels are set to the lowest level where any one of the following three results could occur:

- The exposed individual could die at that level.
- The exposed individual could have serious or irreversible health effects as a consequence of exposure at that level.
- The exposed individual could be unable to escape the area on their own and would require help getting out of the area.

Any exposure that exceeds the IDLH level is a serious matter. Only properly trained and protected workers should enter areas with known or suspected concentrations above IDLH. While it is possible to work safely at levels above IDLH, and while these levels can often be present in emergency situations where the concentration of the hazardous substances tends to be extreme, there are some important rules that must be followed to protect workers who are exposed to these levels. Throughout the book, we will see examples of these rules and will expand on them in several areas, including the use of respiratory protection, response procedures, and personal protective equipment (PPE) requirements.

Another group of exposure values that are important in the discussion of safe levels of airborne hazards relates to those involved when a substance is released and the exposed group is the public. Unlike the OSHA-established exposure limits, which are designed for the average employee, other exposure limits must take into account the effects of a material on the average member of a population. When taking into account the effects of a material on the public, consideration must be given to the effects of an exposure on the youngest and oldest members of the population, who would not be represented by typical employees. In setting these levels, consideration must be given to the effects of the exposure on the youngest and the oldest members of the group. For this reason, the use of the occupational exposure values does not provide this information and other levels must be used.

■ Note
Employee exposure limits are not suitable for use when a release of a hazardous substance involves the population. In such cases, newly developed levels can provide guidance related to evacuation planning and protection programs.

Emergency Response Planning Guideline (ERPG)
ERPG values are established to assist emergency response planning when a release involves exposure of the public to the substance; three levels of ERPG are developed to describe the effects of the exposure at each level

To assist in the protection of the public when exposures occur outside the workplace, such as in the event of a transportation accident involving a hazardous substance or a release from a facility in the community, the **Emergency Response Planning Guideline (ERPG)** values have been established. The ERPG values are intended to provide estimates of concentration ranges where one reasonably might anticipate observing adverse effects as described in the definitions for the three levels used in the system. The levels include ERPG-1, ERPG-2, and ERPG-3. They were developed and are updated by the American Industrial Hygiene Association, a group that studies the effects of airborne exposure. Following is a summary of the effects that occur at each of the levels as a consequence of exposure to the specific substance.

- The ERPG-1 level is the maximum airborne concentration below which it is believed that nearly all individuals could be exposed for up to 1 hour without experiencing other than mild transient adverse health effects or perceiving a clearly defined, objectionable odor.
- The ERPG-2 level is the maximum airborne concentration below which it is believed that nearly all individuals could be exposed for up to 1 hour without experiencing or developing irreversible or other serious health effects or symptoms that could impair an individual's ability to take protective action.
- The ERPG-3 level is the maximum airborne concentration below which it is believed that nearly all individuals could be exposed for up to 1 hour without experiencing or developing life-threatening health effects.

It is important to consider that human responses do not occur at precise exposure levels but can extend over a wide range of concentrations. The values derived for ERPGs should not be expected to protect everyone, but should be applicable to most individuals in the general population.

Safety
ERPG values help to set parameters for evacuation and emergency response planning, but care should be taken because no value will provide protection for all exposed persons.

Health and Toxicity Terminology To more completely understand the issues associated with the toxicity of certain materials, we must have a working knowledge of other important terminology, often found in reference materials, including the MSDS for the particular substance.

Carcinogen Apart from causing death, some materials can produce harm or create illness. One harmful effect of exposure to certain materials is the production of cancer in our bodies from known or suspected carcinogens.

As we noted in Chapter 2, a *carcinogen* is a material that is known or suspected to cause cancer in an organism. Because our study of toxicology is primarily focused on occupational exposures, or exposures to human beings as a consequence of a release, we will further limit our working definition to substances that are capable of causing cancer in human beings. Because cancer is a disease that carries a high mortality rate, exposures to human carcinogens should not be taken lightly. In spite of the importance of the topic, a lot of misinformation is accepted by many as fact.

While considerable study has been done on the topic of cancer and carcinogens, there remains a high degree of confusion on many levels. For those who work where carcinogenic materials are present, it is important to have an accurate understanding of the truth as it relates to these substances. To illustrate some of the *misconceptions* that relate to Carcinogens and cancer, review the following statements and try to determine which of them are true and which are only myths. The results may be very surprising, but will help in understanding the real issues because they relate to this vital topic.

1. Cancer is the leading cause of death in the United States.
2. Almost all chemicals that are produced will cause cancer in humans, and there are thousands of chemicals that are known to cause cancer in humans.
3. Most carcinogenic materials are synthetic (man-made).
4. Over one-third of all cancers are caused by exposure to chemicals on the job.
5. Chronic infections can cause cancer and actually contribute to about one-third of all of the cancers reported worldwide.

While some of these statements are true, a number of facts need to be discussed relative to cancer and our exposure to Carcinogens. These include:

1. Cancer is generally thought to be the second leading cause of death in the United States after cardiovascular disease. The statistical odds of getting cancer in the United States are approximately one in four. The odds are that one in five people (20%) in the United States will die from cancer.
2. While many materials are reported to cause cancer, the data in these reports most often applies to the ability to cause cancer in rodents or rats. Most chemical substances (over 50%) tested on rats have produced cancer when the rat is exposed to high doses. Yet many of these substances have not been shown to cause cancer in humans. Four agencies—OSHA, the National Toxicology Program (NTP), the International Agency for Research on Cancer (IARC), and the ACGIH—list the materials that are generally accepted to be human carcinogens. OSHA requires that if either the NTP or IARC lists a material as a human carcinogen, the MSDS must indicate this. A listing of those substances on the IARC list that are believed to cause cancer in humans is found in Appendix C. Updated information on the list can be found on the IARC Web site located at http://www.iarc.fr/. On reviewing the list, we find that there are not thousands of materials as many believe.
3. Most of the materials on the IARC and other lists are natural and not man-made. Consider asbestos, wood dust, tobacco, hormones, and the sun as examples of common carcinogens not created by man. Never assume that chemicals are the only causes of cancer, because most chemicals do not cause cancer in humans while many natural materials can.
4. Our occupational risk for contracting cancer (statistically) is low. Most studies place it at around 2%. Although the emotion and hysteria that may follow a media report concerning cancer may indicate otherwise, and while the history of exposure to materials such as asbestos is still in our minds, most materials capable of causing cancer are now well regulated in the workplace.
5. Sometimes just the irritation effects of a substance are enough to cause cancer. Asbestos and wood dust may not contain any chemical that causes cancer in humans, but continual irritation by them may contribute to the cancer. Likewise, diseases that remain untreated can also cause cancer. For example, Hepatitis B and Hepatitis C, serious diseases caused by viruses that attack the liver, are leading causes of liver cancer, and infections involving them are listed as human carcinogens by the IARC.

> **■ Note**
> While there are a relatively low number of materials that are known to cause cancer in humans in the occupational setting, it is always prudent to limit our exposure to all materials, especially those that are suspected of causing cancer.

Teratogen
material that can cause harm to the developing fetus as a consequence of exposure by the mother; teratogenic materials can cause birth defects and include items such as lead, mercury, alcohol, and cigarette smoke

Teratogen The word **Teratogen** is derived from a Greek word teratos, meaning "monster" or "monster baby." The Ancient Greeks believed that if a couple had a child who was deformed, it was the result of having angered the gods. Now we know that such deformations can be the result of exposure to a substance while the woman is pregnant. If a pregnant woman is exposed to a Teratogen, there is an increased risk that the developing fetus will suffer some type of health effect. In such cases, the mother seldom suffers any effect from the exposure and does not necessarily have an increased risk of problems with subsequent pregnancies. The drug thalidomide was a major cause of birth defects in the 1950s. Pregnant women were prescribed the drug to combat morning sickness associated with pregnancy. Unfortunately, many birth defects, including deformities of the arms and legs, were attributed to this drug, which was later removed from the market in the United States. There are a number of examples of Teratogens commonly found in today's society, including cigarette smoke, mercury from eating certain types of fish, and alcoholic beverages.

There are numerous other factors related to the toxicity part of the equation. There is little we can do to control the inherent properties of the materials, so it is incumbent on us to learn as much as we can about them. To help further understand other toxicity issues, see the complete Glossary at the back of this book.

Exposure Factors

The second variable in our understanding of our real risk of harm is far less quantifiable and somewhat variable, but it is the one that is most easily controlled. This is both good news and bad news. While we can't fully quantify all of the exposure factors easily, and while the factors do vary, the good news is that our ability to limit exposure is well within our control in many cases. Let us now look at some exposure factors and discuss how they affect our overall risk when we work with or respond to chemical incidents.

Concentration of the Chemical Not all concentrations of substances are capable of producing harm. Take for example some of the forms of acid. Not all acids cause harm because their concentrations vary widely. The phosphoric acid in a soft drink clearly makes the material acidic, but that level of acidity is not harmful to us in most cases. In some cases, the exposure potential may be low because the concentration of the material is low. Battery acid is a weaker or less concentrated form of sulfuric acid. Sulfuric acid in high concentrations can produce serious injury or even death. At the weaker concentration found in a vehicle battery, the result of exposure may be discolored or damaged clothing. In considering the relative hazard, we must consider the relative concentration of the material.

> **■ Note**
> The concentration of the material that we are exposed to plays an important role in the overall hazards we encounter. Higher concentrations can produce significantly higher health effects.

Duration of the Exposure The duration of the exposure, or how long we were exposed to it, will give us information relative to the effects that we can expect. Not all exposures to all materials result in the same level of harm. OSHA considers this in its development of the exposure limits, such as the PEL as the level to which we can be exposed for a full 8 hours. The STEL duration is limited to only 15 minutes because the consequence of longer exposure could be harm. The TLV/C is a level that we would never want to have any exposure to.

Short-term exposures of limited duration are often those that our bodies can recover from, assuming the initial dose is not significant. Long-term exposure to something may cause cancer or other serious health effects. If the exposure persists, and the level of the exposure is high, the overall dose of the material taken in will generally be higher. In these cases, the body does not have time to rid itself of the harmful materials and the consequences of that exposure could be harmful. Generally, the shorter the duration of exposure to high levels of materials, the lesser the effect they will have.

> **■ Note**
> Prolonged exposure to a material even in low concentrations can produce harm as well as exposure to higher concentrations because the body is constantly exposed to the substance and cannot recover.

Uptake Rate The uptake rate refers to the rate at which material is taken into our bodies. It is tied in somewhat to the last two factors, concentration and duration. However, in addition to those factors, the uptake rate involves the actual rate that we intake the materials into our bodies. Essentially, how fast something gets into our bodies is a function of a range of variables. When confronted with an emergency situation, our body is often more at risk from this factor than at other times because our respiratory rate will be higher and our skin will be moist from sweat. These are the times when we need to be careful because we are more vulnerable to an overexposure given how fast the material will get into us. Higher uptake rates generally result in higher doses within the body.

> **! Safety**
> Exposure during emergency response activities can often be more hazardous than other types of exposures because rapid breathing and sweating increases intake of the material. Higher uptake rates generally result in higher doses within the body.

Chemical Interactions Another important consideration is that overall exposure to hazardous substances is a function not just of one material but also the effects of multiple materials that may be present in a particular environment, as can be seen in Figure 3-2. Chemicals and other hazardous materials can often mix together, causing effects that are different from the effects of either material by itself. To safely work with mixtures of materials, it is important to know what the actual effects of the mixture will be. While we have studied and quantified the effects of specific materials, in many cases we do not know what effects the materials might have when they mix or are taken in combination with other materials. Sometimes the resulting mixtures have a totally different effect, one we might have not expected. However, the effects resulting from these mixtures fall into four categories of chemical interactions: additive, synergistic, potentiation, and antagonistic interactions. The following paragraphs discuss these briefly.

Figure 3-2
A release of a material in this setting could involve multiple materials mixing together.

> **Safety**
> Chemicals and other hazardous materials can often mix, causing effects that are different from the effects of any of the materials by itself.

The *additive interactions* that occur when chemicals mix can be described by the equation: 2 + 3 = 5. In this case, the effects caused by an exposure to a mix of two materials will not result in any more of a hazard than exposure to either of the materials by itself. Mixing these materials produces the same result as adding the materials together. With additive interactions, there is no interaction of the materials and the effects are simply the same as taking two different materials. An example of this occurs with certain prescription drugs. Taking a Tylenol tablet for a headache while drinking alcohol may cause additive effects because both are metabolized in the liver. If both are taken in at the same time, the liver must work overtime to break them down, causing an additive effect on the body.

> **Note**
> Additive interactions occur when chemicals mix together. The effect of mixing two materials will not result in any more of a hazard than exposure to either of the materials by itself.

Synergistic interactions occur when chemicals mix, and can be summed up by the equation: 2 + 3 = 20. Synergistic interactions are the most dangerous of all toxic chemical interactive effects because the resulting mixture can have a far more serious effect on the exposed individual. In many cases, we do not fully understand what happens when chemicals mix. The mix may produce chemical effects that are more hazardous than the effects of either of the materials alone or may react more vigorously than expected. Not all synergistic interactions are spectacular or attention-getting events. Some interactions are insidious and may occur without the knowledge of the person affected.

> **Safety**
> We do not fully understand what happens when chemicals mix. Synergistic interactions are the most dangerous of all toxic chemical interactive effects because the resulting mixture can have a more serious effect on the exposed individual.

For example, synergistic interactions occur in people who smoke tobacco and are also exposed to asbestos. The result of this exposure combination is often far worse than would be expected from exposure of these two materials simply being added together, or by either by itself. The effects of each of these are compounded and made much worse by the presence of the other. In the case of smoking and asbestos exposure, while data varies considerably, it is estimated that the chances of developing lung cancer are increased between 50 and 90 times over what would be expected from either smoking or exposure to asbestos alone. Synergistic interactions should always be considered whenever multiple materials are present.

Potentiation interaction effects may result when chemicals mix, and can be summed up by the equation: $0 + 2 = 10$. In this case, one of the materials does not have any adverse effects of its own, but can potentiate, or increase the likelihood of the other material into having far worse effects than it otherwise would. The material that is the potentiator might simply cause the other material to be taken into the body at a higher rate, increasing the uptake rate for that material without the potentiator itself having any direct effect. Such is the case with the popular solvent dimethyl sulfoxide (DMSO). While DMSO might not present any serious problems on its own, its interaction in a mixture often causes other materials to be rapidly absorbed by the body. If DMSO is applied to intact skin, it acts as a catalyst for transporting any other applied chemicals immediately through the skin. Essentially, it opens the door for another substance to enter the body. This potentiates or enables another chemical to cause significantly higher exposures than would otherwise have occurred.

> **Note**
> Potentiation interaction may result when chemicals mix and one of the materials does not have any adverse effects of its own, but can potentiate the other into having far worse effects than it otherwise would.

Antagonistic interactions are sometimes confusing but can be mathematically expressed as $4 + 6 = 8$; $4 + (-4) = 0$; or $4 + 0 = 1$. In certain rare instances, specific chemicals have an adverse effect on each other and cause harmful effects to be reduced. Such is the case of antagonism. With antagonistic chemicals, the harmful effects of one (or both) are decreased or eliminated due to their interaction with each other.

A classic example of this antagonistic effect is the drug naloxone (Narcan) which is given to combat narcotic overdoses such as heroin. In serious overdoses, patients are given the drug naloxone, an antagonist for narcotics. The naloxone competes for the same nerve receptor site as the narcotic does and consequently lessens the bad effects of the narcotic. Naloxone is an antagonist only for narcotics and does not interact with other substances.

> **Note**
> Antagonistic interaction may result when chemicals mix and one causes the harmful effects of another to be lessened.

Routes of Entry
the pathways that materials use to enter the body; the four routes of entry are inhalation, contact, ingestion, and injection

Route of Entry Perhaps the single most important of the exposure factors relates to the ability of the material to enter the body in a dose high enough to create problems. **Routes of Entry** refers to the pathways materials use to enter or come into contact with the body. In simple terms, they are the highways or paths that the material uses to travel into the body, and each material that we work with has to take one or more of these paths in order to adversely affect us. Simply working in an area where a material is present, as either someone who is handling the material or as response personnel, would not necessarily mean that an exposure had occurred. An exposure occurs when the material has the opportunity to take one or more of these routes of entry. These pathways include *inhalation, contact, ingestion,* and *injection.*

Take, for example, the situation in which someone works in an area where asbestos is present. Asbestos can be a significant workplace hazard in certain circumstances. In the example, the worker is in an area where pipe insulation contains asbestos. The asbestos in this example is encapsulated and not able to become airborne. Simply working in this space would not necessarily expose the employee to the hazard because asbestos is most harmful when it enters the body through breathing. Understanding the principle of the routes of entry is the key to safety when dealing with hazardous materials.

■ Note
Materials that have a toxic effect on the body must enter or contact it through one or more of the four routes of entry.

The means by which a chemical enters the body is certainly a major factor in its ability to do harm. Water is a good example of this concept. We can drink it almost without issue. It doesn't permeate the skin and doesn't cause problems, but it could prove fatal if we try to breathe it for any length of time. Certainly, this example shows that the route a material takes in entering the body is an important factor in the measure of its ability to do harm.

Whether as a responder or as a worker at a site, it is vital to understand that the concept of routes of entry is the foundation for much of our study. In fact, knowing the route of entry a chemical may take is one of the most important considerations in selecting an appropriate level of PPE. Each material that we are potentially exposed to may use one or more of the routes to get into us and cause harm.

Inhalation
the route of entry that includes exposure via the respiratory system

alveoli
the plural of alveolus. Microscopic air sacs in the lungs, which are surrounded by a blood supply, where oxygen is delivered to the blood, and where hazardous wastes are exchanged

Inhalation The route that poses the highest level of concern to those who work in the field of hazardous materials handling and response is **inhalation**. This term is used to describe chemical exposure via the respiratory system. In Figure 3-3, we can see that the respiratory system starts at the nose and mouth and continues into progressively smaller passages until it terminates in the lungs.

The respiratory system is quite complex and encompasses a large area within the body. Ultimately, the inspired air terminates in the small air sacs called **alveoli** (the plural of alveolus). These small air sacs number in the millions in the average healthy adult and each is surrounded by its own blood supply. As shown in Figure 3-4, this is where the alveolus and the blood supply meet and oxygen passes from the lungs into the blood. This is also where the waste products such as carbon dioxide (CO_2) in the blood pass into the lungs so they can be excreted with exhaled air.

Because of the large numbers of individual air sacs involved—millions in a healthy adult—the surface area for gas exchange inside the lungs is incredibly

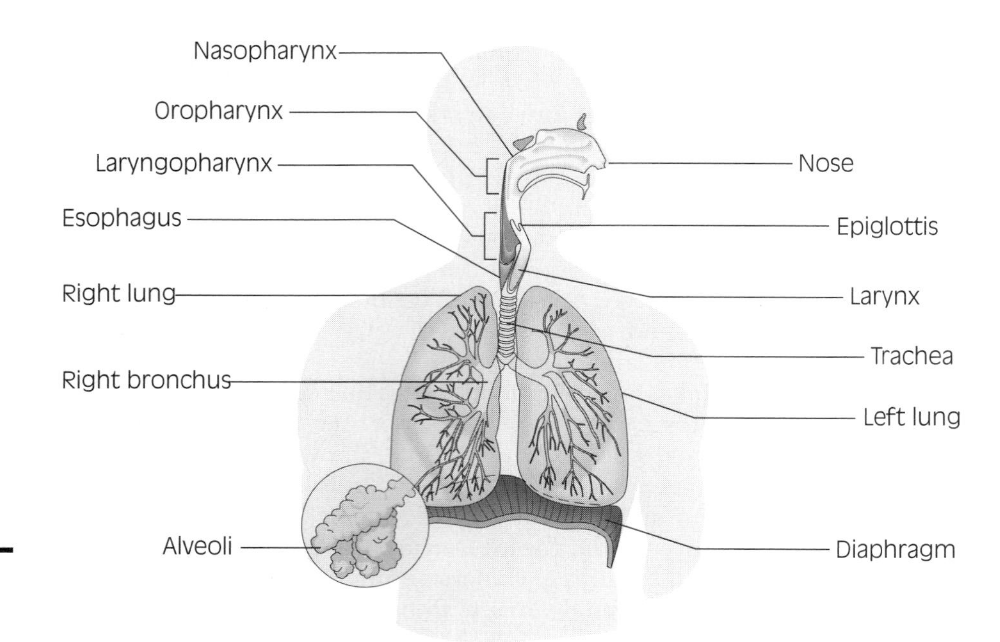

Figure 3-3
Anatomy of the respiratory system.

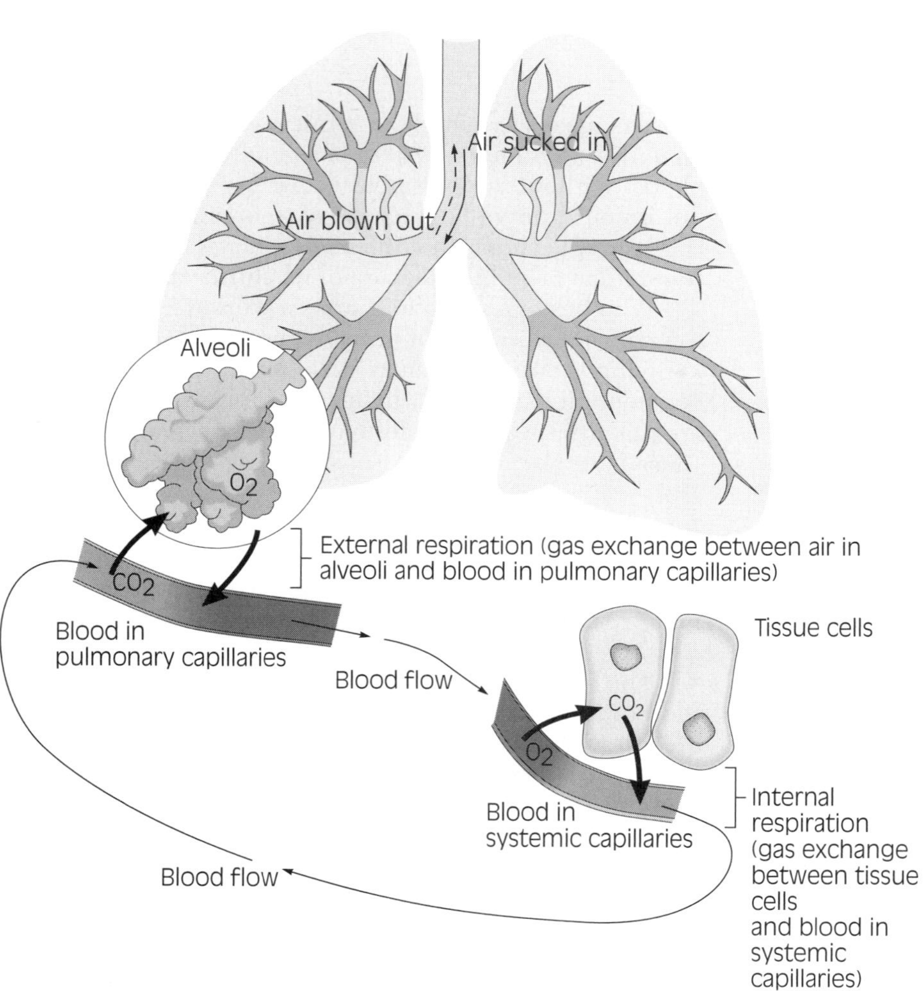

Figure 3-4
Each alveolus is in direct contact with the blood, facilitating transfer between the two.

large. In fact, it is reported that if we were to open up each of the hundreds of thousands of alveoli and measure their surface area, we would have approximately 700 square feet. Additionally, a normal, healthy adult at rest breathes approximately 15 times per minute. While this rate varies with exertion and sleep, we can generally count on the fact that an adult will breathe over 21,000 times in a 24-hour period. Based on the average respiratory rate, it is estimated that the normal, healthy adult breathes approximately 35 pounds of air per day, making the inhalation route a major consideration when the adult works with or around hazardous materials that could use the route as a pathway into the body.

> **■ Note**
> The lungs provide a major potential for materials to enter the body.

In any work in the hazardous materials field, the concern should be to reduce the possibility of exposure through inhalation. Many inhaled materials can cross the alveolar membrane just as oxygen does, including most gases and microscopic dust particulates that enter the alveoli, where they are broken and then cross into blood. Particulates such as dust needs to be smaller than 10 micrometers (microns) to reach into the deepest of the alveoli, because particles larger than 10 microns are filtered out by the tiny hairs and other mechanisms that help protect the respiratory system. Because of this, larger, visible particles are not the ones that present a hazard, and the body can either swallow them or rid itself of them in mucus. Without magnification, it is estimated that the normal human can only see particles down to approximately 25 microns. These would be the visible dust particles that we see floating in the air when bright sunlight shines into a room. Without the light, these tiny particles are invisible despite being 2½ times larger than those that can get into the lungs. So the inability to see dust in the air does not mean that we are not being exposed because it is the less visible particles that can enter the alveoli. Figure 3-5 illustrates the relative sizes of particulates.

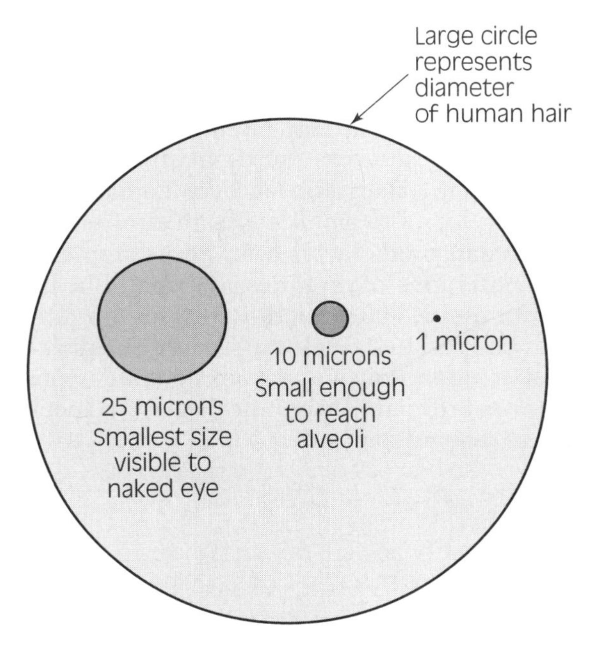

Figure 3-5
The relative sizes of particulates.

■ **Note**
In order to reach the alveoli, a particle must be smaller than 10 microns.

■ **Note**
The eyes cannot see particles small enough to enter the alveoli. Without magnification, a particle smaller than 25 microns cannot be seen.

odor threshold
the level at which the typical person can detect an odor from a particular material

Another similar issue when working with hazardous substances is the presence or lack of an odor for a given substance. This is referred to as the **odor threshold** for a particular material and describes the level at which the typical person can detect an odor. However, whether or not the material has an odor, you can detect it through smell. We should never rely on the presence or absence of odor to determine the presence of a chemical because a number of factors influence this. Some materials have an odor threshold that is higher than the level at which the material can cause harm. In addition to the fact that each of us may detect odors at different levels, the material may not have a noticeable odor, or the odor may be masked by the presence of other materials. The lack of an odor does not guarantee that we are not exposed. Carbon monoxide (CO) gas is an excellent example to illustrate this point. It is produced from almost all types of combustion processes and is colorless, tasteless, nonirritating, and completely odorless. No one can detect its presence by odor alone, and an exposure to levels of CO above 1200 ppm could be fatal. CO is only one of many materials that cannot be detected by their odor at levels lower than those that cause adverse health effects.

In other cases, the chemical may be detected by odor initially, but prolonged exposure deadens the sense of smell. A classic example of this is found with hydrogen sulfide (H_2S), which is a very common material found in a range of applications and one that everyone who works in the field of hazardous materials should be familiar with. At low concentrations, it has a strong odor, much like that of rotten eggs, but at 200 ppm, it can actually knock out a person's ability to smell it through the odor. It is widely accepted that exposures above 200 ppm of H_2S will immediately deaden the sense of smell, causing an inability to detect its presence. Exposure to concentrations above approximately 900–1000 ppm can be immediately fatal. In addition, H_2S is also very flammable and should be considered a dangerous fire risk.

Another problem with detecting chemicals by their odor is that once you smell a particular substance, it may be too late to prevent damage from occurring. Some materials are so toxic that by the time we can smell them, we have already been exposed to a dangerous or even Lethal Dose.

Conversely, the ability to smell a chemical does not necessarily mean that we are exposed to dangerous levels of it. For example, ammonia vapors can be detected in concentrations of as little as 5 ppm, but is not dangerous until it reaches a concentration of 300 ppm, the IDLH for the substance. Numerous other examples of this can be cited, including home cleaning products and perfume. The detection of an odor simply does not indicate exposure to high levels of a material. Without appropriate air monitoring using specialized instruments, we can only guess the level of exposure.

! **Safety**
The presence or lack of an odor does not indicate or rule out hazardous exposure. Air monitoring is required to determine the level of hazard present.

contact/absorption the process by which materials enter the body by passing into or through the skin, eyes, or mucous membranes

Contact A route of entry in which materials enter the body through direct contact with the skin, eyes, or mucous membranes is **contact**. Another term that sometimes is used instead of the term *contact* is **absorption**. However, the process of absorption is present with other routes of entry because materials are absorbed into the blood as they pass through the alveoli when they are inhaled or into the digestive system when they are ingested. Contact, on the other hand, means we must be close enough to physically touch or contact the materials. Generally, contact exposure most often is in the form of a liquid. However, some hazardous materials can be absorbed when they are in gaseous or even solid forms.

Skin contact is the most likely mechanism for contact exposure to take place because skin makes up a large portion of the body's exterior. The makeup of the skin is very complex. It is composed of several layers that provide barriers of protection against many materials, including those that are hazardous. Figure 3-6 shows how the multiple layers of skin create some protection against contact hazards.

The outermost part of the skin, called the epidermis, is the layer of mostly dead cells. This waxy mantle of cells provides the first layer of defense from chemicals and many other hazards.

The lowest level of the skin is the dermis layer. This layer is rich in blood supply. Once chemicals enter the dermis, they can be quickly absorbed by the blood and transferred to other areas of the body, where they can produce harmful effects in addition to any inflicted on the dermis itself.

Generally speaking, the skin is a waterproof barrier that provides a fairly effective protective layer against a wide range of materials. This is partly due to the makeup of the layers and composition of the skin. Even in the case of disease-producing materials, such as germs or pathogens, intact skin is the first and one of the most effective means of preventing these materials from entering the body. Although it is true that the skin will provide a significant barrier of protection against many substances, we must also understand that there are limitations to this shield. It is important to consider that there are certain types of hazardous materials that can and do pose a significant risk if direct contact is made. Two of those that pose this increased risk are Corrosives and Solvents.

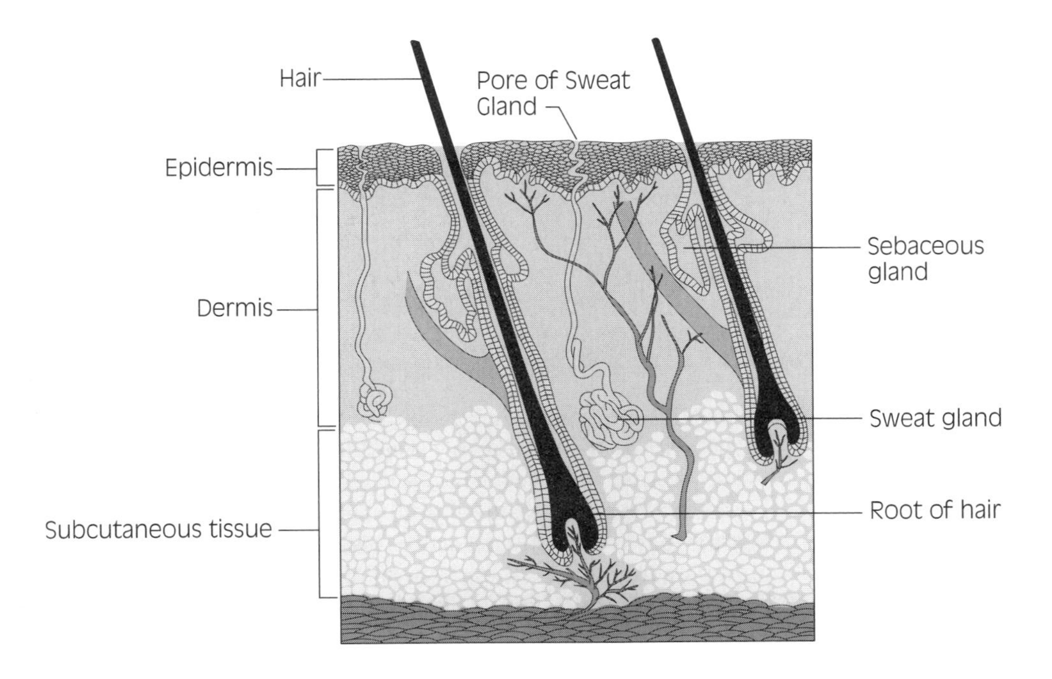

Figure 3-6 *Anatomy of the skin.*

> **Note**
> The skin layers help to provide a high degree of protection against many materials, including disease-causing materials.

Corrosives, by definition, are materials that can cause visible harm or damage to tissue on contact. There are two types of Corrosive materials, acids and bases. Although the effects of the two differ, both acids and bases can significantly damage the skin, as evidenced by the corrosive burn. Some of the common corrosives damaging to the skin are sulfuric acid, hydrofluoric acid, hydrochloric acid, sodium hydroxide, and potassium hydroxide, which all make the skin unable to provide protection from their harmful effects.

Solvents are a class of material similar to Corrosives in their ability to produce very harmful effects on contact. Again, the very term *solvent* indicates that it is capable of dissolving something. In this case, the oils and fats that help to keep us waterproof and provide protection from harmful effects of chemical exposure can also be affected by the solvent. The outer waxy mantle could dissolve when it contacts these materials. Once the protection afforded by the oils, fats, and waxy mantle is gone, the skin becomes an open sieve through which the solvent and other things can enter the body.

> **Safety**
> Corrosives and Solvents are groups of chemicals that can be rapidly absorbed through the skin. The damage they inflict can allow other materials to enter the body through the openings caused by the damage done to the skin.

In addition to the skin, we also need to consider that the eyes are an area of potential exposure through this contact/absorption route. Anyone who has ever exposed their eyes to a mild soap knows how painful the exposure can be. This is due in large part to the fact that the tissue in the eyes is moist, extremely thin, and filled with a large number of blood vessels very close to the surface. Throughout the day, we consistently replenish the moisture in our eyes through blinking. This moisture provides the vehicle that the chemicals can ride into the body. This is particularly true in the case of an **anhydrous material**, which is a material that contains little or no water. Anhydrous ammonia as a gas and sodium hydroxide pellets are examples of anhydrous materials. Each normally exists without water and each combines with the water in the eyes to produce very concentrated forms of those materials, resulting in significant damage.

anhydrous material
material that contains little or no water

> **Note**
> Anhydrous materials can cause serious harm if they contact our eyes where they rehydrate and make very concentrated forms of the material.

The mucous membranes are another area of concern relative to the contact/absorption route of entry. Mucous membranes are specialized tissues that are both moist and **vascular**, and are concentrated in areas around the face and groin. Vascular tissues contain a high concentration of blood vessels, making them particularly vulnerable because the closer the blood system is to the surface of the membrane, the less distance the chemicals have to travel in order to enter the blood and then into the body. As with the eyes, mucous membrane areas are extremely vulnerable to exposure, making absorption a real concern.

vascular
those areas that have a high concentration of blood vessels

Safety
Vascular areas present a higher contact hazard because the blood system is closer to the surface of the membrane, making it less distant for the materials to travel in order to enter the body.

An example of absorption through the mucous membranes would be the use of nitroglycerin tablets to treat chest pain. Patients who take nitroglycerin tablets do not swallow the tablets, instead the tablets are placed under the tongue where they dissolve and are rapidly absorbed into the bloodstream given that the blood supply to this area is high. This effect occurs within just a few minutes because the blood supply is concentrated and very close to the surface of the membrane. The active ingredients in the medicine can rapidly get into the blood stream where, in this case, positive effects will occur. In the case of other harmful substances such as chewing tobacco, a similar action occurs and the nicotine and other harmful materials enter the blood supply and have their harmful effects throughout the body.

Finally, any contact with a hazardous substance requires immediate action. Unlike other of the routes of entry, contact with a material gives us an opportunity in many cases to recognize the potential exposure and take action to reduce the impact of the contact. For this reason, in the case of contact with a hazardous substance, we should take immediate action and reduce the exposure. This is most often done through the use of water. Water should be applied to exposed skin or eyes in the event of a chemical exposure for at least 15 minutes. This involves a complete soaking, removal of contaminated clothing if indicated, and holding back the eyelids to allow complete irrigation to the eyes if they are exposed. Time is critical once the material has touched the body, so the prompt application of water is vital to reducing the exposure. For this reason, OSHA requires that safety showers and eyewash stations such as those shown in Figure 3-7 be available within a 10-second travel distance to areas where corrosive and other irritating materials are present.

ingestion
the route of entry that involves materials entering the digestive system

Ingestion The third route of entry for materials to enter the body involves entry through the mouth into the digestive system, and is known as **ingestion**. This route begins at the mouth and continues some 28–30 feet until it passes through the body at either the rectum or urethra. This path is essentially a long tube composed of specially designed tissue that will extract nutrients and vitamins from the food we eat. The purpose of this type of absorption is to bring energy into the body and eliminate the waste products from the digestion process. However, harmful materials can enter the system through this route and cause considerable harm. Figure 3-8 shows the components of the digestive tract of the average adult and helps to illustrate the complexity of the system.

secondary exposure
exposure that occurs as a result of contact of a material with the hands, which get contaminated and then can transfer the contaminant into the mouth where the material is then ingested

While exposure through this route is not as direct as with the previously discussed routes, the route does present issues to those who work with hazardous materials in that the exposure is often not recognized. In fact, most of the intake of harmful materials via this route of exposure occurs as a result of a **secondary exposure**. A secondary exposure occurs when a material is taken into the body after it contacts some other item, which is then placed into the mouth. This is the case when working with hazardous materials in the course of the day, which subsequently contact the hands. Unless the hands are washed or decontaminated, everything that is touched after the initial contact with the hazardous material is contaminated with it. This could result in a secondary exposure when eating, smoking, or in any way touching skin, eyes, or mucous membranes.

Figure 3-7
Example of an eyewash/safety shower station at a site where hazardous materials are found.

injection
materials getting through the skin through openings or with pressure

pathogen
a disease-causing material

Injection The final route of entry is termed **injection** and is the route by which chemicals pass through the skin or through openings in the skin. In many studies, this route is often discussed when covering the contact/absorption route. However, doing so minimizes the ability of the skin to serve as a blocker for many types of materials. Injection is where the material may go through openings in the skin created if the skin is already broken, has an open wound, or when damaged as a result of a puncture or contact with something contaminated with a hazardous material. The best example of this is contact with a used hypodermic needle, which might be contaminated with blood that contains a **pathogen**. Pathogens can cause disease if they are allowed to enter the body, and while the skin would normally block the entry of pathogens, the needle serves to inject the material through the skin where it can cause harm.

Another, and often overlooked, example of the injection route is that of compressed air forcing chemicals through the skin. Workers inappropriately using the compressed air hose to blow contaminants off their clothing and skin run a risk of exposure. By doing so, the workers may actually be injecting materials that otherwise would not penetrate their skin. Such injection can also occur whenever employees use power sprayers in painting operations. This is the reason that OSHA has established maximum levels of air pressure that are allowed to be used when blowing substances off employees. But even with these low levels, additional hazards are created when the materials are blown back into the air where another chance is presented to breathe in the harmful products.

Protection for the Routes of Entry Identifying the potential route(s) of entry for a chemical exposure is only the beginning of chemical safety. Once we know how the

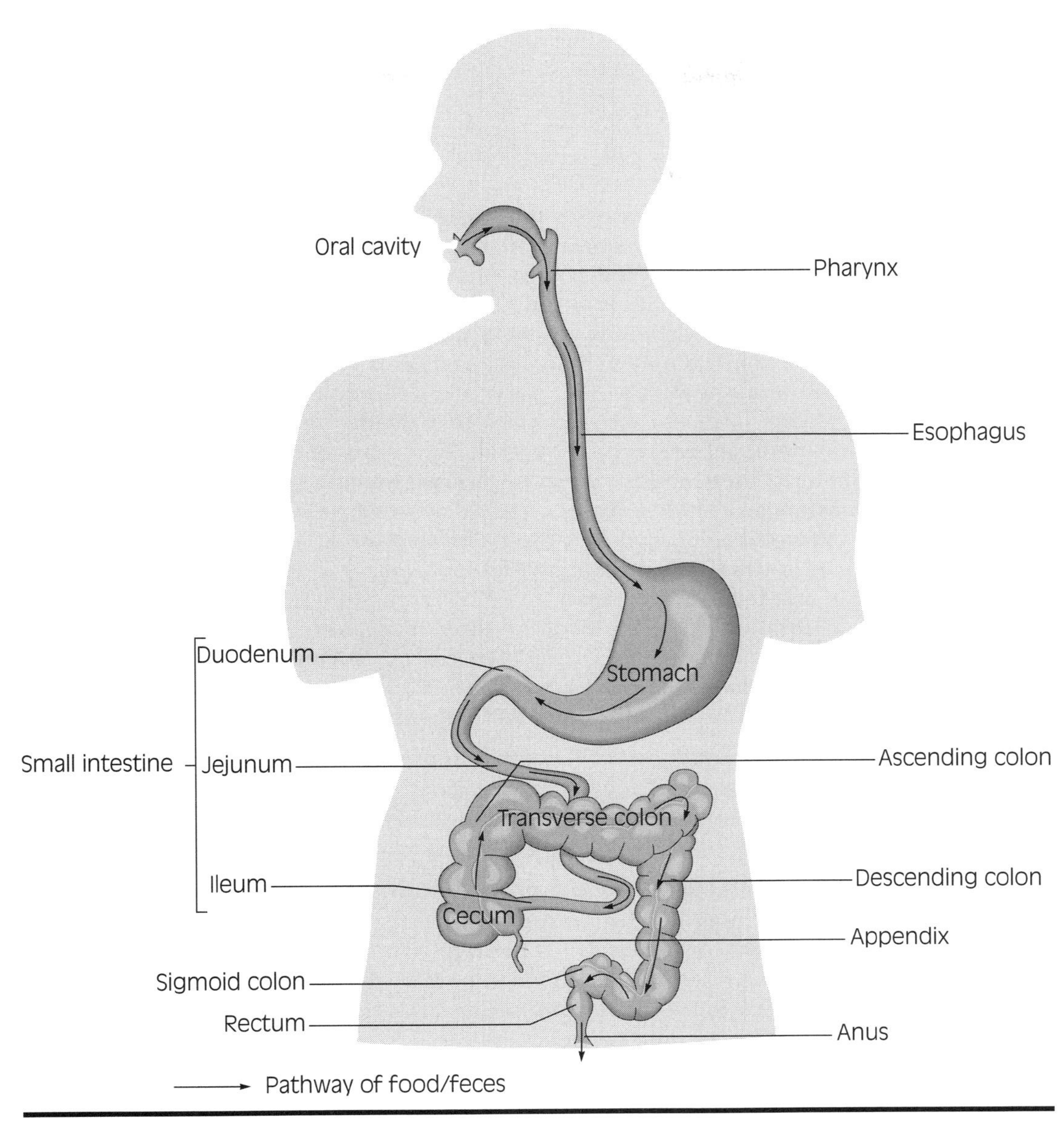

Figure 3-8 *Anatomy of the digestive tract.*

material will enter and harm us, we must also develop a strategy to protect or block these routes so that materials do not enter the body and produce harmful effects. In virtually every case, once we know which route or routes to block, we can provide a means to do so and work safely with the chemicals involved. Keep in mind that it is important to consider the potential that some materials can be harmful through more than one route.

To illustrate how some materials pose a hazard through multiple routes of entry, let us look at the examples of CO_2 and hydrogen cyanide. CO_2 is a gas that is present in the air we breathe. While it is considered to be nontoxic, atmospheres with high concentrations of CO_2 (above 40,000 ppm) may pose a hazard through its ability to **asphyxiate** us by the exclusion of oxygen from the air. Whenever work is performed in areas of high concentration of CO_2, respiratory protection must be worn. CO_2 has no dermal toxicity, so it will not hurt the skin in its gas form. It is not corrosive to the skin and poses no other risk to personnel

asphyxiate
the ability to suffocate through the exclusion of oxygen

than that of asphyxiation—the inhalation risk. It is not hazardous to us if we have cuts or breaks in the skin, and will not pose a problem to us if we ingest it. In fact, we may even swallow CO_2 when we drink carbonated drinks with nothing worse than a strong burp or belch as the result. Essentially, it is dangerous by a single route of entry, and if the inhalation route is protected, the material is safe to work with.

Contrast the properties of CO_2 to those of hydrogen cyanide. Hydrogen cyanide or hydrocyanic acid is a material that is hazardous to us through more than one route of entry. We certainly cannot breathe it because it is toxic by inhalation at approximately 50 ppm. Furthermore, if the material enters the body through injection or ingestion, it can be rapidly fatal. A person suffering a mild exposure to this chemical may exhibit such symptoms as nausea, vomiting, headache, and irregular respirations. As a gas or even dissolved in liquid form, hydrogen cyanide can permeate the skin and cause death. Unfortunately, this material poses considerable high risks through all four routes of entry, making it critical to identify all of them and to take appropriate protective action when working with this substance.

To effectively block the route or routes that a particular substance will take is the key to safely working with the material. From the discussion of hazard and risk assessment in Chapter 2, we know that the method of providing protection is to follow the hierarchy of the three levels of control. Engineering controls are the first line of defense in blocking the route of entry, followed by administrative controls, then the use of appropriate PPE. Table 3-1 is a summary of how these systems come into play once the route of entry has been identified.

> **■ Note**
> Exposure factors are the ones we can control by not allowing materials to enter.

Modes, Methods, and Effects of Exposure It is important to understand the effects that chemicals may have should we allow them to enter the body. This discussion is important for two reasons. First, numerous OSHA regulations require that we discuss the effects of chemical exposure as part of our program, and second, it only makes sense that we understand the consequences of chemical exposure if we are the ones being exposed.

In reviewing the effects that hazardous materials have on us, it is necessary to look at the manner in which they exert their effects. These chemical effects

Table 3-1 *Protection for routes of entry.*

Route	How	Prevention
Inhalation	Vapors, gases, fumes, mists, dusts, and airborne substances	Engineering controls, including ventilation Administrative controls, including distance Respirators (PPE)
Contact/ absorption	Skin contact, eye contact, mucous membranes	Keep materials in closed storage Warnings such as labels to help avoid contact Proper use of PPE
Ingestion	Secondary exposure: Chemicals on hands, food	Proper hygiene/decontamination
Injection	Direct chemical exposure through the skin via open wounds or cuts, or by injection of compressed air	Proper tools, and proper PPE

are accomplished through either an *acute* or *chronic* exposure, or a combination of both. While both of these terms represent exposure modes or methods, they are not the same and occur at two distinctly different levels. It is important to understand that these terms relate only to the method or way that the exposure is received and not the time that it takes for the materials to exhibit their effects on us. The time frames of chemical exposure are discussed later in this chapter.

acute exposure a one-time or relatively brief exposure of short duration

Acute exposure is likely to be the one best known to most response personnel. Acute exposure refers to a one-time or relatively brief exposure of short duration. While there are a number of variations of this definition, it is clear in all of them that an acute exposure is one that is short in duration, and which is related to a single exposure of the material. In general, the acute exposure, in order to exhibit toxic effects, is generally of a higher concentration than we are physically able to deal with at that time.

Examples of acute exposures include the immediate skin destruction that occurs secondary to a dermal exposure of concentrated sulfuric acid, or death that occurs from a relatively brief exposure to a Highly Toxic gas such as hydrogen cyanide. Acute exposures can occur whether we work with the product on a regular basis or as an emergency responder. The key to understanding this method of exposure is that the dose is delivered to us in a single, short-term event. It often will expose us to a higher level of the material with a range of results based on the materials involved.

chronic exposures occur as a result of multiple exposures to the material over a period

Chronic exposure effects occur when we are exposed to materials multiple times. Generally, the levels involved are lower than those associated with acute exposure. Again, depending on the definition used, chronic exposures describe exposures to the same material multiple times and generally to a lower level or dose. Depending on the material involved, chronic exposures may or may not result in serious health concerns to the exposed person.

A classic example of a chronic exposure is when someone smokes cigarettes for a long time and develops cancer or heart disease. Smoking a single cigarette would constitute an acute exposure and generally not result in health concerns. But the multiple doses produce the diseases. Another example is of workers who are exposed to asbestos in low levels for extended periods. A single exposure to the material may not result in health concerns, but the chronic exposure can produce lung disease as a result of the buildup of the asbestos fibers in lung tissue or the possibility of cancer as a result of the chronic doses. Chronic exposure effects are more pronounced in occupations placing workers in contact with the materials on a daily basis, rather than those that involve emergency response.

It should be noted that a material could have both acute and chronic effects. There are a large number of materials that can exhibit effects from a single exposure as well as from repeated doses of smaller amounts. A good example of this concept can be found in some alcohol compounds. Ethyl alcohol is the form of alcohol found in liquor. While it can produce effects from a single and relatively large exposure, it can also produce adverse health effects if ingested for a long time. Intoxication can occur on an acute level, and liver disease is possible in those who ingest alcoholic beverages on a chronic basis.

> **■ Note**
> Health effects can occur as a result of a single exposure called acute, or as a result of multiple exposures termed chronic.

Specific Chemical Effects In addition to the differences between acute and chronic exposure, it is also important to study the overall effects of an exposure to a hazardous substance. When we study this, we find that there are two distinct types of effects that can occur as a result of that exposure. Whether the mechanism of the exposure was acute or chronic, the health effects that result are either *local health effects* and/or *systemic health effects.*

local effects
effects of chemical exposure occur at the site where the chemical contacts the body

Local effects of chemical exposure occur at the site where the material contacts the body. An example of this would be a chemical burn following contact with an acid. Local exposures most often can be identified early and acted upon immediately because the effect is at the point of contact with the material. If the effect is local, the material is generally still in contact at that point and can often be washed off or removed in some other way. Local exposures are more easily recognized than systemic exposures because they occur at the point of contact with the body. The burn injuries are an example of a local health effect because the damage occurs at the point of contact with the material.

systemic effects
occur in a body system as a result of the material entering the body and getting into the blood stream

target organ
an organ adversely affected by the hazardous materials that enter the body

Systemic effects of exposure to hazardous substances are quite different from local effects and can be very serious. **Systemic effects** occur when the substance(s) is taken into the body and then enters the blood where it goes to a body system or **target organ**. Target organs are affected by the material as it enters the body system. This is accomplished when the material is absorbed through the skin, inhaled into the lungs, ingested, or injected through the skin. Because the area of the body affected is not necessarily the area of the body that contacts the material, the symptoms might be dismissed or attributed to something else. Commonly impacted body systems include gastrointestinal, reproductive, respiratory, and cardiovascular systems. However, the most common and often most problematic of the body systems affected is the central nervous system that includes the brain. Once this happens, the exposed person is often confused and unable to even recognize that an exposure has occurred or is occurring. This is one of the reasons OSHA mandates that the buddy system be used when working in hazardous areas. The second person can watch for changes in the behavior of the exposed person and take action to help them get out of the area and treated for their exposure.

> **Safety**
> It is important to realize that the central nervous system is the body system most often affected by chemical exposure.

Chemical Exposure Time Frames Local and systemic effects from either acute or chronic exposures do not necessarily occur as soon as we come into contact with them. In fact, a person may not suffer the effects or symptoms of the exposure for some time following the actual event. As we look carefully at this concept, we find that the effects of chemical exposure time frames fall into three distinct time frames. While determining these time frames is not an exact science, it does provide us with a basis from which to learn. These three areas include immediate, delayed, and long-term effects.

immediate health effects
generally occur within minutes following the exposure

Immediate Effects Some materials we contact might have immediate effects while others may not. **Immediate health effects** occur within minutes of exposure to the hazardous substance. These immediate effects can take many forms depending on the substance and can range from minor irritation to significant injury or even death, depending on the substance, its properties, and the dose that we receive. Immediate health effects most often occur from an acute exposure to the substance and can be either local or systemic depending on the material. The key point with these types of exposure is that the effects are immediate, making us more aware at the time of the exposure. From a safety perspective, these are the best types of effects to have because the exposed individual is aware of the exposure and can take steps to stop, reduce, or immediately treat the symptoms of exposure.

A classic example of this type of exposure would be the result of contact with an acid on the skin. The burning and pain would likely occur in a short time, making us aware of the exposure and allowing us to take the appropriate action.

> **■ Note**
> Immediate effects are generally more recognized because we become immediately aware of the exposure, generally at the time of the exposure.

delayed effects
not seen for up to several hours after the initial event or exposure

Delayed Effects The **delayed effects** of exposure to a chemical are not seen for up to several hours after the initial event or exposure. Because of the time delay between the exposure and the onset of symptoms, delayed health effects are often mistaken for other problems and often are not associated with an earlier occurring chemical exposure. This is a dangerous situation because the exposure to the material may continue for a longer time than with immediate effects.

Examples of this type of exposure include the delayed pain and damage resulting from exposure to hydrofluoric acid, which unlike other types of acid often can penetrate the skin without damage or pain and then migrates to the bones where it reacts with calcium causing serious damage. Another example would be the delayed onset of disease after exposure to pathogens or **pulmonary edema** resulting from an exposure to chlorine gas. Pulmonary edema is a condition that occurs when the lungs are damaged and fill with fluid. In each of these cases, by the time the symptoms are noted, the actual exposure has stopped and therefore cannot be reduced.

pulmonary edema
a condition in which fluids fill our lungs

long-term health effects
do not show until months or years following exposure

Long-Term Effects **Long-term health effects** occur long after the initial exposure. In many cases, the symptoms may not be realized for months or even years after the exposure actually happened. Again, as with delayed effects, by the time the actual symptoms are noted, the exposure will be long over and there may be little or nothing that can be done to stop the health effects from happening.

Long-term health effects vary considerably. Many of these effects fall into three distinct groups of problems that include reproductive harm, organ damage, and the formation of cancers.

The reproductive long-term effects could result in damage to a developing fetus from exposure to a teratogenic substance. This could occur months after the exposure.

Organ damage can show up years after exposure, often as a result of chronic exposure to hazardous substances. Often, the lungs, liver, and kidneys suffer damage from such exposure because these organs are filters in the body. Over time, the organs start to fail due to the continuing exposure to hazardous materials. Other organs are also vulnerable to long-term exposure depending on the specific types of materials that are involved. Again, the effects may not be apparent until years after the first exposure.

> **■ Note**
> Our body filters, including the lungs, liver, and kidneys, are often the ones suffering long-term health effects.

The final long-term health effect is cancer. Cancer is most often caused by exposure to carcinogenic materials, as we discussed earlier. However, the actual disease may not show up for years after the exposure, such as the case with cancers associated with exposure to asbestos.

■ Note
Delayed and long-term effects of exposure to chemicals present unique hazards in that once the effects of exposures are noted, it is too late to go back and take steps to reduce the exposure. Because of this, these exposures should not be taken lightly.

■ Note
Long-term effects of chemical exposure often include organ damage, reproductive damage, or cancer.

The You Factors

Like the toxicity factors, the You Factors are mostly fixed and something that we are unable to do much about. Each of us are the way we are, and there isn't much that we can do about it. We are born a particular sex, our heredity and genetic makeup largely are set as a result, and there is little that we can do to change any of those things. In fact, even our basic patterns of behavior are set by the time we reach adulthood and enter the working world. Each of us is largely different from the other, and yet there are a number of individual differences that are worth noting as we review the toxicological effects that chemicals have on us.

! Safety
It is important to recognize that individuals may respond differently to the same material at the same level.

It is a very well documented fact that men and women can and do respond differently to similar levels of the same chemicals. Differences in the average body fat between men and women influence how a chemical will be tolerated in many cases. A person with a greater mass may tolerate a certain dose of a chemical better than a person with a low body mass. Certain materials that only have an effect on a developing fetus would be of less concern to men than to women who are or might become pregnant.

Even ethnicity can influence some of the ways that a material can affect us. Certain populations have more or less of enzymes and other substances in their livers that help to process materials that enter the body. Darker skin populations have a lesser risk of skin cancer than lighter pigmented population groups. And the list goes on and on.

■ Note
A number of personal factors influence our response to a particular substance.

Other personal factors also influence our response to an exposure to hazardous materials. Individual sensitivities to a material, prior exposure to the material, age, personal health, prior illness, and many other personal factors come into play when we discuss this topic. Based on these compounding factors, each of us will respond differently once the chemical enters our body. But in them we can identify two points that the individual differences have in common. These are the retention factors that each of us has to various materials and our individual metabolism as we process them.

retention factor
the time taken by the body to rid itself of the material

Retention factor is partly a result of the material and its toxicity as well the body's ability to rid itself of the exposure. In some cases, the body is quickly able to deal with the exposure and reduce or eliminate any adverse health effects. In other cases, the body is unable to metabolize the chemical and subsequently stores the substance instead of eliminating it. A person's metabolism has a great influence on his/her retention factor and a number of personal traits change this at various times. The length of time a chemical remains unchanged is the basis of the metabolism factor.

■ **Note**
An individual's metabolism is a factor in how long a material remains inside the body.

CONCLUSION: THE DOSE MAKES THE POISON

The final concept for discussion in the area of toxicology is that of dose. Throughout the chapter we have described terms that relate to the dose. These terms have included descriptions of doses where harm occurs, such as the LD50 or IDLH. In other cases, we have talked about the safe doses of the material. Here we have related the safe 8-hour dose or PEL, or the safe 15-minute dose, which is termed the STEL. In each case we work under the premise that in order to produce harm, most materials must reach a specific threshold or dose level. That level is achieved as materials enter the body through one or more of the routes of entry. This concept of dose is one that has been around for a long time. In fact, one of the early Toxicologists named Paracelsus, who lived in the early 16th century, is credited with the saying that summarizes this concept.

> All substances are poisons; there is none that is not a poison. The right dose differentiates a poison and a remedy.

This concept is one of the most basic in the area of toxicity. To truly understand the concepts of toxicology, we must understand that it is the dose that we receive which creates the resulting harm. All of the health effects that result as a consequence of exposure are based on the concept of dose. In working safely with the materials, and in an effort to protect ourselves in whatever field of work places us in a situation with potential exposure to a hazardous substance, we simply work to maintain a safe level of exposure or dose. While we have little influence on the material itself and its toxicity, and on our personal factors, we can limit our exposure by blocking the routes of entry, which then will reduce the dose to which we are exposed.

SUMMARY

- Toxicology is the science devoted to the study of how materials adversely hurt us. This study includes identifying the chemicals' mode of action and health effects as well as other facets of exposure mechanisms.
- The branches of the science of toxicology include clinical toxicology, descriptive toxicology, mechanistic toxicology, forensic toxicology, environmental toxicology, regulatory toxicology, and industrial toxicology.
- The term toxic refers to the ability of a material to cause adverse effects to living tissues at a low dose. Poison materials can also cause harm, but generally require a higher dose.

- An acute exposure is one that causes a person to exhibit symptoms based on a single and often brief exposure to a high dose of the material.
- Chronic exposures occur as a result of multiple exposures to the material, which is generally in a low dose. Many carcinogens are harmful due to chronic exposures over a long time.
- Local effects of chemical exposure occur at or near the site of exposure. Strong Corrosives such as sulfuric acid cause localized skin damage at the site of contact only.
- Systemic health effects are those that affect the body systems after an exposure. Carbon monoxide is a systemic poison that may be brought into the body through inhalation, but ultimately affects vital organs and the circulatory system.
- Materials enter the body through the four routes of entry: inhale, contact, ingest, and inject.
- Inhalation is a major route of entry for chemicals to enter the body. This is due to the fact that the lungs provide a direct path of access for chemicals to enter. Inhalation hazards pose the highest threat to workers and responders.
- Ingestion is a route of entry whereby chemicals are brought into the body by contaminated food or unwashed hands.
- Injection is a direct route for chemicals to enter the body. If a person has cuts or other open wounds that are unprotected, a chemical can gain immediate access to the circulatory system. Compressed air lines can also inject chemicals through the skin when they are used as a means of decontamination.
- Contact or absorption refers to the route of entry chemicals take when they come into contact with skin, eyes, and mucous membranes.
- Exposures can occur over three distinct time frames. They are immediate, delayed, and long term. Immediate health effects are those that are felt by a person within minutes after exposure. Many corrosives cause pain upon contact. Delayed health effects are those that occur up to several hours after exposure. Hydrofluoric acid, for example, may not alert a person to the fact that they have been exposed. Long-term health effects occur after years of exposure. Cancer is a common long-term health effect that may result from years of exposure to substances like benzene.
- LD50 stands for Lethal Dose, 50% mortality rate to an animal test population. This is an expression of a material that is taken into the body.
- LDlo is the point at which the first subject in the animal test population dies as a result of receiving the dose.
- LDhi is an expression of a dose that is strong enough to kill 100% of the animal test population.
- The LC50, LClo, and LChi follow the same logic as LD50, LDlo, and LDhi. The difference between the two values is that LC, or Lethal Concentration, illustrates an airborne concentration.
- The odor threshold of a chemical is the lowest point at which it can be detected by smell. Some materials like ammonia have a very low odor threshold.
- PEL stands for Permissible Exposure Limit. It is an OSHA term that establishes the limit of a chemical exposure based on an 8-hour workday, 5 days a week for an entire working career.
- TLV–TWA is a term used by the ACGIH to denote an occupational exposure limit. As with the PEL, if exposed to chemicals below these levels, a worker should suffer no ill effects as a result of the exposure. Both the PEL and the TLV values are expressed as ppm, ppb, or mg/m^3.
- The Short Term Exposure Limit or STEL is a term used to describe the occupational exposure limit for a 15-minute work period. The STEL is an expression of an exposure which is higher than the TLV–TWA and the PEL, and is of a short duration. A STEL is limited to 15 minutes with at least an hour break from any contamination. These levels allow for short excursions into chemical atmospheres, but are based on short exposure times.
- IDLH is a term that describes the level that would present a serious exposure for workers. The term stands for immediately dangerous to life or health and indicates atmospheric levels of

exposures above which there could be death, irreversible health effects, or worker's inability to escape the area of contamination.

- An asphyxiant is a material that could cause death by displacing oxygen in the atmosphere. Nitrogen gas is an example of an asphyxiant that could cause the level of oxygen in the air to drop to harmful levels.
- Carcinogens are those materials that are known or suspected to cause cancer in humans.
- Teratogens and mutagens are those substances that may cause birth defects in a developing fetus.
- Additive effects are when two materials combine but do not result in any more of a hazard than each one would pose alone.
- Synergistic interactions are those that occur when two agents are combined with results that are far more dangerous than either substance would pose individually.
- Antagonistic reactions are those that are reduced in severity by the combination of two or more agents. Essentially, one agent would reduce or offset the effects of another.
- Potentiation reactions occur when a material of low hazard interacts with another and is made worse or more dangerous as a result of the reaction.

REVIEW QUESTIONS

1. List and briefly describe the four routes of entry to the body and the means for blocking each of the routes.
2. List three factors that make each person different in their response to a hazardous substance.
3. Identify the difference between a material that is toxic and one that is a poison.
4. Why are systemic exposures more hazardous than local exposures?
5. Define *carcinogen.*
6. Explain the difference between PEL and STEL.
7. List the three factors that help establish the IDLH level.
8. What resource concerning toxicity is required by the Hazard Communication regulation to be made available to employees?
9. List two reasons why delayed health effects present more problems than immediate health effects.
10. What is the difference between teratogenic effects and mutagenic effects?

ACTIVITY

TOXICITY DETERMINATION

Obtain an MSDS for a substance. The material may or may not be one that you work with or have prior knowledge of. For the substance, identify the following:

1. What are the primary routes of entry for the material to enter the body?
2. Do the effects of the substance occur as a result of an acute exposure, chronic exposure, or both?
3. Has the material been listed as being capable of causing cancer in humans?
4. What are the effects of exposure to the material? Are these local effects, systemic effects, or both?
5. Can the material cause teratogenic effects?

HAZARDOUS MATERIALS CLASSES AND PHYSICAL HAZARDS

Learning Objectives

Upon completion of this chapter, you should be able to:

- List the nine major hazard classes as outlined by the U.S. Department of Transportation (DOT).
- List the major hazards associated with each class of hazardous material.
- Identify safe storage and handling practices for each class of hazardous materials.
- Define detonation and deflagration.
- Define brisance and describe the potential effects of blast fronts and fragmentation.
- Define compressed gas, liquified gas, cryogenic liquid, vapor density, diffusion, upper and lower explosive limits, and BLEVE.
- Define oxygen-deficient atmosphere and asphyxiation.
- Define Flammable Liquid, Combustible Liquid, flash point, ignition temperature, miscibility, and specific gravity.
- Define Flammable Solid, Oxidizer, and Organic Peroxide.
- Define pyrophoric.
- Identify the three main groups of Poisons: pesticides, heavy metals, and infectious substances.
- Define Radioactive Materials.
- Identify Corrosive materials.
- Define acid, base, and neutralization.

CASE STUDY

Early one evening, the City of Campbell, California, Fire Department responded to a report of a "chemical fire" with possible injuries at a large chain drugstore in their jurisdiction. On arriving at the scene, fire personnel discovered that the entire store was filled with a white gas with a pungent odor. Numerous people were in the parking lot coughing, exhibiting signs and symptoms of exposure to a hazardous substance. The symptoms included irritation and tearing of the eyes, coughing, shortness of breath, and vomiting. Additionally, a small fire was observed inside the store on one of the shelves.

As the incident progressed, it was quickly determined that the white cloud contained chlorine gas that had evolved from some swimming pool and spa chemicals sold in the store. A total of 25 victims of the exposure were treated at the scene and those with serious exposure were transported to a medical facility for further treatment. The small fire was extinguished by fire and hazardous materials response personnel.

Subsequent investigation by the city Fire Marshal and chemical engineering staff identified the cause as the mixing of two materials that were stored on the shelves at the front of the store. It was surmised that the two materials, dry chlorine and a spa water conditioner that contained glycerin, alcohol, and fragrances, mixed on the store shelves. This interaction between the materials started a reaction that ultimately released significant amounts of chlorine gas and enough heat to start a fire on the shelf holding the materials and the packing materials that contained the substances.

What this incident ultimately demonstrated was the effects of chemical interactions. The materials in this incident were not highly concentrated industrial chemicals—they were materials available to the public for use in homes. But even commonly encountered household materials can, through their interactions, create hazardous and potentially deadly results. The incident shows the importance of understanding the physical effects caused by the interactions of materials, even those that are common. The effects can have serious consequences, and a working knowledge of these types of effects is essential for those who work with or respond to incidents where hazardous materials are present.

INTRODUCTION

In addition to the potentially harmful health effects, there is another way that hazardous materials can pose a threat to those who work with them or who are responsible for handling an emergency involving them in another way. We term this second group of hazards the **physical effects** of the materials. Physical effects are those associated with properties other than the health or toxicological ones discussed in the previous chapter. These include the material's ability to create fires or explosions, hazards from which ordinary measures such as blocking the routes of entry would afford us protection.

physical effects
non-health effects that hazardous materials can produce; examples include fire or explosion

To help in this, the chapter is divided into nine major sections representing the nine hazard classes assigned by the U.S. Department of Transportation (DOT) for all materials that are transported. Each of the nine classes is discussed at length, including the hazards associated with the various classes.

Although each of the hazard classes identifies a potential for a significant risk if not properly managed, in much the same situation found in toxicology, we know that there are ways to protect ourselves from the hazards. Each section in this unit identifies the hazards and presents ways to protect yourself when confronted with a situation involving these materials. Understanding the basic chemical and physical properties of each class is the key to proper handling and response procedures. Some basic hazardous materials chemistry is also included in this chapter to help you gain a broader understanding of how and why things behave as they do.

> **Note**
> Physical properties include the ability of the material to catch fire or create explosions.

EXPLOSIVES

Most people who are confronted with a situation where an explosive material is present have some basic understanding of the relative hazards of these types of materials. Perhaps this is conveyed through movies or television programs where explosions occur on a regular basis or through news programs showing the devastation created by weapons explosions. From these depictions, it is easy to see that these materials have considerable potential to cause damage not just in the immediate area, but also well beyond the area where the explosion occurs. For this reason, those who work with this class of hazardous substance require considerable training and certification. For those who do not work with explosives but who may confront them in cleanup activities or emergency response programs, a basic understanding of the hazards of these materials is essential.

> **Note**
> Personnel who work with explosive materials are required by OSHA and other regulatory agencies to receive training on the specific hazards of these materials.

explosive material
a substance that reacts to produce gases at high temperature and pressure so as to be able to damage surrounding areas

brisance
an expression of the shattering effect of a particular explosive material

An **explosive material** is a chemical substance that can react to produce a gas at a relatively high temperature and pressure and at such a speed as to damage the surroundings. The explosive reaction can be initiated by heat (including that caused by friction) or pressure, and the severity of the reaction depends on the particular materials and the conditions present. In an explosion, a whole range of hazardous events occur, including the following:

- *Projection or fragmentation hazard.* Fragments of the explosive container, its contents, or the physical surroundings are carried outward by the blast front. These fragments can be quite large, depending on what is surrounding the explosive material. Protection against fragmentation can best be accomplished by maintaining a safe distance. This decision on distance relies greatly on the materials involved, but should take into account the high potential for injury or death if any of these materials strike someone.
- *Pressure front propagation.* Explosions are always accompanied by an outward expression of pressure, which is sometimes called a blast front or **brisance**. The pressure created can be enormous (up to several thousand

overpressure
the pressure created over and above the normal or ambient pressure

psi
pounds per square inch

pounds over ambient pressure) and very fast moving. This resulting pressure wave can cause distant windows to break and the ground to shake following a sizable blast. It is interesting to note that it only takes a few pounds **overpressure** (≈5–7 **psi**) to cause damage and even rupture the eardrums of anyone within a short distance of the blast. When explosions occur within buildings, 10 psi overpressure has been significant enough to completely destroy the structure. If a person is caught in the pressure front of a serious explosion, a major injury or death is very likely.

- *High generation of heat.* Think of an explosion as an extremely fast-moving fire. The pressure front as well as the fire front move outward very rapidly and can cause great damage in the process. Temperatures of these fire fronts can be as high as 2000°F and will cause significant damage to personnel as well as most common construction materials.
- *Sound wave production.* Explosions are often identified by their sudden eruption of the sound or *bang* that they produce as the explosion occurs. In some cases, the *bang* of the sound wave will be heard before the actual *whoosh* of the pressure front arrives. In other cases, the sound wave will arrive after the pressure front has passed because the pressure wave of explosion moves faster than the normal speed of sound. Later in our discussion, we will see how this concept helps identify types of explosions.

Safety
Explosive materials create a series of hazards, including fragmentation, pressure, heat, and sound.

Because there is a broad range of explosive materials, this class is subdivided by the DOT into six subclasses based on the type of explosion that results. The subclasses of explosives include the following:

1. Explosives with a mass explosion hazard
2. Explosives with a projection hazard
3. Explosives with predominately a fire hazard
4. Explosives with no significant blast hazard
5. Very insensitive explosives with a mass explosion hazard
6. Extremely insensitive articles

Note
Explosive materials can be grouped into six divisions based on their ability to cause damage.

These six subclasses have been widely studied by those who work with explosive materials and there is a sound rationale for assignment into the six categories. For those of us who may occasionally encounter these materials, it may be simpler to reclassify explosives into three groups to help us better understand how to work safely with or around them. These three groups are:

1. High-order explosives
2. Low-order explosives
3. Blasting agents

High-Order Explosives

high-order explosives materials capable of detonation that can cause significant damage

detonations explosions where the pressure waves move faster than the speed of sound

High-order explosives are those that detonate or move at supersonic speed (greater than 1100 feet per second) throughout the mass of the explosive material. This rapid decomposition occurs instantaneously and self-propagates through the body of the explosive substance. Pressure and temperature waves move outward with the blast front, and the pressure developed in the areas close to the blast can be extreme. In true **detonations**, the pressure waves usually move faster than the heat waves, but often they move at close to the same speed. Dynamite and trinitrotoluene (TNT) are classic examples of high-order explosives that can have these devastating effects on anything close to the location of the explosion. Although dynamite and TNT are both classified as high-order explosives, they are quite different from a chemical standpoint. But at our level of involvement, the distinction is not important to our understanding. It is important to understand some of the basic concepts about handling these types of materials if they are encountered.

Dynamite is a mixture of various absorbent materials saturated with nitroglycerin. This combination creates an intense mixture capable of substantial destruction when properly detonated. Freshly manufactured dynamite is actually quite stable and requires a booster charge in order to detonate. Older dynamite is completely different because it may begin to *sweat* or leak if subjected to heat for a long time. The resulting materials become shock sensitive, which can be recognized by a milky-looking substance clinging to the outside surface of the stick of dynamite. It is critical to our safety that we never move or otherwise disturb older dynamite that is showing such signs, as the slightest disturbance may cause detonation. If you encounter sweating dynamite or other explosive materials, do not disturb the scene and move to a safe distance with caution. Unfortunately, a safe distance may be hard to determine because each incident may vary given a range of variables such as the amount of material present, the temperature and conditions in which it is located, and the time that it has been exposed. Generally, it is wise to start your safe zone at 1500–2000 feet from the area involved. If in doubt, always add more area to provide a broader buffer zone because these materials can produce unpredictable results.

! Safety
If sweating dynamite or other explosives are encountered, do not disturb the scene and move to a safe distance with caution!

TNT is another explosive widely used in the commercial sector as well as by the military. When newly synthesized, it looks like a yellowish sludge and is somewhat crystalline in structure. It is used as a composite explosive with other materials and as a bursting charge and standard demolition charge. As with other explosives, these materials should never be disturbed because their condition may have been impacted by prior handling, accident, or prior exposure to contamination. Additionally, TNT is a threat in terms of its flammability and may detonate if heated above 450°F. It is also considered toxic by ingestion, inhalation, and absorption. So again, leave these materials alone and create a safe zone to isolate the hazard.

! Safety
TNT can create a range of problems, including explosion and toxicity, and like other explosives, it should not be handled without specialized training and protection.

Low-Order Explosives

low-order explosives
materials in which reactions create pressure waves that move more slowly than those created by a detonation

deflagrations
explosions created by low-order explosives with very rapid reactions, but are slower than detonations

Low-order explosives function primarily by **deflagration**. Low-order explosives function primarily by deflagration, which has a slower reaction rate than the high-order explosives, but can be just as devastating. Deflagrations are explosions that move through the explosive material more slowly than the speed of sound and in which the heat wave precedes the pressure wave. Make no mistake, however, that both fronts are propagated within milliseconds and can certainly prove fatal to anyone nearby.

One of the more common examples of these types of materials is black powder. Black powder is one of the oldest explosives known to mankind. It looks like crushed-up charcoal and is commonly stored in small steel cans. It will deflagrate rapidly when exposed to heat, thereby presenting a dangerous and rapidly spreading fire and explosion hazard. Black powder is commonly used in timed fuses for blasting as well as in igniters and primers for propellants and detonators. If timed fuses, igniters, or primers are encountered, it is best to treat them like other high-order explosive materials.

■ **Note**
Deflagrations are essentially rapidly spreading fires.

Blasting Agents

blasting agents
the lowest explosion group, and often used to initiate other explosive charges

Blasting agents are the lowest of the explosive hazard subgroups and include the materials commonly used to initiate the higher order explosives. Materials like dynamite and C-4 (a military explosive) are not easily detonated under normal circumstances unless an explosive train is established. An explosive train is set up to enable low-order explosives to detonate higher order explosives. In other words, a smaller explosion must be used to initiate the bigger explosion created by the high-order explosives. Blasting agents are often used to create these smaller explosions but can be quite hazardous by themselves.

ANFO
ammonium nitrate and fuel oil; an example of a common blasting agent

One blasting agent commonly talked about is a material called ANFO. **ANFO** is an acronym for ammonium nitrate and fuel oil. Ammonium nitrate is a commonly used fertilizer material and fuel oil can be something as simple as diesel fuel. While relatively stable by themselves, these materials can form an explosive mixture that can cause considerable damage. The power of this combination was tragically displayed in the April 19, 1995, bombing of the Murrah Federal Building in Oklahoma City, as seen in Figure 4-1. The nation was shocked by the event and amazed at the destruction caused by such a commonly obtained chemical substance.

■ **Note**
Blasting agents can be formed by mixing other groups of materials together.

hygroscopic
materials that absorb moisture from the air

This low sensitivity explosive must be initiated by a booster charge to be successfully detonated. It is commonly used in composite explosives, where it is combined with a more sensitive explosive partner. Ammonium nitrate is also very **hygroscopic** and must be packed in airtight containers. Hygroscopic materials absorb moisture from the surrounding air, and if exposed to the air for any period of time, the strength of a blasting agent will be greatly decreased. Because of its explosive effects and low cost, ammonium nitrate is primarily used for commercial blasting at construction sites and rock quarries. This material is not suitable

Figure 4-1
Even the lowest order of explosive material can cause significant damage, as seen in the Murrah Federal Building explosion. (Courtesy of FEMA.)

for underwater blasting unless packed in a watertight container or detonated immediately after placement. Table 4-1 illustrates the detonation velocities of some of the more common explosive materials.

Additional Explosive Situations

In addition to the hazard posed by explosive substances like dynamite, there is a danger of explosive conditions caused by Flammable Liquids and Gases. Concentrations of vapors of a Flammable Liquid or a Flammable Gas can build up in confined areas to form an explosive atmosphere that create small explosions when ignited. While these are not technically classified as Explosives, the consequences of the incident can be almost as serious if the concentration of vapors is high, such as in an underground vault or storage room with poor air movement.

Table 4-1 *The detonation velocity of some of the more explosive compounds.*

Name of material	Use	Velocity of detonation (in feet per second)
Black powder	Timed blasting fuse	1,300
Nitroglycerin	Commercial dynamite	5,200
Ammonium nitrate	Demolition charge Composition explosives	8,900
TNT	Demolition charges Composition explosives	22,600
Tetryl	Booster charges Composition explosives	23,200
C3/C4	Demolition charges	25,000
PETN	Detonating cord Blasting caps Demolition charges	27,200
RDX	Blasting caps Composition explosives	27,400

Note
Many types of flammable materials can create explosions if ignited under confined conditions.

It is not hard to imagine the potential consequences of a prolonged release of a material such as propane into an enclosed space. If the leak occurred in a room that was poorly ventilated, the vapors would begin to collect inside the room. Eventually, the area would have enough flammable vapors that an ignition source would ignite the highly concentrated mixture within the enclosed space. Once ignition of the gas occurred, the rapid fire would result in an explosion that would blow out windows and doors, causing the propagation of a fire and pressure front similar to that of other explosive materials. Ignition can come from many sources, including pilot lights, turning on a light switch, electrical arcs from machinery turning on or off, or even static electricity. Structural damage would occur and anyone trapped inside the room would probably suffer serious injury or death. It is for this reason that responders and site workers alike should be greatly concerned with explosive atmospheres and should monitor areas carefully before entering. Constant air monitoring should be done in work areas with suspected flammable atmospheres to ensure that the levels of vapors remain low enough not to present a fire or explosion hazard.

Safety
Explosive situations can be created by materials that are not classified as Explosives. Confined gases that are ignited are examples of this.

Health Hazards Posed by Explosives

As has been discussed, the principal danger posed by explosives is the explosion itself and/or the significant fire dangers that are present. However, as noted with nitroglycerine, some types of materials also pose a health hazard. Substances that are toxic, irritants, and sensitizers must be labeled as such in accordance with OSHA regulations. Hazard warning information must be contained on the label and identified on the MSDS for the material. While it is important to be mostly concerned with the explosive effects that may occur with these materials, the significant fire hazard or toxicity associated with them should not be discounted.

Safety
While the primary hazard of explosives is the physical damage that can result, some explosives can also present a health hazard.

Safe Handling, Use, and Storage of Explosives

When working with explosive materials, it is essential to have a comprehensive fire prevention and storage program. Areas where explosives are handled or stored, or any area where explosive vapors may exist, must be free of ignition sources. Non-sparking tools should be used, and electrical circuits should be properly installed to prevent short circuits and sparking. It is vital to understand exactly what we are working with and what the potential hazards may be. Explosive incidents are relatively infrequent but potentially deadly because of the violent nature of the reactions.

Safety
A strong fire prevention program must be in place to ensure safety in work areas with explosive materials.

Finally, consider that normal personal protective clothing is not likely to be effective against an explosive blast. Structural firefighting gear may provide some protection from the heat and fire but not from the effects of a blast. Chemical protective materials may be suitable for protection against the toxicity of the materials but are inefficient against heat or explosive forces.

Note
Working with explosive materials requires considerable care and additional training.

Emergency Procedures

Planning and training for emergencies or cleanup operations involving Explosive materials should be carried out before any personnel approach the incident scene. Anyone dealing with a situation involving explosives should have a thorough knowledge of the materials and the hazards that may be present. It is generally a good practice to leave an area where materials have been found and to call for more help from those with higher training. Recall from Chapter 1 that persons such as Explosive or demolition experts with specific knowledge of a material can be brought to the scene and used in the handling of an emergency situation without requiring that those persons be HAZWOPER trained.

Medical personnel should also be available to take care of personal injury injuries and must be in place before anyone goes to work at an incident site where explosives have been unexpectedly encountered. Information about particular chemical hazards should be available to them as it relates to the toxicity of the material. This can be found on the MSDS for each of the items found.

Note
Handling explosive materials involved in an emergency may require assistance from others with special knowledge on safe handling of these materials.

GASES

A thorough and complete understanding of the hazards associated with any class of hazardous materials is critical for people working with each of the classes of hazardous materials encountered. This is even more critical for those who work with the second class of hazardous materials, which includes a broad range of gases. Gases can present an array of hazards, including Nonflammable, Flammable, Toxic, Corrosive, Oxidizing, or even Radioactive. Some gases, such as chlorine, have multiple properties, including being toxic, corrosive, a strong oxidizer, and poisonous. Figure 4-2 shows a typical one-ton chlorine cylinder that contains the gas in its liquified state.

Gases present hazards in three ways. These include a primary hazard, a secondary hazard, and a hazard associated with the manner in which they are stored or transported. As we study the hazards, we will identify each of these

Figure 4-2
Example of a liquified gas (chlorine) cylinder.

and cite examples of how these three types of hazards can create more dangerous situations than might otherwise be expected. Additionally, the very nature of the gaseous state of the material makes it more challenging in both handling and response programs. This is due to the fact that once a gas leaves its cylinder or container, it is virtually impossible to get it back in again. In most cases, you will likely be dealing with the effects of the release rather than the release itself. The behavior of a gas is such that once it escapes, it goes in many directions and poses its hazards as it moves.

> **■ Note**
> Gases can present numerous hazards that must be identified in order to safely work with them.

Definition of Gases

The DOT has several technical definitions of Flammable, Nonflammable, and Poison Gases. However, working safely with these materials does not require that we know those definitions if we understand some basic concepts about gases and their properties. Simply stated, a **gas** is a material in a state of matter that at normal temperature and pressure tends to fill the space available to it. Gases are different from solids and liquids, as they are more mobile and less dense at a molecular level. If the gas is confined inside a cylinder, it will expand to fill the whole container. If the gas is released into the air, it will spread out, or **diffuse**, until restricted by some confining feature such as the walls of a room or building. Diffusion is the action of a gas to move from an area of higher concentration to an area of lower concentration. A common example to illustrate this point may be the pleasant smell of food cooking in the kitchen. In this case, the smell is much stronger in the kitchen (area of higher concentration), but you may be able to detect the same odor in an adjacent room (area of lower concentration). This distant detection of the odors is an excellent example of how a typical gas may diffuse when released.

gases
materials that form one of the three states of matter; they move freely when released and can occupy the entire area of release

diffuse
the ability of a gas to move around and occupy the area of release

■ Note
Knowing the basic properties of gases is critical to working safely with them.

Primary Hazards of Gases

The primary hazard of a gas is the one most often associated with it. Many of the hazards of gases are obvious. The gas itself may be classified as a Poison or as a Toxic gas, indicating that it can cause serious harm. Examples of these include chlorine, ammonia, and many gases used as chemical weapons such as Sarin gas, which was released into the subway system in Japan in 1995. Others may be Flammable Gases such as acetylene, propane, or hydrogen. Still others may seemingly present minimal hazards with their classification as Nonflammable Gas. Examples of these include argon, nitrogen, and carbon dioxide (CO_2). To further our understanding of these hazards, we will review each of the major primary hazards.

■ Note
The primary hazards of gases are the obvious hazards most often associated with the gas. However, they are not the only hazards related to the gas.

In some cases, the gases are flammable and can be easily ignited. Unlike most materials that can be catch fire and pose a threat, flammable gases require little effort to ignite. Solid materials and many liquids require some heating and an ignition source in order to catch fire. Gases, on the other hand, are already in an ignitable state and can be ignited with little energy such as from a static discharge if the concentration in the air is correct.

■ Note
Flammable materials that are already in the gaseous state pose the highest danger of fire because they exist as a gas and are easily ignited.

When a flammable gas is released and not ignited immediately, the gas can fill the area, resulting in a considerable hazard once any part of the vapor is ignited. When the gas is finally ignited anywhere within the area, the vapors will flash back to the initial area of the release, catching fire along the way. Anyone caught in the resulting flash fire could be seriously burned or killed. Chemical protective clothing without special flash fire protection will not protect against fire. Flammable atmospheres, such as those illustrated in the Table 4-2, must be carefully monitored and only entered under the most extreme circumstances and by personnel who are adequately protected from the fire hazards. Most of the time, ventilation should be accomplished before personnel enter the area, and activities at the site should be carefully monitored until personnel are out of the hazard zone. Care should be taken to ensure that any mechanical ventilation device does not create a hazard and is operated from a remote area or that **intrinsically safe** equipment is used. Intrinsically safe equipment is equipment that is not subject to creating sparks or other potential ignition sources and that is designed to work in a flammable atmosphere.

intrinsically safe
safe to use in flammable atmospheres and does not create sparks or ignition sources

Safety
Flammable gases that are released and that do not immediately catch fire are likely to create explosive atmospheres.

Table 4-2 *Examples of Flammable Gases that may pose a threat of a fire or an explosion.*

Acetylene
Butadiene
Butane
Cyanogen
Diborane
Ethylene
Hydrogen
Hydrogen sulfide
Methane
Propane
Silane

Additionally, many gases, like those shown in Table 4-3, are poisonous or toxic by multiple routes of entry and can even kill you if you suffer the smallest exposure. Arsine, for example, is a highly toxic gas that can be fatal after only a few breaths. Arsine is widely used in the semiconductor industry as well as fluorine, germane, and other highly toxic gases. Take care to understand what these materials are capable of doing and protect yourself accordingly.

It is important to realize that many gases present multiple primary hazards. A common example is the gas form of ammonia. Ammonia gas or vapor is a poison gas that is flammable and that will create a corrosive solution when mixed with water. Many people fail to recognize ammonia as a Flammable Gas, but make no mistake, it is definitely dangerous and can create a flammable atmosphere if confined, just like propane or acetylene. In addition to the other hazards, we also need to be careful to avoid skin contact with ammonia vapors because sweat or normal body moisture can react and form the corrosive compound called ammonium hydroxide. Ammonium hydroxide or aqua ammonia has a relatively high pH (it is corrosive) and may cause localized skin irritation or burns.

> **Safety**
> Some gases may have multiple primary hazards that need to be considered in order to safely work with them.

Finally, it is also important to remember that when a substance is in the gaseous state, your potential for exposure via inhalation or contact with eyes and

Table 4-3 *Examples of toxic or poisonous gases that may be encountered by workers or emergency responders.*

Ammonia, anhydrous	Germane
Boron trifluoride	Hydrogen chloride, anhydrous
Carbon monoxide	Hydrogen fluoride, anhydrous
Chlorine	Hydrogen sulfide
Cyanogen	Methyl bromide
Diborane	Nitric oxide
Dichlorosilane	Phosgene
Ethylene oxide	Phosphine
Fluorine	Sulfur dioxide

skin is increased. Remember that you breathe about 35 pounds of the air a day and expose your lungs to an enormous surface area. Reduce your exposure and protect that route of entry.

Secondary Hazards of Gases

secondary hazard
ability of gases to create an oxygen-deficient atmosphere

The **secondary hazard** of gases, one that is not as obvious as the primary hazard, is that many gases are capable of creating an oxygen-deficient condition resulting in asphyxiation. OSHA defines oxygen deficiency as an atmosphere that contains less than 19.5% oxygen by volume. While this amount is still safe, exposure to low levels of oxygen can result in serious health effects, including death to the person exposed. This generally occurs when levels reach less than 15%. Additionally, oxygen deficiency may go unnoticed by the person exposed to it because the materials creating the condition may be colorless, odorless, and tasteless.

> **Note**
> An oxygen-deficient atmosphere is one that contains less than 19.5% oxygen.

By their nature, gases that are released will move out of their container and start filling up the area where they are released. This could result in the gas occupying the area and pushing out or excluding the normal atmosphere that was present before the release. This area may become oxygen deficient, resulting in harm to those exposed. This situation most often can occur in closed or confined areas where outside ventilation is limited. From a safety perspective, we must consider that all gases other than those classified as *Toxic* or *Poison*, or oxygen itself, are capable of producing a condition that could lead to oxygen deficiency and asphyxiation. Propane, while flammable, is also capable of killing through its ability to push out the normal atmosphere and create an oxygen-deficient condition.

> **Safety**
> From a safety perspective, we must consider that all gases other than those classified as *toxic* or *poisonous*, or oxygen, are capable of producing a condition that could lead to oxygen deficiency and asphyxiation.

In addition, due to the nature of gaseous materials, some of them do not need to be contained within enclosed areas in order to create the potential for asphyxiation to occur. Many gases can settle into low areas of open trenches or similar locations and push out or exclude the normal air because the gases are heavier than the normal air that would otherwise occupy the space. In such situations, we need to be extremely careful even when the area above us may be open to the atmosphere. This is because of the nature of the gases and their ability to rise or sink.

> **Safety**
> Gases can create hazardous atmospheres without completely occupying the entire area of their release.

vapor density
the weight of the vapor or gas relative to the weight of air

To understand this concept, we must have a working knowledge of the term **vapor density**. Vapor density is defined as the weight of a gas or vapor relative to the weight of air. Air is given the value of 1 and a substance that is heavier

molecular weight
the weight of a given volume of material

than air is assigned a value greater than 1. Materials that are lighter than air have a vapor density of less than 1. Each material has a defined value based on its physical properties. To calculate this, we determine the **molecular weight** of air and divide the molecular weight of the material by the molecular weight of air. Air has a molecular weight of approximately 29. The calculation of vapor density is as follows:

- Air: 29/29 = 1. Molecular weight of air divided by the molecular weight of air
- Acetone: 58/29 = 2. Molecular weight of acetone divided by the molecular weight of air
- Gasoline: 94/29 = 3.2. Molecular weight of gasoline divided by the molecular weight of air
- Ammonia gas: 17/29 = 0.59. Molecular weight of ammonia divided by the molecular weight of air.

Note
Air has a vapor density of 1. Heavier-than-air gases have vapor densities greater than 1. Lighter-than-air gases have a vapor density less than 1.

Note
Vapor density is calculated by dividing the molecular weight of the vapor by the density of air, which is approximately 29.

Vapor density is not the only factor that influences the movement of a gas. However, understanding what vapor density is and how it works is vital in working safely with gases. Gases and vapors that are heavier than air, as most are, can settle into low areas were ventilation is limited. When this occurs, the potential for asphyxiation could exist with little or no warning.

Note
The vapor density of most gases is greater than 1, making them sink when released.

Safety
Heavier-than-air gases pose a high potential for problems in that they can stay low to the ground.

To help understand how this concept must be taken into account, consider the following scenario. Propane, a gas whose primary hazard is its flammability, is released into a closed shed area. The shed has no easily recognized ignition sources and is completely closed. An employee recognizing that the cylinder is releasing its product enters the shed in order to remove the cylinder and open up the windows to air out the area. Because there are no obvious ignition sources, the employee considers that the situation is safe to enter as the flammability concerns are addressed. Walking through the area to open the back windows, the employee is exposed to the asphyxiation hazard and is overcome and passes out. The lack of oxygen created by the displacement of the normal air causes this to occur.

This is not an isolated circumstance. Because most gases are heavier than air, the potential for displacement of the normal atmosphere to occur is considerable. Every open trench or excavation is subject to this. Without adequate ventilation in the area, the vapor density of most materials can contribute to the creation of an asphyxiation hazard. Care must always be taken in any area where gases or vapors can be present because they can create the oxygen-deficient conditions that lead to death. In fact, a simple way to remember which gases are lighter than air is to memorize the mnemonic H-A-H-A-M-I-C-E, which can be seen in Table 4-4. The phrase may be silly, but it will help you remember the majority of gases that are lighter than air. If the gas is not on that list, its vapor density is likely greater than 1 and the gas will be on the ground.

■ Note

The vapor density of most gases is greater than 1, making them heavier than air and more likely to settle into low areas.

While the concept of vapor density is important to our safety and working knowledge of gases, there are two additional factors that must be considered in discussing factors that influence the movement of the gas. These include wind and ventilation systems, and temperature extremes. In fact, these two factors are likely to have a greater effect on the movement of the gas that is released than the material's vapor density.

■ Note

Wind, ventilation, and temperature extremes are three factors that influence the movement of gases in addition to vapor density.

Consider the effects of a laboratory fume hood or other type of ventilation system on gases. Figure 4-3 shows an example of a hood that can be used to eliminate inhalation hazards while working with materials. Such equipment with its strong ventilation system will move the gases into the current created by the fans and other equipment regardless of the vapor density of the material. Other types of ventilation and wind currents can have equal effects and should always be considered when establishing the location of a gas after it is released.

Table 4-4 *A list of lighter-than-air gases using the H-A-H-A-M-I-C-E mnemonic.*

Material	Vapor density
Hydrogen	0.07
Ammonia	0.59
Helium	0.14
Acetylene	0.9
Methane	0.6
Illuminating gases	
Gases used to light up areas, including:	
Neon, used in some types of light bulbs	0.7
Natural gas, used for city streetlights prior to electricity	0.65
Carbon monoxide	0.97
Ethylene	0.97

Figure 4-3
Picture of a laboratory fume hood used to provide protection for those who work with airborne hazards.

> **■ Note**
> Wind and ventilation are the strongest factors in the movement of gases.

Another factor to consider in the discussion of gas movement is that of temperature extremes. Gases that are heated will become buoyant and will rise. Conversely, gases that are cooled may sink independent of their weight in air. Hot air balloons are good examples of this concept, as we see that the heated air is lighter than normal air, causing the balloon to rise.

> **■ Note**
> Temperature extremes will cause gases to rise or sink despite the vapor density of the gas.

Storage Hazards: Compressed, Liquified, and Cryogenic Gases

A third and final concern when working with gases is that they must be stored or transported to the area where they are going to be used. Storage and transport of the gas may add a third hazard to the situation that must be considered if we are to work safely with these materials. Gases are primarily transferred and stored in one of three states. This includes the compressed state, where the gas is squeezed into cylinders or pipelines, transformed into the liquified state, and placed into cylinders, and cryogenic state, where a combination of cooling and pressure is used to create very concentrated forms of the gas. Each of these methods of storage and transfer can present a variety of hazards and problems for all who use, work with, or respond to emergencies involving gases.

> **■ Note**
> The manner of storage can add a third hazard to gases and must be considered.

compressed gases gases stored under considerable pressure

Compressed Gases Gaseous materials that are stored in cylinders (or pipelines) under relatively high pressures are **compressed gases**. The concept is that a gaseous material such as oxygen or helium is forced into a cylinder by applying lots of pressure. One of the basic laws of gases, termed *Boyle's Law*, states that as the pressure exerted on a gas is increased, the volume of that gas will decrease. This act of compression or adding the pressure makes it possible for a lot of gas to be stored in a cylinder. An example of this occurs when an air cylinder is filled for use in underwater diving or SCUBA activities, or when compressed air is used to fill cylinders for firefighting or response to hazardous materials releases. A complete discussion of compressed air used in breathing apparatus will occur in our study of respiratory protection in Chapter 6 of this book.

Note
Compression of a gas will cause its volume to decrease and allow more gas to be stored in the cylinder.

Anyone who has ever used compressed air cylinders to store breathing air realizes that there is a considerable amount of air in a relatively small cylinder as a result of the pressure used to concentrate the gases into the small space. This increases the amount of available air and adds a hazard due to the pressure now present inside the cylinder. This pressure can be thought of as a potential energy source that is independent of the primary and secondary hazard of the gas. This pressure can cause harm if it is not properly managed. Pressure up to 5000 psi can be used to help store gases into a compressed cylinder, and with that much pressure available, the consequences of a cylinder failure can pose hazards far in excess of the materials inside.

Safety
Storage pressures alone can pose significant danger to those working with compressed gases.

To work safely with compressed gas cylinders and to prevent this energy from suddenly releasing, we should follow some basic practices. These include:

- If the cylinder has a protective cover, make sure to secure the cover to the cylinder prior to moving.
- Use a cylinder dolly or cart to move cylinders whenever possible.
- Do not expose compressed gas cylinders to temperatures above 250°F.
- Do not modify the valves or threads of any cylinder.
- In the event cylinders are involved in a fire, treat them like explosives and maintain a safe distance.
- Keep cylinders secured when in use or in storage to prevent unwanted movement. Large cylinders should be secured with chains or strong materials to prevent them from falling over.
- Never assume a cylinder is empty if it is unmarked. Develop a system to designate full and empty cylinders.

Safety
Basic safety concepts for working with compressed gases include proper handling, ensuring that cylinders are restrained when not in transit, and keeping caps on cylinders that are not in use.

liquified gases gases that become liquids through a combination of temperature and pressure

Liquified Gases Gases that are used or stored in the liquified state are **liquified gases.** The process of liquification is accomplished by subjecting the gas to a certain amount of pressure while lowering the temperature inside the storage vessel. The ease with which this is done depends on the properties of the gas. Some gases are easily liquified, given their properties. Examples of commonly found liquified gases include butane, propane, chlorine, and ammonia.

Consider a cigarette lighter containing butane. If you look at the plastic storage container, it is clear that the material is not under a lot of pressure. Unlike compressed gases where the additional hazard of pressure and energy release is possible, liquified gases do not have high pressures but are hazardous in other ways. As you look at the lighter, you will also notice that liquified gases are stored in their cylinder just as the description implies—in a liquid state. The gas remains in a liquid state as long as the pressure is minimal. The pressure is released when the valves open, and the liquid reforms into a gas, as is the case with the butane lighter. Once reformed into its gaseous state, the butane is easily ignited by the ignition source on the lighter.

Note
Liquified gases tend to have low cylinder pressures.

The primary advantage to liquifying gases is that the container can hold significantly more volume when stored in a liquid state. This can be a difficult concept to understand, but remember you have taken a huge volume of gas and condensed it into the liquid state. Liquids are more tightly packed together at the molecular level than gases, so in essence a material has greater volume when it is in a liquid state. Another helpful illustration of this is a small amount of water boiling in a pan on a stove. As the water heats up, it goes from its liquid state to its gaseous state. The gas in this case is the steam that is produced, which is considerably more in volume than the water in the pan. An entire area can be filled with the steam produced from a small pan of water on a stove. So too can an area be filled quickly from a small liquified gas cylinder release.

Safety
Liquified gases contain considerable volumes of the gas because the gas concentrates as it liquifies.

expansion ratio the amount of gas released from a given volume of liquid with a liquified gas

To quantify the expansion potential of a liquified gas, we must understand the term **expansion ratio.** The expansion ratio describes the capacity of a liquified gas to convert to its gaseous form as it is released. Large expansion ratios indicate that the liquids are more concentrated and form much more gas from the liquid volume present. For example, propane has an expansion ratio of approximately 270:1, which means that each volume of propane liquid will convert to approximately 270 volumes of propane gas once it is released from its liquid state and turns back into a gas. This expansion ratio concept is important because a liquified gas cylinder contains a lot more potential gas than is obvious from the size of the cylinder. It is much like trying to contain the genie inside the lamp of Aladdin. The genie is many times larger than the volume of the tiny lamp and once the cork is pulled, watch out!

From a safety perspective, the concept of the expansion ratio adds to the other hazards present. This is because the volume of gas present may be masked by the size of the cylinder present. A gas capable of asphyxiation may be more hazardous when stored as a liquified gas in that the amount of gas available to displace the normal atmosphere is far greater than the size of the cylinder. All of the hazards of the gas are increased due to the volume of gas present from the vaporization of the liquified gas and its corresponding expansion ratio.

> **Safety**
> Because of the larger volume of gas released from a liquified gas, the primary and secondary hazards are increased.

Another hazard associated with liquified gases also relates to the concept of expansion ratios. No gas cylinder, including those that contain liquified gases, can be expected to contain an unlimited volume and pressure increase without failing. Because of the high concentration inside a liquified gas cylinder, cylinder failure is a potential concern. When a cylinder containing a liquified gas fails, we say that a **BLEVE** has occurred. BLEVE is an acronym that stands for Boiling Liquid Expanding Vapor Explosion.

BLEVE
Boiling Liquid Expanding Vapor Explosion

Propane can be used to illustrate the concept of a BLEVE. As with all liquified gases, propane has a particular temperature at which it converts from its gaseous state to its liquid state. This is termed its **boiling point**. The boiling point is the temperature at which a material changes from its liquid state to a gas at standard pressure. Increasing the pressure can alter this and keep materials as liquids at higher temperatures. In the case of propane, the boiling point is −44°F. At normal pressure, propane instantly converts to its gas state at any temperature above −44°F when exiting a cylinder.

boiling point
the temperature at which a liquid becomes a gas

When heat is applied to a cylinder containing liquified propane, the metal of the cylinder will weaken. As this occurs, the propane liquid inside the cylinder will also start to heat up and expand. With propane's expansion ratio of 270:1, this volume increase inside the cylinder will soon become critical. This creates even higher pressures within the cylinder whose shell is weakening due to the heat impingement on the exposed metal. As the propane is heated above its boiling point, more and more gas is evolved and the resulting pressures will exceed the capacity of the cylinder to contain it. Because the cylinder cannot hold the ensuing pressure, a catastrophic container failure and explosion could result. This is technically a BLEVE because the material involved has a high expansion ratio and is heated above its boiling point inside a closed vessel. Figure 4-4

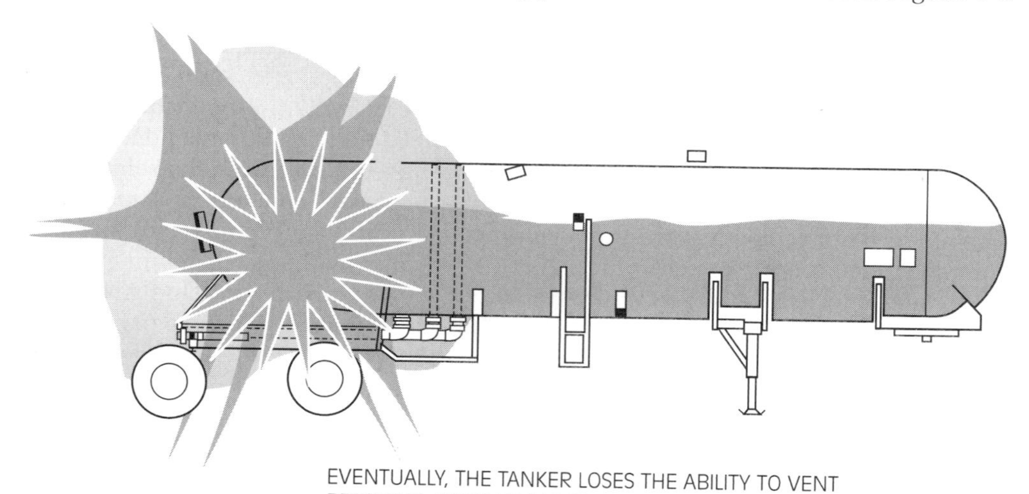

Figure 4-4
A cylinder of liquified gas that fails from a BLEVE.

illustrates a tanker exploding due to its inability to contain growing pressure. In this case, a gas explosion occurs with incredible force and vaporizes anything in close proximity to the blast.

Safety
The potential for a BLEVE must always be considered whenever a liquified gas is exposed to added heat.

However, a BLEVE does not always cause harm, as there are times when the result may be positive. Any type of liquid that is enclosed inside a container can cause this to occur. Take the example of how popcorn is made. Inside each kernel of corn is a little water, a small drop. Water has a boiling point of 212°F, and when it boils, its expansion ratio is approximately 1700:1. For every one volume of water, there are 1700 corresponding vapor volumes (steam in this case). As we heat popcorn, the stage is set for a BLEVE just like in our propane example. In this case, we put the popcorn in the microwave and turn it on. The microwaves heat the water until it reaches its boiling point of 212°F. The water then becomes a boiling liquid inside the kernel and instantly expands according to its 1700:1 ratio. Since the kernel of corn does not have a relief valve, it suffers a catastrophic container failure due to the BLEVE. We achieved the same results with two totally different materials. BLEVEs are not limited to things like propane and butane. Any liquid capable of expanding once it has been taken above its boiling point is susceptible to a BLEVE.

Note
Popcorn popping is an example of a BLEVE.

If the potential for a BLEVE is possible, it is vital that actions be taken to maintain a safe distance from the potential consequences that could result. Establishing safe zones at these incidents will be critical, as huge chunks of the container may be hurled hundreds of yards from the blast site. To ensure that these conditions are not created, it is important to control fire and ignition sources when working around any type of liquified gas because all are subject to this phenomenon.

Safety
Establishing a safe working distance is critical if the potential for a BLEVE is present.

A third hazard to consider when working with liquified gases is that the materials are stored as cold materials. Because the gas is in liquid state due to the pressure that is present, once the pressure is released the gas will be released at or near its boiling point. While the boiling point is not as cold as the Cryogens that will soon be discussed, contact with these materials can produce some harm to the exposed skin. For this reason, care should be taken to avoid any direct contact with the liquid or concentrated vapors as they are released. Additionally, because the vapors are initially cold when released, the coldness will tend to make the gases stay low to the ground independent of their vapor density. This factor must always be considered in emergency response operations because cold gases will stay low to the ground and will be cold enough to cause

ambient temperature the temperature of the surrounding area

harm until they heat up to the **ambient temperature**. Ambient temperature is the temperature of the area where the gas is released. Most often this is above the boiling point for the liquified gas.

Cryogens liquified gases with a boiling point below −150°F

Cryogens The third state in which gases are typically used and stored is called cryogenic. The word *cryogenic* is derived from the Greek word *Kryos*, which means "very cold." In effect, **Cryogens** are a subgroup of liquified gases. The DOT definition of these materials includes liquified gases whose boiling point is lower than −150°F. If the boiling point of a gas that is liquified is above that temperature, we classify it as a liquified gas. If it is lower than −150°F, we classify that material as a Cryogen. Because it takes considerable cooling to get a material to this low temperature, the Cryogen can only be maintained in its liquid state by ensuring a low temperature. This is most often done by mechanical refrigeration.

> **■ Note**
> Cryogens need to be mechanically refrigerated in order to remain cold.

Most Cryogens originate from the atmosphere and are cooled to the necessary temperature by a relatively simple process. Initially, the atmospheric air or gas is drawn into a compression chamber via a dryer to remove as much moisture as possible. The gas is then filtered as much as possible to remove all the contaminants and particulates before it moves through the system. The compressed gas is then cooled inside a chamber by a circulating liquid that is relatively cold. Liquid ammonia is often used, as it can drop the temperature to approximately −28°F. Another of the gas laws that explain the behavior of gases states that as a gas is cooled, its pressure drops. In order to keep a significant pressure (usually 2000 psi) constant, more of the gas gets pumped in and cooled until there is a full chamber of gas at 2000 psi cooled to −28°F. The cold compressed air is then allowed to escape into another vessel that again lowers its temperature. This process is repeated and different sized containers are used as catch vessels to reduce the temperatures to the desired levels. Once the air condenses into a liquid, the desired gases (nitrogen, hydrogen, helium, etc.) are boiled off or distilled at the prescribed temperatures and captured elsewhere.

> **■ Note**
> Most Cryogens come from cooling normal air until the various components reach their boiling point where they convert to a liquid.

The benefit of using Cryogens is that at these temperatures, it is possible to get an even larger amount of gas into a given size cylinder, resulting in even higher concentrations and expansion ratios. Expansion ratios of up to 1445:1 are possible, making these Cryogens economical to both use and store. Table 4-5 shows the list of common Cryogens and their properties.

> **■ Note**
> One major benefit of Cryogens is that large volumes of them can be stored in small containers, as most of the materials have high expansion ratios.

Table 4-5 *List of common Cryogens and their properties.*

Substance	Boiling point	Expansion ratio
Liquid argon	−302°F	840:1
Liquid fluorine	−306°F	980:1
Liquid helium	−452°F	700:1
Liquid hydrogen	−423°F	848:1
Liquid krypton	−243°F	695:1
Liquid natural gas	−289°F	635:1
Liquid neon	−411°F	1445:1
Liquid nitrogen	−320°F	694:1
Liquid oxygen	−297°F	857:1
Liquid xenon	−163°F	560:1

Because of their extremely cold temperatures, Cryogens also behave like liquified gases when they are released, and stay low to the ground independent of their vapor density. Because these materials can be stored at extremely low temperatures, breathing released vapors can cause lung damage and should be avoided. Figure 4-5 shows a cloud of cryogenic nitrogen as it is released and stays low to the ground.

In order to maintain the low temperatures Cryogens require, they are stored in unique vessels. As with liquified gases, the pressures inside cryogenic vessels are also low, generally under a few psi. However, unlike liquified gases, keeping a constant temperature inside the vessel is of utmost importance. Cryogens are therefore stored in cylinders, which are comparable to thermos bottles with insulated outer coverings. These containers have silvered linings and are designed to use the inherent temperature of the material inside to help maintain the extreme cold. In the event temperatures are not maintained, cylinder failure can be rapid, as most cryogenic containers are not at all designed to handle significant pressures. Figure 4-6 shows a typical and commonly used Dewar cryogenic container.

Figure 4-5
A release of cryogenic nitrogen. Note that the cloud is lying low on the ground.

Figure 4-6
Example of a cryogenic Dewar.

> **■ Note**
> Cryogenic containers do not look like typical gas cylinders. They may look and act like a thermos bottle and are often insulated.

From a health and safety perspective, it is vital that we take into account the extreme cold temperatures that are present when working with these materials. Cryogens are readily capable of causing significant damage to anything that comes into contact with them. Never touch the liquid, and do not walk into visible clouds because such a cloud may freeze tools, equipment, and humans, irrespective of PPE worn. If you come in contact with cryogenic materials, severe frostbite and tissue damage could result. Care should always be taken when working around these materials, as no conventional chemical-resistant fabrics or standard gloves are sufficient to protect you against these extreme temperatures. Specially designed and insulated gloves must be worn to provide enough protection when working with these materials. Figure 4-7 shows pipes under a cryogenic tank where the pipes have ice forming on them due to the extreme cold. Care should be taken to avoid contact with items that have been in contact with the cryogenic gases.

> **! Safety**
> Exposure to the extreme cold vapors or liquids associated with Cryogens can cause significant harm independent of the other hazards of the material.

Figure 4-7
Cryogenic materials caused ice to form on pipes.

A final hazard when working with these materials is the expansion ratios. Consider that a release from a small cryogenic vessel inside an enclosed space may result in a large gas release, given the expansion ratios. The volume of gas even in a small cryogenic vessel may be enough to completely exclude the normal atmosphere that was present before the release. This can rapidly result in an oxygen-deficient condition that may go unrecognized because many common cryogenic materials do not have an odor or other detectable properties.

> **Safety**
> The large expansion ratios associated with Cryogens can make the primary and secondary hazards of the material more hazardous by releasing more of the gas.

Safe Handling, Use, and Storage of Gases

Safe handling and use of any gas is always based on limiting contact and on modern engineering practices, including using ventilation systems. This is particularly true with gases that have multiple hazards, as has been discussed. If there is a possibility of skin contact or inhalation of the materials, the activity should be reviewed and analyzed to determine the appropriate measures that need to be taken. The danger of toxic vapor inhalation, reaction with or absorption through moist mucous tissues and the eyes, or absorption through the skin should always be minimized through appropriate engineering controls such as ventilation systems, administrative controls such as exclusion of personnel from the area, and the use of proper protective equipment.

> **■ Note**
> The basic safety principle in working with gases is to limit exposure to them.

> **Safety**
> Ventilation is a major consideration when working with all gases.

segregation of hazards storing materials apart so that they cannot react with one another

The principle of **segregation of hazards** applies to gases as well as to other materials. Segregation of hazards simply means that we store materials away from other materials that could react with them. As gases are some of the most migratory of materials, this principle is critical to ensuring our safety. One example of this involves Flammable Gases, which should be stored separately and away from other materials, especially those classified as Oxidizers and Explosives, because they could react and create a hazardous situation. Poisons and Corrosives should also be stored away from each other and from any material that might cause a breach in their containers.

Emergency Procedures

In the event of a gas release, it is important to get all personnel out of the release area as soon as possible. Move any exposed victims to a source of uncontaminated air. Rinse skin or eyes exposed to corrosive or poison gases with large quantities of water. OSHA requires that washing be done for at least 15 minutes to ensure the material is removed. The location and use of safety showers and eyewash stations should be conveyed to all personnel who might be exposed to these materials because the effects can be rapid. Quick action is essential because the longer the contact, the greater the damage. It may also be necessary to support the victims' breathing in the event of respiratory and/or cardiac arrest. Make sure victims are thoroughly decontaminated before such measures are taken. Care should be taken to avoid becoming another victim through careless exposure while trying to help another exposed person.

> **Safety**
> Flushing with water for at least 15 minutes is the standard approach to any contact with hazardous materials.

Other appropriate emergency response procedures should be found on the MSDS. It is important to know about and to follow the MSDS because it provides some of the most complete information available. Medical assistance should be obtained quickly, if needed, and it is also important to provide the physician with the MSDS so that treatment can be quickly provided.

In the event of a release, other emergency response measures include isolation and containment of the material. Unlike many other substances, gases are difficult to capture. Special cleanup procedures may be required if gases are confined inside a given area, particularly because it may be unsafe to vent toxic gases to the atmosphere. Proper procedures must be followed to make sure that no harmful residues are left in the workplace or are released to the environment.

FLAMMABLE AND COMBUSTIBLE LIQUIDS

Flammable and Combustible Liquids are some of the most commonly encountered hazardous materials. While their hazards might seem clear, there are a number of issues that must be understood when dealing with this class of materials. In addition to posing the threat of fire, many of these materials also present a health hazard for the worker or emergency responder. When working with Flammable Liquids, it is usually safe to assume that they possess more than one property.

Flash Point and Flammability

states of matter
the three forms in which all materials exist; they include gas, liquid, and solid

sublimation
the ability of a material to pass from a solid state to a gaseous state without becoming a liquid

One of the most important concepts relating to the study of flammability involves **states of matter**. All materials exist either as a gas, liquid, or solid at a given temperature, and as the temperature is manipulated, most materials have the ability to change their states. As materials are heated, they may change from a solid state to a liquid state, and as the temperature continues to rise, finally to a gaseous state. Ordinary materials such as water change from a solid state (ice), to a liquid state (water), and to a gaseous state (steam), as the material is heated from 32°F to 212°F. While this is the normal transition, some materials are capable of **sublimation**, where a solid material changes to a gas without becoming a liquid. An example of sublimation would be mothballs that contain naphthalene, a material that converts from its solid form directly to a gas.

To understand flammability, it is critical to accept that, usually, in order for a material to ignite, it must be in a gaseous or vapor state. For the most part, materials that are both flammable and in solid or liquid states can only be ignited after they have been heated and become a gas. The closer to being an ignitable gas a material gets, the more likely it will catch fire in the presence of an ignition source. Because gases are already in a gaseous or vapor state, they do not need any additional heat for conversion to a gas, presenting the greatest concern (assuming the gas is flammable) when it comes to creating a flammable environment. Depending on the specific material, most liquid and solid materials will require some heating in order to release potentially ignitable gases or vapors, and this concept leads us into the discussion of Flammable Liquids.

> **■ Note**
> In order to be ignited, the material must be a gas or vapor.

> **■ Note**
> Solids and most liquids require some heat before they can be ignited.

ignitability
the ability of a material to be set on fire

flash point
the lowest temperature at which a liquid can give off vapors that can be ignited

The ability of a liquid that is capable of being set on fire or **ignited** is largely a function of its **flash point**. The flash point is the minimum temperature at which a liquid gives off vapors that are capable of being ignited. Above the flash point, ignitable vapors are given off, and below the flash point, ignitable vapors are not present. Therefore, the lower the liquid's flash point, the greater the danger of ignition.

To aid in our understanding, we will consider some common materials. Gasoline, for example, has a flash point of –44°F. This is quite low and means that a release of the liquid at most commonly encountered temperatures will result in the release of vapors that can be ignited. Diesel fuel, another ignitable material, has a flash point of approximately 125°F, making it somewhat safer than gasoline because a release of this material at most normal temperatures will not result in the production of ignitable vapors. However, if the diesel is released onto a hot parking lot in the middle of summer, the liquid could heat up and release vapors, creating a fire hazard.

Not all materials have a flash point. Water and sulfuric acid are examples of materials that do not have a flash point because no amount of heating will result in the release of vapors that can be ignited.

> **■ Note**
> Materials with high flash points must be heated in order to release vapors that can be ignited.

Flammable Liquid
a liquid with a flash point less than 100°F

Combustible Liquid
a liquid with a flash point at or above 100°F

Most regulatory agencies classify liquids with a flash point as either a Flammable or Combustible Liquid. To be classified as a **Flammable Liquid**, the material must have a flash point below 100°F (37.8°C). To be classified as a **Combustible Liquid**, the material would have a flash point at or above 100°F. While this is the definition used by both OSHA and the NFPA, it is important to note that not every agency uses the same temperature for the difference between the two types of materials. Additionally, some agencies place an upper limit on the definition of Combustible Liquid. When this is the case, the upper temperature is typically listed as 200°F, and any material with a flash point above that would not be classified as either flammable or combustible. Given that not everyone uses the same definitions, it is important to ascertain whose definition is being used. Regardless of the agency involved, the definition of a Combustible Liquid always places it with a higher flash point than a Flammable Liquid.

Flammable and Combustible Liquids are further subdivided based on their relative hazards related to fire. Table 4-6 shows the classification based on the OSHA and NFPA definitions.

Table 4-6 *Classification of Flammable and Combustible Liquids.*

Classification	Flash point	Boiling point	Example
Class IIIB Combustible Liquid	≥200°F	N/A	Ethylene glycol Glycerine Motor oils
Class IIIA Combustible Liquid	≥140°F and <200°F	N/A	Mineral oil Oil-based paint
Class II Combustible Liquid	≥100°F and <140°F	N/A	Diesel fuel Kerosene
Class IC Flammable Liquid	≥73°F and <100°F	N/A	Mineral spirits Xylene, styrene Turpentine
Class 1B Flammable Liquid	<73°F	≥100°F	Gasoline, acetone Isopropyl alcohol Toluene, ethanol
Class 1A Flammable Liquid	<73°F	<100°F	Ethyl ether, pentane

■ Note
Flammable and Combustible Liquids are broken into subclasses based on their flash point and boiling point.

As can be seen, the ability of a material to release vapors is an important consideration in determining the degree of fire hazard. A material with a low flash point is more hazardous than one with a higher flash point. Additionally, the boiling point of the material is also a consideration relative to its **volatility**. *Volatility* refers to the tendency of the material to change rapidly from its liquid state to gaseous state and is a function of many factors, including the **vapor pressure** of the material. Vapor pressure is the pressure exerted by a saturated vapor above its own liquid *in a closed container*. Essentially, the liquid particles vaporize until the space above the liquid is saturated. The vapor then condenses and some of the particles return to the liquid phase. Figure 4-8 illustrates the concept of vapor pressure. Substances with high vapor pressures will evaporate rapidly, releasing more vapors or gases into the air, resulting in the formation of ignitable atmospheres.

volatility
the ability of a material to rapidly change from a liquid state to a gaseous state

vapor pressure
the pressure at equilibrium that a liquid exerts inside a closed container

■ Note
Flash point is not the sole measure of the degree of hazard for a given material.

If the concept of vapor pressure is still confusing, think of it in another way. Vapor pressure is the *wantability* of a liquid to become a vapor. Certain liquids, like methyl alcohol, ethyl alcohol, and isopropyl alcohol (IPA), have high vapor pressures. They give off considerable amounts of vapors inside a closed container and evaporate quickly when released. This is quite evident if you have ever opened a bottle of rubbing alcohol. You immediately smell it, and if the lid is off for any period of time, the whole room will soon smell like a medicine cabinet. Imagine if 55 gallons of the material was spilled or the lid was taken off a full one-gallon container. Clearly a considerable amount of vapor would be in the air, and the flash point of the material released would help us know the degree to which the atmosphere could be ignited should an ignition source be present.

■ Note
A simple way to understand vapor density is to describe it as the *wantability* of a liquid to become a gas.

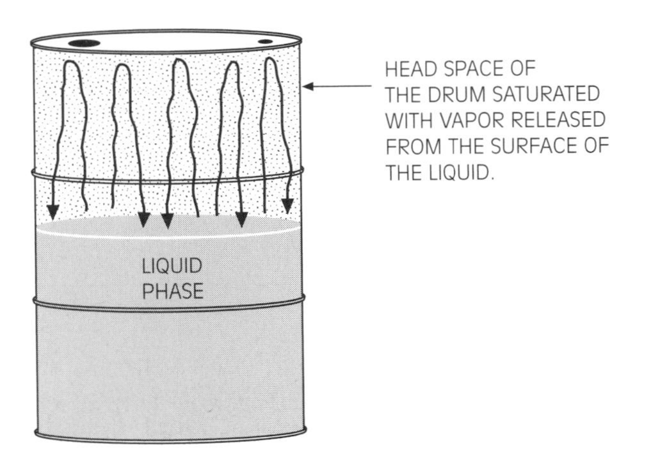

Figure 4-8
Illustration of vapor pressure.

Two additional facts will help in our understanding of the volatility of a material:

1. The vapor pressure of a substance at 212°F (100°C) will always be higher than the vapor pressure of the substance at 68°F (20°C). This means that heated chemicals are more active at the molecular level than cold chemicals, and the eventual vapor production, once released, will be higher.
2. Vapor pressures reported on the MSDS are expressed in millimeters of mercury (**mmHg**). For purposes of a benchmark, 760 mmHg is equivalent to 14.7 pounds psi, which is equal to 1 atmosphere. Materials with vapor pressures above 760 mmHg are capable of exceeding ambient pressure and can be considered vaporous. While this term applies to liquids inside their container, you can be assured that if the material has a high vapor pressure, it will readily evaporate when the lid is removed.

mmHg
the atmospheric pressure that supports a column of mercury, Hg, 1 millimeter high

Note
Heating a material will increase its volatility.

Note
Materials with high vapor pressures will behave more like a gas than a liquid because they release more vapors.

To help illustrate this important concept further, we can again use the rubbing alcohol example. IPA is the primary ingredient in rubbing alcohol and is an excellent example of a material which evaporates rapidly. If you were to spill the alcohol, you would quickly notice that the liquid will evolve great volumes of vapor, which in the case of this material is flammable. While that characteristic does present us with a certain challenge, it also lets us know that the problem is essentially taking care of itself. This means that the liquid product is quickly turning to vapor, which will ultimately diffuse into the atmosphere. High volatility liquids like IPA evaporate quickly on heated pavement and may simply go away before we can get to respond to the leaking material. If an ignition source is present, the fire may occur before we have a chance to react because the flash point for IPA is 53°F. Liquids such as diesel fuel and oils will not evaporate as quickly and present less of an atmospheric hazard and more of a cleanup issue because they are not as volatile.

To determine the volatility of the material and its ability to release vapors that could be above the flash point, we must consult the MSDS or other reference sources. Without this information, we may find ourselves in a hazardous atmosphere that could be toxic or flammable.

Safety
High volatility materials evaporate quickly and are airborne hazards. Materials with low volatility will remain as a liquid for a longer period of time.

Solvents

Solvents
materials that break down other materials

Solvents are often discussed in conjunction with a study of Flammable and Combustible Liquids. While not all Solvents are capable of being ignited, many commonly used ones are. These include materials such as toluene, acetone, methyl ethyl ketone, xylene, and many types of paint thinners. Solvents are materials used to dissolve other materials, most often other hydrocarbon-based substances. Some of them are used as cleaning fluids or for degreasing, while others are used for extracting soluble materials from raw materials.

In addition to their flammability, many types of Solvents also present a health hazard. They can be toxic, or irritating causing damage to the skin, eyes, and mucous membranes. Remember that Solvents work to dissolve fats, oils, and greases, properties that make these materials cause damage to your skin because skin has many of those materials. Solvents can rapidly break through the skin, much as gasoline does. Once you have contact with gasoline, it penetrates your skin, as is obvious from the odor that is released for some time following the exposure. Because of this, it is essential to wear the proper PPE when handling these solvents.

■ Note
Many types of Solvents can permeate into our skin and enter the body.

However, not all Solvents are ignitable. Water is a very common Solvent. It dissolves a wide range of substances, including materials that are soluble or miscible, such as alcohol compounds and others. Because of its ability to break down many materials, water is often regarded as the universal solvent.

■ Note
Water is called the universal Solvent because it has the ability to break down many types of materials.

Health Hazards Posed by Flammable/Combustible Liquids

Various Flammable and Combustible Liquids can present health hazards or combinations of health hazards, because they can be carcinogens, corrosive, highly toxic, irritants, sensitizers, or harmful to specific organs such as the skin. A substance posing one of these health hazards should have a health hazard warning label, and information should be available on the MSDS. Some examples of poisonous and corrosive ignitable liquids are listed in Table 4-7.

Table 4-7 *Examples of poisonous, corrosive, Flammable, and Combustible Liquids.*

Aniline	Dinitrobenzenes	Nitrobenzene
Benzonitrile	Dinitrotoluenes	Nitrotoluenes
Carbon tetrachloride	Ethylene dibromide	Tetrachloroethylene
Chloropicrin	Hexachlorobenzene	Toluidines
Dichlorobenzenes	Nitroanilines	
Corrosive Flammable/Combustible Liquids		
Acetic anhydride	Alkyl amines	Diethylenetriamine
Acetic acid (50% to 80%)	Benzoyl chloride	Ethanolamine
Acetic acid, glacial	Benzyl bromide	2-Ethylhexylamine
Acrylic acid	Butyl amine	1-Pentol

Safety
Flammable and Combustible Liquids can pose additional hazards such as toxicity, ability to produce cancer, and corrosiveness.

Flammable and Combustible Liquids should always be used with adequate ventilation. This is because many of these materials can have effects on the central nervous system, including anesthetic effects leading to dizziness, sleepiness, or unconsciousness. Prolonged exposure to these materials could eventually lead to death.

Note
Ventilation is critical to safety when working with Flammable or Combustible Liquids.

Skin exposure to these materials should be avoided as a matter of common sense. Many materials can remove protective oils from the skin, and some of them can even penetrate the skin. Many can cause systemic or target-organ damage, particularly to the liver and blood-forming organs. Even if the substance is not corrosive or toxic, it is best to avoid all exposure to the material.

Safety
It is important to limit contact with Flammable and Combustible Liquids because many of them can break down the skin and enter the body.

Safe Handling, Use, and Storage of Flammable/Combustible Liquids

The consistent physical danger from Flammable or Combustible Liquids and Solvents is the fire hazard. Vapors may form ignitable or explosive atmospheres and ignite from uncontrolled ignition sources. In most cases, the presence of water will serve to increase the surface area of a fire involving Flammable or Combustible Liquids, making extinguishment more difficult. Foaming agents, such as Aqueous Film-Forming Foams (AFFF), should be utilized for these types of incidents and great care must be taken to properly protect responders from heat and flame. Figure 4-9 shows a flammable liquid fire in a bulk storage facility.

Safety
Never use water on a fire involving Flammable Liquids because it can increase the surface area, vapor production, and size of the fire.

Monitoring for the presence of flammable atmospheres should always be considered in any area where these materials are used. Elimination of ignition sources and the use of intrinsically safe equipment are critical to safety when working with large quantities of materials that readily give off vapors. Ventilation systems in areas where flammable atmospheres could be encountered should be of highest priority and designed to fully eliminate the potential for a flammable atmosphere to exist.

The presence of ignitable vapors in the atmosphere and an ignition source are only part of what is necessary in order to have the potential for fire. The vapors

Figure 4-9
A fire involving Flammable and Combustible Liquids. (Courtesy of FEMA.)

flammable range
the range of concentrations at which a flammable gas or vapor mixed with air will burn if an ignition source is present

themselves must be in a concentration high enough to form an ignitable mixture in air. This level is called the **flammable range**, sometimes referred to as the flammable limits of the substance. It refers to the concentration of the material in air that will create the potential for fire. As with health hazards, this can be thought of as the proper dose of the material necessary to have a fire. A dose of vapors that is too low will not produce the results required to have a fire potential. In reviewing flammability, we must know that each material has its own unique range of vapors in air that will produce the potential for a fire to occur. Anything within this range of vapor-to-air mixture will result in the creation of a flammable atmosphere. Too few vapors in air would be considered to be too lean, or under the required dose. Too much vapor in air would be considered too rich, or over the required dose.

■ Note
Flammable limit is another term for flammable range.

Lower Explosive Limit (LEL)
the lowest concentration of a vapor in air that will form an ignitable mixture

Upper Explosive Limit
the highest concentration of a vapor in air that will form an ignitable mixture

Lower Flammable Limit (LFL) and **Upper Flammable Limit (UFL)**
terms sometimes used in place of the more common LEL and UEL

To help identify the lowest and highest amounts of vapors that would form an ignitable mixture in air for a particular material, the terms **Lower Explosive Limit (LEL)** and **Upper Explosive Limit (UEL)** are used. The LEL is the lowest concentration of vapors in air that are capable of being ignited. Because the vapors are what are required, this would be the least amount present that could be ignited. Conversely, the UEL is the highest amount of vapors that are needed to be present to form an ignitable mixture in air. The difference between the LEL and UEL is termed the flammable range. Sometimes the term *flammable* is used interchangeably with the word *explosive* in the description of the lower flammable limit or lower explosive discussion. **Upper Flammable Limit (UFL)** and **Lower Flammable Limit (LFL)** would therefore replace the terms *Upper Explosive Limit* and *Lower Explosive Limit*, although *explosive* is the more commonly used term.

■ Note
The concentration between the LEL and UEL forms the flammable range for the material.

Table 4-8 lists some flammable ranges for commonly encountered materials. The values are those cited by the NFPA in its standards. Materials with a flammable range are either Flammable Gases, which already exist in a gaseous state, or Flammable and Combustible Liquids, which emit vapors when they are above their respective flash points. Because the term *flammable range* applies to both gases and liquids, both are included in the list.

Safety
Materials with a low LEL pose higher hazards than those with a higher LEL.

Safety
Materials with a wide flammable range pose higher hazards than those with a narrow range.

Some materials listed in Table 4-8 have extremely wide flammable ranges. Acetylene, carbon monoxide, and hydrogen are each capable of ignition in a much wider range than materials such as gasoline. Other materials, such as gasoline and propane, have very low LELs, making them much more likely to have fire issues from smaller releases. Finally, materials such as ammonia and carbon monoxide have a very high LEL, which causes them to present less of a fire hazard because a considerable amount of vapor is required in order to reach the lowest mixture to form an ignitable concentration, the LEL.

Principles of storage and segregation should be based on the recognition of the fire hazard. The storage area should be constructed to be free from ignition sources and able to contain accidental releases by the use of secondary containment. Containing spills to the smallest area is a critical safety component in that limiting the surface area of the spilled material will reduce the amount of vapors released. Additionally, storage containers for these types of materials should remain closed at all times except when materials are being added or removed. This will help to reduce the amount of flammable vapors present in the area. Finally, Flammable and Combustible Liquids should be stored separately from substances that could react with them, such as Oxidizers. While local regulations vary, OSHA, NFPA, and others recommend that a distance of no less than 20 feet be maintained between

Table 4-8 *Common materials and their flammable ranges.*

Material	State	Flash point	LEL (%)	UEL (%)
Acetone	Liquid	0°F	2.5	12.8
Acetylene	Gas	N/A	2.5	100
Ammonia	Gas	N/A	15	28
Carbon monoxide	Gas	N/A	12	74
Diesel fuel	Liquid	125	0.3	10
Ethyl alcohol	Liquid	55°F	3.3	19
Gasoline	Liquid	−44°F	1.4	7.6
Hydrogen	Gas	N/A	4.0	75
Methane	Gas	N/A	5	15
Propane	Gas	N/A	2.4	9.5

the storage of Flammable and Combustible Liquids, and Oxidizers or other incompatible materials.

Safety
Containing the liquid spill to the smallest possible area will limit the amount of vapors produced.

Emergency Procedures

Planning for emergency response to incidents involving Flammable or Combustible Liquids is absolutely critical because of their ability to quickly create fires and other hazardous environments. Even the most carefully developed plans need continual reevaluation because conditions change in any release scenario. Small releases of these materials can present major issues because the materials are hazardous from both health and physical perspectives. A general rule when planning response procedures or for working with any of these materials is to ask "what if" something were to happen. This will ensure that there is a backup plan ready to go should the material suddenly catch fire, and that secondary escape routes have been identified as well as medical treatment made available, should it be required. Other common response practices include the following:

- Ensure that there is adequate ventilation in place to prevent the buildup of the gases and vapors because the accumulation of the vapors creates additional health concerns and can cause fires. Adequate ventilation will ensure that levels do not approach the LEL, the lowest point at which the vapors are capable of igniting.
- Monitor the area with appropriate types of instruments to ensure that the level of vapors never exceed more than 10% of the LEL. Many monitoring instruments that will be discussed in a later chapter are calibrated to sound an alarm when the ignitable vapors reach this level.
- Eliminate ignition sources in the area of the release. This is also a critical element of an emergency response effort. Eliminating ignition sources involves all shutting down equipment, the use of non-sparking tools, etc. If materials are transferred between containers, bonding the two containers together and grounding them will help dissipate static electricity during this process. More details on these procedures will be discussed in Chapter 12.
- Never use water to wash down Flammable or Combustible Liquids release because many of these materials are not soluble in water. Adding water to them will only increase the surface area of the liquid, resulting in the production of more vapors.
- Ensure that personnel are prepared should a fire occur. Appropriate fire extinguishers or foaming agents should be located near the release areas and staffed with properly trained personnel throughout the response effort.

Safety
Proper ventilation, elimination of ignition sources, limiting the use of water, and monitoring for the presence of vapors are the keys to handling Flammable Liquid spills.

FLAMMABLE SOLIDS

Flammable Solids represent a class of material that is relatively uncommon and not frequently encountered in most operations. Unlike ordinary combustible solids such as wood, paper, or cloth, Flammable Solids are considerably more hazardous and present a significant fire hazard. They possess properties that make them more hazardous in their ability to either start fires easily or create fires that cannot be extinguished with ordinary extinguishing agents. The most commonly used extinguishing agent, water, is highly effective on ordinary combustible solids. However, water is incapable of extinguishing most types of Flammable Solids once they catch fire due to the extreme heat that these materials produce.

> **Safety**
> Flammable Solids pose a much higher hazard than ordinary combustible solids and cannot be extinguished with water.

Definition of Flammable Solids

Flammable Solid
a material that is easily ignited or one that presents a serious hazard once ignited

pyrophoric materials
materials with an autoignition temperature less than 130°F

autoignition temperature
the temperature at which a material ignites without an external ignition source

The technical definition from the DOT regulations describes a **Flammable Solid** as a material, other than an Explosive, that is easily ignited or that, once ignited, burns so vigorously or persistently that it presents a serious fire hazard. Unlike ordinary combustible materials, which generally require substantial heating to ignite and once ignited are easily extinguished through the elimination of the heat through the application of water, these materials create considerable problems and all efforts must be directed at preventing ignition.

Flammable Solids can also possess other hazards, further increasing their danger. Some are **pyrophoric**, or spontaneously combustible, and will ignite on contact with air. Pyrophoric materials are those whose **autoignition temperature** is below 130°F. *Autoignition temperature* is the term that describes the lowest temperature at which materials ignite spontaneously in the presence of air without an ignition source. All materials that are capable of ignition, including all Flammable Gases, Flammable and Combustible Liquids, ordinary combustible solids, and Flammable Solids, are capable of being heated to a point where they ignite spontaneously. For most common materials, this temperature is extremely high, as noted in Table 4-9.

Table 4-9 *A autoignition temperatures of some common materials.*

Material	Autoignition temperature
Acetone	869°F
Diesel	425°F
Gasoline	850°F
Combustible solids such as paper	Approximately 451°F
White phosphorous	93°F
Sodium	257°F

Safety
Extreme care should be taken when working with Flammable Solids because they can possess other properties that can increase the hazard present.

Pyrophoric materials have the ability to ignite in air at temperatures well below the normal temperatures cited in the previous paragraph. White phosphorous, a material often used in military ammunition, which helps soldiers see tracer rounds shot from a rifle, is an example of a material with this property. When a bullet is fired, part of the bullet leaves the casing and goes toward the target. If the bullet end is coated with white phosphorous, it will ignite as it leaves the rifle and contacts air. The flame will be visible, showing the soldier where the bullets went, hence the term *tracer round*. Other military ordnance also use this principle and material. Other types of pyrophoric materials are also found in industry.

Note
White phosphorous is an example of a pyrophoric material and is used in military ordnance.

Another property that some Flammable Solids possess is a capacity to react with water in a dangerous manner. Calcium carbide is such a material. In the early days of mining and in some original car headlights water was dripped onto a small piece of calcium carbide, which evolved acetylene gas, which was then ignited and could light up a relatively large area.

Flammable Solids can be found in a number of industrial applications and exist as powders, shavings or filings from a milling process, solid bars or rods, and chips or chunks. Sodium, for example, can come in 55-gallon drums where it is subsequently heated, turned into a liquid, and pumped into a chemical process.

Examples of commonly encountered Flammable Solids include potassium, sodium, magnesium, calcium carbide, red phosphorous, white phosphorous, and lithium.

Note
Some Flammable Solids can be heated and become liquids.

Health Hazards Posed by Flammable Solids

It is relatively easy to protect ourselves from exposure to Flammable Solids, unless they are reacting or on fire. In solid form, their primary hazard is their dusts and finely divided powders, which can be easily inhaled. Once inhaled, some dusts such as sodium will react with the moisture in the lungs to form caustic solutions that can burn the sensitive tissues in the airway. All attempts should be made to avoid exposure of skin, eyes, and mucous membranes because these moist areas are also vulnerable to irritation and chemical burns. If a Flammable Solid is ignited, it is imperative to avoid clouds of smoke because they may include toxic by-products of the burning materials.

> **■ Note**
> Most Flammable Solids present a minimal health concern.

Safe Handling, Use, and Storage of Flammable Solids

Because some Flammable Solids are water reactive, it is often necessary to take special measures in designing containers and storage areas for them. The potential for water contact should always be considered regardless of the types of Flammable Solids involved, as many will slowly react from extended contact with water. For this reason, specially designed storage areas and containers should be present whenever these materials are found in the workplace. Figure 4-10 shows a hazardous storage practice for these materials in which they are exposed to the outside environment and are not protected from water contact.

In addition to water, many other types of hazardous materials can react with Flammable Solids. Both acids and bases may react adversely with this type of material and cause the release of toxic gases. It is generally a good practice to store Flammable Solids in a segregated area, away from other materials, including water, and in sealed containers.

> **■ Note**
> Storage areas for Flammable Solids must be designed to minimize the potential for interaction with other materials.

Emergency Procedures

There are few recommended emergency actions when it comes to response to incidents involving Flammable Solids, given the nature of their hazard. As with the storage recommendations cited in the previous paragraph, keeping these materials from contact with other materials including water is very

Figure 4-10
Example of poor storage practices for water-reactive Flammable Solids.

important. If the material is released in a dust or shaving form, a common tactic is to isolate the released material by covering it with a tarp or heavy plastic. This will reduce the potential for contact with other materials including water, and minimize the amount of dust that could blow off in the event of wind or air currents. Regardless of the material involved, never add any water until proper identification of the specific material is made and it is confirmed that no adverse reaction will occur. Once nonreactive materials are wetted down, they produce less dust and pose less of a hazard because they are less likely to become airborne.

Note
Dusts given off by a Flammable Solid release can be controlled using water if the material is confirmed to not react with water.

If these materials ignite, firefighting is considerably complicated. In most cases, ordinary extinguishing agents such as water, Halon, dry chemical, and CO_2 are ineffective given the high temperature conditions in which these materials evolve. Once ignited, these materials create a serious fire concern and no common extinguishing agent is effective. An example of the hazards of some types of these materials involved in fire can be found in Navy aircraft located on aircraft carriers. If a serious fire occurs in an aircraft composed of many types of Flammable Solids used to make lightweight metal components, the solution is often for the aircraft to be pushed overboard because ordinary firefighting activities may not be successful. Further, the heat generated by a fire involving these materials can result in damage to the deck of the aircraft carrier. If a serious fire occurs involving these materials, the area may need to be evacuated of all personnel, and cleared of combustible materials.

Finally, if the dust or powder form of the material contacts personnel or PPE, care should be taken before water is used to wash it off. In such cases, it is usually best to brush off as much powder as possible using a brush or other appropriate tool. Once this is done, flush the area with copious amounts of water and ensure that all of the material is washed off the skin or PPE. This will reduce tissue damage to the skin and prohibit the creation of a toxic paste being formed on the body or PPE.

Safety
Large fires involving Flammable Solids may require that the material be allowed to burn up because normal extinguishment is not possible.

OXIDIZERS AND ORGANIC PEROXIDES

Oxidizers are a difficult classification of materials to define in that most definitions are written by and for chemists who have a strong understanding of the dynamics of chemical reactions and physics. Some definitions discuss the transfer of minute parts of the atom, the electrons, to other materials to form other compounds, with a secondary receiving of another part of the atom. This is sometimes referred to as an oxidation-reduction reaction and is nice to know, but not overly helpful to the non-chemist. Still others define it simply as a material that readily gives off oxygen, even though not all Oxidizers do this. These definitions do little to inform non-chemists, and often it is more helpful to simply discuss what these materials do and the hazard they present. This increases our understanding of them as they relate to our safety in working with them or in response to an emergency incident involving them.

Oxidizers are very reactive materials, capable of initiating a reaction and helping reactions to continue, often at an accelerated pace. Consider oxygen, one of the most basic Oxidizers. Oxygen is necessary for a fire to occur. Fire is a reaction between the fuel, oxygen, and heat. If enough of each of these materials is present, the reaction (fire) continues. If you remove the oxygen, or any of the other materials, the reaction stops, and the fire goes out. In this case, oxygen is necessary to allow the reaction to continue. But if the oxygen concentration increases, the reaction rapidly accelerates. For example, in an oxygen–acetylene torch the fuel, acetylene, is ignited in the presence of normal atmospheric oxygen. The fire and flame produced from this reaction is rather limited and incapable of cutting metal or welding. However, once additional oxygen is added, the flame suddenly gets brighter and hotter, creating temperatures that allow the gas to cut through several inches of metal. The reactive nature of the Oxidizer causes this. This is true of other Oxidizers in a number of other reactions. Simply stated, **Oxidizers** speed up the rate of the reaction without actually burning themselves. Oxygen, like other Oxidizers, is nonflammable. By itself, an Oxidizer is not flammable, but the flammable effects are significantly more exaggerated in the presence of the Oxidizer.

Oxidizers
materials that speed up reactions

■ Note
Defining an Oxidizer can be difficult for non-chemists.

Safety
Oxidizers are often capable of causing other reactions at an accelerated pace.

In addition to speeding up reactions, Oxidizers are also capable of initiating a reaction and generating the heat necessary to create a self-sustaining reaction, as noted in the chapter Case Study. Ammonium nitrate, a strong Oxidizer, can be used to create explosive mixtures with common materials including most types of **organic materials**. Organic materials contain hydrogen and carbon. Hydrogen and carbon are building blocks for many materials including hydrocarbons, which comprise many types of fuels. Organic materials are typically ignitable, and in the presence of a strong Oxidizer, an ignition can occur simply as the result of combining the Oxidizer with the fuel. Recall from our discussion of ANFO that the combination of the Oxidizer (ammonium nitrate) and the fuel (fuel oil) creates compounds that are capable of considerable destruction.

organic materials
materials that contain hydrogen and carbon

Safety
Oxidizers present serious hazards because of their incompatibility with many other materials.

Oxidizers through a chemical process create conditions that can initiate a reaction or cause other reactions to occur at an accelerated pace. Defining Oxidizers in this way helps us to understand the safety considerations that must be taken into account whenever these materials are present. Their involvement in an emergency incident could start a reaction or cause a reaction to rapidly grow out of control.

However, the ability to react is not unique to exotic Oxidizers, as can be seen from the Case Study at the beginning of the chapter. In that incident, calcium hypochlorite, a type of common swimming pool chemical shock treatment that contains chlorine, mixed with another material on the store shelf. The second material involved in the reaction was simple spa conditioner, a fragrance oil that is added to spa water. In this case, the fragrance contained glycerin and IPA, both serving as fuels for the Oxidizer to react with. What likely occurred was a reaction between the Oxidizer and the fuel when the materials came into contact with each other. This resulted in the formation of heat that created a fire and caused the release of the chlorine into the store.

The incident demonstrated to everyone involved that Oxidizers can self-initiate a reaction, and once started, can accelerate the reaction with potentially deadly results. Similar reactions could occur if concentrated hydrogen peroxide is mixed with wood shavings or when concentrated nitric acid comes into contact with many types of organic materials. The proper handling and storage of Oxidizers is critical to the safety of those who work with these materials.

! Safety
Some ordinary materials are Oxidizers and can create serious hazards if not properly managed.

Because all Oxidizers should be treated with a high degree of care, it is important that their presence be identified. Many times this can be done through proper labeling, as noted in Chapter 5. However, there is a simple set of rules to help the non-chemist identify materials that are potential Oxidizers. Upon reviewing the list in Table 4-10, we can see that many of the materials end with the letters *ite* or *ate*. Additionally, many others have the prefixes *per*, *hypo*, or *oxy* in their chemical names. While not a perfect system, the presence of any of these in the chemical name indicates that the material could be an Oxidizer. When in doubt, it is always best to refer to the MSDS for the material(s) present.

! Safety
It is vital to ascertain the presence of an Oxidizer in order to ensure that proper handling occurs.

■ Note
Many Oxidizers can be recognized by the presence of prefixes *per*, *hypo*, or *oxy* in their name or if their name ends in suffixes *ite* or *ate*.

Table 4-10 *Examples of some common industrial oxidizers.*

Aluminum nitrate	Fluorine	Permanganates
Ammonium nitrate	Hydrogen peroxide	Potassium permanganate
Calcium hypochlorite	Oxygen	Potassium peroxide
Calcium perchlorate	Ozone	Sodium peroxide
Chlorates	Perchlorates	Sodium hypochlorite

Organic Peroxides

A special class of Oxidizers is **Organic Peroxides**. Like Oxidizers, Organic Peroxides are very reactive and can react explosively when involved in a fire. The prefix *per* in the name means many. So *Per*oxide is made up of many oxides, making it significantly hazardous. The reason these substances pose such a high danger is found in their chemical structure. With Organic Peroxides, an organic compound is chemically bonded with some type of Oxidizer. In other words, the Oxidizer and fuel are already bonded together and waiting to react. The concept is used to make many types of Explosives such as TNT.

Methyl ethyl ketone peroxide (MEKP) is an example of an Organic Peroxide. MEKP is a very aggressive solvent and is used extensively in plastic fabrication. It is irritating to the skin and is susceptible to combustion. Another common Organic Peroxide is benzoyl peroxide. You may be acquainted with this substance from the acne medicine commercials. Regardless of how harmless they may sound, Organic Peroxides are challenging because they contain both the Oxidizer to initiate the reaction and the fuel to sustain it. Care should be taken to ensure that these materials are not exposed to heat or other sources that could initiate a reaction.

> **! Safety**
> Organic Peroxides are a subgroup of Oxidizers that present additional hazards because they contain fuels.

Health Hazards Posed by Oxidizers

Oxidizers as a class are not generally considered to present a health hazard. Their primary hazard is physical—the dangers of fire and explosion either from the compound itself or from reacting with other materials. However, some specific Oxidizers are health concerns and contact with them should be avoided. Fluorine is an example of an Oxidizer that is also highly toxic. Many other Oxidizers are skin irritants and dry agents should not be allowed to contact skin, eyes, and mucous membranes where they can react with water. Additionally, because Oxidizers are highly reactive, their presence in a fire situation may cause the release of other toxic compounds, so avoiding smoke or fumes from these reactions is critical. As with any material used in the workplace, it is always best to consult the MSDS for the specific material, as it will identify the hazards and safe work practices required.

> **■ Note**
> Most Oxidizers present minimal health concerns.

Safe Handling and Storage of Oxidizers

Safe handling and use of Oxidizers should be based on segregation of these materials from others, especially fuels. Because these materials are so reactive, storage requirements dictate that a distance of no less than 20 feet be maintained between them and other hazard classes. They are incompatible with most other types of materials and separation by distance or by a noncombustible barrier is required. Even transporting these materials in a cart or on a pallet with other materials should be avoided because the materials can react should they accidentally spill. Figure 4-11 shows a poor storage arrangement with Oxidizers on the top shelves in close proximity to other incompatible materials. If this shelf were to be bumped, materials could mix, resulting in a serious reaction.

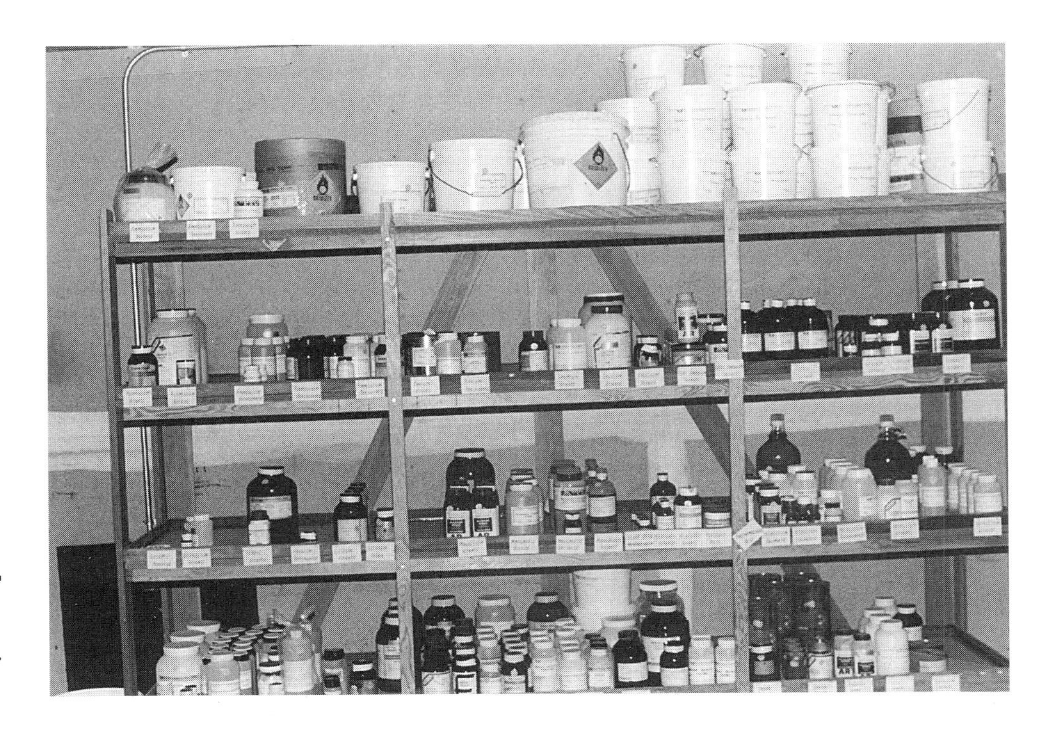

Figure 4-11
Example of poor storage practices for oxidizers.

> **! Safety**
> Oxidizers are required to have a 20-foot separation between them and other materials, especially fuels.

Storage shelves and areas where these materials are present should also be carefully evaluated to ensure that no organic materials are present to form a fuel for the Oxidizer to react with. Figure 4-12 shows Oxidizers stored on wooden shelves, another unsafe practice. If the Oxidizer on the shelf were to leak, a reaction could

Figure 4-12
It is a poor practice to store oxidizers on wooden shelves.

occur, causing the wooden shelves to catch fire and burn. This could result in the shelves failing from the fire, or the other combustible plastic containers could fail, introducing more Oxidizers into the fire. The presence of any organic material in an area where Oxidizers are found is a practice that needs careful monitoring.

> **! Safety**
> Shelving used for storage of Oxidizers should be designed to prevent reactions if the material is released.

To assist in spill control, absorbent materials should be kept in the storage area to limit the spread of the material should there be a release. Care should be taken in selecting absorbents used on Oxidizers because many absorbents are organic based and can react vigorously and cause a fire if exposed to strong Oxidizers.

> **■ Note**
> Proper spill control supplies should be located near storage of Oxidizers to allow for prompt response to releases.

Emergency Procedures

As with many other classes of hazards, planning and training for emergencies involving Oxidizers should be done prior to any emergency response. Oxidizers involved in fires may react with explosive violence. In almost all cases oxygen and heat are released, resulting in a rapidly accelerating fire that is self-sustaining. Oxidizer fed fires may accelerate rapidly and grow beyond the capability of normal fire extinguishers to handle, so preplanning responses to these types of incidents should include additional fire protection systems in the area. Perhaps the most effective means of dealing with large fires involving these materials is to flood the areas with large amounts of water, provided no water-reactive materials are present.

Oxidizers that contact the body must be rinsed off immediately with copious amounts of water. Safety showers and eyewash stations should be present in areas where these and other irritating materials are present. Contact with dry Oxidizers should be handled in a similar manner, as with other dry materials that were discussed previously. This involves using a brush to quickly brush off the bulk of the material prior to the application of the water.

POISONS

Poisons
materials that are capable of causing harm to health

Materials that are classified as **Poisons** are those whose primary hazard is related to damaging health. Numerous definitions and terms are used to describe this class of materials, including the use of the words *toxic*, *poisonous*, or *lethal*. From our study of toxicology in Chapter 3, we know that most materials are capable of causing harm if the dose is high enough. Recall that "the dose makes the poison," which implies that all materials can be poisonous in the right dose. However, to safely work with a material that is classified as a Poison, we need to establish a working definition that identifies the real threat these materials pose.

Poisons are substances that are capable of causing death or serious injury if they are swallowed, inhaled, injected, or come in contact with the skin, eyes, or mucous membranes. This definition can include materials that present other hazards as well, such as being flammable, corrosive, or reactive, so our definition is limited to those materials whose *primary* hazard is to kill or cause serious health effects. In some cases, the Poison interferes with the proper functioning of the body by affecting oxygen distribution in the bloodstream or by blocking nerve impulses. These health effects are usually acute in nature and may ultimately result in death. Poisons are broken down into three main groups of materials that fit this broad definition and include pesticides, heavy metals, and infectious substances.

■ **Note**
Poisons are divided into three main classifications. These include pesticides, heavy metals, and infectious substances.

pesticides
materials that kill various forms of pests

Pesticides are some of the most common types of Poisons. The suffix *cide* at the end of the word literally means "to kill." The word *pesticide* literally means "to kill pests." Other similar types of materials share this identifier such as Rodenticides, substances that are used to kill rodents, Herbicides that kill vegetation, and Fungicides that kill fungus. However, not all Poisons are as easily recognized and the use of an MSDS is critical to understanding whether the material is poisonous to us or not.

■ **Note**
Cide at the end of a word often means to kill.

heavy metals
metals that are poisonous

The second group of materials classified as Poisons are **heavy metals**, such as mercury, beryllium, cadmium, lead, and arsenic. Heavy metals are a class of poisonous materials that negatively affect health in certain doses. Unlike pesticides that kill from an acute exposure, most types of heavy metals require multiple or chronic exposure in order for the poisonous effects to be realized. Regardless of the mechanism of exposure, the end result of exposures to these materials can be health effects that cause harm or even death.

■ **Note**
Heavy metals are mostly toxic through chronic exposure.

infectious substances
materials capable of producing disease

The third group of materials classified as Poisons are **infectious substances**. These materials are capable of causing disease and are sometimes referred to as Pathogens, Biohazards, or Etiologic Agents. Examples of these include the group of Hepatitis A, B, and C viruses, the Human Immunodeficiency Virus (HIV), Tuberculosis (TB), and a host of others. These materials can be transmitted through contact with blood or body fluids (bloodborne pathogens), through the air (airborne pathogens), through drinking water (waterborne pathogens), through eating food (food-borne pathogens), and through exposure to a vector. Vector-borne pathogens are transmitted by an intermediate host such as an animal. Bats are vectors for Rabies and mosquitoes are vectors for a number of diseases such as Malaria and *West Nile Virus*.

■ **Note**
Infectious materials can be bloodborne, airborne, waterborne, food-borne, or vector-borne.

Health Hazards Posed by Poisons

The degree of danger posed by a Poison depends on a number of factors covered in our study of Toxicology. Each type of Poison may have its own unique method of causing harm depending on the material, the route through which we are exposed to it, and our unique sensitivity. Following is a discussion of each of the major groups of Poison materials and how their poisonous properties adversely affect us.

Neurotoxin
a poisonous material that affects the nervous system

Pesticides as a group are materials that are classified as neurotoxins. A **Neurotoxin**, as the name implies, is a substance that is harmful to the nervous system and that can interfere with the ability of the nervous system in the body. Because the nervous system is extremely complex, disruption of any of its functions can lead to considerable harm. Most types of pesticides work in a standard manner and cause the nervous system to malfunction. Often this results in disruption of the normal function of regulation that the nervous system offers to our bodies. While there are three subgroups of pesticides, including Organophosphates, Organochlorides, and Carbamates, each works by attacking the nervous system. To illustrate common effects of exposure to pesticides, we will use common Organophosphates as an example.

> **■ Note**
> Most pesticides are neurotoxins and affect the nervous system.

> **■ Note**
> There are three major groups of pesticides—Organophosphates, Organochlorides, and Carbamates.

Organophosphates kill by paralyzing muscles and affecting nervous system activity. Parathion, a common type of Organophosphate, interferes with nervous system transmissions at neuroeffector junctions and synapses. Neuroeffector junctions are areas in the body where nerves meet the organs. Synapses are junctions between nerve bundles. Essentially, the nervous system is much like the electrical wiring in a house. The central brain is like a main breaker panel with multiple runs of wiring going outward in all directions. Some of these runs of wiring are interrupted by switches that can be turned on or off depending on the need for power to a particular area. The wiring of the body originates in the brain, and the spinal cord is the main wiring that spreads the remaining wiring throughout the body via an intricate framework of nerves. These nerves are also interrupted by switching stations that allow certain body functions and muscular activity to be turned on and off. These switching stations are the synapses and neuroeffector junctions, as discussed earlier. This on–off switching is done by chemical reactions that occur in fractions of a second, instantaneously and unconsciously.

> **■ Note**
> Many pesticides kill by interfering with the ability of the nervous system to regulate itself.

Between each nerve and its point of contact, there is a tiny space or gap. This gap is bridged by chemical impulses that finish the job of nerve transmission.

When the brain says "jump" (any type of organ activity—movement of muscles, tearing of eyes, beating of the heart, etc.), a lightning-fast impulse is sent down the nerve much like electric current through the household wiring. When the impulse gets to the junction, a chemical is secreted by the nerve, which then travels across the gap to find its specific receptor site. This receptor site is the lock that only accepts a very specific chemical profile or key. This lock-and-key relationship is vitally important to the overall success of the nervous system because it is the final step in impulse transmission. Figure 4-13 illustrates the physiology of a typical synapse as the chemicals cross the gaps.

All nerve terminals within the major organs, muscles, arteries, and so on have a number of locks that accept certain keys. Some locks, called Alpha 1 receptors, are stimulated by *go* keys, which cause constriction of arteries and the bronchioles inside the lungs. Other locks are stimulated by *go slow* keys such as Beta 2 receptors, which oppose the Alpha 1s and cause dilation of the same arteries and bronchioles. It is not important to fully understand all the mechanisms of various receptors in our very complex nervous system, but one concept is crucial to our understanding of the effects of neurotoxins. Chemical reactions are a key part of nervous system impulses and organ function, and the body functions as a result of multiple biochemical reactions. In the event the chemistry is not correct, these transmissions are compromised and the body systems may be affected.

Organophosphates work by inhibiting the breakdown of certain chemical compounds that stimulate the action of certain parts of our nervous system. They essentially stop the function of parts of the nervous system by interfering with the chemicals that are required to regulate it. A good illustration of this is the typical symptoms of overexposure to an Organophosphate. The mnemonic S-L-U-D-G-E describes the symptoms of a victim of the overexposure:

*S*alivation—production of significant amount of saliva with the inability to stop drooling

*L*acrimation—production of tears in an uncontrolled manner

*U*rination—inability to stop urinating

*D*efecation—loss of bowel control

*G*astric disturbances—significant stomach distress and pain

*E*mesis—profuse and uncontrolled vomiting

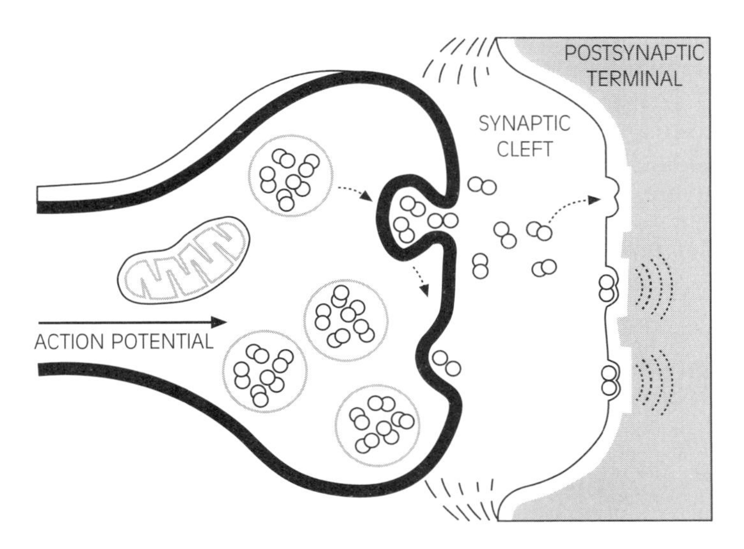

Figure 4-13 *Illustration of neurotransmitter impulse.*

Note
Symptoms of exposure to many types of pesticides can be remembered by the mnemonic S-L-U-D-G-E.

Without rapid treatment, exposure to these materials can be fatal because the victim's heartbeat will start slowing to dangerous levels and convulsions will occur. The condition is very serious and requires the administration of specific drugs such as Atropine to counter the action of the Organophosphate. Atropine will introduce chemicals to help the nervous system to turn off and reduce the symptoms of exposure for a short duration.

Safety
Serious exposure to pesticides requires rapid action because the materials can quickly overwhelm the nervous system.

secondary exposure
results from the material contacting another source that is then taken into the body; an example would be eating food with hands contaminated by a substance

Heavy metals are also neurotoxic, although their toxicity is most often slower acting than those of pesticides. The most common mechanism of exposure to these types of materials is primarily through ingestion. The ingestion is often the result of what we term a **secondary exposure**. Secondary exposures occur when someone works with a material that gets on their hands or food and is later ingested when the hands pick up food to eat. Gradually, the level of material accumulates in the body until a toxic dose is reached. This generally takes a longer time, as seen through the ingestion of lead or mercury with the resulting effects taking much longer to affect the body. In most cases, chronic exposure to these materials is required before their effects on our central nervous system are realized.

pathogens
materials capable of producing disease

Infectious substances present unique challenges given the range of materials and their mechanisms of exposure. Additionally, exposure to these types of materials rarely results in immediate health effects. This fact may contribute to a much larger exposure occurring given that the person exposed does not recognize the hazard at the time. When there is the potential for exposure to these types of materials in the workplace, OSHA requires that additional programs be developed to identify the hazards, train the workers who might be exposed, and develop procedures to prevent exposure. Each specific **pathogen** creates the potential for disease or infection to occur if the organism is allowed to get into the body in a high-enough dose. The good news is that as with other types of hazardous materials, recognition of the hazard and blocking the route of entry will reduce the potential for exposure and allow us to work safely with these materials.

The actual mechanism of exposure to pathogens can be illustrated by a chain of transmission, as seen in Figure 4-14. The chain starts with the actual pathogen that is hosted in a reservoir. The reservoir is often a body fluid or a vector for the material. Blood is often the reservoir for bloodborne pathogens, although any other body material that is contaminated with blood can serve in that capacity. Next, the pathogen must leave the host and be transmitted in some way. This can be done through bleeding or by the presence of blood on surfaces or materials.

Note
In order to have an exposure, all links in the chain of transmission must be present.

Clearly, the first few links in this chain of transmission are not something that we have much control over. It is only when we get to the last two links in

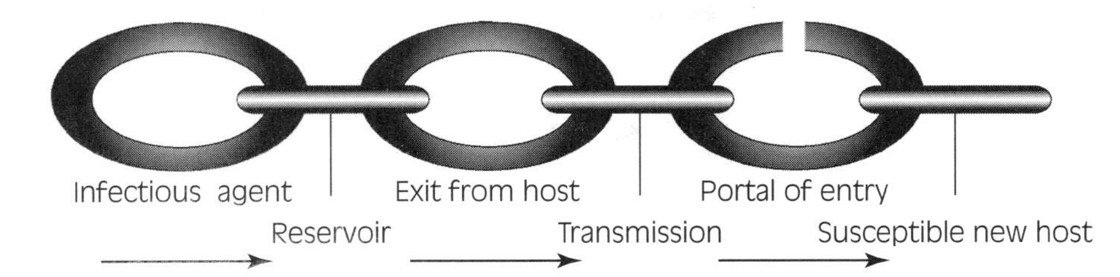

Figure 4-14 *The chain of transmission for exposure to pathogens.*

universal precautions actions that are required to be taken to prevent the potential for infection

the chain that we have some control over the potential for exposure. The control we have for protection from potential exposure occurs when we block the route of entry for the material to enter our bodies. While intact skin does provide some protection, wearing appropriate pathogen-resistive gloves and other barriers is often required. **Universal precautions** should be taken when protection is required from these types of material. Universal precautions are actions that we take whenever the potential for transmission of diseases is present. They include the following:

- Assume that all body fluids from all persons other than yourself are contaminated.
- Always wear protective barriers whenever working with the materials or persons that have potential for transmission of pathogenic materials.
- Never touch body fluid from any other person without an appropriate barrier, such as CPR barrier masks when performing mouth-to-mouth resuscitation.
- Wash your hands thoroughly after dealing with any potentially infectious substance. This is applicable even if gloves are worn.
- Take appropriate precautions to ensure that your health is properly maintained. A strong immune system will help prevent possible infection.
- Make sure to receive all appropriate vaccinations that will help reduce the potential for infection.

The final link in the chain of transmission involves the last of the universal precautions listed. In addition to blocking the route of exposure to the potential pathogenic substance, we can make our bodies less susceptible to disease transmission by getting vaccinations that will keep us from becoming a potential host for the disease. Common examples of this are vaccines for Tetanus that prevent infection, and Hepatitis B vaccinations that almost eliminate the potential for infection with that pathogen. Reducing the potential for being a susceptible host requires that we avail ourselves of proper health care and vaccinations.

> **Safety**
> Blocking the route of exposure to pathogenic substances is one of the two methods of preventing exposure.

> **Safety**
> It is important to have appropriate vaccinations to reduce the potential to be a susceptible host.

Safe Handling, Use, and Storage of Poisons

While the study of Poisons can be unnerving, the good news is that simply working with a material or being in the same area as a material classified as a Poison does not mean that we have been exposed. As we know, exposure is a result of allowing the material to get in or on us in a high-enough dose to cause harm. Once the presence of these materials is known, steps can be taken to eliminate the potential or minimize the exposure through the implementation of controls. Following are some standard practices for working with these types of materials:

- Establish a controlled area where Poisons are processed, used, handled, or stored.
- Restrict access to authorized personnel only.
- Keep the materials in closed containers when not in use.
- Require that all personnel wash hands, forearms, face, and neck after leaving the area.
- Do not allow any eating or drinking in the controlled area.
- Do not allow the application of cosmetics in the controlled area.
- Prohibit smoking and chewing gum in the controlled area.
- Maintain a reduced air pressure in controlled areas to provide for a negative pressure preventing materials from leaving the area. Ensure that the airflow is always inward.
- Prohibit dry sweeping or mopping in the area.
- Use proper labels and warning placards.

Safe handling and use of all Poison or Toxic materials should be based on eliminating contact with the materials, and work practices should employ closed systems where possible. If there is a possibility of contact, it should be reviewed and analyzed for possible routes of exposure. The danger of toxic vapor inhalation, reaction with or absorption through moist mucous tissues, or absorption through the skin may exist. Safety showers and eyewash stations should always be available in any area where the potential for skin or body contact is possible.

Emergency Procedures

Emergency response to incidents involving the release of this class of material can be complicated, depending on the material involved and the circumstances present. A small release of an infectious substance that is hazardous through contact may be relatively easy to handle. For example, covering the material with a rag soaked in household bleach may be all that is required to mitigate the hazards presented by the release. In other instances, it may be necessary to evacuate large areas if the release is airborne and the substance highly toxic. In all cases, the key to safe response involves the rapid and proper identification of the material(s) involved, determining which route(s) of entry is hazardous, and protecting those routes from the effects of exposure. Symptoms of exposure should always be identified to help protect persons working in the area. If any symptoms are noted, prompt action should be taken.

■ Note

Small releases of infectious materials can be handled using bleach-soaked cloth placed on the spill.

Until the specific material(s) is known, isolation of the area where the release occurs is critical. Only after a complete identification of the material, its properties, and the symptoms of exposure, can appropriate action be taken. Additionally, proper procedures must be followed to make sure that no harmful residues are left in the workplace or are released to the environment following completion of the emergency response.

> **Note**
> As with many other types of materials, identification of the substance is critical to proper handling and response.

If contact with any poisonous material occurs, it is critical to immediately rinse the materials from the skin with large quantities of water (the location and use of safety showers and eyewash stations should be discussed). The longer the material is in contact, the higher the potential for exposure. Keep in mind that powders are best brushed off before the application of water.

> **Safety**
> Immediate removal of any type of poisonous substance from the body is critical in reducing the potential for a serious exposure.

If the Poison has been swallowed, medical attention must be sought immediately. It is not always the best course of action to induce vomiting or to administer materials by mouth. Corrosives may cause additional damage coming up as they did going down, or may cause the damaged esophagus to rupture with the added pressures from vomiting. Petroleum distillates and most hydrocarbons should receive special attention in the event of ingestion. Do not induce vomiting with these materials because it may cause further health problems. Aspiration of vomit contaminated with hydrocarbons may cause a condition called **chemical pneumonia**. This results when the material is drawn into the lungs and causes the tissue in the air passages to weep fluids. This chemical pneumonia has a high mortality rate and is oftentimes more fatal than the original exposure. Always have the phone number of the local Poison Control Center available because the experts from that center will be able to provide direction as to what to do when materials are ingested. Finally, it is critical to provide emergency medical personnel and hospital staff with as much information about the chemical as possible. This is easily accomplished by sending the MSDS to the hospital along with the victim. The more that is known about the material, the more quickly the appropriate treatment can be given.

chemical pneumonia
a condition caused when hazardous materials are drawn into the lungs causing fluid to build

> **Safety**
> Poison Control Centers can be critical in helping manage exposure to poisonous materials, especially those that are ingested.

RADIOACTIVE MATERIALS

For most people, the thought of handling Radioactive Materials creates considerable concern. Thoughts of detonating atomic weapons, mushroom clouds, and the damage that can be produced by these materials are now compounded by new concerns regarding "dirty bombs" and other 21st century issues. While there is cause for concern, for the most part the general fear that is expressed by many is actually quite exaggerated. Like other types of hazardous materials, Radioactive Materials can be safely handled once we understand some basic principles of safety and learn a little bit about the different types of Radioactive Materials and their hazards.

radiation
energy that is released as waves or particles

Radiation has been in existence since the beginning of time. Radiation is energy emitted in the form of waves or particles. It is present in many forms and its presence is something that modern society has come to rely on. Used correctly, radiation can provide a clean source of energy, and is the foundation for many medical advances in diagnosing and treating illness. However, its potential to cause harm is often what captures the attention of the media as well as the average person. Yet, all too often those concerns are largely unfounded.

> **■ Note**
> Radioactive Materials can be safely handled by those who have some basic understanding of the hazards and properties.

Radiation has been in use for a considerable amount of time. In 1885, a German physicist named W. K. Roentgen was doing some research that ultimately led to the discovery of X-rays. They were called X-rays because "X" is the mathematical symbol for an unknown and Roentgen did not fully understand where these rays were coming from. Later in his research, he learned a great deal about this new type of energy and ultimately won the Nobel Prize in 1901 for his efforts. Since then, generations of scientists have completed exhaustive research on this topic and have given us volumes of incredibly technical information. Nuclear radiation is one of the most studied environmental hazards in the entire world. It affects life as we know it and will continue to do so until the end of time.

matter
anything that has mass and occupies space

To understand the topic, we first need to understand some basic concepts of physics and **matter**. Matter is anything that has mass and occupies space. Air is made up of different bits of matter, water molecules are matter, and this book and even you are made up of countless tiny pieces of matter. It is helpful to think of matter as the building blocks that make up everything on earth. If it exists, it is made up of matter.

atoms
small particles that make up all matter; composed of protons, neutrons, and electrons; can be stable or unstable based on the relationship between particles

Tiny as it is though, matter is not the smallest piece of our environment because matter is actually made up of smaller components called **atoms**. Atoms are even smaller divisions of matter and are the key players in the formation of millions of chemical compounds. Atoms are very complex and dynamic, and are composed of three smaller types of particles called the *proton, neutron,* and *electron.* Two of the particles, the proton and neutron, are found in the nucleus of all atoms. The nucleus is the center of the atom where all the weight of that particular atom is concentrated. Protons carry a positive charge and neutrons hold a neutral charge. The electrons are negatively charged particles that orbit the nucleus and are instrumental in how and why certain atoms bond with certain other atoms. The bonding of the electrons is the basis for the creation of many compounds. Although the illustrations are sometimes helpful in our understanding of the process, the electrons do not circle the nucleus like planets orbiting the sun as seen in most textbooks. The electron movement is actually much more chaotic and is believed to be like an *electron cloud* that surrounds the nucleus. The orbiting electrons in the cloud are essentially weightless, with each one exhibiting its own negative electrical charge.

stable atom
the number of protons is equal to the number of electrons

Atoms can be described as either stable or unstable on the basis of the ratio of the particles to one another. A **stable atom** is an atom where the number of protons inside the nucleus equals the number of electrons orbiting it. This provides an equal number of positive and negative charges that create a balance. Figure 4-15 illustrates a stable carbon atom and a stable chlorine atom with their respective amounts of protons and neutrons. Note that the protons and electrons are equal in number.

The protons make up approximately one half of the atom's total weight and are found in the nucleus of every atom. Each material has a different number of protons in the nucleus, which distinguishes one material from the other. The carbon atom in Figure 4-15 has six protons in the nucleus. All carbon atoms have six protons in the nucleus and atoms with other than six protons could not be carbon. Chlorine has a total of 17 protons in its nucleus, making it a different material.

Neutrons play a key role in keeping the nucleus together. The number of neutrons in a stable atom should equal the number of protons in the nucleus. When a material becomes radioactive, the number of neutrons is usually much larger than the number of protons it possesses. Essentially, a large imbalance of neutrons to protons in the nucleus prompts an atom to give off energy. Because electrical stability is the goal of all atoms, the unstable nucleus will emit this energy until stability is achieved. The energy that is given off is referred to as radiation and is the result of an **unstable atom**. Unstable atoms are radioactive and they in turn emit radiation.

unstable atom
the numbers of protons and neutrons do not equal the number of electrons; emits radiation

radioactive materials
emit radiation as a result of the presence of unstable atoms

radioactive isotope
the unstable form of an atom

The term **radioactive** is used to describe the spontaneous emission of electromagnetic particles from the nucleus of an unstable atom, and radiation is the energy emitted. This is critical to the understanding of a **radioactive isotope** and the ensuing radiation from its instability. An isotope is a form of an unstable atom. The atom has all the properties of the stable form except that it differs in mass and has a different number of neutrons in the nucleus. An example of this can be found by comparing the stable carbon atom and its radioactive isotope called carbon-14. The carbon-14 is a form that has more neutrons in the nucleus and a greater atomic weight than its stable counterpart. A radioactive isotope is an isotope of its stable counterpart that emits radiation. All elements on the periodic table have at least one radioactive isotope.

All atoms have the potential to become unstable and decay. The process of decay is the way that the atom attempts to achieve stability. In essence, it is dealing with an internal instability usually resulting from the imbalance of protons and neutrons. The nucleus is described as undergoing decay and its decomposition is a process of emitting energy. When this imbalance is substantial, as in

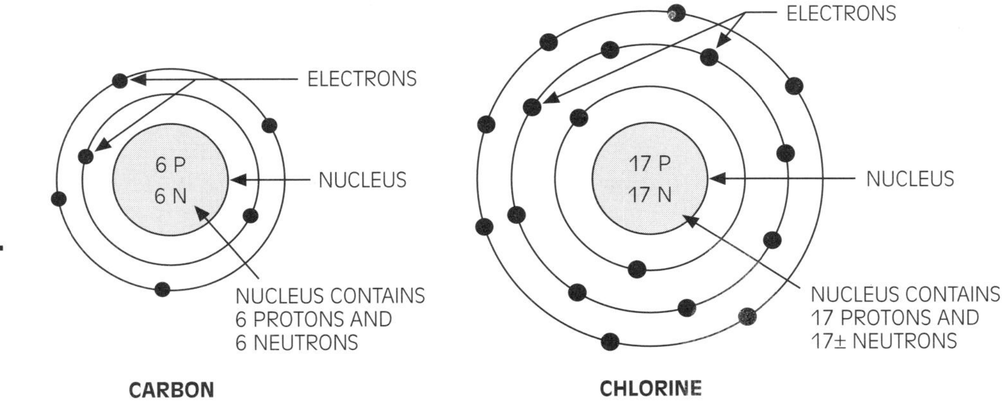

Figure 4-15 *Atomic configurations of carbon and chlorine.*

the example of uranium (atomic weight of 238, atomic number of 92), the atom releases its excess energy from the nucleus, thereby creating electrically charged particles and/or **electromagnetic radiation**. In some cases, the emissions are actual particulates that have mass. In other cases, they can be in the form of electromagnetic radiation, a massless, nonelectrical force that travels from the nucleus in the form of "waves."

electromagnetic radiation
the type of radiation that moves in the form of waves and does not have mass

As the atom struggles to attain stability, energy is released. Depending on the specific material involved, this process can take considerable time. Many reference sources refer to this energy release duration as the **half-life** of the isotope. The half-life is the time it takes for a radioisotope substance to undergo enough changes in an energy state to reduce its original mass by half. Some half-lives are as short as millionths of a second, while others can be as long as millions of years. Carbon-14 has a half-life of approximately 5700 years, while Cobalt-60 has a half-life of approximately 25 years.

half-life
the time required for a radioactive substance to lose one half of its energy

Regardless of the half-life of the material, the material comes off as either a particle or as pure energy. If the material coming off this process is in the form of charged particles, we call them either Alpha or Beta particles depending on their makeup. If the energy comes off in the form of waves or rays, the pure energy (electromagnetic radiation), is called Gamma rays. Other types of electromagnetic radiation include microwaves and visible light, but Gamma radiation is different because it is of higher energy and originates in the nucleus of unstable atoms.

Ionizing Radiation: Alpha, Beta, and Gamma

Alpha radiation is the result of a charged particle that is emitted from the nucleus of an unstable atom. This type of particle is quite heavy, atomically speaking, and is the most highly charged of the three forms. But because the Alpha particle uses up most of its energy before it travels very far from the nucleus, safe distances from Alpha radiation can be as little as just a few feet.

Alpha radiation is one type of particulate radiation that is generally emitted from elements with heavy nuclei such as radium and plutonium. Elements with atomic numbers greater than 82 are regarded as heavy and can be suspected radioactive elements. These particles can become airborne and be inhaled by an unsuspecting worker. Once the particle has entered the lungs and released its energy, everything within a few inches is affected. Because these are particles, specific types of respirator filters can be used to protect the lungs from Alpha particles and should be used whenever contaminated dusts are suspected. Respiratory protection will be covered in Chapter 6.

Alpha radiation
emission of a charged particle from the nucleus of an unstable atom

> **Safety**
> Alpha radiation is primarily an internal hazard that must be blocked from entering our bodies.

In **Beta radiation**, particles are much smaller than Alpha particles and have the ability to travel greater distances from the nucleus. They are able to move faster but have less energy than Alpha particles. One major difference between the two types of radiation is the ability of Beta particles to travel into the skin and cause skin damage or burning, but they are not able to go deep enough to reach the vital organs. However, as with the Alpha particles, Beta particles are hazardous if they enter the body by being inhaled or even ingested. Drinking water contaminated with either Alpha or Beta particles could cause radiation sickness or cancer, or some other type of fatal organ dysfunction. Keeping the

Beta radiation
particulate radiation that is much smaller than Alpha particles, moves faster, and has the ability to travel greater distances from the nucleus

materials from entering our bodies is the goal because these materials are hazardous when taken internally.

> **Safety**
> Like Alpha radiation, Beta radiation is primarily an internal hazard that must be blocked to prevent contact with the body.

Gamma radiation
radiation that is most often caused by a fission reaction and is pure energy, not a particle

fission
the splitting of an atom's nucleus that results in the release of considerable amounts of energy

The third type of radioactive decay is **Gamma radiation**. This is the type of radiation most often reported in the media and what most people think of when they consider the topic of radiation. Gamma radiation can be caused by several activities, but most often it is caused by a fission reaction. **Fission** is the splitting of an atom's nucleus, resulting in the release of an awesome amount of energy. Gamma radiation is quite different from Alpha and Beta in that it is not a particle. It is a form of pure energy that can travel through space in the form of electromagnetic waves similar to our concept of X-rays. The energy moves at the speed of light (186,000 miles per second) and can travel deep into the body affecting organs and body tissues with extreme exposures, causing severe burns and rapid death. The energy is strong enough to go through most common materials, including chemical suits, firefighters' turnout gear, the standard walls of a building, or any other common construction materials.

ionizing radiation
radiation that interacts with the atoms of the somatic cells within the body, creating ions within the cell

ion
a neutral atom with an imbalance in electrical charge as a result of the gain or loss of an electron

cations
atoms that have gained an electron

anions
atoms that have lost an electron

All three forms of radiation are known as **ionizing radiation**. Ionizing radiation is far more hazardous than non-ionizing radiation because it interacts with the atoms of the *somatic cells* within the body. Somatic or body cells are neutral in charge unless they are hit by an Alpha or Beta particle. This collision can knock an electron out of its natural orbit, thereby creating an electrical imbalance in the atom. If this happens, an **ion** is created inside the body's cells. An ion is a neutral atom with an imbalance in electrical charge as a result of the gain or loss of an electron. If the atom gains an electron, it is called a **cation**. If it loses an electron, it is called an **anion**.

radiation sickness
a group of symptoms resulting from an excessive radiation exposure to large parts of the body where there is active cell division

Under most circumstances, the body is capable of repairing this kind of damage on a regular basis and life goes on without problems. In the event the body responds negatively to the newly created ion, or does not respond at all, biological changes can occur. These changes may manifest themselves as **radiation sickness** or various types of cancer. Radiation sickness is a group of symptoms resulting from an excessive radiation exposure to large parts of the body, most often in areas where there is active cell division. Some of the symptoms include nausea and vomiting, hair loss, loss of energy, and hemorrhage. Length of exposure, intensity of the source, biological factors, and a variety of other conditions influence just how bad the symptoms will be.

Sources of Radiation

mrem
unit of absorbed dose of radiation taken into body tissues, representing one-thousandth of a rem

We are bombarded daily by a number of radioactive sources, including cosmic radiation from the sun, that enter the atmosphere and rain down on us. While the dose from this is relatively low ($\approx$30 **mrem** per year), it is still present and is part of the background radiation from which we cannot escape. An mrem is a unit of absorbed dose of radiation taken into body tissues and represents one-thousandth of a rem, which is the dose of ionizing radiation that causes the same effect as one roentgen of X- or gamma-radiation. The rocks and soil that make up the earth are also sources of radiation and contain uranium, thorium, and polonium, minerals that produce radiation energy. Your own body also emits radiation because of the presence of elemental isotopes within soft tissues and organs. We certainly cannot escape this source because we cannot survive without carbon, potassium, and other vital elements.

In addition to these naturally occurring sources, we can receive radiation through many man-made sources. Radiation is used to treat cancer and other diseases. X-rays and fallout from weapons testing are also man-made sources from which we cannot escape. Nuclear power is widely used around the world and adds to the amount of radiation in the environment, although incidents involving these facilities are rare and releases carefully monitored.

Note
Our bodies are exposed to many types of natural radiation and man-made types. Most often, the levels present little potential for harm.

Emergency Procedures

Responding to incidents involving the potential release of Radioactive Materials can be a cause of concern for even the most seasoned veteran. This is due in part to the fact that such incidents are extremely rare. Even when Radioactive Materials are involved in a transportation incident, such incidents are often relatively minor because the packaging and transportation requirements for radioactive substances are considerably more comprehensive than for other materials. Containers for transporting these materials are designed well above the normal DOT standards. Shipments and packaging must be capable of withstanding normal handling practices. While glass bottles can be used to transport many types of seriously hazardous substances, packaging for Radioactive Materials must be capable of protecting the material from accidents. In the event of a transportation accident involving one of these containers, significant releases are unlikely.

Safety
Containers used to transport highly Radioactive Materials are specially designed to prevent leakage should they be involved in a transportation accident.

However, there is always a possibility of spilling liquids or container breakage and the spread of radioactive contamination when materials are handled in the workplace. Fixed facilities where Radioactive Materials are used are required to be carefully monitored, and require special licensing, written plans, and appointment of a **Radiation Safety Officer** who is to oversee the program. A Radiation Safety Officer is a specially trained site member who is responsible for ensuring that all radiation licensing requirements for the site are maintained. When a source is identified and thought to be leaking, it should be isolated from all personnel and the Radiation Safety Officer notified immediately. For all routine handling and emergency response activities involving Radioactive Materials, all personnel should be familiar with the protection concepts of time, distance, and shielding.

Radiation Safety Officer a specially trained site member who is responsible for ensuring that all radiation licensing requirements for the site are maintained

The concept of time simply refers to the fact that the longer a person is exposed to the Radioactive Material, the greater the exposure that results. Time spent working with radiation should be limited and carefully monitored. The time exposure is linear, meaning that if you are there twice as long, your exposure will be twice as much. Reducing the time will significantly reduce the exposure. Rotating personnel in both emergency and routine exposures is a good method of reducing potential harm because each group of personnel will spend less time exposed to the hazard.

> **Safety**
> The longer the time we are exposed to radiation, the higher the potential for radiation damage to occur.

Distance is also very effective and, in many cases, the easiest method of radiation protection. Technical research shows that radiation intensity falls off very rapidly as distance increases. The inverse square law for reduction of radiation applies to most sources of radiation. The law simply states that doubling the distance from the source of radiation decreases the intensity of the radiation by a factor of four. Therefore, it is important to stay as far from the source of radiation as possible. This is not always possible for those who work directly with Radioactive Materials, but it can definitely be employed in emergency situations.

> **Safety**
> Maintaining adequate distance from a radiation source is an effective way to reduce exposure given that exposure decreases radically when distance is extended.

Shielding should be designed so that no large amount of radiation is released except through the intended portal. This is useful at fixed facilities but once again may not be practical in emergency response. For these situations, it may be necessary to protect personnel with physical barriers. The more material used to make the barrier, and the greater its density, the more radiation it will stop. Lead is an example of a very dense material that is effective in limiting the spread of radiation. While it is impossible to eliminate all emitted radiation, it can be reduced to levels that are considered acceptable.

> **Note**
> Shielding from Radioactive Materials must be done using barriers with extremely dense materials.

CORROSIVES

Corrosives are the class of hazards that most people working with hazardous materials will likely encounter. Corrosives are some of the most widely used substances in industry. One type of Corrosive, sulfuric acid, continues to be the most widely produced and used industrial chemical in the United States. Sulfuric acid is so important to industry that some economists can gauge how technologically advanced a nation is by identifying how much sulfuric acid it uses. The reason for this is that sulfuric acid is used in just about every segment of industry, including petrochemical plants, semiconductor fabricators, biotechnology, agriculture, and plastics manufacturing. With such a wide demand, involvement with this material and other Corrosive materials is likely.

■ Note
Corrosives are some of the most widely used materials.

Definition of Corrosives

Corrosives
materials that can cause visible destruction to tissue and/or steel during a predetermined period

While there are a number of definitions that can be used to define a Corrosive, the ones that provide the best information relative to safety are those used by OSHA and the DOT. **Corrosives** are defined by the DOT as materials that can cause visible destruction to tissue and/or steel during a predetermined period. For most cases, this testing occurs over a 4-hour duration. From an occupational safety and health point of view, the definition of a Corrosive results from its effects on living tissue rather than on metal plates. Specific tests have been prescribed to be performed on test animals, and Corrosives are compounds that cause, in the words of OSHA, a "visible destruction of, or irreversible alterations in, living tissue by chemical action at the site of contact" during the 4-hour test period. The definition certainly helps us understand the primary hazards of these materials. It is the health effects of Corrosives that cause the class of materials to be grouped together. However, not all Corrosives are the same, and in fact, Corrosives are divided into two categories that include acids and bases.

Acids

acid
material that contains loosely held hydrogen ions

In its simplest definition, an **acid** is a material that contains a loosely held hydrogen ion. Materials that loosely hold hydrogen are more likely to give it up when they react with other materials. This loosely held hydrogen is a key factor in determining the strength of the acid. It is also an indicator of how much potential damage could be caused by a skin exposure.

You should know that all things called acids are not strong or necessarily harmful. Boric acid is used as an eye wash solution, carbonic acid makes the fizz in a soda, and dilute acetic acid is known as vinegar. Information on whether a compound is a true corrosive capable of skin destruction must ultimately be based on experimentation and is recorded on the MSDS for the substance. However, it is generally a good idea to treat all acids as strong and capable of damage until proven otherwise.

pH scale
describes the relative strength of a Corrosive; 7 on the scale is neutral, numbers lower than 7 are acids, and numbers higher than 7 are Bases

The relative strength of a Corrosive material is expressed by the **pH scale.** In simple terms, pH stands for the "power of hydrogen." This scale is shown in Figure 4-16 and has a range of 0–14. Acids occupy the lower half of the scale with their pH between 0 and 6. The other half of the scale is reserved for bases, which have a pH in the range from 8 to 14. A pH of 7 is considered neutral and is the point we strive for when neutralizing substances. Because the scale is used to show the relative strength of the material, a pH at either end of the scale would indicate a strong Corrosive, either *acid* at the low end or a *base* at the high end. Strong acids are considered to have a pH of 2 or less. Strong bases have a pH of 12.5 or more.

■ Note
Acids are materials with a pH less than 7.

ionization
the process whereby an acid releases the hydrogen ion

The relative strength of an acid is determined by how easily the material releases its hydrogen ion. The term for this process is **ionization,** and Inorganic Acids (those not containing carbon) tend to be stronger than Organic Acids. This is due to the fact that Inorganic Acids will release their hydrogen ions more easily. Some examples of common Inorganic Acids include hydrochloric acid,

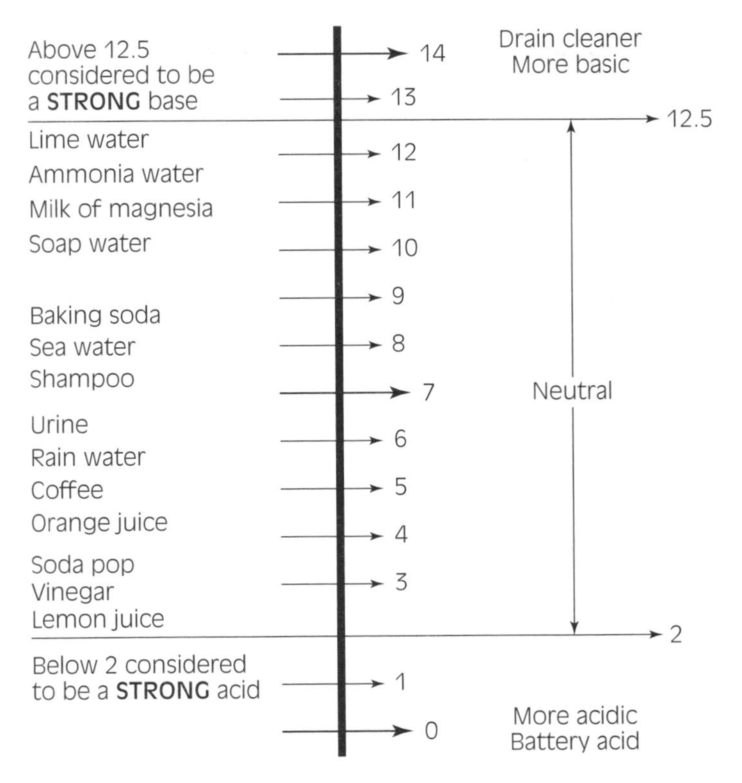

Figure 4-16 *The pH scale, showing pH values for common materials.*

solvation the process that occurs when an acid mixes with water; hydrogen ions are released, which then bond with some of the water molecules to become H_3O

phosphoric acid, and sulfuric acid. This action of ionization is then our basis for defining pH as the power of hydrogen.

The technical definition of pH is a bit more complex. An important point is that when an acid mixes with water (H_2O), hydrogen ions are released and subsequently bond with some of the water molecules. When this happens, the H_2O gets an extra hydrogen ion and becomes H_3O. The process even has a very descriptive definition. It is said that the extra hydrogen ion (H^+) has *solvated* to the water (H_2O), creating a hydronium ion (H_3O). Figure 4-17 shows the process of **solvation**.

Figure 4-17 *Example of the solvation process.*

Table 4-11 identifies some of the more commonly encountered acids in use in a range of industries. It is easy to see from the chemical formulas that the "H" (hydrogen) is present in all of them. This H^+ is ultimately released to the water and causes the pH paper to change color (usually red for acids and blue for bases). The free H^+ also causes the damage to skin and metals. The H^+ is the hazard, and the more H^+ that goes into a solution, the stronger the acid. The strength of an acid is based directly on how readily the material gives up its H^+ to water.

concentration
how much of an acid is mixed with water

Another concept that is important to understanding Corrosive materials is that of **concentration**. This term is often confused with the strength of an acid, but it is really a separate issue. The strength of an acid is identified by how readily it ionizes, while concentration simply refers to how much of the acid is mixed with water. Sulfuric acid is traditionally used in 97% concentration. This means that the mixture contains 97% sulfuric acid and the other 3% of the volume is water. Water is the other part of an acid solution and its presence is always implied when discussing concentration.

Safety
It is important to note that concentration is different from the strength of the acid.

At the 97% concentration of sulfuric acid mentioned in the previous paragraph, we could say that we have a concentrated, strong acid. If the concentration of the sulfuric acid were reduced to 10% (the other 90% being water), we would have a strong, dilute acid. Strength is not related to concentration because it does not affect the ability of an acid to ionize. Conversely, it is possible to have a concentrated version of a weak acid. Such is the case with acetic acid, an acid that does not easily give away its H^+ to water, making it a weak acid. Even if the concentration were 70%, it would still be considered weak because of its poor ionization potential. So in this case, we would have a concentrated, weak acid. Again, if we drop the concentration to 10%, we would have a dilute, weak acid. Remember, ionization potential is not affected by a change in the concentration of an acid.

Bases

base
a material with a pH greater than 7

The other type of Corrosive material is a **base**, also called caustic or alkali. Using any of these terms refers to the same material with the same properties. Bases are common materials and are found in many ordinary household products, including drain cleaners, which contain sodium hydroxide, baking soda which

Table 4-11 *Examples of inorganic acids that ionize readily and are considered strong.*

H_2SO_4	Sulfuric acid
HF	Hydrofluoric acid
HCl	Hydrochloric acid
HNO_3	Nitric acid
H_3PO_4	Phosphoric acid
$HClO_4$	Perchloric acid
HBr	Bromic acid
HI	Hydriodic acid

contains sodium bicarbonate, and many cleaners such as ammonia. While we earlier defined an acid as a substance that gives up its H^+ to water, a base is a material that gives up its OH^- into the surrounding solution. The process is conceptually the same as with acids, the main difference being what is given off, in this case the OH^- or hydroxide ion. The OH^- is negatively charged and naturally attracted to anything positively charged.

Another difference between an acid and a base is the damage they do when they contact skin. Unlike an acid, which burns into the skin from the top down, contact with a base will not always result in direct skin destruction because the base tends to break down fats and oils that are deeper under the skin. The result of such contact can be more damage and a deeper burn to the tissue. This can be shown using the example of an exposure of the skin to products such as drain cleaners that contain sodium hydroxide, a strong base. Contact with this material can often result in a slippery feeling on the skin because of the dissolution of the fats and oils under the skin by the base. The process is called **saponification** and is a common result of an exposure to a base that melts away the fat pockets under the skin.

saponification process in which a base dissolves fat and turns the fat into water-soluble materials

Safety
Contact with a base can result in deeper damage to our bodies as the base tends to dissolve fats that are located under our skin.

Bases are mostly used and stored in a liquid or solid state. They are common substances, and will likely be encountered by anyone working in the field of hazardous materials. Table 4-12 lists some commonly encountered bases.

Sodium hydroxide and potassium hydroxide are very common materials used in a number of industrial processes. They are generally used and stored as concentrated liquids or solid pellets. The solid pellets pose problems if they are allowed to become airborne dusts. These dusts are anhydrous, and only need to find water to become strong liquid bases capable of destroying tissue. When they are in dust form, they can react with the water in the eyes, nose, armpits, or any other wet or sweaty part of the body.

Safety
Concentrated bases in powder form present a hazard if inhaled or if they contact moist tissue.

Bases are classified as strong or weak, and concentrated or dilute, in the same manner as acids. Concentration varies depending on the need, but remember that it does not relate to strength.

Table 4-12 *Examples of some commonly encountered bases.*

NaOH	Sodium hydroxide
KOH	Potassium hydroxide
NH_4OH	Ammonium hydroxide
$Mg(OH)_2$	Magnesium hydroxide
$CaCO_3$	Calcium carbonate
$NaHCO_3$	Sodium bicarbonate

Neutralization

neutralization the process of deactivating a strong acid using an appropriate amount of a base at an appropriate concentration, or vice versa

The concept of neutralization is very important for anyone who works with a Corrosive material. **Neutralization** is the process of deactivating a strong acid using an appropriate amount of a base at an appropriate concentration, or the process of deactivating a strong base with an appropriate type of acid. Neutralization is based on the antagonistic relationship of the two classes of materials, acid and base. The goal of neutralization is to reach a neutral pH so the material can be safely handled and then properly disposed.

Neutralization is important because of the relative strength that an acid or a base possesses based on its pH. The pH scale is logarithmic in its numbering system, meaning that to change the pH of a material with water alone, a neutral material, requires 10 times the amount of water found in the original Corrosive. For example, one gallon of acid with a pH of 0 would require 10 gallons of water to change the pH to 1. Changing that one gallon of acid to a pH of 2 would require 100 gallons of water. To change the pH to 7, where the acid becomes neutral, would require over one million gallons of water, far more than is practical. This property applies whether you are trying to adjust the pH up or down, making it nearly impossible to use water and requiring the use of neutralization. Because of this, one of the great benefits to neutralizing a Corrosive is that it does not add appreciably to the overall volume of material left for cleanup and disposal.

The process of neutralization is demonstrated in the following example.

1. A Corrosive is spilled on a flat surface where a puddle forms. The material is identified and referenced for chemical properties.
2. The appropriate neutralization agent is selected. If the spill is an acid, a moderate pH base is selected for the neutralization. If the spill involves a base, a mild acid is chosen. The use of weak bases for strong acids and weak acids to neutralize a strong base is done to limit the possibility of a violent reaction between the two incompatible materials. When two strong opposites are allowed to contact each other, the resulting reaction can cause splattering of Corrosive materials, the release of various gases, and extreme heat. Commercial neutralizing agents are available for this purpose and work well to minimize the reaction between the two materials.
3. The neutralizing agent is slowly added to the material being neutralized, taking care to check the pH of the mixture. While this process can take some time, it is important to do it slowly because it is possible to overneutralize the material, resulting in a pH that is the opposite hazard.
4. The resulting mixture then can be evaluated for proper disposal but will be considerably less corrosive than it was prior to the neutralization process.

Safety
Never neutralize a strong acid with a strong base or a strong base with a strong acid. Serious reactions can occur.

The neutralization reaction is simple in concept and is shown in the following chemical equation involving a neutralization of hydrochloric acid and sodium hydroxide.

$$HCl + NaOH = NaCl + H_2O + \text{Heat that is generated from the reaction}$$

This reaction leaves water and sodium chloride, a salt compound formed by the sodium combining with the chlorine. Other neutralization reactions happen in a similar manner with the resulting material containing water and a salt, which is neutral and poses less of a problem for disposal. Before any material is thrown away, it should be evaluated because some neutralized solutions contain other hazardous wastes that require special disposal.

Health Hazards Posed by Corrosives

The primary hazards of Corrosives are associated with their ability to cause tissue damage. The degree of danger posed by these materials depends on a number of factors, including the strength of the material, its concentration, the area of the body affected, and the duration of the exposure. Some of these materials have a slower corrosive action on the skin but are respiratory hazards because they are more vaporous and release poisonous or toxic vapors. Other Corrosives are more dangerous through their ability to cause skin destruction within a short duration. Weaker Corrosives may only be mildly corrosive and are classified as **irritants**. Irritants are materials that cause a reversible inflammatory effect on living tissue by chemical action at the site of contact. Weak forms of corrosives can produce the irritation effects.

irritants
materials such as mild Corrosives that cause a reversible inflammatory effect of tissue

It should be noted that some Corrosives can present multiple health hazards. The possibility of inhalation toxicity is present with some Corrosives, including hydrochloric acid and hydrofluoric acid. If contacted, these materials can also cause skin or other body damage. Hydrocyanic acid is an excellent example of a material that presents multiple hazards including the ability to kill rapidly. Once it is absorbed by the skin, local burning occurs, but that is the least of the issues. The material can become rapidly systemic and produce death if inhaled, ingested, or injected, or if enough skin area is exposed.

Safety
Knowledge of the specific Corrosive is critical because some Corrosives are also toxic.

Safe Handling, Use, and Storage of Corrosives

Because of the reactive nature of many types of Corrosives, it is necessary to ensure that the containers that hold them and the areas where they are stored are free of materials that could initiate a reaction. Acids and bases corrode many types of metals, requiring them to be shipped and stored in plastic and glass containers. Figure 4-18 shows typical polyethylene (plastic) drums and other containers used to store Corrosives.

Many other materials can react with Corrosive materials, causing reactions that evolve heat. Contact between some acids and metals can release hydrogen gas and heat, a potentially lethal combination. Because of the range of possibilities, it is always important to consult the MSDS for the material. The MSDS will provide information relative to the types of incompatible materials for the specific Corrosive. Storage methods and handling procedures must always take into account the potential for a reaction between materials. Additionally, because of their ability to react with each other, the different types of Corrosives, acids and bases, should be kept separated so that accidental mixing cannot occur.

Finally, to help ensure protection for those who might be exposed to a Corrosive material, it is required that areas where Corrosives are used, stored, or handled be provided with a safety shower and eyewash station to allow for the immediate flushing with water should contact occur. Recall that showers should be placed so that the exposed employee can get to it within 10 seconds of

Figure 4-18 *Example of polyethylene drums and totes used to store Corrosive Liquids.*

recognizing the contact. Areas around these stations should be carefully maintained because the exposed worker may not be able to see clearly if the material contacted their eyes.

> **Safety**
> Safety showers and eyewash stations are required to be present where Corrosives are used and stored.

Emergency Procedures

Response to releases of Corrosive materials requires some knowledge of the type of materials and properties present. Many strong bases are solid in form, making their release less mobile than a typical liquid release. Yet, airborne dusts can certainly complicate the response if winds or rain are present. Conversely, some Corrosives can be stored in gas form, making them very hazardous and mobile. Liquid releases can also result in the production of vapor and the possibility of reactions with materials found in the environment. All of these factors make proper identification a necessity when it comes to responses involving Corrosive materials.

Once the material is known, Corrosive Liquid spills are handled in much the same manner as with other liquids. Proper PPE should always be worn and the initial goal of the response efforts should be to stop the flow and reduce the surface area as quickly as practical. Specific types of absorbent pads and materials can be used to contain the spill and keep it from spreading. A release that is not contained and that is moving poses a hazard in that the Corrosive material may begin to react with materials present in the area.

Safety
It is vital to stop the movement of a Corrosive spill to prevent any reactions with other materials.

The use of water with Corrosive releases should be prohibited except under the most extreme circumstances. Recall that adding water will simply increase the amount of Corrosive material, causing it to become more mobile and spread. In addition to the increase in the volume of material resulting from the application of water, adding water to the Corrosive can cause a reaction that produces splattering of the material and possible release of gases because the water produces heat on contact with the Corrosive. The exception to the rule of not using water on Corrosive materials occurs when the Corrosive contacts personnel or their PPE. In such cases, it is always best to immediately apply large amounts of water in an attempt to dilute the product and stop the tissue damage. Because of the potential for a release, safety showers and eyewash station locations should be identified in a fixed facility where Corrosives are found. In cases of transportation emergencies, charged hoses and other decontamination procedures should be in place, making sure that the water is immediately available if needed. In the event of an exposure, quick action is essential, because the longer the contact, the greater the damage.

Safety
Water should not be used to dilute a Corrosive spill unless there is contact with the person or the PPE he is wearing.

Lastly, because the Corrosive materials can be easily neutralized, personnel who are required to handle releases of the materials should be thoroughly familiar with the process and materials used for neutralization. Untrained personnel can make the situation worse, so only properly trained and protected personnel should ever attempt to neutralize a material. Figure 4-19 shows a worker neutralizing a Corrosive following a release.

Figure 4-19
Worker neutralizing the release of a Corrosive material.

Neutralization should only be attempted by properly trained and protected workers.

MISCELLANEOUS MATERIALS

While not a hazard class, the ninth class of material that the DOT recognizes is termed Miscellaneous. This group is technically not a class of material but is a group of materials that are hazardous, but do not have the specific qualities to be grouped into one of the other eight classes. When confronting a Miscellaneous material, the appropriate action should be directed at determining the specific material because it can possess numerous properties including some degree of toxicity, corrosiveness, or ignitability.

SUMMARY

- Physical effects are associated with properties other than the health or toxicological ones and include a material's ability to create fires or explosions, hazards from which ordinary measures such as blocking the routes of entry would afford us protection.
- An explosive material is a chemical substance that can react to produce a gas at a relatively high temperature and pressure and at such a speed as to damage the surroundings. The explosive reaction can be initiated by heat (including that caused by friction) or pressure, and the severity of the reaction depends on the particular materials and the conditions present.
- Deflagrations are explosions created by low-order explosives and are very rapid reactions, slower than detonations. Deflagrations are essentially rapidly spreading fires.
- Gases can present an array of hazards, including Nonflammable, Flammable, Toxic, Corrosive, Oxidizing, or even Radioactive. Some gases such as chlorine have multiple properties, including being toxic, corrosive, a strong Oxidizer, and poisonous. Gases can present numerous hazards that must be identified in order to safely work with them.
- From a safety perspective, all gases other than those classified as toxic or poisonous, or oxygen are capable of producing a condition that could lead to oxygen deficiency and asphyxiation.
- The vapor density of most gases is greater than 1, making them heavier than air and more likely to settle into low areas.
- In addition to vapor density, the movement of gases is influenced by wind, ventilation, and temperature extremes.
- A BLEVE is a Boiling Liquid Expanding Vapor Explosion that could occur as a result of a liquified gas cylinder failing.
- Because Cryogenic gases can be stored at extremely low temperatures, breathing released vapors can cause lung damage, and contact with the skin can cause serious tissue damage and should be avoided.
- In order to ignite, most materials must be in a gas or vapor state. For the most part, materials that are in solid or liquid states can only be ignited after they have been heated and become a gas. The closer a material is to being an ignitable gas, the easier it is to have a fire.
- The flash point is the minimum temperature at which a liquid will give off vapors that are ignitable. The lower the liquid's flash point, the greater the danger of ignition. Most regulatory agencies classify liquids with a flash point as either a Flammable or Combustible Liquid on the basis of their flash point.

- Many types of Solvents present a health hazard and can be toxic, irritating, and can damage the skin, eyes, and mucous membranes. It is important to remember that Solvents work to dissolve fats, oils, and greases, properties that make these materials damage the skin. Water is the universal solvent and has the ability to break down many types of materials.
- The consistent physical danger from Flammable or Combustible Liquids and Solvents is the fire hazard. Vapors may form ignitable explosive atmospheres and ignite from uncontrolled ignition sources. Never use water on a fire involving Flammable Liquids because this will usually increase the surface area and size of the fire.
- Containing spills to the smallest area is a critical safety component in that limiting the surface area of the spilled material will reduce the amount of vapor released.
- Flammable Solids pose a much higher hazard than ordinary combustible solids because most of them cannot be extinguished with water.
- Oxidizers are very reactive materials capable of initiating a reaction and of helping reactions to continue, often at an accelerated pace. Oxidizers speed up the rate of the reaction that is occurring without actually burning themselves. Oxidizers are also capable of initiating a reaction and generating the heat necessary to create a self-sustaining reaction.
- It is vital to ascertain the presence of an Oxidizer in order to ensure that proper handling occurs. Many Oxidizers end with the letters *ite* or *ate* and others have the prefixes *per*, *hypo*, or *oxy* in their chemical names.
- Organic Peroxides are challenging because they contain both the Oxidizer to initiate the reaction and the fuel to sustain it. Care should be taken to ensure that these materials are not exposed to heat or other sources that could initiate a reaction.
- Oxidizers are required to have a 20-foot separation between them and other materials, especially fuels. Shelving used for storing Oxidizers should be designed to prevent reactions if the material is released.
- Poisons are substances that are capable of causing death or serious injury if they are swallowed, inhaled, injected, or come in contact with the skin, eyes, or mucous membranes. Poisons are divided into three main classifications—pesticides, heavy metals, and infectious substances.
- Exposure to infectious substances rarely results in immediate health effects and may contribute to a much larger exposure because the person exposed does not recognize the hazard at the time. OSHA requires that additional programs be developed to identify the hazards, train workers potentially exposed, and develop procedures to prevent exposure if these materials are present in the workplace.
- Universal precautions are required actions taken to prevent infection and include hand washing, wearing of gloves, and proper disposal procedures.
- Rapid removal of any type of poisonous substance from the body is critical in reducing the potential for serious consequences from an exposure.
- Alpha radiation is a particulate radiation. These particles can become airborne and be inhaled. The particle enters the lungs, releasing its energy and affecting everything within a few inches. Because Alpha radiation are particles, specific types of respirator filters can and should be used to protect the lungs whenever dusts contaminated with Alpha particles are suspected.
- In Beta radiation, the particles are much smaller than Alpha particles, move faster, and have the ability to travel greater distances from the nucleus. Like Alpha radiation, Beta radiation is primarily an internal hazard that must be blocked to prevent contact with the body.
- Gamma radiation is a form of pure energy that can travel through space. It can travel deep into the body, affecting organs and body tissues, with extreme exposures causing severe burns and rapid death. The energy is strong enough to go through most common materials.
- Time, distance, and shielding are the most effective ways to protect against radiation exposure.
- The definition of a Corrosive, "visible destruction of, or irreversible alterations in, living tissue by chemical action at the site of contact," describes the primary hazards of these materials.

- One difference between an acid and a base is the type of damage they do when they contact skin. An acid burns into the skin from the top down. A base tends to break down the fats and oils that are deeper under the skin, resulting in a deeper burn.
- Neutralization is the process of deactivating a strong acid using an appropriate amount of a base at an appropriate concentration, or the process of deactivating a strong base with an appropriate type of acid. However, a strong acid should never be used to neutralize a strong base or a strong base to neutralize a strong acid because serious reactions can occur.
- To ensure protection for those who might be exposed to a Corrosive material, it is required that areas where Corrosives are used, stored, or handled be provided with a safety shower and eyewash station to allow for immediate flushing with water should contact occur.

REVIEW QUESTIONS

1. Explain the difference between a detonation and a deflagration.
2. List, in order, the three factors that influence the movement of gases and vapors.
3. Identify the additional hazards created when gases are stored.
4. Describe the process involved in a BLEVE.
5. List the gases identified in the H-A-H-A-M-I-C-E mnemonic.
6. Identify the hazards posed by Oxidizers.
7. Isopropyl alcohol (IPA) has a flash point of 53°F. The outside temperature is 68°F. While this product is being unloaded from a truck, several cases fall to the floor and the product spills. Several large pieces of machinery and a welding operation are going on nearby. Is there a potential danger from fire in this scenario? Justify your response.
8. Identify the three types of Radioactive Materials and list their properties and hazards.
9. List the three methods used to provide protection from Radioactive Materials.
10. Describe the difference between concentration and strength relative to acids.

ACTIVITY

GAS RELEASE

Review the case study given at the beginning of the chapter. Review an MSDS for chlorine gas and answer the following questions:

1. You are with the fire response company. After arriving on the scene of the incident, you believe from the distinctive odor that the cloud present is chlorine. You are assigned to assist the people suffering symptoms of acute exposure to the gas.
    a. What symptoms can you expect from the inhalation of the gas?
    b. What first-aid treatment is recommended for those exposed?
    c. Are other routes of entry of concern with exposure to the hazard?
2. There is a small fire in the building. However, the MSDS says that chlorine gas is not flammable.
    a. Is there some property of chlorine that is related to the fire?
    b. What is that property, and why is it significant?
3. Absent the heat from the fire, would the gas sink to the floor or rise to the ceiling? What information given in the MSDS provides you with the answer to that question?
4. Would people exposed to the gas be at risk from other health concerns after their initial exposure?
5. The MSDS lists some incompatibilities for chlorine gas. What properties of the spa materials may have contributed to the reaction that occurred?

IDENTIFICATION SYSTEMS

Learning Objectives

Upon completion of this chapter, you should be able to:

- Identify the basic use of and need for identification systems.
- Define what identification systems can tell a responder or waste site worker.
- List the advantages and disadvantages of the major identification systems.
- Describe and interpret the DOT system of labels and placards.
- Define the conditions and situations when the DOT identification system is used.
- Describe and interpret the NFPA 704 identification system.
- Describe the conditions and situations when the NFPA 704 is used.
- Describe the Hazardous Materials Identification System (HMIS).
- Identify the uses and types of shipping papers.
- Describe the uses of Uniform Hazardous Waste Manifests.
- Demonstrate an ability to use the *Emergency Response Guidebook* (ERG).

CASE STUDY

A truck or vehicle carrying hazardous materials was involved in an accident near a major freeway interchange in the San Francisco Bay area. The accident occurred near a toll plaza at the north end of the Benicia-Martinez Bridge and involved several vehicles in addition to the tank vehicle. As a result of the accident, the tank's contents spilled onto the roadway and ignited many types of combustible materials in the area, including automobiles that were also involved in the accident. Emergency response personnel from nearby jurisdictions responded.

First arriving response personnel were quickly overtaken by the magnitude of the situation. The tank vehicle did not have any placards or warning signs indicating the contents of the shipment, making approach to the vehicle cab impossible. Response personnel worked feverishly to free some of the drivers of the vehicles and extinguished the fires that resulted from the accident and contact with the tank car cargo. At the time of the accident, fog blanketed the area, and the moisture in the fog reacted with the materials released in the accident. It took over 90 minutes to definitively determine the exact content of the shipment, which turned out to be molten sulfur used in nearby refineries. The fatalities and exposures that resulted from the accident could have been prevented through the prompt identification of the cargo. Changes in the DOT placarding system followed, but these were too late for those involved in this incident.

INTRODUCTION

The amount and types of hazardous substances that are transported, stored, and used in the United States are staggering. Since the end of the Second World War, the amount and variety of chemicals and other hazardous substances used in industry has risen dramatically. Daily production and consumption involve products as diverse as TNT, pesticides, gases of all types, and fuels such as gasoline. All of these, by one means or another, have to be transported throughout the country and indeed the world.

Those whose jobs place them in potential contact with the material, such as the worker who handles the materials at a site, someone responsible for the safe shipping of the material, or a responder dealing with the release of a hazardous substance, need to know what they are handling and exposed to at all times. Today, standardized systems in use across the United States and around the world help in the proper identification of materials being shipped. Without knowledge of the presence and type of hazardous substance involved, workers in contact with the materials would have no way of knowing what they were dealing with.

While there are a number of systems used, two identification systems have become the standards for identifying materials in shipment and in storage at sites across the United States. These two systems are the DOT system, which has also been adopted by the United Nations and which covers identification of products being transported, and the NFPA system, which is used to identify hazardous substances at fixed facilities where they are stored and handled.

> **■ Note**
> Two major identification systems are currently used in the United States. These are the DOT and NFPA systems.

The DOT identification system is legally mandated to be used on the shipment of all hazardous materials or, as termed in the DOT regulation, "dangerous goods." The regulations are quite comprehensive and require an entire volume of the Code of Federal Regulations (CFR). Title 49 of the CFR gives the specific rules that must be followed when hazardous substances are placed into the transportation system. Among other things, the regulation requires that hazardous substances be identified, labeled, transported with proper documentation, and handled by trained workers. The labeling of these materials is so effective that OSHA references it in its Hazard Communication regulation as one means of helping employees identify the hazardous substances in the workplace.

■ Note
The DOT system of identification is mandated by law to be on all shipments of hazardous materials.

■ Note
The DOT identification system is a part of the OSHA Hazard Communication program.

The NFPA system was developed by the agency, which primarily is involved in the development of standards that are often incorporated into law by various regulatory agencies. The NFPA labeling system is also known as NFPA 704, indicating that it is standard number 704. The standard has been adopted by many agencies across the United States and incorporated into regulations such as those found in many of the locally adopted Fire and Building Codes. The system is used to identify hazardous substances that are used and stored in aboveground storage tanks, buildings, or areas outside of buildings.

■ Note
The NFPA 704 labeling system is found on fixed facilities and storage tanks.

Across the United States and around the world, other systems of identification may also be used. Many times, these are locally adopted by a state, county, or local agency. In some cases, a specific site may also use its own system for identification of its products. Some other systems used but not as common as the DOT and NFPA systems include:

- Pipe labeling standards recommended by the American National Standards Institute (ANSI). This standard recommends that all pipes that carry hazardous substances be properly labeled to show the material that the pipe carries as well as the direction of flow of the material. Where used, it is quite effective in identifying the materials carried in the piping system. An example of pipes labeled with this system is shown in Figure 5-1.
- The Hazardous Materials Identification System (HMIS) was developed by the National Paint and Coating Association to help identify the hazards of various materials used by workers in that industry. The system is similar to the NFPA system but is found on actual containers of hazardous materials.
- Military identification systems are used to identify the hazards associated with military ordnance and other materials. Figure 5-2 shows an example of some commonly used military systems of labeling.

Figure 5-1
Example of a pipe labeling system.

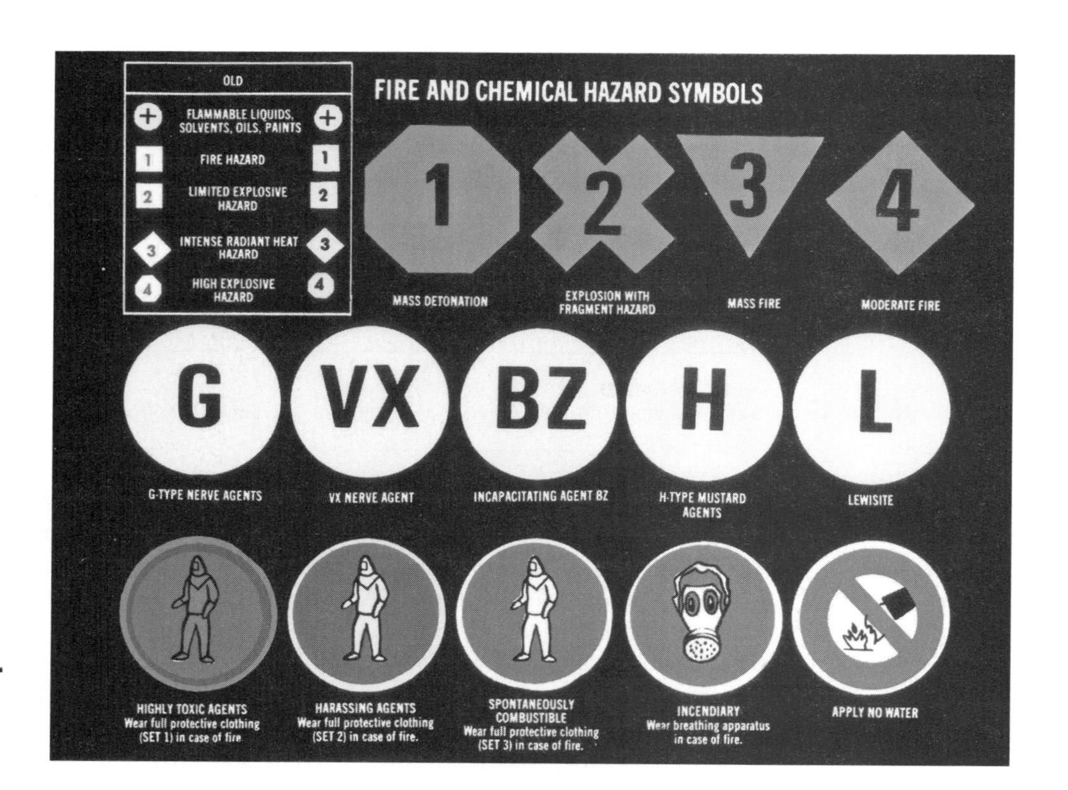

Figure 5-2
Example of military labeling system.

Regardless of the system used, to be effective, all identification systems must be properly used and understood by all. A system that is not widely known is of little value to those who need the information when working with or responding to a release of a hazardous substance. For this reason, we will focus our discussion on the two systems that are most common and most likely to be present.

THE DOT IDENTIFICATION SYSTEMS

In 1974, the Hazardous Materials Transportation Act was enacted by the DOT to develop regulations for identifying hazardous materials in transit. This was to be accomplished through requirements for standard labeling and placarding of hazardous materials while they are being transported, and the use of shipping papers to accompany shipments of hazardous materials. As part of the Act, the DOT defined a hazardous material in 49 CFR, Part 171.8 as "a substance or material that the Secretary of Transportation has determined is capable of posing an unreasonable risk to health, safety, and property when transported in commerce, and has designated as hazardous under section 5103 of Federal hazardous materials transportation law (49 U.S.C. 5103). The term includes hazardous substances, hazardous wastes, marine pollutants, elevated temperature materials, materials designated as hazardous in the Hazardous Materials Table (see 49 CFR 172.101), and materials that meet the defining criteria for hazard classes and divisions in part 173 of subchapter C of this chapter." This broad definition covers most types of hazardous substances, including hazardous wastes.

Shipments of hazardous materials can be done in many forms, including air shipments by aircraft, water shipments by boat or barge, and ground shipments by car, truck, or train. Within the regulation are specific rules for each form of transportation and for each type of hazardous material. Depending on the amount of material carried, the type of material carried, and the method used to transport the material, the rules for each shipment can change.

> **■ Note**
> Shipments of hazardous materials whether by ship, air, truck, or train are all regulated by the DOT regulations.

One of the most helpful aspects of the DOT regulations is the requirement that hazardous materials and hazardous wastes that are in shipment must be appropriately labeled. This is necessary to alert those who handle the packages or respond to accidents where the materials are involved, to be made aware of the presence of a hazardous substance. Within the DOT system, this is done using a combination of **labels** that are affixed to the specific package, box, or cylinder containing the hazardous materials, and through the use of a larger label called a **placard** on the outside of the transporting vehicle. Both the label and the placard are diamond-shaped identifiers. Labels are smaller versions of the placard. Labels are placed on two sides of a typical package within the transport vehicle. The placard, when required, is commonly found on all four sides of train cars or trucks that contains the hazardous cargo. From a safety perspective, the presence of a placard on the side of a vehicle involved in an accident can often alert a responder to the presence of a hazardous material in shipment and allow the response efforts to factor in the hazards associated with the material(s) identified. However, most hazardous materials in transit are shipped in vehicles that do not contain a placard. This is a very important point to consider when approaching any vehicle involved in an incident because the DOT regulations do not require a placard unless the amount of material being shipped is greater than 1000 pounds, or is one of the hazard classes identified in a table given in the DOT regulations. The table is most commonly known as Table 1 and includes the following list of materials:

labels
standard means of identification required to be used by the DOT on packages containing hazardous materials

placards
larger versions of labels and are placed on vehicles used to ship hazardous materials

- Certain classes of Explosives—Classes 1.1, 1.2, and 1.3 (highly hazardous types)

- Poison/Toxic Gases—Class 2.3
- Water-Reactive Flammable Solids—Class 4.3
- Some Organic Peroxides—Class 5.2 (Organic Peroxide, Type B, liquid or solid, temperature controlled)
- Some Poison Liquids—Class 6.1 (material poisonous by inhalation)
- Some types of Radioactive Materials—Class 7 (Radioactive Yellow III label only)

> **■ Note**
> In the DOT system, placards and labels are used to help identify the hazardous materials. Labels are affixed to the individual containers, while placards go on the transport vehicle.

Ground transportation of any amount of a material given in Table 1 would require that the vehicle be placarded. For the hazard classes not on the list, it is generally required that the shipment contain greater than 1000 pounds of these materials. For a more detailed description of these requirements, refer to 49 CFR, Part 172.504.

> **■ Note**
> In the United States most types of hazardous materials are shipped in vehicles that may not have a visible placard on the outside indicating that the shipment inside contains hazardous materials.

> **■ Note**
> Shipments of any quantity of items found in Table 1 of the DOT regulations must be placarded.

Knowing this, it is easy to see that many types of vehicles may contain a hazardous material even though there is no outside visible indication of this. Assuming that the shipment is in compliance with the DOT regulations, the lack of the outside placard indicates that the shipment does not involve anything on the Table 1 list because those materials, in any amount, all require a placard, and also that the total amount being transported is less than 1000 pounds of other hazardous materials subject to the DOT regulation.

To understand the information provided on labels or placards, it is important to know how the system is applied and what information the specific placards or labels indicate. This is done by reviewing the labels and placards.

DOT Labels and Placards

Once it is determined a hazardous material being transported requires identification, the appropriate label or placard must be properly affixed to the package containing the material, and if required, to the vehicle carrying the shipment of hazardous materials. Whether this involves a 40-ton container or a cardboard box, labeling and placarding must be done in the correct manner. The specific requirements for the type of label or placard to be used are given in the DOT regulations. The regulations are specific as to the exact color, letters, sizes of letters, symbols, numbers, and language that is required for each classification.

> **■ Note**
> DOT labels and placards are standard in their colors, symbols, numbers, etc. to ensure that they are readily recognized.

A label is a diamond-shaped identifier that is usually about four inches square. The label is required to be placed on two sides of the specific package or container that is being identified as carrying a hazardous material. The DOT regulation specifies the type of packages to be used, including cardboard boxes, gas cylinders, and drums. Most types of packages are required to be tested, approved by the DOT, and contain information on the package to identify its approval for specific materials. Examples of approved packages are contained in Figure 5-3.

> **■ Note**
> The DOT requires that some of the packages used to ship hazardous materials be tested and listed for this use.

subsidiary label
a second DOT label placed onto the package of specific types of hazardous materials to identify a second hazard of the material

As a general rule, each package carrying a hazardous material that is regulated by the DOT has a single label placed onto one side. Most materials regulated by the DOT have only one hazard class assigned and only a single label is present. In some cases, a secondary label, called a **subsidiary label**, is also placed onto the package to denote a second hazard assigned to the material being transported in the package.

A placard is a larger version of the label and is placed on four sides of the vehicle used to ship the hazardous material. The DOT regulations require that the placard be placed in an area where it can be clearly seen and it must contrast with the area where it is placed to cause it to stand out. The placards are typically 10¾" × 10¾" and diamond shaped, just like the label. Placards are most often identical to the label. However, there are a couple of exceptions to this general rule. Because labels are placed directly on the package itself, more types of

Figure 5-3
Examples of DOT-approved packages.

labels are required to identify the specific contents of the package. An example would be the use of one of three types of Radioactive labels that could be used on the specific packages that are contained in a truck shipment of Radioactive materials. The specific label will help identify the type of Radioactive material within each package. The vehicle carrying the various labeled packages would only show the standard Radioactive placard, but the individual packages would be labeled with one of three specific labels as shown in Figure 5-4.

■ Note
There are more types of DOT labels than placards to help show the range of materials in transit.

Using the DOT labeling and placarding system is not difficult if we remember a few important points. Seeing a DOT label or placard should first and foremost alert us to the presence of a hazardous substance. Without knowing anything about the system, its very presence should make us slow a bit and consider that there is something hazardous present.

Once we identify the presence of a label or placard, we then can look for one or more of the four key indicators present on the identifiers. These include the following:

1. The symbol or picture at the top of the label or placard
2. The color or color combinations of the label or placard
3. The hazard class number located at the bottom of the label or placard
4. The plain language name on the label or placard and/or the four-digit identification number required for some placards

■ Note
Four indicators are used on a DOT label or a placard to help identify the hazard of a primary hazard classification of the material.

In many cases, simply seeing any one of these can help us identify the primary hazard class that has been assigned to the material by the DOT. However, a working knowledge of all four components can be even more useful and give us further information. To help with this, it is necessary to review each of the indicators individually and then combine the information that is provided on them.

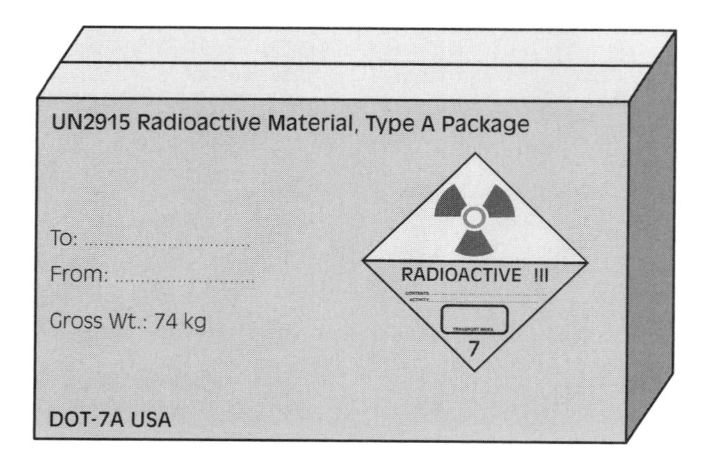

Figure 5-4
Example of one of the three radiation labels that differs from the radiation placard.

Symbols used in the system on both the label or placard are standard and most often clearly indicate the property of concern to the DOT. Table 5-1 lists the major symbols currently used in the DOT system.

The most conspicuous of the identifiers is the color of the label or placard. As previously discussed, the specific colors to be used are assigned by the DOT in the regulation. While the colors may fade or get dirty, they start off as one of the standard colors because each color or color combination on a label or placard symbolizes a specific hazard classification. Variation of these colors is not allowed within the system as it can create confusion in this critically important system. Table 5-2 presents a list of the colors and color combinations that define their respective hazard classes.

> **■ Note**
> The DOT system uses standard colors and color combinations on the labels and placards to indicate the hazard of the shipment.

Colors used in the system do have some disadvantages. The person looking at the placard could be colorblind, or the color can be difficult to determine under adverse light conditions, making it difficult to identify the hazard class in some cases. Two other conditions that often impact the identification of the color are the presence of direct sunlight glaring off a metal placard and looking at

Table 5-1 *Descriptions of the symbols used on DOT labels and placards.*

Symbol/Pictogram	Hazard
Flame	Flammables: gas, liquid, or solid
Cylinder	Nonflammable gas
Flaming O	Oxidizer or organic peroxide
Skull/crossbones	Poison
Melting bar/hand	Corrosive
Trifoil	Radioactive
Black stripes	Miscellaneous
Wheat w/stripe	Foodstuff hazard
Explosion	Explosive

Table 5-2 *A list of the colors used on DOT labels and placards.*

Background color	Hazard classification
Orange	Explosive
Green	Nonflammable gas
Red	Flammables/combustibles: gas, liquid, or solid
Yellow	Oxidizer/organic peroxide
Blue	Dangerous when wet materials
White/black	Corrosive
White/yellow	Radioactive
White	Poisons: gas, liquid, or solid
Red/white	Flammable solid

orange and white placards under sodium vapor lights at night. If either of these situations is encountered, it is important to look for other confirming information present on the label or placard.

> **■ Note**
> Poor lighting and other conditions can limit the ability to detect the colors used on DOT labels and placards.

The third of the four identifying items in the system is the use of a number at the bottom of the label or placard indicating the primary hazard class assigned for the material by the DOT. The numbers are assigned based on the nine hazard classes assigned by the DOT, which have been discussed in Chapter 4. The number, usually a single digit, is located at the bottom of the label or placard. Generally speaking, the DOT system works under the premise that the lower the hazard class number, the higher the hazard. For example, Class 1 is assigned to Explosives that are considered to be the most hazardous types of materials in transportation, and Class 8 is assigned to Corrosive materials that are usually less hazardous when shipped. However, there are some exceptions to this rule as found in the regulations, but it is worth noting that lower numbers generally mean higher hazards when the materials are shipped. The major hazard class numbers are as follows:

1. Explosives
2. Gases including flammable, nonflammable, and toxic/poisonous
3. Flammable and Combustible Liquids
4. Flammable Solids
5. Oxidizers and Organic Peroxides
6. Poisons—liquids and solids
7. Radioactive Materials
8. Corrosives
9. Miscellaneous Materials not otherwise classified

> **■ Note**
> There are nine hazard classes identified in the DOT regulations. The number of each class is shown on the bottom of a label or placard.

The final way to identify the hazard is by reading the plain language name (such as the word Oxidizer) on both labels and placards, or by identifying the four-digit identification number that could be used in lieu of the hazard class name for certain placards. In order to ensure the consistency of the system, only specific words are allowed to be used on the placard or label. Specific terms are well defined in the regulation and are the only ones that are allowed to be displayed on either a label or a placard.

In some cases, large volumes of some materials are shipped. When this occurs, the DOT regulations require that additional information be made available to those who respond to emergency situations or accidents involving the shipment. For this reason, the DOT has established a system to identify what is termed a **bulk shipment** of a hazardous material. Bulk shipments are those meeting any of the following criteria:

bulk shipment
a shipment of specific amounts of hazardous materials above the limits found in 49 CFR, Part 171.8

- Contain a maximum capacity greater than 450 L (119 gallons) as a receptacle for a liquid

- A maximum net mass greater than 400 kg (882 pounds) and a maximum capacity greater than 450 L (119 gallons) as a receptacle for a solid
- A water capacity greater than 454 kg (1000 pounds) as a receptacle for a gas

In bulk shipments, the wording used on the placard is required to be replaced by a four-digit identification number that is given in the DOT regulations for the specific material to be shipped. The rationale for the use of this numbering system is to alert emergency response personnel to the presence of a larger shipment of material and to be able to obtain more specific information on the material involved. For example, if a shipment that contains 1000 gallons of sulfuric acid, all in one-gallon containers, is involved in an accident, it is unlikely that each of the entire 1000 individual containers would be damaged and release their products. In this case, the use of the standard Corrosive placard is required. If the same 1000 gallons of the Corrosive material is transported in larger containers, such as in 250-gallon totes, then the shipment requires the use of the four-digit identification number for the material based on the specific type of Corrosive material present. In this case, it is possible that a larger volume of the material would be released in the event of an accident involving the materials, and emergency response personnel would need to be provided with additional information because of the increased hazard.

Figure 5-5 shows a picture of a truck with several placards. The placards indicate that the truck is carrying a variety of materials in the same load. Note that the general hazards are identified without specifically identifying the materials present in the shipment.

> **■ Note**
> Bulk shipments of hazardous materials require the use of the four-digit identification number in lieu of the standard wording.

The presence of the four-digit identification number is important because the responders can gain additional critical information that will help in their response efforts. However, the presence of the four-digit identification number

Figure 5-5
Example of a vehicle carrying a mixed shipment and bearing multiple placards.

on the placard does not necessarily identify the specific material because one four-digit identification number can be used for several different materials that have similar properties. But with the identification number, more information may be obtained using a standard Guidebook commonly used throughout the transportation and emergency response industry.

The Guidebook has had a number of name changes since its introduction. The HAZWOPER regulation that came into effect in 1990 termed it the *United States Emergency Response Guidebook* (*ERG*). Introduction to the *Guidebook* is mandated for all levels of emergency responders under the HAZWOPER regulation. The *Guidebook* is updated and republished about every 3–4 years. In 1996, it became known as the *North American Emergency Response Guidebook,* with the introduction of the North American Free Trade Agreement or NAFTA, which opened up shipments of hazardous materials on a more international basis. In 2000, it was again updated and became simply known as the *ERG*, adding other countries outside of North America to its list of agencies using it. In 2004, the book was updated again without a change in its title. Regardless of what it is called, the use of this book is an important part of HAZWOPER training, and while it is not perfect or complete, it is often the most universal tool to help add information on shipments of hazardous materials if through nothing more than listing what all of the four-digit identification numbers used on a placard indicate. A more thorough discussion on the use of the Guidebook is found later in this chapter.

> **■ Note**
> The *Emergency Response Guidebook* will help response personnel learn more about the types of materials in transit if the material has a four-digit identification number on it.

Limitations of the DOT System

While helpful in our need to identify the hazardous materials present, the DOT system is not without its drawbacks and it is critical for the safety of those who use this system to thoroughly understand the limitations. As we have already discussed, one of the primary problems with the system is that it may not always be present. An overturned truck may contain 1000 pounds of a hazardous substance and not require a placard. Such a shortfall could give responders to the incident a false sense of security if they don't understand the 1000-pound exemption for most types of hazardous materials.

A second of the major problems with the system is that it will only identify one hazard of the material even though many hazardous materials possess more than one property. It is important to know that the hazard assigned by the DOT is that which it believes to be the most problematic while the material is in transit. In many cases, a material may have more than one hazard. Take for example the substance acrylonitrile, a material that is both a Flammable and a Toxic Liquid. When this material is transported, only one of the hazards can be listed because only a single placard will be placed on the outside of the vehicle used to transport the material. In this example, a Flammable Liquid (Class 3) placard is required and not the Poison or Toxic Liquid (Class 6) placard. The shipment in this case would display the Flammable Liquid (Class 3) placard similar to that used for gasoline. When this happens, responders may incorrectly assume that the placard is more indicative of a substance such as gasoline and not one similar in toxicity to cyanide.

> **■ Note**
> The DOT system most often identifies a single hazard of the material. The system may not alert personnel to all of the hazards of the material.

> **■ Note**
> The hazard listed by the DOT system may not be the primary hazard in every case.

A third point regarding the limitations of the system can also be seen with the example of acrylonitrile previously discussed. Not only does the DOT system not list *all* of the hazards of the material, but it also may not identify the *highest* of the hazards present. In some cases, the use of the material in a fixed facility could make one of the hazards a higher concern than the other. The DOT system is designed to identify the primary hazard for the material when it is transported. This hazard could be different than the one identified for the material when it is used in the workplace.

A fourth limitation of the system is that it does not identify the exact material present. The use of terms such as Oxidizer or Nonflammable Gas do not tell us what types of materials are present. Even when a four-digit number is present, many times that number is used to identify several materials with similar properties. For example, when we look up the identification number 1993 in the current edition, we find that it is used to identify a number of materials including the following:

- Flammable Liquid, n.o.s.
- Combustible Liquid, n.o.s.
- Compound, cleaning liquid (flammable)
- Compound, tree or weed killing, liquid (flammable)
- Diesel fuel
- Fuel oil
- Medicines, flammable, liquid, n.o.s.
- Refrigerating machine

The term *n.o.s.* in this list stands for *not otherwise specified*, which means numerous materials that meet that criteria are included as well. From a short review of the list, we can see that there are a number of materials assigned the 1993 number.

A final limitation of the system relates to the politics of regulations and the influences that can and do occur. Most people recognize that there are groups with special interests in specific regulations and the DOT regulations are certainly a big component of that. To illustrate the point, consider the product anhydrous ammonia. As we know from the discussion on toxicology, an anhydrous substance is one without water. In this case, this is the gaseous form of ammonia. Ammonia in its gaseous form is the most dangerous. It is an extreme inhalation hazard, is corrosive when it contacts moist tissue, and can burn in twice as wide a concentration as gasoline. The material has essentially three hazard classes—Poison Gas, Flammable Gas, and Corrosive. So which of the DOT placards is assigned in the regulations? The answer may be surprising in that it is classified as a Nonflammable Gas. This is the hazard class assigned to relatively harmless materials such as nitrogen, argon, and helium. But ammonia is certainly not harmless. In fact, exposures to the material can kill us in a number of ways.

So why is this the case? In this case, the regulation is heavily influenced by the agricultural industry within the United States that uses significant amounts of the material as a fertilizer. By making the definition of a Flammable Gas (Class 2.1) narrow enough, ammonia can be excluded from that classification and be assigned the next hazard class, Class 2.2 or Nonflammable Gas. It can be argued that it is difficult to ignite ammonia gas in most settings, especially those conditions typically found in agriculture where the ammonia gas cylinder is on wheels and stored adjacent to the field where it will be used. But if those wheels came off the cylinder and the material were stored in a fixed facility, the DOT regulations would not apply and other codes would require that it be used, stored, and handled as one that is poisonous, flammable, and corrosive.

Anyone who relies on the DOT system must understand that it has limitations. It must be simple enough to use; otherwise even more mistakes could be made. In making it simple enough to use, limitations can and do occur.

> **■ Note**
> The DOT identification system works best by alerting us to the presence of a hazardous material. It does not give us all of the information that we need to know, and other sources of information should always be consulted.

Major Changes to the DOT System—Global Harmonization

Given the global nature of transportation of hazardous materials, it is becoming more and more clear that international standards must be developed to help ensure consistency among those who ship hazardous materials. After more than a decade of work and discussions, the United States and many other countries throughout the world have developed a Globally Harmonized System for the Classification and Labeling of Chemicals (GHS). After years of technical work and negotiation, a United Nations Economic and Social Council Subcommittee adopted the Globally Harmonized System for Classification and Labeling (GHS) and recommended that it be disseminated throughout the world. By promoting common, consistent criteria for classifying chemicals and developing compatible labeling and safety data sheets, the GHS is intended to enhance public health and environmental protection, as well as reduce barriers to trade. Countries lacking systems for hazard classification and labeling are to adopt the GHS as the fundamental basis for national policies for the sound management of chemicals; countries that already have systems will align them with GHS. This includes the United States, which will phase in the program as required.

The purpose of the GHS is to promote common, consistent criteria for classifying chemicals according to their health, physical, and environmental hazards, and encourage the use of compatible hazard labels, MSDSs for workers, and other hazard communication information based on the resulting classifications. The major classification that is proposed for each of the groups of materials and examples of the labeling requirements is contained in Appendix D.

NFPA IDENTIFICATION SYSTEM

Once a hazardous material has reached its destination, and is no longer in the transportation system, most of the DOT regulations no longer apply. When materials are being stored or used at a fixed facility, the NFPA 704 system of identification often comes into play. Like the DOT system, it is designed to be easy to use and simple in identification of the hazards of the specific material present, and helps to alert response personnel to the presence of hazardous materials. Unlike the DOT system, it is only a recommended guideline and must be adopted as the local code by a government authority such as a state, county, or city government and enforced at that level. In some areas, the NFPA 704 placard may be in wide use, but in other areas, its use may be severely curtailed or it may not even be used at all. Figure 5-6 shows a site where the system is used to alert response personnel as they enter the area.

> ■ **Note**
> Unlike the DOT system, the NFPA 704 placard may not be required in all jurisdictions.

Despite not being required to the same level as the DOT system, the NFPA system is widely used to provide helpful information and is often found on an MSDS for the material seen in Figure 5-7, making it a useful tool to quickly determine some of the hazards of the material.

The primary purpose of the NFPA system is to provide an early warning to fire department and other emergency response personnel to the presence and hazards of the materials in the building, area, or storage tank. It does this by classifying each of the various materials present on the site on the basis of their individual hazard in three key areas. The NFPA 704 diamond is generally displayed prominently at entrances to the site so that responding personnel will easily see it and quickly determine the hazard located in the area. In many situations, such as a small factory or manufacturing building, the hazards will be identified by one placard located on the outside of the building or even the entrance to the company parking lot. It is generally placed in an area where the emergency responders are likely to see it when they enter by the main route of approach. If a larger building or facility is involved, the placards may be placed in several areas to identify the specific hazards in that one area of the complex.

Figure 5-6
An NFPA 704 diamond on a building, indicating multiple hazards at that site.

Material Safety Data Sheet Collection

Genium group inc.

1171 RiverFront Center, Amsterdam, NY 12010
(518) 842-4111

Issue Date: 2006-06

Phosphine
PHO3520

Section 1 - Chemical Product and Company Identification 61

Material Name: Phosphine **CAS Number:** 7803-51-2
Chemical Formula: H_3P
Structural Chemical Formula: H_3P
EINECS Number: 232-260-8
ACX Number: X1003336-8
Synonyms: CELPHOS; DELICIA; DETIA; FOSFOROWODOR; GAS-EX-B; HYDROGEN PHOSPHIDE; PHOSPHINE; PHOSPHORATED HYDROGEN; PHOSPHORETTED HYDROGEN; PHOSPHORUS HYDRIDE; PHOSPHORUS TRIHYDRIDE; PHOSPHORWASSERSTOFF
General Use: Used as grain fumigant, rodenticide when produced by action of a moist atmosphere on aluminum phosphide, zinc phosphide or magnesium phosphide.
Such baits are highly dangerous if wet with water.
Phosphine is corrosive to copper and fumigation may severely damage electrical equipment.
Used as doping agent for solid state electronic components.
Electronic grade phosphine may be diluted and compressed with nitrogen, the hazard potential is unchanged.
Phosphine may be liberated during the following industrial processes.
-dismantling of aluminum melting furnaces [ALCOA] -acid leaching of certain arsenic bearing ores (particularly cobalt) -action of moisture on dross produced by the refining of certain nonferrous metals (tin and cobalt, for example) -action of moisture on ferrosilicones -action of moisture on impure calcium cyanamide and calcium carbide

Section 2 - Composition / Information on Ingredients

Name	CAS	%
phosphine	7803-51-2	99.999

OSHA PEL
TWA: 0.3 ppm; 0.4 mg/m^3.

ACGIH TLV
TWA: 0.3 ppm; STEL: 1 ppm.

EU OEL
TWA: 0.14 mg/m^3 (0.1 ppm);
STEL: 0.28 mg/m^3 (0.2 ppm).

NIOSH REL
TWA: 0.3 ppm (0.4 mg/m^3);
STEL: 1 ppm (1 mg/m^3).

IDLH Level
50 ppm.

DFG (Germany) MAK
TWA: 0.1 ppm; PEAK: 0.1 ppm.

Section 3 - Hazards Identification

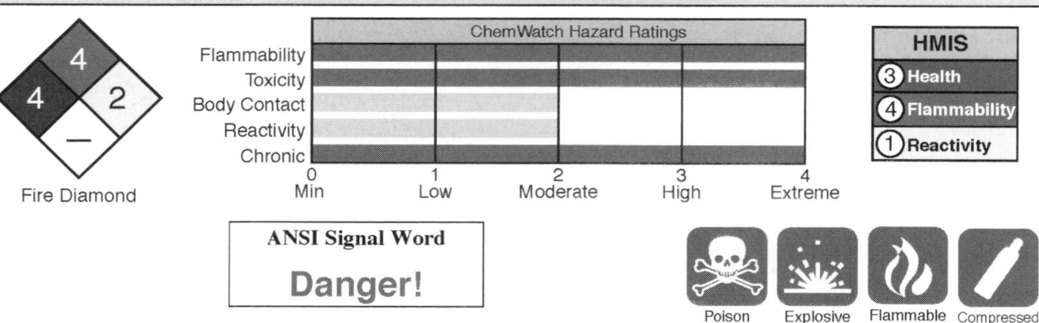

☆☆☆☆☆ Emergency Overview ☆☆☆☆☆

Colorless gas; garlic odor. Stored as a compressed gas which can cause frostbite. Poison. Other Acute Effects: coughing, wheezing, pulmonary edema, convulsions, coma. Chronic Effects: injury to bone/kidney/CNS. Flammable. Pyrophoric. Explosive.

Potential Health Effects

Target Organs: skin, eyes, respiratory system, kidneys, central nervous system (CNS)
Primary Entry Routes: inhalation, skin contact

Figure 5-7 *Portion of an MSDS showing the NFPA system on the document.*

2006-06 **Phosphine** **PHO3520**

Acute Effects

Inhalation: DANGER. Highly Toxic.
The gas is highly discomforting and may be fatal if inhaled.
Reactions may not occur on exposure but response may be delayed with symptoms only appearing many hours later.
The only signs during exposure may be mild respiratory irritation although some victims report dyspnea, weakness, tremor and convulsions.
Overexposure may cause tightness of chest and cough, headache, dizziness, nausea, vomiting, tremor, loss of coordination, diarrhea.
More severe poisoning may result in pulmonary edema, cardiovascular collapse, cardiac dysrhythmias, myocardial injury, disordered liver function. Mortality from severe poisoning is high. Death has resulted from exposure to 8 ppm phosphine for 1-2 hours per day over several days.
Asthma and inflammatory or fibrotic pulmonary disease will be aggravated.
Acute phosphorus poisoning results in fatality rates of almost 50% once hypoglycemia, azotemia, hepatomegaly or delirium appear. The mean-time to death is 5 to 6 days. Phosphorus is eliminated in exhaled air, urine and feces and death results from gastroenteritis, hepatic and renal failure and, in some cases, acute myocardial infarction.

Eye: The gas is highly discomforting to the eyes and is capable of causing pain and severe conjunctivitis.
Corneal injury may develop, with possible permanent impairment of vision, if not promptly and adequately treated.

Skin: The gas is moderately discomforting to the skin.

Ingestion: Not normally a hazard due to physical form of product.
Considered an unlikely route of entry in commercial/industrial environments.

Carcinogenicity: NTP - Not listed; IARC - Not listed; OSHA - Not listed; NIOSH - Not listed; ACGIH - Not listed; EPA - Class D, Not classifiable as to human carcinogenicity; MAK - Not listed.

Chronic Effects: Chronic phosphorus intoxication is characterized by anemia, cachexia, bronchitis and skeletal necrosis.

Section 4 - First Aid Measures

See DOT ERG

Inhalation: Avoid becoming a casualty. Remove victim to uncontaminated site.
Remove to fresh air.
Lay patient down. Keep warm and rested.
If available, administer medical oxygen by trained personnel.
If breathing is shallow or has stopped, ensure clear airway and apply resuscitation. Transport to hospital or doctor, without delay.

Eye Contact: Immediately hold the eyes open and flush with fresh running water.
Ensure irrigation under the eyelids by occasionally lifting upper and lower lids. If pain persists or recurs seek medical attention.
Removal of contact lenses after an eye injury should only be undertaken by skilled personnel. Symptoms of exposure may be delayed.
Immediately transport to hospital or doctor. DO NOT delay.

Skin Contact: Immediately remove all contaminated clothing, including footwear (after rinsing with water).
Wash affected areas thoroughly with water (and soap if available).
Seek medical attention in event of irritation.
Symptoms of exposure may be delayed.
Immediately transport to hospital or doctor. DO NOT delay.

Ingestion: Not normally a hazard due to physical form of product. DO NOT delay. Immediately transport to hospital or doctor.

After first aid, get appropriate in-plant, paramedic, or community medical support.

Note to Physicians: For severe acute or short-term repeated exposures to phosphine:
1.There is no antidote. Clinical manifestations include headache, fatigue, nausea, vomiting, cough, dyspnea, parethesias, jaundice, ataxia, intention tremor, weakness and diplopia.
2. Care is supportive and all obviously symptomatic patients should be monitored in an intensive care setting. Watch for dysrhythmias. Replace fluids/electrolytes. Follow blood chemistries (calcium, phosphorus, glucose, prothrombin time, CBC) at least daily. Follow renal and hepatic function at least daily. Avoid any alcohol intake.
3. The risk of pulmonary edema after severe exposure requires observation for 24-48 hours but can appear several days later. Initial X-ray may be useful in assessing development of edema. If edema develops, nurse with trunk upright and administer oxygen at atmospheric pressure. Diuretics, morphine, theophylline derivatives are of little benefit since edema is exudate rather than transudate.
Bronchodilators by nebulizer or metered aerosol may reduce bronchospasm and dyspnea. Where immediate respiratory symptoms suggest lower airway exposure, steroids may be beneficial, with intravenous injection of methylprednisolone up to 30 mg/kg body weight initially with subsequent smaller doses. Prophylactic antibiotics are indicated in all but mild cases.
Intermittent positive pressure ventilation with bronchial toilet and suction may be important elements of treatment.

Figure 5-7 (*Continued*)

2006-06 **Phosphine** **PHO3520**

Section 5 - Fire-Fighting Measures

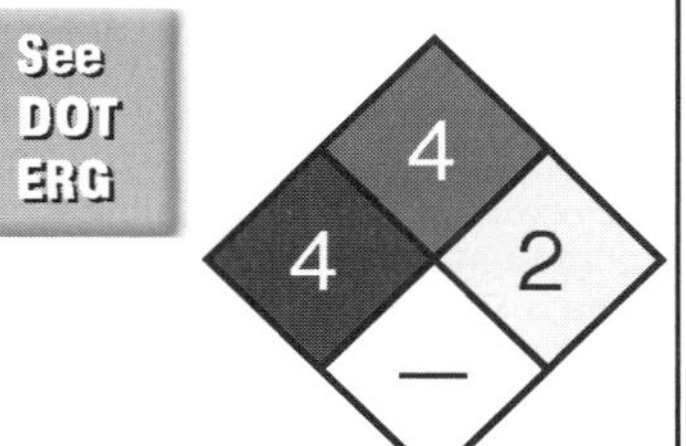

Fire Diamond

Flash Point: Flammable gas
Autoignition Temperature: 100 to 150 °C
LEL: 1.79% v/v
UEL: 1.79% v/v
Extinguishing Media: Water spray or fog; foam, dry chemical powder, or BCF (where regulations permit).
Carbon dioxide.
General Fire Hazards/Hazardous Combustion Products: EXTREME HEALTH HAZARD. Flammable gas. May emit poisonous fumes.
Severe vapor explosion hazard, when exposed to flame or spark.
Vapor may travel a considerable distance to source of ignition.
Heating may cause expansion or decomposition leading to violent rupture of containers.
Decomposes on heating and produces acrid and toxic fumes of phosphorus oxides (PO_x).
Fire Incompatibility: WARNING: May decompose violently or explosively on contact with other substances.
This substance is one of the relatively few compounds which are described as "endothermic" i.e. heat is absorbed into the compound, rather than released from it, during its formation.
The majority of endothermic compounds are thermodynamically unstable and may decompose explosively under various circumstances of initiation.
Many but not all endothermic compounds have been involved in decompositions, reactions and explosions and, in general, compounds with significantly positive values of standard heats of formation, may be considered suspect on stability grounds.
Explosion hazard may follow contact with incompatible materials.
Avoid contact with oxidizing agents, oxygen gas, fluorine, chlorine, nitrates, alkalies and alkali metals e.g. sodium, potassium, lithium.
Fire-Fighting Instructions: EXTREME HEALTH HAZARD. Contact fire department and tell them location and nature of hazard.
May be violently or explosively reactive. Wear full body protective clothing with breathing apparatus. Prevent, by any means available, spillage from entering drains or waterways. Consider evacuation.
Do not extinguish burning gas.
Fight fire from a safe distance, with adequate cover.
If safe to do so, switch off electrical equipment until vapor fire hazard is removed.
Use water delivered as a fine spray to control the fire and cool adjacent area. Water spray or fog may be used to disperse vapor.
Do not approach cylinders suspected to be hot.
Cool fire-exposed containers with water spray from a protected location.
If safe to do so, remove containers from path of fire.
Equipment should be thoroughly decontaminated after use.

Section 6 - Accidental Release Measures

See DOT ERG

Small Spills: EXTREME HEALTH HAZARD. Clear area of personnel and move upwind.
Restrict access to area.
Contact fire department and tell them location and nature of hazard.
May be violently or explosively reactive. Wear full body protective clothing with breathing apparatus. Prevent, by any means available, spillage from entering drains or waterways. Consider evacuation.
No smoking, bare lights or ignition sources. Increase ventilation.
Stop leak if safe to do so. Water spray or fog may be used to disperse / absorb vapor.
Do not exert excessive pressure on valve; do not attempt to operate damaged valve.
Remove leaking cylinders to a safe place. Fit vent pipes. Release pressure under safe, controlled conditions by opening valve. Burn issuing gas at vent pipes.
Use only spark-free shovels and explosion proof equipment.
After clean-up operations, decontaminate and launder all protective clothing and equipment before storing and reusing.
Large Spills: EXTREME HEALTH HAZARD. Clear area of personnel and move upwind.
Restrict access to area.
Contact fire department and tell them location and nature of hazard.
May be violently or explosively reactive. Wear full body protective clothing with breathing apparatus. Prevent, by any means available, spillage from entering drains or waterways. Consider evacuation.
No smoking, bare lights or ignition sources. Increase ventilation.
Stop leak if safe to do so. Water spray or fog may be used to disperse/absorb vapor.
Do not exert excessive pressure on valve; do not attempt to operate damaged valve.

 Page 3 of 6

Figure 5-7 *(Continued)*

2006-06 **Phosphine** **PHO3520**

Remove leaking cylinders to a safe place. Fit vent pipes. Release pressure under safe, controlled conditions by opening valve. Burn issuing gas at vent pipes.
Use only spark-free shovels and explosion proof equipment.
After clean-up operations, decontaminate and launder all protective clothing and equipment before storing and reusing.
Regulatory Requirements: Follow applicable OSHA regulations (29 CFR 1910.120).

Section 7 - Handling and Storage

Handling Precautions: Used in closed pressurized systems, fitted with safety relief valve.
Vented gas is flammable, denser than air and will spread. Vent path must not contain ignition sources, pilot lights, bare flames.
Atmospheres must be tested and O.K. before work resumes after leakage.
Obtain a work permit before attempting any repairs.
Do not attempt repair work on lines, vessels under pressure.
Handle and open container with care.
Avoid all personal contact, including inhalation.
Wear protective clothing when risk of exposure occurs.
Use in a well-ventilated area. Prevent concentration in hollows and sumps.
DO NOT enter confined spaces until atmosphere has been checked.
Avoid smoking, bare lights, heat or ignition sources.
When handling, DO NOT eat, drink or smoke.
Vapor may ignite on pumping or pouring due to static electricity.
DO NOT use plastic buckets. Ground and secure metal containers when dispensing or pouring product. Use spark-free tools when handling.
Avoid contact with incompatible materials.
Keep containers securely sealed. Avoid physical damage to containers.
Always wash hands with soap and water after handling.
Work clothes should be laundered separately.
Use good occupational work practices. Observe manufacturer's storing and handling recommendations. Atmosphere should be regularly checked against established exposure standards to ensure safe working conditions. Technical material is transported as liquified gas under pressure. Packed as liquid under pressure and remains liquid only under pressure. Sudden release of pressure or leakage may result in rapid vaporization with generation of large volume of highly flammable/explosive gas.
Recommended Storage Methods: Cylinder. Ensure the use of equipment rated for cylinder pressure.
Ensure the use of compatible materials of construction.
Valve protection cap to be in place until cylinder is secured, connected.
Cylinder must be properly secured either in use or in storage.
Cylinder valve must be closed when not in use or when empty.
Segregate full from empty cylinders.
WARNING: Suckback into cylinder may result in rupture.
Use back-flow preventive device in piping.
Check that containers are clearly labeled.
Packaging as recommended by manufacturer.
Regulatory Requirements: Follow applicable OSHA regulations.

Section 8 - Exposure Controls / Personal Protection

Engineering Controls: Areas where cylinders are stored/used must have discrete, controlled exhaust ventilation.
Operators should be trained in correct use and maintenance of respirators. Local exhaust ventilation usually required.
If risk of overexposure exists, wear NIOSH-approved respirator.
Correct fit is essential to obtain adequate protection.
Provide adequate ventilation in warehouse or closed storage area.
Personal Protective Clothing/Equipment:
Eyes: Close fitting gas tight goggles and DO NOT wear contact lenses.
Contact lenses pose a special hazard; soft lenses may absorb irritants and all lenses concentrate them.
Hands/Feet: Wear chemical protective gloves, eg. PVC. Wear safety footwear.
Respiratory Protection:
Exposure Range >0.3 to 15 ppm: Supplied Air, Constant Flow/Pressure Demand, Half Mask
Exposure Range >15 to <50 ppm: Supplied Air, Constant Flow/Pressure Demand, Full Face
Exposure Range 50 to unlimited ppm: Self-contained Breathing Apparatus, Pressure Demand, Full Face
Other: Overalls. PVC apron. PVC protective suit may be required if exposure severe.
Eyewash unit. Ensure there is ready access to a safety shower.
Rescue gear: Two sets of SCUBA breathing apparatus, rescue harness, lines, etc.

 Page 4 of 6

Figure 5-7 *(Continued)*

2006-06 | Phosphine | PHO3520

Section 9 - Physical and Chemical Properties

Appearance/General Info: Flammable colorless gas at normal temperature and pressure.
EXTREME HEALTH HAZARD. Pure phosphine has no odor and is flammable. So-called phosphine odor of Technical grade gas, i.e. decaying fish, is from contaminants i.e. up to 5% diphosphine in the gas.
Phosphine released from fumigants may contain diphosphine, ammonia, methyl phosphine, arsine.

Physical State: Liquefied gas
Odor Threshold: 0.03 ppm
Vapor Density (Air=1): 1.17
Formula Weight: 34
Specific Gravity (H_2O=1, at 4 °C): 0.746 liquid
Evaporation Rate: Very fast
pH: Not applicable

pH (1% Solution): Not applicable.
Boiling Point: -87.7 °C (-126 °F)
Freezing/Melting Point: -133 °C (-207.4 °F)
Volatile Component (% Vol): 100
Decomposition Temperature (°C): 375
Water Solubility: 0.26 vol at 20 °C

Section 10 - Stability and Reactivity

Stability/Polymerization/Conditions to Avoid: Avoid any contamination of this material as it is very reactive and any contamination is potentially hazardous. Gas containing diphosphine may be PYROPHORIC, may ignite spontaneously or accumulate and explode without a source of ignition.
Product is considered stable under normal handling conditions.

Storage Incompatibilities: WARNING: May decompose violently or explosively on contact with other substances.
This substance is one of the relatively few compounds which are described as "endothermic" i.e. heat is absorbed into the compound, rather than released from it, during its formation.
The majority of endothermic compounds are thermodynamically unstable and may decompose explosively under various circumstances of initiation.
Many but not all endothermic compounds have been involved in decompositions, reactions and explosions and, in general, compounds with significantly positive values of standard heats of formation, may be considered suspect on stability grounds. Segregate from oxidizing agents, ammonia, alkalies, chlorine, nitrates and alkali metals e.g. sodium, potassium, lithium.
Materials of construction restriction: Incompatible with aluminum, copper, copper alloys. Do not use with natural rubber, neoprene, polyethylene, PVC.

Section 11 - Toxicological Information

Toxicity
Inhalation (human) LC_{Lo}: 1000 ppm/5m
Inhalation (rat) LC_{50}: 11 ppm/4h

Irritation
Nil reported

See *RTECS* SY7525000, for additional data.

Section 12 - Ecological Information

Environmental Fate: No data found.
Ecotoxicity: Toxicity to microorganisms: Bacillus subtilis growth inhib. EC_{50} 2.7

Section 13 - Disposal Considerations

Disposal: Recycle wherever possible. Consult manufacturer for recycling options.
Follow applicable federal, state, and local regulations.
Incinerate residue at an approved site.

Section 14 - Transport Information

DOT Hazardous Materials Table Data (49 CFR 172.101):

Shipping Name and Description: Phosphine
ID: UN2199
Hazard Class: 2.3 - Poisonous gas
Packing Group:
Symbols:

Label Codes: 2.3 - Poison Gas, 2.1 - Flammable Gas
Special Provisions: 1
Packaging: **Exceptions:** None **Non-bulk:** 192 **Bulk:** 245

Figure 5-7 *(Continued)*

2006-06	Phosphine	PHO3520
Quantity Limitations: **Passenger aircraft/rail:** Forbidden	**Cargo aircraft only:** Forbidden	
Vessel Stowage: **Location:** D **Other:** 40		

Section 15 - Regulatory Information

EPA Regulations:
RCRA 40 CFR: Listed P096
CERCLA 40 CFR 302.4: Listed per RCRA Section 3001 100 lb (45.35 kg)
SARA 40 CFR 372.65: Listed
SARA EHS 40 CFR 355: Listed
RQ: 100 lb
TPQ: 500 lb
TSCA: Listed

Section 16 - Other Information

Disclaimer: Judgments as to the suitability of information herein for the purchaser's purposes are necessarily the purchaser's responsibility. Although reasonable care has been taken in the preparation of such information, Genium Group, Inc. extends no warranties, makes no representations, and assumes no responsibility as to the accuracy or suitability of such information for application to the purchaser's intended purpose or for consequences of its use.

Page 6 of 6

Figure 5-7 (*Continued*).

> **■ Note**
> NFPA placards are found on buildings and areas where hazardous materials are present, and also on above-ground storage tanks.

The placard itself is a diamond that is composed of three smaller colored diamonds and a white diamond located at the bottom. Each of the colored diamonds of the placard represents one of the specific hazard types. On the far left is a blue diamond used to identify the health hazards. On the top is a red diamond used to describe the extent of the flammability or fire hazards. On the right side of the diamond is a yellow diamond used to identify the extent of the reactivity hazards for the material present. Finally, a white diamond on the bottom identifies other specific or special hazards that are not apparent by the use of a number and that are not related to the health, fire, or reactivity hazards identified by the colored diamonds.

> **■ Note**
> The colors used in the NFPA system indicate three key hazards: health, flammability, and reactivity.

In each of the blue, red, and yellow diamonds, there are numbers ranging from 0 to 4. These numbers are assigned by the NFPA committees for each specific material based on its hazards in the areas denoted by the colors. The higher the number assigned in any given area, the higher the hazard relative to that area. So the number 0 in any of the given sections indicates that there is little or no hazard relative to the area of concern. For example, a 0 in the fire diamond would indicate that the material will not burn and fire is not a concern with this material. A 4 in the fire diamond would indicate the most extreme of fire hazards.

> **■ Note**
> Numbers ranging from 0 to 4 are used to indicate the degree of hazards for each of the hazards identified for a particular material.

Although it may be clear, it is nevertheless important to note the similarities and differences between the NFPA color and numbering system and that used by the DOT system. There is one common color used in each system, red, which is used to identify a flammable or fire hazard in each of the two systems. However, each of the remaining colors and numbers means different hazards in each of the two systems, and it is critical that we do not confuse the two. For example, blue in the NFPA system indicates a health hazard. In the DOT system, the color blue is used to classify a Class 4 Flammable Solid that is dangerous when wet. Confusing the two systems can be deadly!

> **■ Note**
> The colors used in the DOT and the NFPA are not consistent in what they identify. Care should be taken to know each system individually.

The white diamond located on the bottom of the NFPA 704 system is used to identify other specific hazardous conditions or properties. If the material has any specific hazards that are not represented by a number in the other colored diamonds, a symbol or lettering that relates to the hazard will be shown in the white portion. An example of this would be a Radioactive Material where the radioactive hazard associated with the material is not identified in the colored boxes. In this case, the standard "Trifoil" or "spinning propeller" that is commonly used to identify materials is placed into the white section of the diamond. Other examples of special hazards include the use of the letters *oxy,* which denotes an Oxidizer, *cor* to identify that the material is Corrosive, and *W* with a slash through it to warn that the chemical reacts with water.

> **■ Note**
> Additional hazards of a material can be shown in the white section of the NFPA placard.

In order to fully appreciate the information provided through the use of the NFPA system, it is necessary to understand the criteria used for the assignment of numbers in each of the areas. In the NFPA standard, there is some helpful information that is summarized in the following subsections.

NFPA 704 Health Hazard Assignments

The NFPA system assigns its rating of the health hazards based on the consequences of a single exposure that could range from a few minutes up to an hour. A single or acute exposure is an exposure that is likely to occur with involved personnel in case of a release (i.e., responders). The health effects may be compounded in cases where the worker is in contact with the material on a chronic basis, so the use of this system to indicate those types of health effects is not appropriate. This information is helpful in determining the level and type of PPE that may be suitable for a particular material. The NFPA ratings of the health hazards are as follows:

0: Materials whose exposure would offer no hazard beyond that of ordinary combustible materials. They are not considered to be acutely toxic. Examples of materials with this ranking include diesel fuel and oxygen gas.

1: Materials that are only slightly hazardous to health. As with the previous group, these materials are not considered to be acutely hazardous. Examples of materials with this rating include gasoline, nitrogen gas, and propane.

2: Materials with this rating are hazardous to health, but which you can be exposed to safely with appropriate eye and respiratory protection. Examples of these materials with this rating include benzene, hydrogen peroxide, toluene, and polychlorinated biphenyls (PCBs).

3: Materials that are extremely hazardous to health but to which personnel could be exposed when using a self-contained breathing apparatus and firefighter's protective clothing, including coat, pants, helmet, boots, gloves, and bands around the arms, legs, and waist. No skin area should be left exposed because some irritation or burning could occur. Examples of materials with this rating include anhydrous ammonia, sulfuric acid, cryogenic materials, and sodium hydroxide.

4: Materials that are too dangerous to the health of personnel to risk exposure without considerable protection. A few breaths in an exposed individual to the material could cause death or the vapors could penetrate clothing, including that worn by the average firefighter. The normal clothing and breathing apparatus worn by firefighters will not provide protection from these materials and often specialized chemical protective clothing will be required. Examples of materials in this highest class include the gases chlorine, fluorine, germane, and hydrogen cyanide.

■ Note
Health hazard ratings for the NFPA system are based on an acute exposure. Exposure on a chronic basis can produce different results and should be considered.

■ Note
A higher number in each of the colored boxes shows a higher hazard for the material.

NFPA 704 Flammability

The NFPA system assigns degrees within this category according to how easily a fire would occur. This most often involves a review of the flash point of the materials coupled with the method of combating a fire. Following is a brief description of the flammability hazard for each number and an example of a material with the rating:

0: Materials that will not burn or ignite under ordinary conditions. Examples of materials with this rating include oxygen, water, chlorine, and sulfuric acid.

1: Materials that must have significant preheating before they can be ignited. In most cases, the materials have a flash point greater that 200°F. If materials in this group are involved in fire, water may be effective in extinguishing them because it can cool the material below its flash point. Examples of materials with this rating include ethylene glycol, most types of oils such as motor oil, lubricants, and related products.

2: Materials that must be moderately heated before ignition can occur. They have a flash point between 100°F and 200°F. Water may be effective in extinguishing these materials because it can cool the materials below their flash point. However, it should be used with extreme caution and only by those properly trained to combat these types of fires. Examples of materials with this rating include kerosene, diesel fuel, and formaldehyde.

3: Materials that can be ignited under almost all normal temperatures. The flash point for a material in this category is below 100°F. Water is generally ineffective on these materials because it cannot cool the material below its flash point. Examples of these types of material include gasoline, alcohols, and most flammable solvents.

4: Materials such as Flammable Gases and very volatile Flammable Liquids. Extinguishing of fires involving these materials should be directed toward shutting off the flow of the material and protecting other exposed areas. Examples of materials with this rating include acetylene, propane, hydrogen, and other Flammable Gases.

■ Note
Fire hazards are largely assigned on the basis of the flash point of the material.

NFPA 704 Reactivity Hazards The NFPA system assigns the degree of reactivity based on the susceptibility of the materials to release energy either by themselves or in combination with water. The following is a brief description of the hazard for each rating and includes examples:

0: Materials that (in themselves) are normally stable even under fire exposure conditions and which do not react with water. They present minimal or no hazard in this area. Examples of commonly encountered materials in this group include gasoline, oxygen, and most common materials.

1: Materials that (in themselves) are normally stable but might become unstable at elevated temperatures and pressures, or might react with water with some release of energy. Generally, these are minor reactions posing relatively minor hazards. Examples of materials in this class include moderately concentrated hydrogen peroxide, sodium hydroxide, and magnesium.

2: Materials that (in themselves) are normally unstable and readily undergo violent chemical change but do not detonate. This class includes materials that can undergo chemical change at elevated temperatures and pressures, may react violently with water, or may form potentially explosive mixtures with water. Examples of these materials with this rating include sulfuric acid, stabilized acetylene, and sodium.

3: Materials that (in themselves) are capable of detonation, explosive decomposition, or explosive reaction, but usually require a strong initiator or heating under confinement before initiation. This class includes materials that are sensitive to temperatures and pressures or that react explosively with water without requiring heat or confinement. According to the NFPA, fire-fighting activities involving this class of materials should be done from an explosion-resistant location. Examples of these include ammonium nitrate, unstabilized acetylene, and silane gas.

4: Materials that (in themselves) are readily capable of detonation or of explosive decomposition or explosive reaction at normal temperatures and pressures. This class includes materials that are sensitive to mechanical or thermal shock. In the case of a major fire involving these materials, the NFPA recommends the area should be evacuated. Examples include fluorine gas, nitromethane, and picric acid.

■ Note
Materials with a high reactivity number are capable of explosion or rapid release of energy. Extreme care is required when working with these materials.

Limitations of the NFPA System

As with any of the systems used for identification, the NFPA system has its limitations. While it gives us considerably more information on the various hazards, identifies more than one hazard, and also tells us the degree of hazard of the materials identified, like the DOT system, it does not give us the name of the material or materials involved. With the NFPA system we only know the relative hazards of the material as they relate to the emergency response personnel. Care should be taken to try to find out the name of the exact substances involved in any release situation.

> **■ Note**
> The NFPA system, like the DOT system, does not give the specific name of the material to be identified.

Additionally, in many cases, a fixed facility will have more than one hazardous material present. Each chemical at the site may have an NFPA diamond on its individual container, but the facility entrance may have only one diamond on the front gate to show the highest hazards present at the site. When this occurs, the main entrance diamond must convey the worst-case scenario in each of the four assigned areas. This can often lead to confusion, as the following example illustrates.

As an example, a particular facility uses and stores materials such as diesel fuel, motor oil, gasoline, chlorine, sulfuric acid, and acetylene. When the appropriate hazard classes of these materials are determined, the individual materials are first classified as follows:

Diesel Fuel: Rated as 0 (health), 2 (flammability), and 0 (reactivity)

Motor Oil: Rated as 0 (health), 1 (flammability), and 0 (reactivity)

Gasoline: Rated as 1 (health), 3 (flammability), and 0 (reactivity)

Chlorine: Rated as 4 (health), 0 (flammability), and 0 (reactivity). It also has an "OX" in the white section

Sulfuric Acid: Rated as 3 (health), 0 (flammability), and 2 (reactivity). It also has "W" with the slash through it indicating it can react with water in the white section

Acetylene: Rated as 1 (health), 4 (flammability), and 2 (reactivity)

In this case, the NFPA placard on the front entrance would read 4 (health), 4 (flammability), and 2 (reactivity). The additional "W" with the slash and an "OX" would be found in the white diamond. In a review of the material on site, we find that there is no single material that is a 4-4-2 with the additional special hazards. Yet that is the information provided to the response personnel who may be hesitant to enter the site given the high hazards identified. If the incident involved a tank of diesel fuel, the response may be slightly delayed until more detailed information is made available.

> **■ Note**
> When multiple materials are stored in an area, each material is classified and the highest number for each category is used on the NFPA placard for the entire area.

Another problem with the system is similar to what we found in the study of the DOT system, the placards can be obscured by physical factors such as smoke, fog, darkness, and distance. Additionally, in many cases, the lettering of the NFPA 704 placards is black, which can be especially difficult to read against the blue (health) background. Or, because they are posted and exposed to the elements on an ongoing basis, the information may become faded, making it difficult to see.

> **■ Note**
> All identification systems can be hidden or not seen depending on a number of factors. It is always wise to carefully look for identification systems in every emergency response.

HAZARDOUS MATERIALS IDENTIFICATION SYSTEM

A system similar to the NFPA 704 system is found on the package or container of a hazardous substance. This system is known as the HMIS and is shown in Figure 5-8. As with the NFPA 704 placards, the HMIS uses the same combination of blue, red, and yellow colors and also the same numbering system (0–4) for classifying the hazards in each of the same areas.

However, the system is different in three key areas. First, the system is not required by most federal, state, or local agencies and its use is most often voluntary. Because of this, many people do not know of its presence and often confuse this system with the NFPA 704 system.

> **■ Note**
> The HMIS labels are used on individual containers of hazardous materials. Use of this system is mostly optional.

Second, the assignment of numbers indicating the hazard levels within the health, fire, and reactivity boxes may not be the same for any specific substance because the HMIS labels are designed to provide information to workers who use the materials on a regular basis rather than in the event of an emergency. As we have studied, acute exposures are what the health hazard classification is based on in the NFPA system, while the HMIS takes into account the chronic exposure issues and assigns a higher number in the same colored box for the exact same material. This can lead to confusion for both emergency response personnel and persons who use the HMIS ratings to apply to their required NFPA diamond.

> **■ Note**
> While the HMIS system uses a similar numbering and color system as the NFPA system, the assignment of individual numbers for the same material may be different.

The final difference between two systems is found in the white section. As we know, the white section of the NFPA 704 diamond gives us additional hazards of the substance being described. With the HMIS, the white section is used

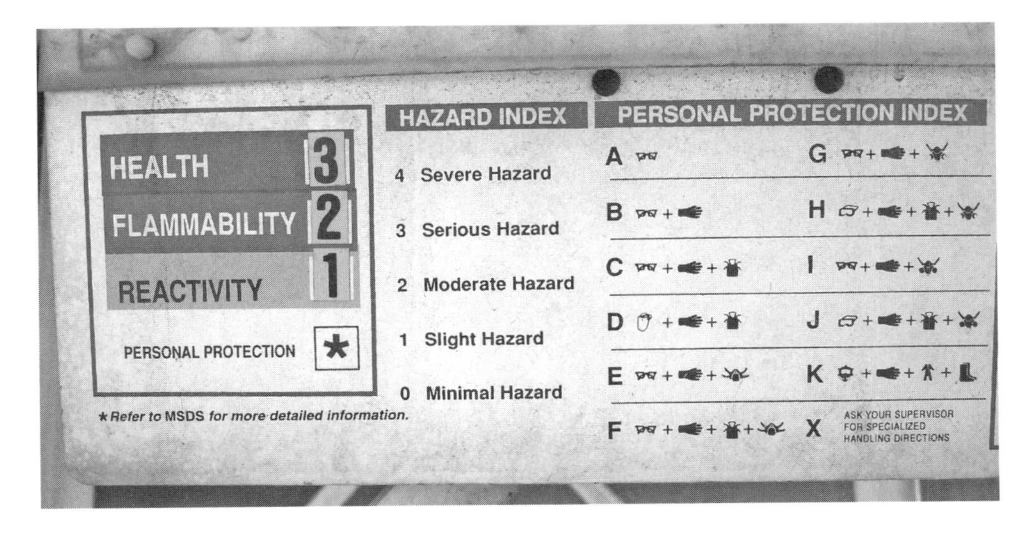

Figure 5-8 *An HMIS label on a container of a hazardous material.*

to define the types of PPE that are required to be used when handling the material. This is done by using a series of letters that run from "A" through "K" and then "X." Figure 5-8 also shows an example of what these letters indicate.

As seen, higher letters in this system denote that a higher level of PPE is required. While this sounds like a good idea, the letters in this system conflict with the standard letters used to describe levels of protection. A detailed study on this will be covered in Chapter 7. In that chapter, we will find that the standard letter designation of levels of protection is used by OSHA, NFPA, and EPA. With the HMIS system, the letter "A" in the white box denotes a pair of safety glasses should be worn. With the standard levels of protection described by the other systems, the letter "A" describes the use of a totally encapsulating gastight chemical protective suit with supplied-air respiratory protection. There is a significant difference between the two types of protection, one that could lead to considerable confusion if overlooked.

> ■ **Note**
> The HMIS letter "A" through "D" system conflicts with the OSHA, EPA, and NFPA standards of chemical-protective clothing. Never confuse the two systems.

SHIPPING PAPERS AND HAZARDOUS WASTE MANIFESTS

While not technically an identification system, there is another useful source of information that will help emergency response personnel to identify the substances involved in transportation incidents. As discussed previously, the DOT requires that shipments of most types of hazardous materials be appropriately managed to ensure their safety on the roadways. This is accomplished in part through the use of the labeling and placarding system. Another requirement within the DOT regulations is that with limited exception, hazardous materials shipments are required to be accompanied by shipping papers and/or a Uniform Hazardous Waste Manifest if the shipment is a hazardous waste. Figure 5-9 shows a common type of shipping paper.

Bill of Lading
a standard shipping document carried in trucks that ship hazardous goods

While shipping documents come in several forms, one of the most common is the **Bill of Lading**. This is the type most often used in the trucking industry when materials are carried in trucks or vans. It is often stored in the map pocket of the truck door, making it immediately available as the driver exits the vehicle, and to emergency responders who can approach the cab in the event of a release or accident. No matter what mode of shipping and what type of shipping paper is present, these documents have the advantage of informing the emergency responder what exact materials are present, how much is being carried, and who to contact for further information.

> ■ **Note**
> Shipping papers are carried in the cab of trucks and available to alert the emergency responder as to the type and amount of hazardous materials being transported.

Uniform Hazardous Waste Manifest
a standard shipping document required by the EPA for shipments of hazardous wastes

Another form of a shipping paper is the **Uniform Hazardous Waste Manifest**. The EPA regulates the handling, processing, and storing of all hazardous wastes and requires that they be removed from the site where they were generated within a specified amount of time and taken to a TSDF. While

CONTAINS HAZARDOUS MATERIALS

STRAIGHT BILL OF LADING – SHORT FORM – Original – Not Negotiable

Shipper's No. ________

(Carrier) ________ SCAC. ________ Carrier's No. ________

Received, subject to the classifications and tariffs in effect on the date of this Bill of Lading:

at ________, date ________ from ________

the property described below, in apparent good order, except as noted (contents and condition of contents of packages unknown), marked, consigned, and destined as indicated below, which said company (the word company being understood throughout this contract as meaning any person or corporation in possession of the property under the contract) agrees to carry to its usual place of delivery at said destination, if on its own road or its own water line, otherwise to deliver to another carrier on the route to said destination. It is mutually agreed, as to each carrier of all or any of said property over all or any portion of said route to destination, and as to each party at any time interested in all or any of said property, that every service to be performed hereunder shall be subject to all the conditions not prohibited by law, whether printed or written, herein contained (as specified in Appendix B to Part 1035) which are hereby agreed to by the shipper and accepted for himself and his assigns.

TO: (Mail or street address of consignee for purposes of notification only.)	FROM:
Consignee	Shipper
Street	Street
Destination Zip	Origin Zip
Route:	

Delivering Carrier	Trailer Initial/Number	U.S. DOT Hazmat Reg. Number

No. of packages	HM	Description of articles, special marks, and exceptions	Hazard Class	I.D. Number	Packing Group	*Weight (subject to correction)	Class or rate	Labels required (or exemption)	Check column

Remit C.O.D. to:
Address:
City: State: Zip:

COD AMT: $

Subject to Section 7 of conditions, if this shipment is to be delivered to the consignee without recourse on the consignor, the consignor shall sign the following statement:
The carrier shall not make delivery of this shipment without payment of freight and all other lawful charges.

(Signature of consignor)

C. O. D. FEE: Prepaid ☐ Collect ☐ $

*If the shipment moves between two ports by a carrier by water, the law requires that the bill of lading shall state whether it is "carrier's or shipper's weight".
Note. – where the rate is dependent on value, shippers are required to state specifically in writing the agreed or declared value of the property.
The agreed or declared value of the property is hereby specifically stated by the shipper to be not exceeding ________ per ________

Charges Advanced $ ________

FREIGHT CHARGES ☐ Prepaid ☐ Collect

This is to certify that the above-named materials are properly classified, described, packaged, marked and labeled, and are in proper condition for transportation according to the applicable regulations of the Department of Transportation.
Per ________

PLACARDS REQUIRED

PLACARDS SUPPLIED ☐ YES ☐ NO - FURNISHED BY CARRIER
DRIVER'S SIGNATURE:

SPECIAL INSTRUCTIONS:

SHIPPER: ________ CARRIER: ________

PER: ________ DATE: ________ PER: ________ DATE: ________

EMERGENCY RESPONSE TELEPHONE NUMBER: ( )
Monitored at all times the Hazardous Material is in transportation including storage incidental to transportation (§172.604).

Permanent post office address of shipper

8-BLS-C3 (Rev. 5/95)

CONTAINS HAZARDOUS MATERIALS

Figure 5-9 *Example of common shipping papers used for hazardous materials.*

in transit, the materials overlap into the DOT regulations and the Manifest becomes the shipping document that accompanies the load. As with the Bill of Lading and other shipping papers, the Uniform Hazardous Waste Manifest will give you information on the shipping name, the hazard class of the wastes being shipped, and the total quantity being transported. The form should be readily accessible, but separate or distinguishable from other shipping papers. Figure 5-10 shows the new Uniform Hazardous Waste Manifest introduced in late 2006.

In addition to the Bill of Lading and the Manifest, there are other commonly used shipping papers. The type and location of the shipping papers depend on the mode of transportation and are generally found at locations mentioned in Table 5-3.

Please print or type. (Form designed for use on elite (12-pitch) typewriter.) Form Approved. OMB No. 2050-0039

UNIFORM HAZARDOUS WASTE MANIFEST | 1. Generator ID Number | 2. Page 1 of | 3. Emergency Response Phone | **4. Manifest Tracking Number**

GENERATOR

5. Generator's Name and Mailing Address — Generator's Site Address (if different than mailing address)

Generator's Phone:

6. Transporter 1 Company Name — U.S. EPA ID Number

7. Transporter 2 Company Name — U.S. EPA ID Number

8. Designated Facility Name and Site Address — U.S. EPA ID Number

Facility's Phone:

9a. HM	9b. U.S. DOT Description (including Proper Shipping Name, Hazard Class, ID Number, and Packing Group (if any))	10. Containers No.	10. Containers Type	11. Total Quantity	12. Unit Wt./Vol.	13. Waste Codes
	1.					
	2.					
	3.					
	4.					

14. Special Handling Instructions and Additional Information

15. **GENERATOR'S/OFFEROR'S CERTIFICATION:** I hereby declare that the contents of this consignment are fully and accurately described above by the proper shipping name, and are classified, packaged, marked and labeled/placarded, and are in all respects in proper condition for transport according to applicable international and national governmental regulations. If export shipment and I am the Primary Exporter, I certify that the contents of this consignment conform to the terms of the attached EPA Acknowledgment of Consent.
I certify that the waste minimization statement identified in 40 CFR 262.27(a) (if I am a large quantity generator) or (b) (if I am a small quantity generator) is true.

Generator's/Offeror's Printed/Typed Name — Signature — Month Day Year

INT'L

16. International Shipments ☐ Import to U.S. ☐ Export from U.S. Port of entry/exit: ______
Transporter signature (for exports only): Date leaving U.S.:

TRANSPORTER

17. Transporter Acknowledgment of Receipt of Materials

Transporter 1 Printed/Typed Name — Signature — Month Day Year

Transporter 2 Printed/Typed Name — Signature — Month Day Year

DESIGNATED FACILITY

18. Discrepancy

18a. Discrepancy Indication Space ☐ Quantity ☐ Type ☐ Residue ☐ Partial Rejection ☐ Full Rejection

Manifest Reference Number:

18b. Alternate Facility (or Generator) — U.S. EPA ID Number

Facility's Phone:

18c. Signature of Alternate Facility (or Generator) — Month Day Year

19. Hazardous Waste Report Management Method Codes (i.e., codes for hazardous waste treatment, disposal, and recycling systems)

1. 2. 3. 4.

20. Designated Facility Owner or Operator: Certification of receipt of hazardous materials covered by the manifest except as noted in Item 18a

Printed/Typed Name — Signature — Month Day Year

EPA Form 8700-22 (Rev. 3-05) Previous editions are obsolete.

DESIGNATED FACILITY TO DESTINATION STATE (IF REQUIRED)

Figure 5-10 *Example of the new uniform hazardous waste manifest.*

Table 5-3 *Types of shipping documents and their locations.*

Mode of transport	Name of document	Location
Truck	Bill of Lading	Driver's side door map pocket or on driver's seat if unoccupied
Rail	Way bill, consist bill (wheel report)	Engine cab manned by engineer
Airplane	Air Bill	Cockpit or by main passenger door to airplane
Ship	Dangerous cargo manifest vessel stow plan	Wheelhouse, ship's office, or in the possession of chief mate

In order to be useful, the shipping documents must be available to response personnel and are required to be in the locations listed in Table 5-3. A relatively common problem is that the set of shipping papers available may not exactly agree with what is actually present due to constant loading and offloading of material from the mode of transport. Although shipping papers must be current by law, it is sometimes very difficult to accurately list real-time locations of every hazardous material. For example, a train in a transfer yard is being put together or taken apart, or a large container ship at a dock is constantly loading and discharging cargo. These situations are not usually the case with trucks in transit, so a higher level of accuracy may be expected.

In addition to providing shipping papers with information on the amount and type of hazardous material in transit, every shipper must make previous arrangements with an emergency agency for technical assistance should the material shipped be involved in an accident. The DOT regulations state that the site to contact for additional information must be staffed for the entire time the material is in transit. One commonly used agency that is available 24 hours a day is the Chemical Transportation Emergency Center (CHEMTREC). The phone number for CHEMTREC is 1-800-424-9300. Agencies that ship hazardous materials can contract with CHEMTREC to provide the assistance required. In all cases, a phone number such as the one for CHEMTREC or another emergency contact must appear on the shipping papers along with the following information:

- Material description (technical name)
- Immediate health hazards
- Fire or explosion risk
- Immediate response actions
- Immediate methods for handling fires
- Initial actions for handling spills or leaks
- Preliminary first aid in case of exposure

Sometimes some shippers attach a copy of the appropriate MSDS onto the shipping paper to avoid reprinting all of the required information. If the responder is able to access this document, they will have some of the most complete information relative to the material and its hazards.

CONTAINER PROFILES

While not a formal system of identification, a study on identification systems would be incomplete without a discussion of the types of containers that are used in the storage, handling, and shipment of hazardous materials. Often, the

container profiles
a form of identification that matches the types of containers with the types of materials commonly transported in them

type of container, or **container profile**, will give us valuable information relative to the hazards associated with a particular material. In fact, the type of container may be the only indication as to the hazards of the material and is what we will need to rely on until more complete information becomes available.

> **■ Note**
> The type of container used to ship or transport the hazardous material may be helpful in determining the type of material that could be present.

Containers used for storing and shipping hazardous materials and their wastes fall into two major groups. The first type includes the portable containers such as drums, jars, boxes, and cylinders. The second type is the actual transport vehicle. Following is a short discussion on how to use this information to help in the identification process. Keep in mind that in some cases, materials may have been placed into containers that are not suitable for them. In such cases, reliance on the container profile alone may expose workers or response personnel to unnecessary hazards. The use of the container profile is one source of information that should be confirmed as soon as possible by identifying the placards, labels, shipping papers, or NFPA ratings for the particular material.

> **■ Note**
> Containers used to transport hazardous materials include the package used or the vehicle.

Small containers are portable ones generally handled manually. The primary types of these include the range of drums. Drums can be composed of a range of materials and can either be open at the top or closed, with access only by small openings called bungs. Depending on the substance, some drums are more suited to specific materials than others. The drums are primarily composed of metals such as steel, plastics such as polyethylene, and composite materials such as fiberboard. Each of these types of drums has different uses, which are described in the following paragraphs.

Metal drums are generally constructed of low-carbon steel with a mechanically seamed side. They typically have two rolling rings along the vertical body of the drum to help seal the top and body of it. The top of the drum is referred to as the head. The top and bottom lips of these drums are often called chimes, and rolling a drum on the bottom ring is termed **chiming a drum**. This is an often used method of moving them short distances. Additionally, the drums are painted with an array of colors, most commonly blue, red, black, or white. It should be noted that there is no accepted standard as to the colors used, so the color is not generally an indication of the contents.

chiming a drum
describes moving a drum by rolling it on the bottom ring

> **■ Note**
> Drums come in a variety of colors. Color should not be used as a primary means to identify what is in a drum because a color system is not standard.

The tops of these metal drums often contain small openings referred to as the bungholes. The bottom of the drum may also have these as well in cases where the product needs to be pumped out. The bungs vary in size but typically include a large bung, which is approximately two inches in size, and a smaller one that is approximately three-fourths of an inch. Sometimes the smaller hole is used to help vent the drum as it is being loaded or emptied. The caps that seal the bungholes are easily removed with standard bung wrenches often made of non-sparking materials such as brass. Figure 5-11 shows a standard type of drum and the use of a bung wrench to loosen the bung caps. Closed-top drums with the bungs are most often used to store liquid materials.

> **■ Note**
> The openings in most closed-top drums are called bungs and require a specialized tool for removal.

> **■ Note**
> Closed-top drums most often are used for storage of liquids.

overpack drum
used to place one drum inside another

In some cases, steel drums may be completely open at the top to allow other materials or even other drums to be placed inside them. Figure 5-12 shows an open-head drum being used to transfer another smaller drum into it for disposal. This is sometimes referred to as an **overpack drum** or salvage drum. With the open-head configuration, the drum is composed of a similar material as ones with the bungholes, but the entire top of the drum is a one-piece lid that is placed onto the drum and sealed with a removable ring. Once full, the lid is placed onto the top and the ring is tightened to secure it to the drum. Open-head drums are used to hold solid materials rather than liquids.

> **■ Note**
> Open-top drums are often used to store solid materials or to overpack other drums.

Figure 5-11
Bung wrench being used to tighten the bungs on a drum.

Figure 5-12
A salvage drum being used to overpack a leaking drum.

Another commonly used type of drums are made of a plastic material. These drums are often made of polyethylene, which is rigid enough to carry the weight of the material contained in the drum. Like metal drums, the colors of these drums vary considerably and do not have a nationally recognized system of color coding. Most often, the drums are black or blue in color. As with the metal drums, these drums can be either the open-head type or the closed-head type with standard bung configurations. When working with plastic drums that have had significant exposure to the sun, care should be taken because these drums can be damaged and weakened by the exposure.

> **Note**
> Exposure to sunlight can damage plastic drums and make them subject to failure.

The third type of drum is made of fiber materials. These drums may look like they are composed of paper or cardboard and in fact are made of tightly bound cardboard materials that are often coated with a waxy substance to help resist and repel liquids. These drums are typically open-head type and many of these have a plastic bag liner to help contain the material that is placed into them. These drums most often store solid materials even with the liner. The method for closing these drums is similar to the method for closing other of open-head type drums, with a fiberboard top and a metal ring that is tightened to hold the lid in place.

Regardless of the materials that they are made of, drums tend to be of standard sizes. Most often, these include 33-gallon small drums and the more standard 55-gallon type. Larger overpack drums can be used to help to place the smaller drums inside as shown in Figure 5-12. In this case, a standard 55-gallon drum is being placed into a larger overpack drum. Overpack drums up to 95 gallons are commonly found.

Identification of the type of drum material and whether it has an open or closed head can help us know something about the material that is contained inside. Following is a summary of the major types of drums and the products most often contained:

- *Closed-head steel drums.* These are frequently used in the storage and shipment of Flammable or Combustible Liquids because the metal will help provide protection against exposure to fire and the possibility of drum failure as a consequence.
- *Open-head steel drums.* These are most often used in the storage and shipment of combustible greases, lubrication materials, or sludge. They are also commonly used to overpack other drums, which may be leaking as previously noted, or for use as a **lab pack** for other hazardous materials or wastes. A lab pack operation involves the placement of smaller containers, which are surrounded by loose absorbents that help to hold the smaller containers in place, into larger containers. Several smaller containers can be placed into the open-head drum, surrounded by the absorbent packing material, and then sealed inside when the lid is secured through the placement of the ring.
- *Closed-head plastic drums.* Most often, plastic drums are used for the transport or storage of Corrosive materials and for other nonflammable or noncombustible liquids.
- *Open-head plastic drums.* As with the open-head metal drums, these containers are often used to lab pack other materials or to contain solid or semisolid materials such as absorbents that have been used to clean up spills.
- *Open-head fiber drums.* These drums most frequently contain solid materials such as sodium hydroxide pellets, soap flakes, or other nonmetallic solids.

lab pack
a process of placing smaller containers such as glass jars inside an open-head drum for proper shipment or disposal

Note
Determining the drum type can often help identify the types of materials that could be present.

Another type of container used for storage and shipping of hazardous substances is the transportation vehicle itself. As with any other type of approved shipping container, the DOT has specific requirements for the vehicles that are allowed to be used in shipping. Figure 5-13 shows the commonly used types or tank cars and tank trucks with their DOT numbers and the types of materials that they carry.

While looking at these, it is easy to see that their design often can alert us to the types of materials contained inside.

EMERGENCY RESPONSE GUIDEBOOK

No discussion of identification systems would be complete without an understanding of the *ERG*. Figure 5-14 shows the cover of the book currently used. The book is updated approximately every 3–4 years and is a useful resource for those involved in the transportation of hazardous materials or in the response to a transportation incident.

ROAD TRAILER IDENTIFICATION CHART*

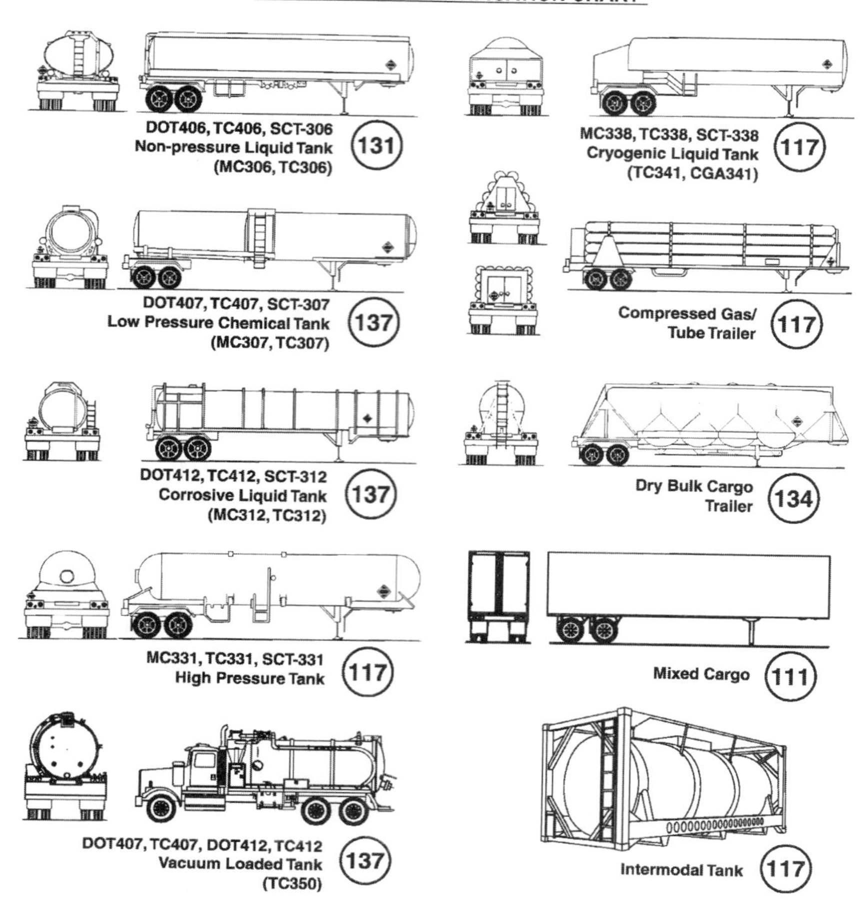

CAUTION: This chart depicts only the most general shapes of road trailers. Emergency response personnel must be aware that there are many variations of road trailers, not illustrated above, that are used for shipping chemical products. The suggested guides are for the most hazardous products that may be transported in these trailer types.

* **The recommended guides should be considered as last resort if the material cannot be identified by any other means.**

Page 19

Figure 5-13
Examples of DOT-approved tank cars and trucks identified in the ERG.

The *ERG* is designed for the first responder for initial response action only and should be used in conjunction with other reference sources. In the first section of the book, we find the following description: "It is primarily a guide to aid first responders in quickly identifying the specific or generic hazards of the materials(s) involved in the incident and protecting themselves and the general public during the initial response phase of the incident." The information should be used with caution because it is a guide for initial assessment and identification only. Many who receive training on the *ERG* never fully understand this and often use it beyond the intent for which it was developed. The book cautions users to its limitation to transportation incidents primarily and to the fact that the information is general in nature and not specific to the material(s)

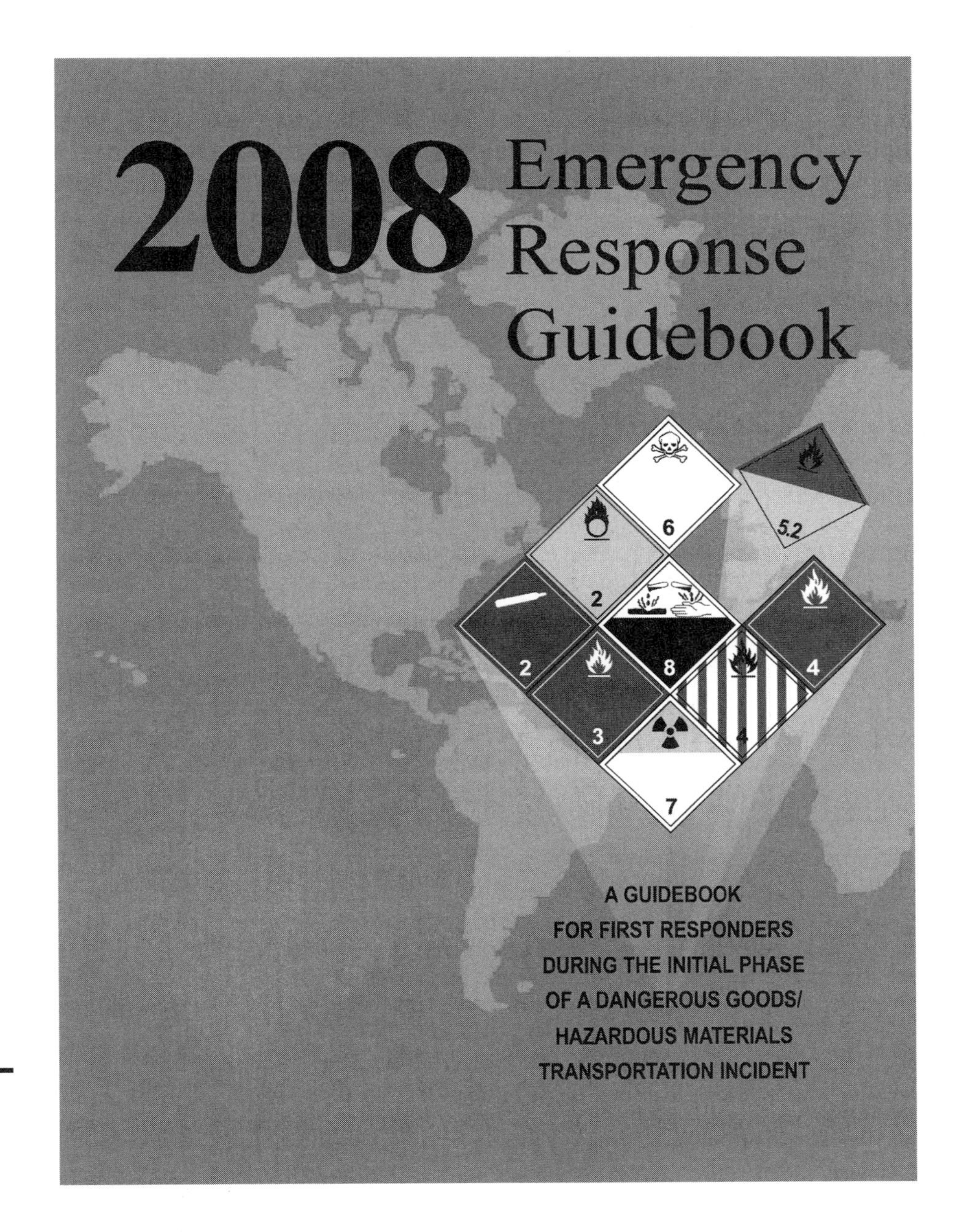

Figure 5-14 *The Emergency Response Guidebook.*

involved. However, in many situations, the guidebook will help us to make an initial identification of a hazardous substance being transported. The *ERG* is simple to use and designed to provide basic information until more detailed information becomes available.

To begin using the *ERG,* it is important to note that the pages are color-coded to break the book into sections. Following is a summary of the colors and their use:

- *Orange section (see Figure 5-15).* While the orange section is located in the middle of the book, it should be discussed first because it provides the specific response information for each group of materials and is referenced by the other colored sections. Within each of the guides contained

GUIDE 119 GASES - TOXIC - FLAMMABLE ERG2008

POTENTIAL HAZARDS

HEALTH

- **TOXIC; may be fatal if inhaled or absorbed through skin.**
- Contact with gas or liquefied gas may cause burns, severe injury and/or frostbite.
- Fire will produce irritating, corrosive and/or toxic gases.
- Runoff from fire control may cause pollution.

FIRE OR EXPLOSION

- Flammable; may be ignited by heat, sparks or flames.
- May form explosive mixtures with air.
- Those substances designated with a **"P"** may polymerize explosively when heated or involved in a fire.
- Vapors from liquefied gas are initially heavier than air and spread along ground.
- Vapors may travel to source of ignition and flash back.
- Some of these materials may react violently with water.
- Cylinders exposed to fire may vent and release toxic and flammable gas through pressure relief devices.
- Containers may explode when heated.
- Ruptured cylinders may rocket.
- Runoff may create fire or explosion hazard.

PUBLIC SAFETY

- **CALL Emergency Response Telephone Number on Shipping Paper first. If Shipping Paper not available or no answer, refer to appropriate telephone number listed on the inside back cover.**
- As an immediate precautionary measure, isolate spill or leak area for at least 100 meters (330 feet) in all directions.
- Keep unauthorized personnel away. • Stay upwind.
- Many gases are heavier than air and will spread along ground and collect in low or confined areas (sewers, basements, tanks).
- Keep out of low areas. • Ventilate closed spaces before entering.

PROTECTIVE CLOTHING

- Wear positive pressure self-contained breathing apparatus (SCBA).
- Wear chemical protective clothing that is specifically recommended by the manufacturer. It may provide little or no thermal protection.
- Structural firefighters' protective clothing provides limited protection in fire situations ONLY; it is not effective in spill situations where direct contact with the substance is possible.

EVACUATION

Spill

- See Table 1 - Initial Isolation and Protective Action Distances for highlighted materials. For non-highlighted materials, increase, in the downwind direction, as necessary, the isolation distance shown under "PUBLIC SAFETY".

Fire

- If tank, rail car or tank truck is involved in a fire, ISOLATE for 1600 meters (1 mile) in all directions; also, consider initial evacuation for 1600 meters (1 mile) in all directions.

Page 184

ERG2008 GASES - TOXIC - FLAMMABLE GUIDE 119

EMERGENCY RESPONSE

FIRE

- **DO NOT EXTINGUISH A LEAKING GAS FIRE UNLESS LEAK CAN BE STOPPED.**

Small Fire

- Dry chemical, CO_2, water spray or alcohol-resistant foam.

Large Fire

- Water spray, fog or alcohol-resistant foam.
- **FOR CHLOROSILANES, DO NOT USE WATER**; use AFFF alcohol-resistant medium expansion foam. • Move containers from fire area if you can do it without risk.
- Damaged cylinders should be handled only by specialists.

Fire involving Tanks

- Fight fire from maximum distance or use unmanned hose holders or monitor nozzles.
- Cool containers with flooding quantities of water until well after fire is out.
- Do not direct water at source of leak or safety devices; icing may occur.
- Withdraw immediately in case of rising sound from venting safety devices or discoloration of tank. • ALWAYS stay away from tanks engulfed in fire.

SPILL OR LEAK

- ELIMINATE all ignition sources (no smoking, flares, sparks or flames in immediate area).
- All equipment used when handling the product must be grounded.
- Fully encapsulating, vapor protective clothing should be worn for spills and leaks with no fire.
- Do not touch or walk through spilled material.
- Stop leak if you can do it without risk.
- Do not direct water at spill or source of leak.
- Use water spray to reduce vapors or divert vapor cloud drift. Avoid allowing water runoff to contact spilled material.
- **FOR CHLOROSILANES**, use AFFF alcohol-resistant medium expansion foam to reduce vapors.
- If possible, turn leaking containers so that gas escapes rather than liquid.
- Prevent entry into waterways, sewers, basements or confined areas.
- Isolate area until gas has dispersed.

FIRST AID

- Move victim to fresh air. • Call 911 or emergency medical service.
- Give artificial respiration if victim is not breathing.
- **Do not use mouth-to-mouth method if victim ingested or inhaled the substance; give artificial respiration with the aid of a pocket mask equipped with a one-way valve or other proper respiratory medical device.**
- Administer oxygen if breathing is difficult.
- Remove and isolate contaminated clothing and shoes.
- In case of contact with substance, immediately flush skin or eyes with running water for at least 20 minutes.
- In case of contact with liquefied gas, thaw frosted parts with lukewarm water.
- In case of burns, immediately cool affected skin for as long as possible with cold water. Do not remove clothing if adhering to skin.
- Keep victim warm and quiet. • Keep victim under observation.
- Effects of contact or inhalation may be delayed.
- Ensure that medical personnel are aware of the material(s) involved and take precautions to protect themselves.

Page 185

Figure 5-15 *Example of the orange section of the* ERG.

in the orange section, information is given relative to the flammability or toxicity of the substances for which the guide is assigned. Additional information on the proper response procedures, first aid protocols for treating exposed persons, methods for control of leaks, and firefighting is given. While there are thousands of materials referenced in the *ERG*, there are only 62 guides that are used, making the information somewhat generic.

- *Yellow section (see Figure 5-16).* The yellow section lists materials in order by their identification number and then refers to a guide number for the recommended response procedures to an incident involving these materials.

ID No.	Guide No.	Name of Material
——	**112**	Ammonium nitrate-fuel oil mixtures
——	**158**	Biological agents
——	**112**	Blasting agent, n.o.s.
——	**112**	Explosive A
——	**112**	Explosive B
——	**114**	Explosive C
——	**112**	Explosives, division 1.1, 1.2, 1.3, 1.5 or 1.6
——	**114**	Explosives, division 1.4
——	**153**	Toxins
1001	**116**	Acetylene
1001	**116**	Acetylene, dissolved
1002	**122**	Air, compressed
1003	**122**	Air, refrigerated liquid (cryogenic liquid)
1003	**122**	Air, refrigerated liquid (cryogenic liquid), non-pressurized
1005	**125**	Ammonia, anhydrous
1005	**125**	Anhydrous ammonia
1006	**121**	Argon
1006	**121**	Argon, compressed
1008	**125**	Boron trifluoride
1008	**125**	Boron trifluoride, compressed
1009	**126**	Bromotrifluoromethane
1009	**126**	Refrigerant gas R-13B1
1010	**116P**	Butadienes, stabilized
1010	**116P**	Butadienes and hydrocarbon mixture, stabilized
1011	**115**	Butane
1011	**115**	Butane mixture
1012	**115**	Butylene
1013	**120**	Carbon dioxide
1013	**120**	Carbon dioxide, compressed
1014	**122**	Carbon dioxide and Oxygen mixture
1014	**122**	Carbon dioxide and Oxygen mixture, compressed
1014	**122**	Oxygen and Carbon dioxide mixture
1014	**122**	Oxygen and Carbon dioxide mixture, compressed
1015	**126**	Carbon dioxide and Nitrous oxide mixture
1015	**126**	Nitrous oxide and Carbon dioxide mixture
1016	**119**	Carbon monoxide
1016	**119**	Carbon monoxide, compressed
1017	**124**	Chlorine
1018	**126**	Chlorodifluoromethane
1018	**126**	Refrigerant gas R-22
1020	**126**	Chloropentafluoroethane
1020	**126**	Refrigerant gas R-115
1021	**126**	1-Chloro-1,2,2,2-tetrafluoroethane
1021	**126**	Chlorotetrafluoroethane
1021	**126**	Refrigerant gas R-124
1022	**126**	Chlorotrifluoromethane
1022	**126**	Refrigerant gas R-13
1023	**119**	Coal gas
1023	**119**	Coal gas, compressed
1026	**119**	Cyanogen
1026	**119**	Cyanogen gas
1027	**115**	Cyclopropane
1028	**126**	Dichlorodifluoromethane
1028	**126**	Refrigerant gas R-12
1029	**126**	Dichlorofluoromethane
1029	**126**	Refrigerant gas R-21

Page 27

Figure 5-16 *Example of the yellow section of the* ERG.

- *Blue section (see Figure 5-17).* The blue section lists the materials by name in alphabetic order and refers to a guide number for the recommended response procedures to an incident involving these materials. If the name of the substance is available, this is the best way to determine which guide to use until additional information is found.
- *Green section (see Figure 5-18).* The green section is the last of the colored sections of the book and is found near the end. It is called the "Table of Initial Isolation and Protective Action Distances" and identifies some of the materials presenting a significant risk when involved in a transportation accident and that are not on fire. A major release of one of the materials identified in the green section could require large

Name of Material	Guide No.	ID No.	Name of Material	Guide No.	ID No.
Hydrazine, aqueous solution, with not more than 37% Hydrazine	**152**	3293	Hydrofluoric acid and Sulphuric acid mixture	**157**	1786
Hydrazine, aqueous solutions, with more than 64% Hydrazine	**132**	2029	Hydrofluorosilicic acid	**154**	1778
Hydrazine hydrate	**153**	2030	Hydrogen	**115**	1049
Hydrides, metal, n.o.s.	**138**	1409	Hydrogen absorbed in metal hydride	**115**	9279
Hydriodic acid	**154**	1787	Hydrogen, compressed	**115**	1049
Hydriodic acid, solution	**154**	1787	Hydrogen in a metal hydride storage system	**115**	3468
Hydrobromic acid	**154**	1788	Hydrogen in a metal hydride storage system contained in equipment	**115**	3468
Hydrobromic acid, solution	**154**	1788	Hydrogen in a metal hydride storage system packed with equipment	**115**	3468
Hydrocarbon gas, compressed, n.o.s.	**115**	1964	Hydrogen, refrigerated liquid (cryogenic liquid)	**115**	1966
Hydrocarbon gas, liquefied, n.o.s.	**115**	1965	Hydrogen and Carbon monoxide mixture	**119**	2600
Hydrocarbon gas mixture, compressed, n.o.s.	**115**	1964	Hydrogen and Carbon monoxide mixture, compressed	**119**	2600
Hydrocarbon gas mixture, liquefied, n.o.s.	**115**	1965	Hydrogen and Methane mixture, compressed	**115**	2034
Hydrocarbon gas refills for small devices, with release device	**115**	3150	Hydrogen bromide, anhydrous	**125**	1048
Hydrocarbons, liquid, n.o.s.	**128**	3295	Hydrogen chloride, anhydrous	**125**	1050
Hydrochloric acid	**157**	1789	Hydrogen chloride, refrigerated liquid	**125**	2186
Hydrochloric acid, solution	**157**	1789	Hydrogen cyanide, anhydrous, stabilized	**117**	1051
Hydrocyanic acid, aqueous solution, with less than 5% Hydrogen cyanide	**154**	1613	Hydrogen cyanide, aqueous solution, with not more than 20% Hydrogen cyanide	**154**	1613
Hydrocyanic acid, aqueous solution, with not more than 20% Hydrogen cyanide	**154**	1613	Hydrogen cyanide, solution in alcohol, with not more than 45% Hydrogen cyanide	**131**	3294
Hydrocyanic acid, aqueous solutions, with more than 20% Hydrogen cyanide	**117**	1051	Hydrogen cyanide, stabilized	**117**	1051
Hydrofluoric acid	**157**	1790	Hydrogen cyanide, stabilized (absorbed)	**152**	1614
Hydrofluoric acid, solution	**157**	1790			
Hydrofluoric acid and Sulfuric acid mixture	**157**	1786			

Page 126

Figure 5-17 *Example of the blue section of the* ERG.

Page 312

ID No.	NAME OF MATERIAL	SMALL SPILLS (From a small package or small leak from a large package) First ISOLATE in all Directions Meters (Feet)	Then PROTECT persons Downwind during- DAY Kilometers (Miles)	NIGHT Kilometers (Miles)	LARGE SPILLS (From a large package or from many small packages) First ISOLATE in all Directions Meters (Feet)	Then PROTECT persons Downwind during- DAY Kilometers (Miles)	NIGHT Kilometers (Miles)
1929 1929 1929	Potassium dithionite **(when spilled in water)** Potassium hydrosulfite **(when spilled in water)** Potassium hydrosulphite **(when spilled in water)**	30 m (100 ft)	0.1 km (0.1 mi)	0.2 km (0.1 mi)	30 m (100 ft)	0.3 km (0.2 mi)	1.1 km (0.7 mi)
1931 1931 1931	Zinc dithionite **(when spilled in water)** Zinc hydrosulfite **(when spilled in water)** Zinc hydrosulphite **(when spilled in water)**	30 m (100 ft)	0.1 km (0.1 mi)	0.2 km (0.1 mi)	30 m (100 ft)	0.3 km (0.2 mi)	1.1 km (0.7 mi)
1953	Compressed gas, flammable, poisonous, n.o.s. (Inhalation Hazard Zone A)	100 m (300 ft)	0.6 km (0.4 mi)	2.5 km (1.5 mi)	800 m (2500 ft)	4.4 km (2.7 mi)	8.9 km (5.6 mi)
1953	Compressed gas, flammable, poisonous, n.o.s. (Inhalation Hazard Zone B)	30 m (100 ft)	0.2 km (0.1 mi)	0.8 km (0.5 mi)	400 m (1250 ft)	1.9 km (1.2 mi)	4.8 km (3.0 mi)
1953	Compressed gas, flammable, poisonous, n.o.s. (Inhalation Hazard Zone C)	30 m (100 ft)	0.1 km (0.1 mi)	0.3 km (0.2 mi)	300 m (1000 ft)	1.3 km (0.8 mi)	4.1 km (2.6 mi)
1953	Compressed gas, flammable, poisonous, n.o.s. (Inhalation Hazard Zone D)	30 m (100 ft)	0.1 km (0.1 mi)	0.2 km (0.1 mi)	150 m (500 ft)	0.7 km (0.5 mi)	2.7 km (1.7 mi)
1953	Compressed gas, flammable, toxic, n.o.s. (Inhalation Hazard Zone A)	100 m (300 ft)	0.6 km (0.4 mi)	2.5 km (1.5 mi)	800 m (2500 ft)	4.4 km (2.7 mi)	8.9 km (5.6 mi)

Figure 5-18 *Example of the green section of the* ERG.

evacuations. Not all materials listed in the *ERG* are covered in the green section. Only those materials that are "highlighted" in either the yellow section or the blue section of the *ERG* are covered in the green section. Once a material is identified by its name or the identification number and is highlighted in either section, we know to check into the green section for additional information on further actions related to evacuation.

In addition to the Table of Initial Isolation and Protective Action Distances, the green section also contains a subsection entitled "Table of Water Reactive Materials That Produce Toxic Gases." This list shows the materials that produce specific types of toxic gases when they are spilled in water. The list includes a table that identifies what toxic materials are produced when the substance contacts water.

Finally, there is a lot of additional information in the white pages of the Guidebook, including sections that show all types of placards used, a diagram of the types of DOT-approved truck and rail cars, and a glossary of terms. Specific information on the proper use of the book and instructions for using each of the colored sections are also found in the white pages of the Guidebook. New information relative to terrorism and weapons of mass destruction are found in these pages.

Procedures for Using the Guide

Using the Guidebook properly requires some basic information on the material. The *ERG* requires that the name of the material, the identification number (four-digit), or one of the generic placards be known. If any of these are known, the *ERG* will direct you to the appropriate orange section, which will provide the appropriate guide to use. Following is a description of how this is used.

- *Name of material is known.* If the name of the material is known, use the blue section of the *ERG* to refer you to the appropriate orange guide number for use until additional information becomes available.
- *Four-digit identification number is known.* If the four-digit identification number is known, such as is the case with a bulk shipment, use the yellow section of the Guidebook to refer you to the appropriate orange guide number to use until additional information becomes available.
- *Generic placard is known.* If the only indication found is the presence of one of the generic hazard class placards on the outside of the shipment—for example, Flammable, Poison, Oxidizer—the *ERG* directs us to use the pictures of the placards found in the front section of the Guidebook. Here we find that the placards will direct us to one of the orange guide numbers for use until additional information becomes available. It is important to know that the guides used for generic placards are often not as specific as those used when the material is identified by name or identification number. Figure 5-19 shows the generic placards found in the current *ERG* and the recommended guide to be used.

To properly use the *ERG,* some practice is required. Exercises given at the end of this chapter will show its use in a range of simulated conditions.

Finally, it is critical to remember that the Guidebook, while helpful, is also very limited in what information it can provide. It has a number of limitations, including the following:

- Remember that even if the four-digit identification number is known, the *ERG* will not always identify the one specific material that corresponds to that number. Recall the example of the numerous materials that display the 1993 identification number.
- Each of the orange guides describes the actions to be taken for a large number of materials and not specifically the single material that we are interested in. They must be somewhat generic in their information because they are useful only for the initial actions described.
- The *ERG* is primarily used for transportation incidents, as it clearly states. It cautions users that the information may be of "limited value in its application at fixed facility locations."
- The green section providing guidance on the evacuation distances can be confusing. It divides spills into "large" and "small" but does little to define these terms. Small spills are defined as those "from a small package or a small leak from a large package" and large spills are those "from a large package or from many small packages." Obviously this can present problems in applying the information in an emergency situation.
- The green section provides the guidance on evacuation distances, but cautions the user that the recommended distances only apply to the first 30 minutes following the release. The cautions are contained in the small print in the front of the green section and are often missed by users of the guidebook.

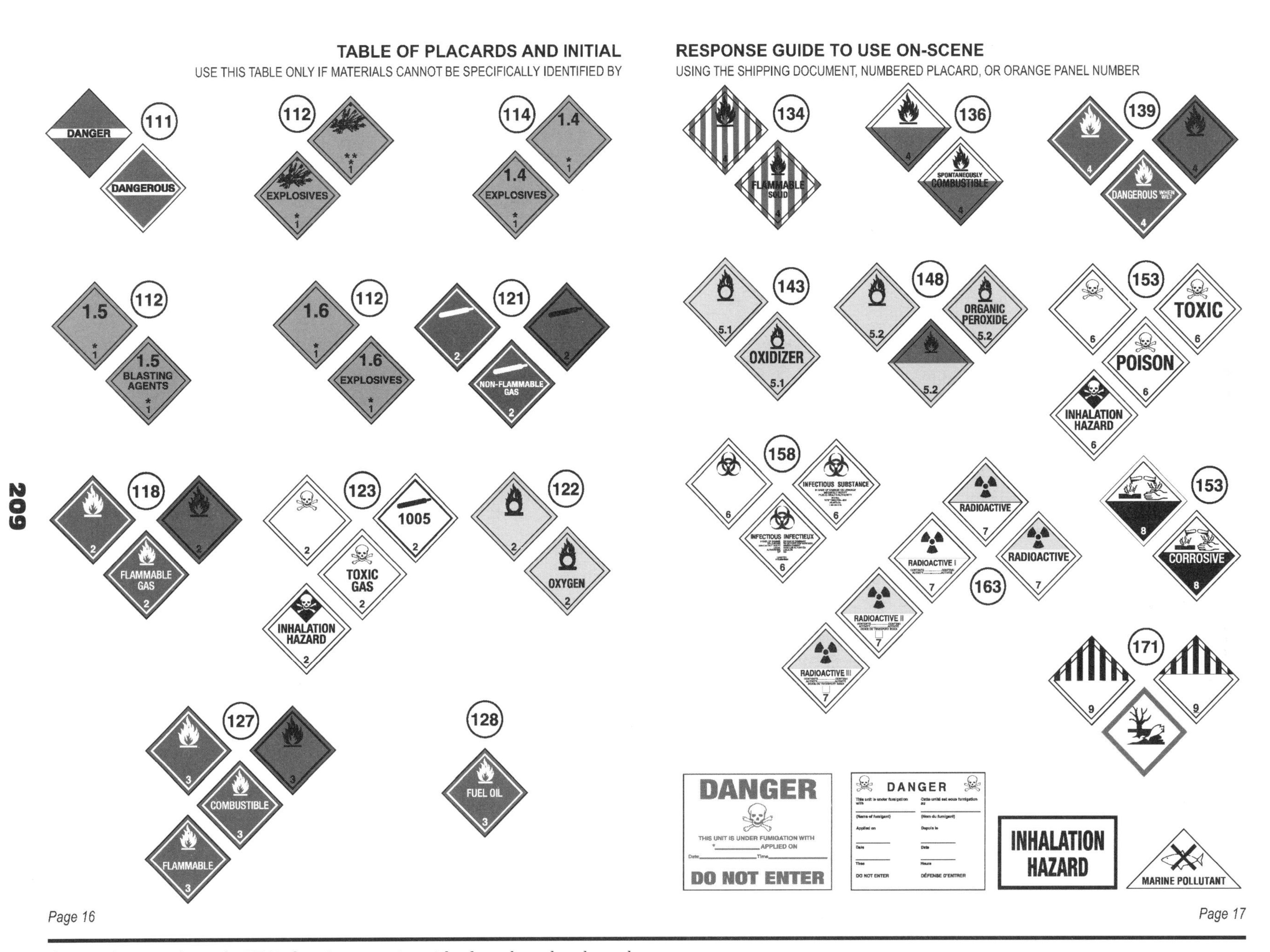

Figure 5-19 *Page from the ERG showing generic guides based on the placard.*

SUMMARY

- Although there are many types of identification systems in use, the two most common are the DOT system and the NFPA 704 system. Never confuse the markings of one system with another.
- Most identification systems do not identify the specific material present, but rather identify one or more of the potential hazards posed by the material.
- Any identification system in use is only effective if it is implemented properly and understood by those who use it.
- On a DOT label or placard, four factors are to be used in determining the hazard: color, identification number, generic hazard class name or a four-digit identification number, and the symbol. All four factors can be used together or separately to determine what material is present.
- Proper labeling and placarding requirements for hazardous materials and hazardous wastes can be found in 49 CFR.
- The DOT system is used for hazardous materials being transported, and the NFPA system is used when hazardous materials are being stored at a fixed facility.
- The DOT system is mandated by federal law, and the NFPA 704 system is a recommended guideline that can become law if adopted by a local governmental agency.
- The NFPA 704 system has the advantage of being easy to use, but can be limited in some situations because it does not inform responders how much, how many, and where chemicals are located.
- Shipping papers and Uniform Hazardous Waste Manifests must accompany hazardous materials during shipment and must be accurate and readily accessible.
- The *Emergency Response Guidebook (ERG)* is for initial action only and is one of many available reference resources. Isolation and protective action distances are effective for only about 30 minutes.
- The *Emergency Response Guidebook* lists chemicals by identification number (yellow section), name (blue section), basic action to take in case of spill (orange section), and initial isolation/protective action distances (green section).
- The Hazardous Materials Identification System (HMIS) is similar to the NFPA 704 diamond, but the white section denotes levels of PPE that may conflict with other standards.

REVIEW QUESTIONS

1. What are the two most common identification systems and where are they used?
2. What other identification systems do you use or have come across?
3. List three disadvantages of the DOT identification system.
4. Describe the difference between a hazardous material and a hazardous waste and cite which agency regulates each one.
5. List the nine standard DOT hazard classes.
6. If you are shipping 10 pounds of 1.1 Explosives, is labeling or placarding required? If so, why and what other materials require placards in any amount?
7. According to the *Emergency Response Guidebook* (*ERG*), what is the difference between a large and a small spill?

ACTIVITY

EMERGENCY RESPONSE GUIDEBOOK EXERCISE

Obtain a copy of the latest version of the *Emergency Response Guidebook* or review the online version and use the information to answer the following questions. The *ERG* online can be found at http://hazmat.dot.gov/pubs/erg/gydebook.htm.

1. As part of an emergency response to a large factory, a worker reports that several containers

have been knocked over by a forklift on the loading dock. He reports seeing a placard with the identification number 1830 on the truck carrying the material. What are its health effects and what is the isolation distance with a 75-gallon spill?

2. A second container is also found in the area. The second container has an all yellow diamond label on the side. What type of material is likely found in that container?
3. A material found alongside a road is determined to be acrylonitrile. Acrylonitrile is a Flammable Liquid and would ordinarily have a generic Flammable Liquid (Class 3) label. Identify which guide number would be used if you knew the specific name of the material. Following this, identify the guide number that would be used if only the Flammable label was seen. List three actions that are different between the two guides.
4. A release of a material with the identification number 1017 occurs. The release is a large one and occurs during the day. How far is the initial evacuation distance and how far should you move others who may be downwind? If the spill occurred at night, how far would the distances for evacuation be?
5. If you confront a material leaking from a tank vehicle and the only indication is that it is a type DOT 412 tank truck, what is likely in the tank and what guide should be used until other information becomes available?
6. A third container lying next to the above container had a red and white label with a "flame" on it. What type of material is likely to be in that container?

RESPIRATORY PROTECTION

Learning Objectives

Upon completion of this chapter, you should be able to:

- Identify the five types of respiratory hazards likely to be found in situations involving hazardous materials and wastes.
- List the elements of a Respiratory Protection Program required by Federal OSHA regulations.
- Identify the two classes of respiratory protective equipment.
- List the advantages and disadvantages of each class of respiratory protective equipment.
- Select the appropriate type of respiratory protective equipment for specific situations likely to be encountered in the hazardous waste field.
- List the maintenance and inspection procedures for each type of respiratory protective equipment.
- Describe the steps necessary to effectively don and doff an air-purifying respirator.
- Identify the components of a supplied-air (umbilical) system.

CASE STUDY

Approximately 40,000 workers were called into service to assist in the rescue and recovery efforts as a result of the attacks on New York on 9/11. On the five-year anniversary of the event, a report was issued by the National Institute of Environmental Health Sciences, which is part of the National Institutes of Health, U.S. Department of Health and Human Services. The report, entitled "The World Trade Center Disaster and the Health of Workers: Five-Year Assessment of a Unique Medical Screening Program," included information on the health of traditional first responders such as police, fire, and medical personnel, as well as others involved in the recovery program, including construction, utility, and other public sector employees.

The data on the level of workers' respiratory damage that led to the study's conclusion was shocking. The results showed that 69% of those evaluated in the years after the disaster showed a worsening of respiratory symptoms. Further, of those who did not have respiratory problems before 9/11, approximately 61% developed respiratory symptoms while performing the work at the site. Respiratory symptoms were reportedly worse for those at the site who arrived early in the response effort.

What caused these tragic results and are they preventable in future disasters? The report concluded that the exposure of the workers to the "caustic dusts and toxic pollutants" caused or contributed to the conditions found in the study. Were these exposures preventable? The report's conclusion stated that the "lessons learned should guide future responses to civil disasters." Those lessons include many related to the assessment of hazards, the use of controls to reduce the exposure, and the proper use of respiratory protective equipment that form the basis of the next section.

INTRODUCTION

From the discussion of toxicology, it is clear that inhalation hazards are a commonly encountered problem when working with many types of hazardous substances. In Chapter 3, we learned that the inhalation route of entry, which includes the respiratory system, is the body system most vulnerable to the harmful effects of chemicals. Because of this, it is vital to anyone who works with hazardous substances in the workplace, or those who respond to emergencies involving the release of these materials, understand the need for proper respiratory protection. It is also critical that proper selection and use of respiratory protection be an important element of any training program related to the safe handling and response to releases of hazardous substances.

> **■ Note**
> The respiratory system is the most vulnerable system in the body and is subject to harm from a range of hazardous substances.

RESPIRATORY HAZARDS

Respiratory hazards are present when there is a hazardous level of atmospheric contamination. The types of hazardous conditions vary considerably and include a broad range of specific materials, the types of hazardous atmospheric conditions can be divided into five major groups:

1. Particulates including dusts, fumes, and mists
2. Gases and vapors
3. Combinations of particulates and gases
4. Oxygen-deficient atmospheres
5. High or low temperature areas

The first group of respiratory hazards is actually comprised of several components that ultimately end up as particulates, some visible and some invisible in the atmosphere. Particulates are materials that have an actual weight or mass. Unlike vapors, which are most often invisible, particulates can be visible with the naked eye in some cases. Some particulates are small enough to be both invisible and able to enter the deepest portion of the lungs, the alveoli. Recall from the discussion on inhalation hazards that the particulate must be extremely small in order to get into the alveoli.

dusts
solid particles in the air created through operations such as grinding, sanding, demolition, or drilling

fumes
particles created through the heating and cooling of solid materials, such as happens during welding

mists
suspended liquid droplets in the air created through pressure such as with spraying operations

atomizing
creating small mist droplets through the introduction of pressure

gases
nonparticulate forms of matter that move freely through air

vapors
similar to gases but are formed when a liquid evaporates and the liquid portion of the material enters the air

evaporation
the process of a liquid entering the air as it dries out; this is caused by a number of factors, including heat and the liquid itself

■ Note

In order to get into the deepest part of the lungs, a particle must be smaller than 10 microns in size.

One of the more commonly encountered particulate types is dust. **Dust** can be composed of a variety of materials and is created when solid materials are broken down by operations such as drilling, grinding, demolition, or sanding. Generally speaking, the smaller the dust, the longer it hovers in the air and the greater the likelihood that it will be inhaled. Also, smaller dusts may not be visible and therefore more difficult to detect.

Fumes are another example of particulates and are produced when metal or plastic solids are heated and then cooled quickly. This creates very fine particles that drift into the air. For example, in welding operations some of the welding fume may be visible in the form of smoke, or invisible in the form of particles that are too small to see but small enough to enter the lungs.

Another common particulate type is mists. **Mists** are tiny liquid droplets usually created during operations involving pressure or **atomizing** of the liquids. Examples of mist-generating activities include spraying operations and processes involving aerosols or high pressure.

Gases are nonparticulate substances that are airborne at ambient temperature and pressure. Gases can travel far and fast in many cases. Because they are very mobile, gases can pass immediately through the airway and enter the lungs, where they most often are absorbed into the bloodstream. Systemic health effects are quite common with the inhalation of a range of materials entering the lungs and then passing into the blood. Unlike particulate materials, gases tend to be difficult to see and detection devices are usually required to identify their presence.

Vapors are similar to gases but are actually the gaseous form of a solid or liquid created by the process of **evaporation**. Evaporation occurs when liquids are exposed to air and the liquid portion goes into the atmosphere because of a number of factors such as heat. Like gases, vaporous materials are mobile, can enter the respiratory system very quickly, and cause systemic effects through their entry into the blood steam. They are most often invisible to the naked eye, and like gases, they may be difficult to detect without proper air monitoring procedures.

Some atmospheres can contain a combination of particulates and vapors/gases. An example of this would be a fire where smoke caused by unburned particulates is present along with gases such as carbon monoxide. In conditions such as these, the respiratory system can quickly become overwhelmed if not protected from this range of hazards.

> **Note**
> Some airborne hazards can be both particulate and gaseous at the same time.

oxygen-deficient atmosphere
defined by OSHA as one that contains less than 19.5% oxygen

Hot Work activities
activities that generate heat or open flame, such as welding and cutting torch operations

The next major type of inhalation hazard is oxygen deficiency. This occurs when the level of oxygen within the environment drops to dangerously low levels. Normal atmospheres contain approximately 20.9% oxygen. The remainder of the makeup of the air is nitrogen, which composes about 78%, and other trace gases such as argon and carbon dioxide, which make up the remaining 1%. An **oxygen-deficient atmosphere** is defined by OSHA as an atmosphere where the percentage of oxygen in the air drops below 19.5% by volume. While there are a number of causes of oxygen-deficient conditions, most often they occur in confined areas where the oxygen is displaced by other gases. Oxygen can also be used up during welding operations or other types of **Hot Work activities**, rusting, and even by the bacterial decomposition process because bacteria may use up the available levels of oxygen in the decomposition process. Hot Work involves operations such as welding or using a cutting torch to cut through metals.

Regardless of the cause, oxygen-deficient conditions can be rapidly fatal and must be identified. Relying on our normal sense of smell will not let us know that the levels of oxygen in the air are safe to breathe. An oxygen-deficient atmosphere most often does not have any characteristic odor to alert people that a serious condition is present. This, coupled with the symptoms produced by this type of exposure, makes oxygen deficiency one of the most serious issues confronting protection of the respiratory system. It is only through the use of special detection devices that we can know oxygen levels may be dropping.

The effects of oxygen deficiency on the body encompass a range of symptoms. Initially, the condition affects the brain, where loss of judgment, confusion, or dizziness can occur. These symptoms may mask the condition to the exposed person, making this a disastrous combination—a confused worker in a deadly environment. Often, the next reaction to low oxygen levels is that the exposed individual will start to breathe more rapidly and the heart rate will increase. This may then result in a higher demand for the limited supply of oxygen. Following this, motor function becomes impaired and the exposed individual may require help to leave the area. Eventually, if the condition continues, death can often be the end result. A summary of these effects is given in Table 6-1.

Table 6-1 *The effects of low levels of oxygen on the body.*

Percentage of oxygen	Effects on body
21%	Normal atmospheric air
	Normal respirations: no problems
<19.5	OSHA: O_2 deficiency
15%	Beginning of confusion
	Increased respiratory rate
	Slight loss of coordination
12%	Fatigue: extreme in some cases
	Decreased level of consciousness
	Rapid respiratory rate
9%	Coma or death

■ Note
Early symptoms of lack of oxygen are generally related to brain function and involve confusion, poor judgment, and dizziness.

Additionally, regardless of what is or is not present in the air we breathe, if the air is extremely hot or cold, respiratory damage can occur. Very hot or cold air can damage tissue in the nose, mouth, airway, and lungs, interfering with normal breathing. If these conditions are found in the air we breathe, steps need to be taken to remove the condition or otherwise protect the respiratory system.

■ Note
Hot and cold atmospheres can cause significant lung damage.

RESPIRATORY PROTECTION FUNDAMENTALS

Because of the dangers presented by inhalation hazards, OSHA has implemented a range of regulations related to exposure. As we know, OSHA and others have established safe levels of airborne contamination from various sources. These are found in the PEL, STEL, and TLV/C levels previously discussed. Even ordinary dust has a listed PEL. When exposures are expected to exceed any of the exposure limits, OSHA requires that the employer develop a Respiratory Protection Program. Included are a number of components to ensure that workers in areas with high exposure potential are properly protected by respiratory protective equipment and training. The key Federal OSHA Respiratory Protection regulation is found in 29 CFR, Part 1910.134. This regulation is also referenced in the HAZWOPER regulation to further illustrate its importance in this field.

■ Note
OSHA has an entire regulation on the subject of respiratory protection. It is found in 29 CFR, Part 1910.134.

Written Respiratory Protection Programs

OSHA requires that any workplace that has the need for respiratory protection must have a written program outlining its rules and application to the worksite. As with other programs mandated by OSHA, a Respiratory Protection Program is required to be in writing and available for employees to review as part of their worker protection system. Because airborne contaminants are a significant hazard for those who work with hazardous wastes or respond to hazardous materials emergencies, the components of such a program are important to the personal safety of the workers and should be understood by everyone involved.

■ Note
Employers whose employees wear respiratory protection are required to develop a written Respiratory Protection Program.

The first part of any Respiratory Protection Program is that respiratory hazards are best handled by elimination. In Chapter 2, we identified the three methods of control in the workplace that are mandated by OSHA. OSHA requires that, if possible, the hazardous environment or conditions be eliminated through the use of engineering controls, and airborne contamination is certainly subject to this. Many inhalation hazards can be easily controlled with effective ventilation or wetting materials down to make them less likely to become airborne. In fact, the regulation in 29 CFR, Part 1910.134 (a)(1) reads as follows:

> *In the control of those occupational diseases caused by breathing air contaminated with harmful dusts, fogs, fumes, mists, gases, smokes, sprays, or vapors, the primary objective shall be to prevent atmospheric contamination. This shall be accomplished as far as feasible by accepted engineering control measures (for example, enclosure or confinement of the operation, general and local ventilation, and substitution of less toxic materials). When effective engineering controls are not feasible, or while they are being instituted, appropriate respirators shall be used.*

■ Note

OSHA requires the use of engineering controls initially to reduce or eliminate an atmospheric contamination.

If the respiratory hazards cannot be eliminated, the next step in the Respiratory Protection Program is to establish a program for use of respiratory protective equipment. This is accomplished through the inclusion of several key areas. They include:

- Procedures to identify the proper selection and use of respirators, including specific requirements regarding the types required for the hazard or combination of hazards likely to be encountered. Respiratory protective equipment used must be compliant with NIOSH guidelines.
- Training program requirements for those who will wear respiratory protection. The number of hours required for this training program is not established in the regulation; however, employees must receive training in several key areas as outlined in the regulation. These include the elements of the written Respiratory Protection Program; training in types of conditions requiring a respirator; procedures for selection of the appropriate respirator; and the use, care, and maintenance of respiratory protection equipment.
- Inspection, maintenance, and sanitation procedures must be specified. Regular inspections should be conducted and maintenance performed in accordance with the manufacturers requirements. Requirements and more details on these procedures are given later in this chapter.
- Storage procedures for each piece of respiratory protection equipment must be specified in order to maintain them in proper working order.
- Medical surveillance practices need to be outlined to include a physical examination by an approved medical provider to determine whether the person is capable of wearing respiratory protective equipment. Respirators can place extra demands on the body. Additionally, the types of operations requiring respirators can cause harm to some individuals. For this reason, OSHA mandates that anyone who wears a respirator obtain medical clearance prior to use. At a minimum, this requires the administration of a lengthy medical questionnaire that is contained as a mandatory appendix to the regulation. On the basis of the information provided to the medical

evaluator, medical testing and a physical examination could be done to ensure that the employee has no pre-existing conditions such as asthma or chronic lung disease that would create additional danger for the employee while wearing a respirator. Follow-up medical examinations are required on a regular basis as determined by the examining medical practitioner.

- Program monitoring and work area condition surveillance, with special attention to potential workers. If monitoring shows deficiencies, appropriate action is indicated to correct them prior to resumption of work.
- Mandatory recordkeeping for all components of the program. This includes employee exposure and medical screening records, maintenance records, air quality records, and training programs.
- If a tight-fitting respirator is worn, fit testing on the specific respirator masks used by employees must be provided. **Tight fitting** refers to masks that are strapped directly to the face and a seal is formed when the mask is pulled tightly to the face. Because the sealing of the mask to the face is critical to ensure proper fit and adequate protection, a test to confirm a proper fit is required. Fit testing is the process where the individual type and size of each type of tight-fitting respirator worn is tested on the person who will wear it. This is done to ensure that the respirators selected fit the individual faces of those who will use them and that each respirator fits tightly to the face. Fit testing is mandated to be conducted at least once each year or as facial conditions change, such as through dental work or other changes. No facial hair is allowed to be present on the individual being fit tested. OSHA clearly states that any facial hair that comes between the sealing surface of the respirator and the face is cause to not test the individual and to not allow him to wear the respirator in the course of his work. The rationale for this is that facial hair can create gaps between the mask and the skin, allowing for the introduction of contaminants into the mask.

tight fitting
the type of respirator that is strapped to the face of the user and a seal is formed between the mask and the skin

■ Note
NIOSH is the agency prescribed by OSHA to approve respiratory protective equipment.

■ Note
There are numerous elements of a Respiratory Protection Program, including required training programs, fit testing, medical surveillance, and recordkeeping.

Mandatory procedures for conducting fit testing for tight-fitting respirators are given in the appendix to the regulation. Two types of fit testing procedures and various agents used in the fit testing process are identified. The two standard types of fit testing include qualitative, which simply establishes that there is some degree of fit, and quantitative, which measures the degree of the fit.

Qualitative fit testing generally involves the individual wearing a mask to enter an area where a test agent with significant warning properties is present. Most often, this is accomplished by placing a hood over the person wearing the mask. He is then subjected to materials in the air that would normally be filtered out by the respirator filter. The approved agents involved are Stanic Chloride (irritant smoke), Isoamyl Acetate (banana oil), Saccharin, and Bitrex. These materials are sprayed into the hooded area and the person is required to do various exercises, including moving the head in various ways, bending over, talking, and breathing deeply. If the person wearing the mask detects any odor or taste, he advises the person conducting the test of the leak and a new mask is found and the test repeated. This is necessary because the material is likely entering the mask through a break or gap in the seal.

> **■ Note**
> Fit testing is required for everyone who uses a tight-fitting respirator for each of the masks that are worn. Fit testing can be done in a number of ways as outlined in the Respiratory Protection regulation.

Quantitative fit testing is more complex and typically uses a computerized system that is adapted to the mask and which takes samples of the air from both inside and outside the mask. Once collected, the two samples are compared and a factor is determined to indicate the difference in the amount of particulates present. If the difference between the amount of particulates outside the mask is high enough relative to the amount measured inside the mask, the mask is determined to have an adequate fit. If the mask is not properly sealed, particulates will enter the mask and a lower value would be present inside. At this point, the mask would need to be changed for one that provides a higher **fit factor**. The fit factor is the ratio of the amount of particles measured inside the mask to those measured in the outside air. An example of quantitative fit testing is found in Figure 6-1. In the picture, you can see that the mask is connected by a tube to a counter that analyzes the samples.

fit factor
the ratio between the levels of particulates inside the mask versus those measured outside the mask

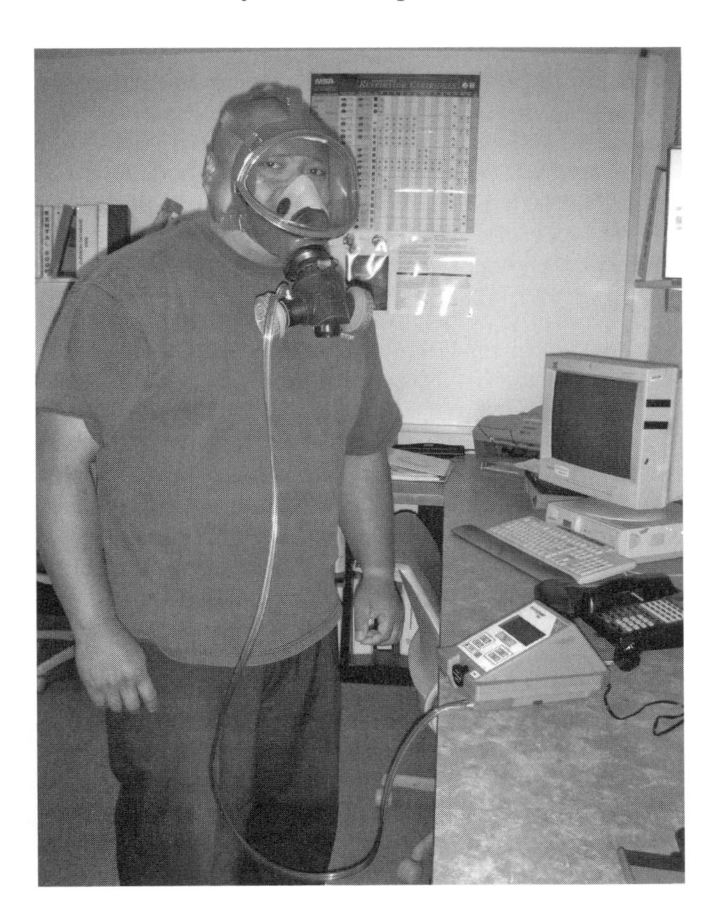

Figure 6-1
Quantitative fit testing being conducted.

RESPIRATORY PROTECTION EQUIPMENT

If it is not possible to eliminate or reduce to a safe level any of the five types of respiratory hazards through the process of engineering controls, respiratory protective equipment is necessary. For this reason, training on the proper use, selection, and maintenance of respiratory protective equipment is an important component of any HAZWOPER training program. Respiratory protective equipment that is used in emergency response, or in work involving exposure to hazardous levels of airborne contamination, is divided into two main categories—air-purifying systems and supplied-air systems.

> **■ Note**
> There are two major types of respirators—the air-purifying system and the supplied-air system.

Air-Purifying Systems

An air-purifying respirator (APR) has specially designed filters that remove specific types of particulates (e.g., dusts, mists, fumes) from the air, or which protect the wearer from specific gases and vapors that might be present. The filters are most often referred to as cartridges or canisters because they are contained in cases that attach to the mask. This type of equipment utilizes either a respirator mask that covers the entire face, called a full-face respirator, or one that only covers the nose and mouth, called a half-face mask. In both cases, the units are equipped with the specific filter or filter combinations to help eliminate the materials as the inhaled air is drawn through them. Hence, they simply take out the harmful materials and *purify* the air. An example of cartridges that have been cut to show the materials inside is shown in Figure 6-2.

The mechanism for filtering the range of types of contaminants varies, depending on the specific type of contaminant present. In the case of particulates, the inhaled air is drawn through paper or another material that acts as a barrier to

Figure 6-2
A cross section of APR cartridges.

block particles of various sizes from entering the mask. As the size of the particles decreases, the filtering capability of the material must be effective enough to handle the tiniest microscopic contamination.

It is important to recognize that not all filtering materials used to trap particulates are the same. In order for the respirator to be effective, it is critical that the proper type of material and filter be used, and that the limitations of each type be known to the user.

> **■ Note**
> Filters used in an APR system are not all the same in their ability to remove materials from the air.

NIOSH approves filters for particulate materials on the basis of two major factors. These include their ability to withstand exposure to oily atmospheres where the particulates are mists or oily in nature. Dry particulates such as dusts would not need to have this level of rating. To classify the ability of the filter relative to oils, the following designations are used:

- N—An N rating on a filter indicates that it is *N*ot resistive to oily environments. The N rating would be suitable for use with dry dusts and fumes.
- R—An R rating on a filter indicates that it is somewhat *R*esistive to oily environments. R ratings are becoming less and less common because most applications where oily atmospheric conditions are encountered are best handled with P-rated units.
- P—A P rating on the filter indicates that it is largely oil *P*roof. Misty conditions where the atmospheric contamination may contain oils require this level of protection. The P level is also useful for non-oily conditions, making it the filter with the broadest range of protection.

> **■ Note**
> Particle respirators are rated to show their resistance to oily atmospheres. An N rating specifies that it is not resistive, an R rating shows that it is resistive, and a P rating shows it is oil proof.

The second factor related to approval by NIOSH is the ability of the respirator to trap particles of a specific size. To show this, the NIOSH ratings also include a number following the letter designation. These numbers are 95, 99, and 100. These numbers indicate the percentage of particulates that the unit will remove based on an exposure to particulates that are 0.3 micron in size. Recall that to reach the alveoli, the particulates must be smaller than 10 microns, so a filter that is capable of filtering down to the 0.3 micron is quite protective for many types of particles. In fact, actual testing shows them to be even more protective than the rating indicates because they are composed of not a single layer of protection, but rather a system of filters through which the air moves randomly and chaotically, which traps many smaller particles in the filtering media. Additionally, electrostatic differences between the particles and the filter materials also traps many particles smaller than the 0.3 micron rating.

> **■ Note**
> NIOSH approves filtering respirators based on their ability to block a particular sized particle and their ability to trap oily materials.

■ **Note**
NIOSH-approved filters are rated to remove particles down to 0.3 micron in size.

Given this, an N95 rating on the respirator indicates that it is rated to remove 95% of the non-oily particulates from the air that are greater than 0.3 micron in size. A 99 rating would indicate that it is capable of removing 99% of them, and the 100 rating would indicate that it is rated to remove 99.97% of them. Using all of the possible letter and number combinations, we find that we can have nine different types of particulate respirators. They include N95, N99, N100, R95, R99, R100, P95, P99, and P100. Of these, the P100 filter is the most protective because it gets both oily and non-oily materials at the highest concentration. Because of this, P100 is the designation that is now replacing the previously used term Highly Efficient Particulate Air (HEPA) Filter for most types of particulate respirators.

■ **Note**
NIOSH assigns numbers to the approved filters to identify the percentage of particulates that they will filter.

As these air-filtering units are used, they may become clogged like any other type of filter that removes particulates from the air. Like a filter on a ventilation system, the filters need to be changed at intervals in order to provide the most efficient and effective filtering of the contaminants. With filtering respirators, the first indication that the filter may be reaching its useful life expectancy is when the user recognizes that it is more difficult to draw air through the filter. This decrease in airflow through the unit is most often recognized by the user as difficulty in breathing. The user works harder to breathe through the clogged filter, resulting in the user expending more energy than normal. If this continues, the user may experience more rapid breathing and fatigue. Once this occurs, the user should leave the contaminated area and change the filter.

■ **Note**
Filter respirators must be changed when the process of breathing becomes difficult.

absorption
the process by which the material inside a cartridge traps the airborne contaminant by taking it into the medium

adsorption
the process by which the material inside a cartridge traps the airborne contaminant by attaching to the contaminant

Another type of air-purifying system is the one used for removing airborne contaminants such as vapors or gases. Unlike the filters that block the passage of the particulates from reaching the user, these types of airborne contaminants are filtered from the atmosphere with substances such as activated charcoal. This kind of filter *absorbs* or *adsorbs* the contaminants and traps them before they are inhaled. **Absorption** is the process of the chemical medium in the filter taking the contaminant into itself much like a sponge takes the water inside. **Adsorption** is a similar process where the contaminant attaches itself to the outside of the medium and does not pass through. While these two processes are different, they are also similar in that the filtering process is often the result of a chemical reaction to the electrical attraction between the medium in the filter and the airborne contaminants. In one case, the activated charcoal carries a positive or negative electrical charge depending on the filter type. The activated charcoal in acid gas filters carries a positive charge in order to attract the negatively charged ions that exist in acidic vapors. Due to the age-old concept of "opposites attract," the charged charcoal particles attract and hold the contaminants that are drawn past them. Depending on the purifying media and the materials involved, these types of systems can be extremely effective.

Because there is no one type of filtering media that can be used in the air-purifying cartridges, the type of contaminant that is in the atmosphere must be known. Each type of contaminant may require a different media in order for it to be removed. It is vital to understand that one of the primary limitations of an APR is that if the hazard is not fully known, this type of system is not to be used. Similar to playing Russian Roulette, using an APR when the hazard is not fully known means risking exposure to hazardous atmospheres which may not be removed by the particular purifying media. If a material has no odor, such as the case with carbon monoxide and a range of other materials, the users would be breathing contaminated air and working under the premise that they were safe. At no time can an air-purifying system be used in the presence of an unknown atmosphere.

> **Safety**
> In order to select the cartridge, the type of material in the atmosphere must be known.

Even when the specific material is known, there are other limitations to this type of respiratory protective system. While a whole range of cartridges is available for a host of substances, there are a number of airborne contaminants for which there is no approved cartridge. In addition to the respirator itself being approved, all cartridges must also be approved by NIOSH. OSHA mandates that all air-purifying cartridges be color coded regardless of the manufacturer, to describe with the color the types of air contaminants for which the cartridge is protective. Standard colors also help to show whether the user has selected the cartridge correctly. The list in Table 6-2 describes some of the more common types of cartridges by color with the materials they are rated for.

Table 6-2 *The standardized colors of commonly used cartridges.*

Type of cartridge:	Color	Typical protection against:
Organic vapors	Black	Organic vapors such as solvents
Acid gases	White	Chlorine, hydrogen chloride, sulfur dioxide, or formaldehyde
Organic vapors/ acid gases	Yellow	Combination of black and white cartridge materials: organic vapors, chlorine, hydrogen chloride, sulfur dioxide, or chlorine dioxide, hydrogen fluoride, or hydrogen sulfide (escape only)
P-100: Dusts, mists, fumes	Magenta (purple)	Dusts, fumes, and mists having a TWA less than 0.05 mg/m^3, including asbestos-containing dusts and mists, radon daughters attached to these dusts, and radionuclides
Organic vapors/ P-100	Black/ magenta	Organic vapors and dusts, fumes and mists with a TWA less than 0.05 mg/m^3, asbestos-containing dusts, mists, radon daughters attached to these dusts, particulate radionuclides, and pesticides
Organic vapors/ acid gases/ P-100	Yellow/ magenta	Organic vapors, acid gases and dusts, fumes, and mists having a TWA less than 0.05 mg/m^3 including asbestos-containing dusts and mists, radon daughters attached to these dusts, radionuclides, and pesticides
Metallic mercury vapors/chlorine	Orange	Effective against metallic mercury vapors or chlorine. Mercury vapor cartridges usually are equipped with an ESLI (end-of-service-life indicator) to provide adequate protection from mercury

■ Note
Standard colors are used to illustrate the types of material that a respirator cartridge is rated for.

Some newer types of cartridges are combination units that will be useful for more than one class of hazardous air contaminant. The most common example of this is the filter cartridge, P-100, stacked on top of a chemical filter. These stacked units are most often used in atmospheres containing a combination of particulates such as aerosols or dusts, and a vapor or gas. Newer gas cartridges also combine two or more classes of media in a single cartridge, such as when acid gas (white cartridge) and organic vapor (black cartridge) are combined into a unit that is yellow. As technology improves, the combination cartridges are becoming able to pick up a broader range of contaminants. At this point, however, there is not any one cartridge that would be useful for all possible combinations of airborne contaminants.

! Safety
There is no single cartridge that will protect the user from all types of airborne contaminants.

In addition to cartridge selection and availability, other limitations to these types of systems must be understood if we are to use them correctly. Even if we have the approved cartridge and have identified the specific contaminants, the level of hazard in the air can make the use of a cartridge respirator unsafe. Consider that the media in the cartridge will remove the contaminants until the media is used up or saturated. Once saturated with the contaminant, the media does not have the ability to remove additional materials from the air. If those levels of exposure were above the IDLH level, and if the cartridge were to become saturated, the air coming through the cartridge would be at an extremely high and unsafe level. For this reason, APRs cannot be used in areas where the airborne contaminant levels are above the IDLH level for the specific material present.

! Safety
APRs cannot be used when the level of contamination reaches IDLH.

Another limitation of this type of respiratory protection is the fact that APRs only remove materials from the air through filtering or purifying. They do not add anything to the air, so in an oxygen-deficient atmosphere the use of an APR would result in exposure to low levels of oxygen. As we know, there is no way for the normal senses to detect such unsafe conditions, so the atmosphere where the APR will be used must be evaluated to determine if an oxygen-deficient condition could be encountered. Often, this will require the use of air monitoring equipment, which will be studied in Chapter 11. It is vital to ensure that the level of oxygen in the air be above 19.5% before use of an APR can be considered.

! Safety
APRs cannot be used in oxygen-deficient atmospheres.

A final consideration in choosing an APR system is the status of the cartridge itself. In previous years, OSHA allowed certain approved cartridges to be used until the media became saturated from exposure. Media saturation was most often determined by the APR user, who detected, by sensing the **warning properties** inside the mask, such as by smell or irritation, that the contaminant had broken through the filter. Most often, the user then determined that the media was saturated and the cartridge needed to be replaced. When OSHA allowed this, the only approved cartridges were ones designed for materials that had warning properties. The warning properties alerted the user that it was time to change the cartridge. If the material did not have warning properties, the user would be exposed. With the updates to the Respiratory Protection regulation, OSHA no longer allows cartridges to be used until they are saturated, and now requires that one of two methods be used. These methods are outlined in the regulation as follows:

warning properties
properties such as odor, taste, or irritation that alert us to the presence of a material

- The presence of an **End-of-Service-Life Indicator** (ESLI): The ESLI is placed on the cartridge by the cartridge manufacturer and indicates when the media is reaching its saturation point. ESLI units may indicate this by a change in color on the outside of the cartridge. Once the color changes, the unit is required to be replaced.
- Absence of ESLI: OSHA now requires that "the employer implements a change schedule for canisters and cartridges that is based on objective information or data that will ensure that canisters and cartridges are changed before the end of their service life. The employer shall describe in the respirator program the information and data relied upon and the basis for the canister and cartridge change schedule and the basis for reliance on the data." This language, taken directly from the regulation, mandates that the cartridges be tracked by individual users to ensure that the time allowed by the employer's requirements is not exceeded. It also implies that individual cartridges must be protected from inadvertent exposure to materials present in the air. This may require that the partially used cartridges be kept in tightly sealed containers to ensure that they are not exposed to airborne contamination while in storage. It is never a good idea to allow a cartridge to be stored outside of a sealed container if that cartridge is to be used again because the media in the cartridge will continue to absorb or adsorb the materials that it is designed to.

End-of-Service-Life Indicator (ESLI)
a system that warns the user that the respirator cartridge is reaching the end of its useful life and needs to be changed

■ Note

When no ESLI is present, objective criteria must be developed to advise the user that the cartridge must be changed.

To help users determine how long a particular cartridge will be protective, some manufacturers and even OSHA have programs on their Web sites where an employer can insert information into a matrix that will help determine how long a particular cartridge can be used. These programs allow the employer to identify the conditions that will be encountered, such as temperature, humidity, physical exertion anticipated, levels of contamination, and air pressures present. Once the employer has inputted these factors, he then identifies the expected levels of contamination present and the type of cartridge to be used. Based on these variables, the evaluation program will give an estimate of how long the unit will be effective. Given their limitations, the assessments tend to be very conservative, requiring a cartridge to be changed out well before the potential for exposure can occur.

■ Note

Help in setting up a change out schedule for cartridges is available online at the Web sites of the OSHA as well as many of the major respirator manufacturers.

It should be noted that very few respirator cartridges have an ESLI. This means the responsibility for determining how long each cartridge can be used usually rests on the employer. For this reason, many employers will often default to using the cartridge once and then disposing of it. Given the variables in emergency response activities, this is a very appropriate protocol.

Regardless of the circumstances and programs used to determine the expected life of a particular cartridge, all employees should be instructed that detecting any odor or irritation to the eyes, nose, or mouth while wearing an APR is evidence of a problem indicating they should immediately leave the area. Such detection could indicate that the cartridge has reached its saturation level or that the APR mask system has developed a leak. In most APR equipment, a negative pressure is created inside the mask as the user inhales. This negative pressure draws air from the environment through the filters and into the mask, where it becomes available to the user. In the case of a poorly sealed mask or the failure of a mask component, the inhaled air may enter the mask through the openings or gaps in the mask seal, bypassing the filter media because that is the most direct way into the mask itself. This is why the user should be aware of any potential warning properties that the material may have, or be able to recognize symptoms of exposure to the material, and immediately leave the area! To stay would be to risk almost certain exposure to the material.

■ Note

A negative pressure is created inside the APR mask as the user inhales. This allows any leaks in the mask seal to accommodate the introduction of contamination inside the mask.

powered air-purifying respirator (PAPR)
a type of APR with a small motor that draws air through the filters and provides a positive pressure into the mask or hood

One variation of the standard APR, sometimes referred to as a positive pressure mask, is the **powered air-purifying respirator (PAPR)**. PAPR units work similar to standard APR units, with one major addition. Instead of user inhalation creating the negative pressure inside the mask that pulls the air through the cartridges, PAPR units are equipped with a small motor that draws air through the media and into the mask. Often the units are powered by a small battery pack that is worn on the belt with the motor assembly. The combined unit pulls the air through the cartridges and into the mask, creating positive pressure inside the mask. This positive pressure helps to reduce the potential for contaminants to enter the mask, as is the case with standard negative pressure APR units. Additionally, the filters used in some PAPR systems are often larger and have a greater capacity than smaller ones worn directly on the mask of the respirator, as is the case with standard APR masks. Some PAPR units also pump air into a hood system that is not sealed directly on the face. This allows the use of respiratory protection equipment in cases where a standard, tight-fitting mask cannot be used. Such would be the case where someone has facial hair or some condition that precludes the use of a standard respirator. An example of a PAPR unit is found in Figure 6-3.

Finally, because of their limitation to only filter specific contaminants, it is clear that the use of any type of APR system would not be appropriate when extreme thermal conditions are encountered. Because the mask of an

Figure 6-3
A PAPR unit being worn with a loose-fitting (hood) respirator.

APR only filters the specific chemical hazard, air-purifying systems do little to provide thermal protection when extremely hot or cold environments are encountered.

Despite their limitations, APR systems do have some distinct and important advantages, and their use may be the most appropriate in many types of circumstances. Some of the advantages of APRs include the following:

- Air-purifying systems are small, lightweight, and allow freedom of movement far beyond what other types of respirators allow. Because of their size, APRs don't prevent the wearer from entering tight areas, nor do they weigh the wearer down or change the wearer's center of gravity.
- APRs require far less training than is needed to use other types of respirators. This is particularly true when another higher-trained person specifies the type of cartridge required.
- Air-purifying systems need considerably less maintenance than other types of respirators. Often the maintenance involves simply cleaning the mask and replacing the cartridge.
- An air-purifying system costs are significantly less than an air-supplying system. This cost includes both initial purchase and ongoing maintenance.

Types of Air-Purifying Systems

While all the APR systems are similar in the way they work, there are some variations. These variations provide additional flexibility to those who use the APR systems.

Half-Face Air-Purifying Respirators The first type of APR is called the half-face respirator because as the name describes, it covers only part of the face. An example of one of these units is found in Figure 6-4. As can be seen, the half-face respirator provides a mask covering just the mouth and nose.

Figure 6-4
Worker wearing a half-face APR.

protection factor
a number OSHA assigns to respiratory protection equipment to illustrate the level of protection offered

Like the other types of respiratory protection, a half-face respirator is attached to the face by straps that go around the head. Unlike full-face respirators, the straps of the half-face respirator do not ensure as tight a fit, which limits its use to conditions where the atmospheric hazards are no higher than 10 times the PEL. This limitation is termed the **protection factor** and varies among the different types of respiratory protection equipment in our study. Protection factors are assigned by OSHA for specific types of respiratory protection equipment and represent the minimum anticipated protection provided by a properly functioning respirator or class of respirators to a given percentage of properly fitted and trained users. The half-face APR is assigned the lowest protection factor because it does not fit the face as securely as other units.

While it does have some limitations, the half-face respirator offers a number of advantages over the full-face type. These include the following:

- Half-face respirators are quite easy to use and require less training than other respirators.
- Half-face units are some of the most comfortable to wear because they are not as tight as full-face units, do not limit visibility, and are usually softer and more pliable than other types of units. Because of these advantages, people are more likely to wear them.
- Because half-face units do not cover the eyes, peripheral vision is unrestricted and fogging is not a problem.
- The half-face mask allows the user to wear standard eyeglasses without any additional equipment or accommodation.

That half-face systems are not without problems must be understood prior to selection of these types of units. These disadvantages include the following:

- Half-face respirators do not provide protection for the eyes or face. This limitation may mean the user must wear some other form of eye protection because many types of respiratory hazards are also hazardous to the eyes and mucous membranes.
- The straps on a half-face respirator do not allow the user to fit the respirator as tightly to the face as is possible with full-face units. Because of this, the protection factor for these units is less than for other types of respiratory protection equipment.
- Because the half-face respirator is an air-purifying type, it has the same limitations as other air-purifying systems.

Full-Face Air-Purifying Respirators The full-face APR attaches to the head and face more securely than the half-face units, with the mask covering the entire face from forehead to chin. It typically utilizes either four or five straps to attach the mask securely to the head. With this arrangement of straps, the full-face unit fits tightly on the face of the wearer. Figure 6-5 shows a worker wearing a typical full-face APR unit.

Many full-face units have an additional nose cup built into the mask. The nose cup looks a little like the half-face mask but provides two important functions. The addition of the nose cup provides a second seal within the mask, which helps limit contamination and reduces the potential for fogging. Many full-face APR units are the negative pressure type, meaning fogging is a common problem,

Figure 6-5
Worker wearing a full-face APR.

as the moist exhaled air and sweat from the user condenses on the inside of the mask. The nose cup helps channel the high moisture content from the expired air into the exhalation valve and out of the mask. Once fogging occurs inside a full-face negative pressure mask, there is no effective means of defogging the mask other than to remove it and clean it manually. If the full-face mask is part of a PAPR system, there will be far less fogging because the PAPR system creates air movement inside the mask, which forces some of the moisture out. Advantages of the full-face air-purifying masks over the half-face models include the following:

- The full-face APR units have tighter seals, providing a higher protection factor. This allows the typical tight-fitting APR mask to be worn in levels of exposure that are up to 50 times the PEL for the substance encountered.
- Some full-face APR systems have the added advantage of being part of a PAPR system that provides positive pressure inside the mask and further increases the protection factor along with other advantages of the PAPR system.
- A full-face system provides full protection for the face and eyes, eliminating the need for other types of facial protection. In many cases, protection of the eyes and face when wearing a full-face mask is superior to that provided by typical safety glasses, goggles, and face shields.
- Some manufacturers have conversion kits to adapt the full-face air-purifying units to their supplied-air units. Such conversion kits reduce the need to have two types of masks and to perform fit testing with two different masks.

Disadvantages of the full-face units are primarily problems associated with air-purifying systems in general. Other disadvantages of the full-face over the half-face units include the following:

- Because of the way the mask is configured, full-face units are often prone to fogging unless they are part of a PAPR system.
- A loss of peripheral vision is common in a full-face mask. The amount of loss varies with the design, but can be significant with some types of masks.
- Full-face units are more expensive than comparable half-face models.

Supplied-Air Respirators

supplied-air respirator (SAR)
supplies clean air from an outside source to the user through either a tank containing air or a hose line connected to an air source

A **supplied-air respirator (SAR)** provides the highest level of respiratory protection when an inhalation hazard is present. Supplied-air units derive their name from their function. These units *supply* air to the wearer, and like APR systems, they have two major types. These include the self-contained breathing apparatus (SCBA), which is shown in Figure 6-6, and the air-line or umbilical system, which is shown in Figure 6-7. Each of these units works in a similar way in that the air that is being breathed is air that comes from outside the contamination area. In each case, the air supplied to the mask comes from a tank and hose configuration, air compressor, or a combination of the two. Because the units provide clean air directly to the user, they do not have the same limitations as air-purifying systems. In either type of SAR, the air breathed by the wearer is independent of the atmospheric conditions of the environment.

To ensure the quality of the breathing air, OSHA specifies that air provided to a SAR system meet the specifications for Grade D breathing air as described in ANSI/Compressed Gas Association Commodity Specification for Air, G-7.1-1989. The breathing air that is supplied to the user comes directly or indirectly from an air compressor. The air provided must be free of harmful

Figure 6-6 *Worker with an SCBA unit.*

levels of contaminants that exist when compressors are used to supply air either directly to the wearer or into a tank. Examples of potential contaminants from a compressor include carbon monoxide and vaporized oils or lubricants liberated from the operating compressor. NFPA 1404 states that the air quality of all breathing-air systems should be tested at least every 3 months by a qualified laboratory. This is not specified in the OSHA regulation.

> **■ Note**
> All breathing air is required to meet national standards in order to be used in a SAR system.

A properly functioning SAR of either type provides a number of advantages over the air-purifying type of respirator. These advantages include:

- SARs provide protection from all types of dusts, fumes, mists, gases, vapors, and particulates. They are not product specific and can be used when the atmospheric hazards are unknown.
- The concentration of the contaminant in the air can be unknown. Contamination levels above IDLH are not an issue for a properly fitted and functioning SAR. In fact, OSHA mandates the use of a SAR when contamination levels are expected to be above the IDLH levels.
- The oxygen level in an area does not rule out the use of a SAR. Because SARs supply their own air, the amount of oxygen is known to be at safe levels even when worn in oxygen-deficient atmospheres.

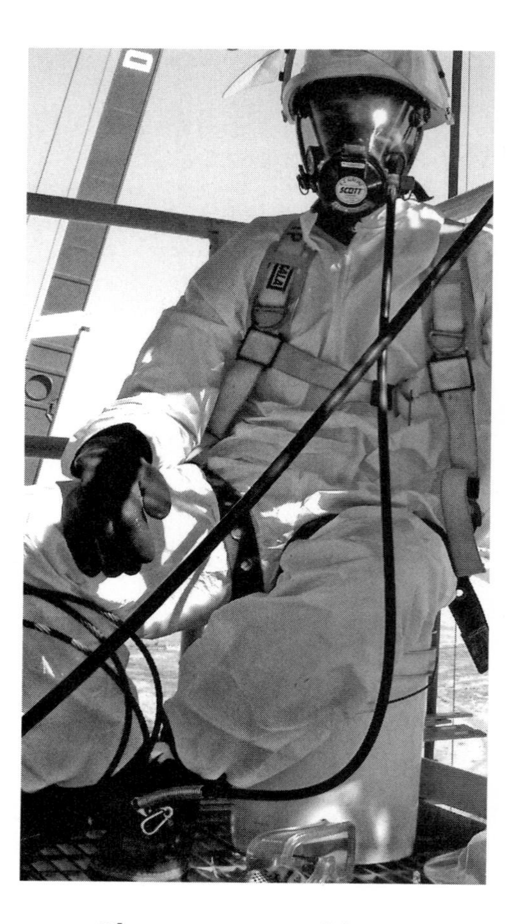

Figure 6-7 *Picture of an air-line system set up for use.*

- The temperature of the air in the hazard zone is of little concern to the user of a SAR. The air in the tank or lines is typically not subject to heating if these units are used in extremely hot (or cold) environments. Firefighters use these units where temperatures exceed 1000°F.
- A SAR provides positive pressure in the mask. This positive pressure increases the protection provided to the user because there is less chance that hazardous contaminants can be drawn into the mask even if a leak is present.
- A SAR provides the maximum level of respiratory protection against the five types of respiratory hazards. This is the highest level of protection available and is mandated for use when the airborne hazards are unknown or extreme.

> **■ Note**
> SAR systems provide protection against all five major types of airborne hazards.

> **■ Note**
> A SAR will provide positive pressure in the mask. SAR systems are the highest level of respiratory protection available.

While SARs are well suited for almost all types of hazardous atmospheres, the decision to use a SAR is not automatic. These units have some distinct disadvantages that limit their application. While they are the *highest* level of protection, the disadvantages do not always make them the *safest* level of protection in all circumstances. Safety is a function of many factors and it is generally best to try to use the *lowest* level of respiratory protection that will provide the necessary degree of safety. This concept will become clear in the Chapter 7 discussion of the levels of PPE. More protection is not always safer because the addition of more equipment will increase weight, limit time, and cause other factors to come into play.

> **■ Note**
> While SARs are the highest level of respiratory protection, they are not necessarily the safest. Many factors must be considered before using a SAR.

Examples of some of the limiting factors with a SAR include the following:

- A SAR is awkward to wear and limits the mobility of the user. This can limit the user's ability to escape in the event of an emergency.
- The use of either type of SAR will add weight to the user. The additional weight can cause fatigue and further reduce the effectiveness of the worker. For example, SCBA units are heavy, some weighing up to 35 pounds. This weight can take a toll on the wearer, especially when added to the weight of other protective equipment.
- SAR equipment is more complex than APR units. This complexity requires the user to have considerably more training on the use, care, and maintenance of the units.
- Because of their complexity, SAR units have higher maintenance requirements than APR units. This increased maintenance results in increased down time, higher costs, and increased potential for failure of the units.
- The use of any type of SAR will result in a reduction in mobility and access in some cases. The user wears a large tank on their back or has to pull a hose line around, which can restrict or even prevent entry into some areas where the simple APR mask would not.
- The use of an SCBA will significantly limit the time during which the user can work. Most commonly used SCBA units are rated to provide between 30 and 60 minutes of air, although some longer-lasting units are available. Although the units are rated to provide this amount of time, the actual amount of time that they provide is between 20 and 40 minutes. This is diminished even further when positive pressure units are used and small leaks occur.
- All types of SAR systems are expensive to purchase and maintain. A single SCBA unit can cost up to $3000. Air-line systems often cost considerably more.
- With the air-line systems, the air supply could be interrupted due to outside factors such as the loss of a compressor or broken equipment. Additionally, there may be a maximum amount of air line that can be used with the system. That maximum is defined by the manufacturer.

> **■ Note**
> SAR units add weight and bulk to the user that can limit access to areas and reduce the mobility of the user.

> **■ Note**
> SAR systems are considerably more complex, cost more, and require more maintenance than APR units.

Self-Contained Breathing Apparatus Perhaps the most common of the two types of SAR systems is the SCBA unit. With an SCBA, the air is stored in a tank worn by the user. Everything that is needed is contained on the pack assembly, hence the term **self-contained units**. The tank holding the air is carried on the back and held in place with a system of straps. The tank supplies air through hoses and pressure regulators to the mask of the user. All of the air breathed comes directly from the tank with no appreciable air coming into the mask from the outside atmosphere. Everything needed to ensure that the air is safe to breathe is contained on the wearer.

self-contained units
breathing apparatus where the user carries all components required for the system

While there are some major differences in the types of systems available, all SCBA units share one common feature: a low-pressure warning device. The low-pressure warning device or alarm alerts the user when the amount of air left in the system gets low. In most cases, activation of the low-pressure alarm indicates that approximately 5 minutes of air remain. Once the warning device activates, the user should leave the contaminated area. Depending on the manufacturer, the warning device can be a bell, a whistle, or a constant vibration of the regulator, indicating that the system is low on air.

■ Note

A low-pressure warning is standard on SCBA units to alert the users that the air left in the system is low and that they need to exit the area.

SCBA units come in a variety of configurations and models from a host of manufacturers. Generally, they are divided into two major subgroups. The two groups are differentiated by the manner in which the air is provided to the user and are called either open-circuit or closed-circuit systems. In the open-circuit type, Grade D compressed air is held in the cylinder and available for use. Typical pressure in the cylinder ranges from 2216 pounds per square inch (psi) to 4500 psi. The pressure and size of the tank provide the required amount of breathing air to the user for the time limits of the system. A standard size cylinder with 2216 psi contains approximately 47 cubic feet of air. This provides the user with approximately 30 minutes of work time, although actual use varies depending on a number of factors.

If the pressure in the cylinder is increased, more air will be available, increasing the time that the unit will provide protection. A doubling of the pressure in a standard cylinder will roughly double the time the unit can be used. In some cases, the pressures are doubled and the size of the cylinder is decreased. This smaller cylinder weighs less yet can provide the user the same amount of time because it contains the same volume of air. In addition to weighing less and reducing the fatigue of the user, smaller cylinders can allow access into more confined areas. Because the time rating of the system is dependent on the amount of cubic feet of air in the cylinder, the smaller tanks can have the same rating as larger ones. Typically, a 60-minute tank will be similar in size to a 30-minute tank, but will have double the pressure.

■ Note

SCBA cylinders can be low or high pressure and be of different sizes. The combination of the size of the cylinder and the pressure inside determines the amount of breathing air available.

Whether high or low pressure, the basic SCBA unit provides air to a pressure regulator or series of regulators, which govern the necessary airflow into

demand systems supply air into the mask as the user inhales and creates a negative pressure inside the mask

pressure-demand systems supply air into the mask as the user inhales and monitor for leakage; if the system detects a leak, it will provide additional air and create positive pressure inside the mask

open-circuit unit allows exhaled air to leave the mask and not be reused

closed-circuit system reuses the exhaled air by recycling it through filters inside the backpack where more oxygen is added to it

the mask. Regulated air enters the mask under proper pressure, usually when the user inhales, because the system senses the pressure decrease in the mask. These types of units are termed **demand systems** because they only provide air when *demanded* by the users as they inhale. The unit will monitor the pressure inside the mask and fill it when the demand is there. Additionally, the unit will also sense a leak in the seal and fill the mask with positive pressure to prevent the user from inhaling hazardous substances through the leaks. When combined with a demand system, these units are termed **pressure-demand** SCBA units.

With either high-pressure or low-pressure systems, when the user exhales, air leaves through an exhalation valve usually located at the bottom of the mask. This is termed an **open-circuit unit** because the air simply flows outside of the unit once it has been used. While some oxygen is still present in the exhaled air, the system simply allows it to go out of the mask and not be reused in any way.

The second major type of SCBA system is called a **closed-circuit** system. These types of units have been in use for a number of years and their favor with various groups comes and goes. A closed-circuit respirator, also referred to as a rebreather system, is composed of a backpack assembly that contains a small cylinder of oxygen, filter system, and reservoirs. In a closed-circuit system, air is taken into the mask in a similar manner as the open-circuit systems. The difference between the two systems occurs when the exhaled air leaves the mask. With the closed-circuit system, exhaled air is captured as it leaves the mask and is reused as it returns from the backpack assembly. Inside the backpack, the exhaled air that still contains oxygen is filtered to remove carbon dioxide, additional oxygen is added from a small cylinder located in the backpack, and air is recycled back into the system. In contrast, the open-circuit unit uses only compressed air with no additional oxygen.

The primary advantage of the closed-circuit system is that time ratings of up to several hours are possible. This increase in time is largely due to the recycling of expired air and is similar to units used in the space program by NASA.

■ Note

Closed-circuit systems are capable of providing several hours of use.

While such advantages would seem to be significant, closed-circuit units have largely disappeared from the hazardous materials industry. The closed-circuit units are costly, require considerable maintenance, and create concern due to the introduction of a strong oxidizer (oxygen) into potentially hazardous atmospheres where the units are worn. While they do have the added advantage of increasing the amount of time the user can work with a self-contained unit, the reality is that a cylinder change often provides a much-needed break for the user. Allowing someone to work in a hazardous atmosphere with additional PPE can create significant stress on the user, which can be alleviated during a break to change the cylinder.

■ Note

Cylinder changes allow the user to get a break from the activities.

Whether open-circuit or closed-circuit, SCBA units are the leading type of SARs. SCBA units provide users maximum flexibility when entering hazardous areas because their breathing air is contained on their backs and no additional equipment or personnel are required. SCBA units can be placed into service in a relatively short period and provide the maximum level of respiratory protection available.

> **■ Note**
> SCBA units provide a high level of protection and flexibility.

However, SCBA units are not without their disadvantages. These include:

- Most commonly used types of SCBA units are limited in the amount of time they provide because of the air supply limitations. Most often, work using an SCBA unit is limited to less than 1 hour.
- SCBA units are some of the heaviest types of respiratory protection equipment. Their weight can increase the physical demands on the user and cause the user to have a higher center of gravity, leading to loss of balance.
- Because of its size, an SCBA unit can limit a user's access to certain areas.
- SCBA units are complex and require considerable maintenance and service.

> **■ Note**
> SCBA units add considerable physical stress to the users, and the amount of time that they can be used is limited.

Air-Line Respirators An air-line respirator is the second major type of SAR and involves the use of a supply line that is continuously attached from the air source to the user. With this system the air line is pulled behind the worker and acts as an umbilical cord, supplying a constant flow of air into the mask. Unlike the SCBA unit, these units do not require the user to carry the air supply on their back. The air supplied to these units is located outside the hazardous area and is supplied to the user through the air hose that follows the user. The air line is attached to the air supply, which can either be a compressor with rated and approved air, or a series of large bottles of breathing air. The other end of the hose is attached to a regulator usually worn on the belt of the user. The regulator provides air directly into the mask. Because time limits are not generally a concern, many air-line respirators supply a constant flow of air into the mask, creating a positive pressure inside the mask. The constant air movement in the mask also cools the user's face and limits heat stress. Air-line systems are always open-circuit systems.

A typical air-line system is shown in Figure 6-7. In the picture, we see the only hardware carried by the user is the small regulator where the line attaches to the belt of the user, the facemask, and a small cylinder of compressed air, also attached to the belt. The small air cylinder, known as the **escape pack,** provides an emergency source of air if the primary supply is compromised, or if escape from the area is necessary. OSHA requires that an escape pack be included with an air-line system in an area where the levels of contamination are above IDLH value.

escape pack
a small cylinder worn on the belt of the user of an air-line system to allow the user to disconnect from the air line in the event of an emergency

> **■ Note**
> OSHA requires that an escape pack be included with an air-line system when the level of contamination reaches IDLH.

The escape pack is essentially a smaller version of an open-circuit SCBA unit in that it provides air directly from the cylinder into the mask. Its use is indicated when the supply of air is cut off, most often due to hose failure, kinking, or mechanical failure of some part of the system. It is also used when the worker needs to leave an area by a different route than the one used to enter. When the escape pack is needed, the user is required to turn on the small cylinder that provides air into the regulator and into the mask. Once the cylinder is activated, the user disconnects the air line from the regulator and is free to leave the area. It should be noted that the escape bottle should only be used in the event of an emergency or unforeseen circumstance. It should not be used to simply get free from the air line and go into areas where access may be limited, even for a minute. Most commonly, the escape bottles provide only a 5-minute air supply, giving the user just enough air to evacuate the area and get to a safe area for decontamination.

> **Note**
> Escape packs should only be used in the event of an emergency or other unplanned event.

There are some advantages and disadvantages of the air-line systems over the SCBA units. Like SCBA, air-line systems provide the highest level of respiratory protection, and with the air supply from outside the hazardous area, they can do this for the longest time. No other type of respirator can provide the highest level of respiratory protection while offering an almost unlimited work time. Additionally, these systems do not have the disadvantage of adding a great deal of bulk or weight on the worker as the SCBA units do.

While it may seem that this type of unit is the ultimate answer, there are limits to their use and effectiveness. The major disadvantages include the following:

- Use of these types of units requires that the area to be entered be largely clear of obstacles and open to allow the air line to be pulled behind the user without becoming entangled in items found in the area.
- The units are generally limited in distance, with air lines typically less than 200 feet long.
- Use of these units typically involves including another person to watch the air supply. This **bottle watch** is a person who monitors the outside air supply, whether from a compressor or series of air tanks, and ensures that a constant supply of air is maintained.
- The complexity of the units causes them to be considerably expensive, both in the initial purchase and in the ongoing maintenance. All of the equipment used must be cleaned and serviced after each use. If the hose is used in an area high in contamination, it will require considerable decontamination and inspection prior to reuse.

bottle watch
a person who monitors the outside air supply when an air-line system is in use

RESPIRATOR SELECTION AND USE

The type of atmosphere or contaminant that will be encountered dictates the type of respiratory protection that must be worn. Although we have highlighted some of the conditions for using an APR, the following criteria should reinforce the concepts. Whether you are dealing with a hazardous material spilled on the roadway, or a waste site requiring cleanup, the rules for selecting the proper respiratory protection are the same. Using the wrong type of equipment can be more harmful than using no protection at all if it leads to a false sense of security.

It may seem that simply wearing a SAR unit will resolve potential problems in any hazardous atmosphere because a SAR is the highest level of respiratory protection available. This reasoning is flawed because more protection is not always the best choice when less protection will provide the necessary degree of safety. For example, using an SCBA unit in a situation where no respirator is necessary, or where a half-face APR would provide adequate protection, means unnecessarily giving up a degree of safety and comfort, and adds cost. As we know, all types of respiratory protective equipment have disadvantages, and there is no perfect system. Recall that wearing an SCBA unit will restrict mobility, upset the center of gravity, and require significant care, maintenance, and training. Also, SCBA units have serious limitations relative to the utility time. It is critical to understand that the safest level is not always the most protective level. So the selection of the equipment should be based on choosing the lowest level of respiratory protection that is appropriate for the hazard and not to simply default to the highest level. Professionals working in the field have an understanding of the hazards and choices in equipment. They use this knowledge to select the proper equipment that is most appropriate to the hazards encountered. When encountering a situation where the hazards are not fully known, or when the type of equipment to select is uncertain, then clearly the use of the higher levels of protection may be indicated. But before any selection of equipment is made, a thought process of the hazards should be undertaken.

> **■ Note**
> The use of more protection than we need can add hazards to the operation.

The selection of the proper equipment is not a difficult process. While it is not a perfect system, the following series of questions provide a good starting point from which to begin the decision process for the selection of the proper type of respiratory protection equipment. Again, if there is ever any doubt regarding the answer to any of these questions, it is always best to be more protective. However, an accurate answer to the questions will provide you with the best choices when it comes to which type of protection is best.

1. Is there any type of respiratory hazard present in the work area?

   The first thing is to ascertain the presence of an atmospheric hazard in the area where work will be performed. This could be an area in a facility where the materials are used or stored, or at the scene of a hazardous substance release. Not all situations where hazardous substances are present pose a threat to the atmosphere. Recall that not all substances are inhalation hazards and therefore do not always require the use of respiratory protection.

   > **■ Note**
   > The first issue of respirator selection is ascertaining whether an inhalation hazard is present.

   The evaluation of the area should be done from outside the contaminated area. Visible indicators and a review of the circumstances most often will alert us to the possibility of a hazard. Visible clouds, odors, and dusts are often the most obvious clues that some type of respiratory protection may be required. Any type of release of a hazardous substance inside a building is a greater concern than a release in an open area, even though outside activities can also present hazards. An indoor release with poor ventilation can be highly hazardous due to the accumulation of vapors compounded by the potentially limited air movement in an enclosed area. The evaluation should be as objective as possible and include the use of visible indicators, a review of the potential materials involved and their ability to create an atmospheric hazard, and even monitoring equipment. Remember that not all materials can be detected with the nose alone because the odor threshold for a material may be lower than the PEL. Just because an odor is not present does not mean that hazardous materials are not present.

> **■ Note**
> It is critical to determine the presence of a hazard from outside the area before entry is made.

> **■ Note**
> Vapors are indicators of a release, as are odors. Monitoring equipment should also be used if available.

Once we have reason to believe that a hazardous atmosphere is present, we need to determine if the level could pose a hazard. Air monitoring equipment may be required to accurately determine which airborne contaminant is there and if the levels of airborne contamination are hazardous. Work on the assumption that the levels could pose a hazard until good air monitoring data becomes available. To do otherwise might expose you to levels that could cause harm. However, if the air monitoring shows that a respiratory hazard is present, but the levels of hazard are below safe limits such as the PEL, it is wise to consider that the most appropriate level of respiratory protection may in fact be no respirator at all. The hazard could still be one that would require the use of other types of PPE, but if there are no respiratory hazards present, no respiratory protection is necessary and the decision process can stop at this point. Remember that the presence of an odor may not indicate the presence of a hazardous level.

> **■ Note**
> If levels of airborne contamination are lower than the PEL, the safest level of respiratory protection may be no respirator at all.

However, if the answer whether a hazard is present is "yes," or you do not know for certain that a respiratory hazard is present, proceed to the next question to determine the proper type of respirator to be worn.

2. What specific material(s) may be present in the area to be entered?

The second question involves determining what specific materials may be present at the work site. Keys to making this determination include identification systems such as labels and placards, the location of the activity and its relation to specific hazardous materials, shipping papers or manifests, the type of containers, and information obtained from witnesses or workers at the site. In fixed facilities, we can often obtain a great deal of information by talking to the people who work in the area, or even those who have worked in those areas previously. Often, people who work in the area have specific knowledge regarding the materials that are or could be present at the time of the release. In other instances, we can obtain information about the materials when we study the type of processes that ordinarily take place in the area or when we review the records of the previous uses of the site. This is particularly helpful in the cleanup of an abandoned waste site. A review of the records often reveals the processes or materials used.

Note
A number of sources can indicate what material(s) may be present. These include labels, placards, shipping documents, containers, and witnesses or workers at the site.

If in reviewing the materials you can specifically identify the substance, you can then proceed with the rest of the decision process and answer the remaining questions to determine if an APR is still an option. While the exact material or product does not need to be known, the specific type or class of material is essential in the selection of the appropriate cartridge. If, on the other hand, you cannot positively identify the material classes, you must use a level of respiratory protection that is not product specific, the SAR.

Safety
If you cannot determine what is present, you must default to a SAR.

3. Is an approved APR cartridge available for the expected contaminant?

   In order to use an APR, you must have an approved cartridge. Find out from the equipment manufacturer whether there is an approved cartridge for the hazard present. All of the cartridges allowed are listed and approved by NIOSH and can be utilized if the factors for use are favorable. If there are no approved respirator cartridges available, you can effectively exclude the APR from our choices. On the other hand, if an approved cartridge is available, you can proceed with the next question in the decision process.

Note
A NIOSH-approved cartridge must be available for the material(s) present in order to use an APR.

4. Is the amount of airborne contamination present above the IDLH level?

   Before using the approved cartridge, confirm that the amount of airborne contamination is below the IDLH level. This step could require using air monitoring equipment or other evaluative measures to determine how much contamination is present. Determining the level of hazard present is relatively easy if air monitoring equipment is available and samples of the atmosphere can be obtained from outside the hazardous areas. Often, remote sensors may already be present in some buildings or areas, which can help determine the level of hazard. In other cases, remote readings might be obtained by instruments that can be dropped or extended into the area.

■ Note
Air monitoring can confirm the levels of airborne contamination.

In some situations, the lack of potential for an IDLH atmosphere can be determined by a review of the material involved and the circumstances of the release. For example, an accident that results in the release of sulfuric acid from a 55-gallon drum in an open area might be determined to be well below IDLH levels because of the openness of the area, the inability of the material to release vapors under the conditions encountered, and wind conditions that could allow personnel to work with the wind directing any vapors away from them.

■ Note
The location of the release and type of material present can also help determine that IDLH conditions are not present.

If it can be determined through analysis of the conditions or monitoring results that the level of airborne contaminants are above the IDLH for the material, a SAR system is required. If, however, the levels are below the IDLH, you can still consider using an APR and proceed to answer the next question in the decision process.

! Safety
APR units cannot be used when the level is above IDLH.

5. Is there any type of thermal extreme that would prohibit the use of an APR?

   While not commonly encountered, the potential for a thermal condition to create an atmospheric hazard should always be considered. While an APR is suitable to remove many contaminants from the atmosphere, they do little or nothing to remove the hazards associated with high or low temperatures, making a supplied-air system necessary. If high or low temperature extremes are encountered, APR units should not be used.

■ Note
High or low temperatures will restrict the use of an APR.

6. Are oxygen levels in the area to be entered adequate?

   Recall that the normal atmosphere contains approximately 20.9% oxygen. Levels below 19.5% are classified as oxygen deficient by OSHA and require the use of a SAR. The potential for oxygen deficiency present is enough to warrant using the SAR because the consequences of exposure to low atmospheric oxygen levels can be rapidly incapacitating and result in death. An APR will not add oxygen to the air, and cannot be used in atmospheres that are oxygen deficient. In these circumstances the SAR must be selected.

The use of a SAR is required when oxygen-deficient conditions exist.

Respiratory Protection Specifics

No discussion of respiratory protection would be complete without additional detail on some of the topics that have been presented in this chapter. The information is important because it will provide expanded details on parts of the respiratory protection requirements that often can be confusing or unclear. Without a full understanding of these areas, you could very easily experience a false sense of security and risk exposure to hazards that are avoidable.

donning
putting a respirator on

doffing
taking a respirator off

Using an APR The terms **donning** and **doffing** are often used to describe putting on (donning) and taking off (doffing) a respirator. Although there are many manufacturers, most APR masks are similar in design and mostly standard in donning and doffing procedures. The following procedures are generally accepted. Despite the standardization of equipment, it is always appropriate to consult the information sheets that come with the equipment at the time of purchase. In all cases, the manufacturer will have the most complete and most accurate information, which will help in donning and doffing. Figure 6-8 illustrates standard procedures for using a typical full-face respirator mask.

An important step, shown in Figure 6-8d, involves using a field fit test to confirm that the mask has a good seal on the face of the user. This step is required to be performed for all types of respiratory protection equipment that have a tight-fitting mask. Testing involves either a positive pressure mask check or a negative pressure mask check. In some cases, depending on the type and manufacturer of the mask worn, only one of these tests can be performed. In other cases, it may be possible to perform both. Some agencies require that both tests be done, if possible. While there is some variation in procedures for various masks, the standard method is as follows.

> **■ Note**
> Field fit testing is required whenever any type of mask is worn. It helps ensure that a tight fit is present.

To conduct a positive pressure fit test, secure the mask to your face and tighten the straps in the approved manner—generally the bottom ones first, working your way up.

1. After the mask is secure, place your hand over the exhalation valve cover, usually located at the bottom of the mask. In some SARs, the intake for the air supply may also have to be sealed to perform this test.
2. With your hand over the exhalation valve cover, exhale with enough force to slightly push the mask away from your face. Once the mask has lifted away from the face, but still sealed to your skin, stop blowing and watch to see if the unit stays pushed away from your face by the positive pressure created in the mask, or it returns to its relaxed position.
3. If the mask returns to its position on the face, it could indicate a leak in the seal and require that the mask be repositioned and/or that the straps be tighten further. Once the mask and/or straps have been repositioned, the test should be repeated.

> **■ Note**
> Positive pressure field fit testing involves exhaling into the mask after sealing the exhalation valve.

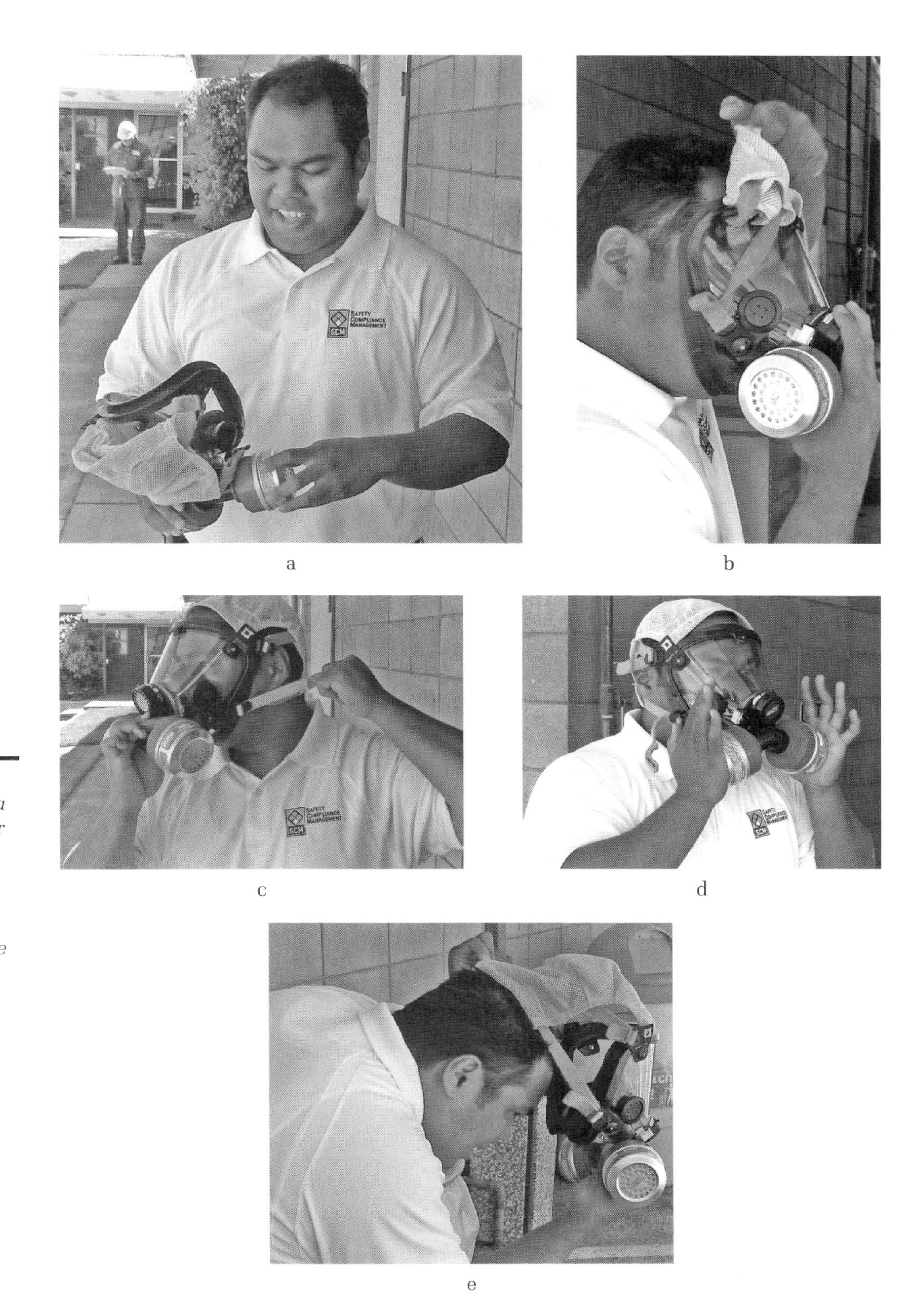

Figure 6-8
Steps in donning a full-face respirator mask: (a) secure the appropriate cartridge to the respirator and tighten it; (b) place the mask to your face, chin first; (c) tighten the straps, beginning with bottom ones, to ensure a secure fit; (d) perform a field fit test; and (e) doff the mask while holding it outward and away from your face to reduce contamination.

To conduct a negative pressure fit test, secure the mask to your face and tighten the straps in the approved manner.

1. Once the mask is secured on your face, seal the inlets for the air to enter the mask with your hand(s). In the case of an APR, the inlets for the air are the areas where the cartridges attach. With an SCBA unit, seal the air

intake line or cover the hole in the mask where the regulator attaches. Some types of SCBA units do not allow this test to be conducted in this way. In those cases, information provided by the manufacturer will describe the appropriate procedure.

2. Once the inlets are sealed, create negative pressure inside the mask by sucking air in through your mouth and nose as you would taking a breath. After the mask pulls into your face, stop sucking and watch to see if the mask stays sealed to the face.
3. If the mask pulls away from your face, additional tightening or readjusting will be required. If it stays tight on your face, the seal is adequate and the respirator is ready for use.

■ Note

A negative pressure field fit test is done by sealing the intakes for air into the mask, and by sucking in. The mask should draw tightly on the face and stay there after sucking stops.

Doffing of most types of masks is relatively simple. The procedure usually requires that you loosen the bottom strap or straps, pull the unit away from your face, and pull your head out. As shown in Figure 6-8e, do not pull the mask above your head while doffing but rather outward and away from your face. This will minimize the amount of any contaminants present on the outside of the mask raining down on the your face as you remove the mask. This should be done even if a formal decontamination procedure has been performed because any contamination that gets onto the face can expose you to both inhalation and contact hazards.

! Safety

When removing a mask, take care to not pull the mask above the face because contaminants present can fall onto your face and enter into your respiratory system.

Using a Supplied-Air Respirator Unfortunately, using SARs is not as easy or as standard as using the simpler APR systems. Earlier we mentioned that using SARs often requires considerably more training than using APR units. The complexity and variation of these units clearly demonstrates that without additional training, it would be difficult to fully understand each type of unit and its operation. For this reason, anyone who is preparing to use an SCBA or air-line system must consult the manufacturer's instructions and information regarding the proper donning and doffing procedures. The placement, removal, and field fit testing are similar between the two types of systems and can be done with the SAR mask.

■ Note

Procedures for using a SAR vary considerably and require that the manufacturer's instructions be followed.

MAINTENANCE, STORAGE, AND RECORDKEEPING

The best equipment and programs available will not provide the needed protection without some type of system of ongoing maintenance and care. Proper maintenance of respiratory protection equipment starts with a commitment to follow the requirements of the OSHA regulation and the manufacturer's instructions. Each of these requires that the units be regularly cleaned and disinfected to ensure that they do not expose the user to unnecessary harm. Respirator masks used by more than one worker must be thoroughly cleaned and disinfected after each use and before anyone else is allowed to use the equipment. Cleaning and disinfecting procedures for each type of equipment are available in the manufacturer's instruction and maintenance manuals provided with the equipment. Additionally, the OSHA Respiratory Protection regulation has a mandatory appendix that provides general information on the proper cleaning and maintenance of equipment if the manufacturer's information is not available.

■ Note
All respiratory protective equipment must be maintained in accordance with the manufacturer's requirements.

! Safety
Masks used by more than one person must be properly disinfected to prevent the spread of germs and disease.

General cleaning involves washing the equipment with a mild cleaning solution and disinfectant, particularly the mask. Once cleaned, all respirator masks should be thoroughly rinsed with clean water and allowed to air dry. Drying the mask with cleaning towels can introduce small pieces of lint into the mask that could interfere with the operations of the delicate inhalation and exhalation valves. Even tiny specks that prevent valves from sealing could be hazardous to the user. It is also prudent to never wipe the face piece of the mask with paper towels or other abrasive fabrics, as small scratches can be formed which will reduce visibility. Finally, disinfecting the mask or its components with alcohol is never appropriate. While alcohol is an effective disinfectant, it can also damage or dry out the rubber and plastic components, allowing them to become brittle and subject to failure.

■ Note
Never use regular towels to dry a mask because small particles of lint can interfere with the valve sealing.

■ Note
Never use paper towels or abrasive materials to wipe the face piece of a respirator because this could scratch it.

■ Note
Never use alcohol on the components of a respirator as the alcohol could cause the rubber and plastic to dry out and crack.

Respirators used routinely should also be inspected during cleaning. Worn or deteriorated parts should be replaced by qualified personnel. Respirators for emergency use are required to be thoroughly inspected at least once a month and after each use. A comprehensive program of inspections is critical to reducing the potential for failure when the unit is in use. While the OSHA regulation provides the basic requirements on this topic, it is also important to follow the manufacturer's instructions for inspection and maintenance of each unit because there can be subtle differences among the various types.

> **■ Note**
> Masks and other components should be inspected on a regular basis.

> **■ Note**
> Properly trained personnel are the only ones authorized to repair respiratory protection equipment.

Basic inspection of the mask portion of a respirator includes a thorough examination for any evidence of damaged, broken, or worn parts. All valves should be clean and tested to ensure that they are in proper working order. All straps and/or fasteners should operate to fully open and closed positions. All rubber or plastic parts should be inspected for signs of deterioration. After cleaning and inspection, the straps should be left in a fully open position, allowing users to quickly don the equipment and tighten the straps to the proper position.

> **■ Note**
> All components of a respirator should be checked regularly to ensure that they function.

Inspection of an SCBA or other air-supplying system is considerably more complex. In addition to the inspection of the mask as noted above, the cylinders need to be examined and refilled with air as needed. OSHA regulations require that air pressure in the cylinder be maintained at 90% or greater of the rated capacity of the cylinder. All fittings should be checked for tightness, and the hoses and straps should be carefully inspected to ensure proper operation. Additionally, all valves should be operated to check for adequate and unrestricted airflow. Finally, the low-pressure warning system should also be tested to ensure that it is within the manufacturer's guidelines.

> **■ Note**
> SAR units require greater care and maintenance.

> **■ Note**
> SCBA bottles are required to be maintained at 90% or greater of their rated capacity.

After cleaning and drying, all types of respirators should be stored in a convenient, clean, and sanitary location. When possible, the mask should be placed in a sealed plastic bag to prevent damage and contamination. One important consideration in storing the equipment is that it should not be in an area exposed to direct sunlight. Direct sunlight can damage the rubber and or plastic components and make them less supple and more brittle. Most manufacturers provide this and other warnings in the user information provided with the units. Again, it is always a good idea to review this material for specific storage requirements and recommendations.

> **Note**
> Respiratory protective equipment should be stored away from direct sunlight exposure and in a clean and sanitary location. Sealed bags can be used to prevent contamination.

Cylinders of compressed breathing air are required to be tested and maintained as prescribed in the *Shipping Container Specification Regulations* of DOT (49 CFR, Part 173 and Part 178). This regulation requires that each steel cylinder be hydrostatically tested every five years, while fiberglass/composite wrapped (full or hoop) cylinders be tested every three years. Additionally, the fiberglass wrapped cylinders have a maximum service life of 15 years regardless of their ability to pass the three-year tests. As part of the inspection program, the service life of each cylinder should be reviewed to ensure that the cylinder is within its useful life expectancy. Cylinders that go beyond the service life should be removed from active service and be destroyed to prevent their unauthorized reuse. The date of service is located on each fiberglass cylinder.

> **Note**
> All compressed air cylinders used in a respirator must meet applicable DOT regulations regarding testing and service life.

There is one addition to the inspection program. OSHA requires that persons who need prescriptive eyewear or glasses be provided with a special kit that allows for the placement of the kit with their glass prescription into the mask. The spectacle kits, as they are often called, are a smaller version of standard glasses, minus the arms that go over the ears. These kits can either be specifically designed for a particular brand of mask or generic enough to fit into several types of masks. Regardless of the type used, a final element of the maintenance program should include a review of the kit to ensure that it is able to fit into the mask and that the lenses on the kit are free from damage. If someone is required to wear glasses while wearing a respirator, the OSHA regulation requires that the employer provide these kits to the employee at no charge. Figure 6-9 shows a typical spectacle kit.

> **Note**
> Spectacle kits are available for persons who need glasses while wearing a respirator.

A final element of the program is the records that are kept. Recordkeeping is an important element of a Respiratory Protection Program in that it documents that the written Respiratory Program is maintained. Records are required to be kept of all inspections, maintenance, and repairs to each respirator used. Proper recordkeeping can also ensure that maintenance schedules are kept. This

Figure 6-9
Picture of a mask with spectacle kit.

includes documentation of the monthly inspections of emergency units, minor repairs performed by authorized employees, and any major service or checking by factory-authorized service technicians.

> **■ Note**
> OSHA requires that all parts of the Respiratory Protection Program be documented.

SUMMARY

- The five types of respiratory hazards you may encounter include particulates, gases and vapors, a combination of particulates and gases, oxygen-deficient atmospheres, and thermal extremes.
- Oxygen deficiency occurs when ambient oxygen content drops below 19.5%.
- Engineering controls are the best method to provide for the control of airborne contamination.
- Elements of a Respiratory Protection Program include the respirator selection procedure; training for those who wear respirators; the inspection procedure; maintenance and sanitation procedures; storage procedures; and medical surveillance practices. Additionally, the program should provide for atmospheric monitoring, recordkeeping, and fit testing for all employees who routinely wear respiratory protection.
- Medical examinations are required for those who wear respirators. OSHA requires the use of a standard medical questionnaire given in the regulation.
- The OSHA standard for Respiratory Protection Programs can be found in 29 CFR, Part 1910.134.
- Respiratory protection devices can be divided into two main categories: air-purifying systems and supplied-air systems.

- An APR is a respirator that has special filters designed to remove particulates and/or vapors from the air. The units have canisters or cartridges that either physically or chemically purify the air when the user inhales it. These units do not add supplemental oxygen when used.
- In order to safely use an APR, the following conditions must be met:
  - The type of material in the area must be known.
  - The cartridge must have an ESLI, or a change out schedule must be present.
  - The airborne contamination levels must be below IDLH values.
  - The atmosphere must have oxygen levels at or above 19.5%.
  - There can be no thermal extremes in the environment to be entered.
- OSHA has established values that are referred to as protection factors, which denote the acceptable level of the airborne contaminants for which various types of respiratory protective equipment may be used. Such protection factors used by OSHA are based on the PEL of the material.
- APRs are usually found in the full-face and half-face models.
- The questions to ask prior to selecting a respirator include:
  - Is any type of respiratory hazard present in the workplace?
  - Do you know what specific materials are present in the area to be entered?
  - Is there a suitable APR cartridge for the contamination present?
  - What are the airborne contamination levels?
  - Are there any thermal extremes?
  - Is there adequate oxygen in the work atmosphere?
- SCBA and air-line systems are the two types of SARs.
- SARs provide the maximum level of protection against the five types of airborne contamination.
- SARs have the potential to provide positive pressure within the mask. Such positive pressure further increases the protection afforded to the user because there is less chance that hazardous contaminants can be drawn into the mask even when a leak is present.
- All air-line systems must have an escape cylinder included in the system if the area entered is above IDLH. These cylinders typically provide 5 minutes of breathing air to the user.
- OSHA mandates that the quality of the breathing air shall meet at least the requirements of the specifications for Grade D breathing air as described in ANSI/Compressed Gas Association Commodity Specification for Air, G-7.1-1989. In order to verify the quality of air as Grade D, the air source should be tested regularly. NFPA 1404 states that the air quality of all breathing-air systems shall be tested at least every 3 months by a qualified laboratory.
- All components of a respirator should be inspected regularly and maintained in accordance with the manufacturer's requirements.

REVIEW QUESTIONS

1. List the five types of respiratory hazards.
2. List four elements of an OSHA-compliant Respiratory Protection Program.
3. What are the two main types of respiratory protective equipment?
4. Define oxygen-deficient atmosphere.
5. List four conditions that must be present before a decision can be made to use an air-purifying respirator.
6. Before using a respirator, what types of controls should be implemented?
7. List three advantages of an APR.
8. List three advantages of a SAR.
9. Identify the major differences between an open-circuit SAR and a closed-circuit SAR.
10. Describe how to conduct positive pressure and negative pressure field fit tests.

ACTIVITY

RESPIRATORY PROTECTION SELECTION

You are responsible for determining the appropriate type of respiratory protection for the following assignments. In each case, list the steps that should be taken to determine what specific respirator type is going to be selected. If an APR system is used, determine the appropriate color of the cartridge from the information provided in the chapter. Also list the rationale for your selection and identify why it is the appropriate type for the project.

1. The project involves sanding fiberglass materials in an open area of the site. Personnel working on the project will be exposed to fiberglass dust as part of conducting their activities.
2. Personnel are working in a confined area where nitrogen tubing is present. The project involves entering the area to replace some other materials. Additionally, personnel suspect that there could be a leak in the nitrogen system and need to determine where the leak(s) is located. The system needs to be operational while they determine the location of the release.
3. Personnel are working with some highly hazardous materials, including hydrochloric acid, which releases vapors, and some organic solvents. The materials are being sprayed into an area. The levels of exposure have been monitored and found to be above the PEL for each material, but below the IDLH. Oxygen levels in the area are above 19.5%.
4. Personnel need to respond to a release of a material that is commonly found at the site. The material is in an area where personnel typically wear APRs with a yellow cartridge when they work with the material that is leaking. The released material is coming from a 55-gallon drum that is mostly full.

PERSONAL PROTECTIVE EQUIPMENT

Learning Objectives

Upon completion of this chapter, you should be able to:

- Identify the importance and need for personal protective equipment (PPE).
- List the three major types of PPE.
- List the conditions for use and components of Level A protection.
- List the conditions for use and components of Level B protection.
- List the conditions for use and components of Level C protection.
- List the conditions for use and components of Level D protection.
- Identify the importance and use of compatibility charts.
- Explain the differences between permeation, penetration, and degradation.
- Identify the health considerations of wearing chemical protective clothing (CPC).

CASE STUDY

A Northern California fire department was called to an industrial park in response to a hazardous materials release. As they arrived, they found a railcar that was releasing some type of vapor in the area behind the site. There were no visible indicators on the railcar to identify the exact substance in the railcar, and they determined that a Level A entry would be required to try to locate the exact nature of the release.

A team was organized and personnel entered the area of the release in large, heavy butyl rubber suits. As the team approached the railcar, the face shield in one of the suits suddenly shattered into small pieces, much like automobile glass breaking. The entry team members quickly exited the area and other crews working in the area immediately decontaminated them. The person inside the suit was wearing appropriate breathing apparatus and did not suffer any significant inhalation exposure.

This incident illustrates a number of critical issues that must be considered every time PPE is used. Personnel realized that despite choosing the PPE with the highest level of protection, the materials in the suit were not compatible with all substances. There is no material that is compatible with all substances, and even though the suit material resisted the released substance, the laminated shield on the suit was subject to failure when an exposure occurred. Another critical lesson that was reinforced related to the release itself. When proper identification cannot be obtained, entry teams must use extreme care to stay out of the area of the concentrated release. PPE alone is the last line of defense and must be used with other safe work practices such as approaching spills and releases from upwind whenever possible.

INTRODUCTION

Although there are many different types of protective clothing and equipment, it is important to understand that not all are equal in providing the necessary protection to the user. Some types of protective clothing are more suitable for protection against fire and heat. Others protect against extreme cold. There is also clothing that provides protection against some specific types of hazardous materials but not others. Clearly which type of protective clothing will be used is a major decision in activities involving potential exposure to a hazard. The decision about what to wear is even more critical, knowing that the use of PPE is the last line of protection after other methods to control or eliminate the hazards have not been fully successful.

> **■ Note**
> PPE selection is a critical aspect of safety because it is often the last line of defense from hazards.

PPE SELECTION

The selection and use of PPE is a critical component of any program dealing with worker protection. While engineering and administrative controls are the most effective ways to reduce or eliminate the potential for exposure, PPE is often also needed to provide the necessary safety around hazardous materials. The proper use and selection of PPE is required in a range of circumstances, including routine handling of materials, cleanup following a release, and emergency response to incidents involving releases. As we know, the hazards created are far reaching and include both physical and health concerns, making decisions regarding the selection and use of PPE even more complicated. Making these decisions requires a thorough understanding of the hazards that are faced, coupled with a process of asking the appropriate questions regarding PPE selection. The questions include:

1. What material is involved and what are the primary hazards?
2. Are the hazards health concerns or physical hazards?
3. Which of the four levels of protection is required for the health hazard(s) present?
4. Which specific clothing types will provide protection against the hazard(s)?

Once these questions have been answered, it becomes a matter of donning the equipment and going to work. Keep in mind, however, one of the most critical points regarding PPE. Because it is literally your last line of defense, you must be careful when you use PPE not to intentionally expose yourself to the materials present and, in fact, you must work to reduce any potential exposure. No one type of PPE is completely protective, and circumstances can develop in which the PPE could fail. Intentionally exposing yourself to the hazard by walking through a spilled material, placing your hands in standing liquids, or working downwind of a release can lead to catastrophic results. It is always best to think of the PPE as protection against unintentional exposures, not those that are planned. While it may be necessary to stand in a spill area or to touch highly contaminated items, you should always consider additional measures before taking any action to ensure that the risk of the exposure is worth the gain and that there are no other practical ways to accomplish the tasks at hand. When such intentional exposure is initiated, it is also important that other appropriate safeguards be in place to ensure safety for those involved. These additional measures could include working in teams, having backup personnel available, and making sure that a thorough decontamination program is in place and that protocols are followed as soon as possible after the exposure. Many of these measures will be reviewed in later chapters covering decontamination and incident management.

No one type of PPE will protect against all hazards.

Safety
PPE protects against accidental exposures but should not be thought of as all-protective.

Safety
When PPE is used to provide protection, additional safety measures should also be in place.

What Is Present and What Are the Hazards That Will Be Encountered?

There is a wide range of hazardous properties to which workers could be exposed. Sometimes, a single hazard may be all that is present. For example, the material may simply be a corrosive liquid that is only hazardous through direct skin contact. In other cases, the hazards may be more complex and include both inhalation and skin contact, as could happen with chlorine. In these circumstances the selection of PPE is more complicated given the need to protect multiple routes of entry. Then there are those circumstances where the potential hazards include both physical and health concerns and require protection from contact with the material, and from its effects such as flammability. In such cases, the choice of PPE can be very complicated and require a thorough assessment of the activities that will be conducted so that the risk of this exposure can be controlled.

■ Note
PPE selection can be complicated if the material is hazardous by multiple routes of entry.

Because of the range of hazards and the fact there is no one type of PPE that is protective from all hazards, the selection of the proper PPE starts with an in-depth assessment of the hazards present and the work to be done. Understanding the types of hazards present becomes the starting point in our PPE selection process, and identification systems discussed in Chapter 5 are a critical component in this. When the hazards are known, an informed choice can be made to ensure that the proper PPE is used and that all other necessary safeguards are implemented. But what happens when the hazards are unknown or competing hazards are present? Often in an emergency response the hazards are not immediately known or the material present is both harmful to health and flammable. In such circumstances, extreme care should be exercised and appropriate measures be taken to limit the risks to all response personnel.

Safety
Identification of the material is critical in the PPE selection process.

Safety
Additional measures must be in place to ensure that responders confronting an unknown material are properly protected.

Identification of the materials and their properties is critical for two reasons. First, no material is truly "chemical proof." There is no such thing as "chemical proof" given that various materials can *resist* the effects of a chemical, but over time and given a range of circumstances, the chemical can break down the resistance and cause harm. Second, chemically resistive materials are generally not resistive to the effects of heat and flame. Many chemically resistive materials are rubberized or made of plastics, which can protect against specific hazardous chemicals, but often the rubber or plastic will melt or break down when exposed to heat or flame.

Safety
No type of PPE is "chemical proof."

Safety
The chemical resistance of the PPE does not afford protection from heat or fire contact.

Have I Chosen a Type of Protection Appropriate for the Hazard?

Once the substances are known, the second question that must be answered relates to the type of hazard the material presents. As we know, materials present a health concern, a physical concern, or a combination of these. While a large range of PPE is available to provide the protection needed, PPE can generally be classified into three basic types. These include structural firefighting clothing, high-temperature protective clothing, and chemical protective clothing (CPC). Most of these are available to the employee whose job involves working with various hazards. Others are for use by emergency response personnel whose job involves attempting to stabilize an emergency incident. Regardless of the use, selection of the type of PPE is a critical step in the PPE selection process because each of these has its advantages and limitations.

> **■ Note**
> The three major types of PPE are structural firefighter clothing, high-temperature clothing, and chemical protective clothing.

Structural Firefighting Clothing Perhaps the most familiar PPE is that worn by the average firefighter. A common example is shown in Figure 7-1. This equipment is designed to protect against extremes of temperature, hot water, hot ash and other particulates, abrasion, and other hazards encountered in firefighting. Protection from these hazards is accomplished through specific types of fabrics and

Figure 7-1 *An example of firefighter protective clothing. (Courtesy of San Ramon Valley Protection District.)*

the design of the system, which includes layering materials to form a protective ensemble. The outer layer is a fire-retardant fabric such as Nomex or Polybenzimidazole (PBI). These fabrics are both durable and fire-resistive, providing protection from many of the major hazards encountered in firefighting. Often, the fabric is reinforced with leather to prevent abrasion at key areas, including the elbows, knees, and shoulders. Under the outer fabric are several additional barriers to protect against steam and moisture, and to insulate against heat.

Safety
Firefighter clothing is designed to protect against the hazards normally encountered in firefighting operations and not hazardous materials response.

The layered system affords the firefighter the necessary protection from the hazards commonly encountered in firefighting. However, the design and materials used may actually increase hazards in some cases. For example, when this type of PPE is worn in areas of potential chemical exposure the fabrics and leather can actually serve as wicks and absorb a number of chemicals into the garment. This is particularly true with the gloves worn by most firefighters, which are composed primarily of leather. Leather is prone to absorbing liquids, and is difficult to decontaminate. Other than the boots, firefighter protective clothing does not provide protection against chemical hazards and should not be worn when potential exposure is present.

Note
Firefighter clothing protects by layering various materials, each designed to protect the user from hazards including flame, heat, moisture, and steam.

High-Temperature Protective Clothing High-temperature protective clothing, the second type of PPE, is designed to provide thermal protection for a short period of time and is usually made of layers, like firefighter PPE, coated with a reflective material. Some of these PPE are termed **proximity suits** because they allow personnel to come into close proximity with high temperatures. Such conditions can be encountered during fires involving flammable liquids or in aircraft crash firefighting and rescue operations.

proximity suits
specialized high-temperature clothing worn when personnel need to enter areas of extremely high temperatures

Note
High-temperature clothing is designed to offer protection from exposure to extreme temperatures such as those found in aircraft crash firefighting.

Like firefighter clothing, high-temperature PPE is not designed to provide chemical resistance and may be seriously degraded by a number of common materials including corrosives and strong solvents. Anyone using this type of PPE should be fully aware of the limitations of the garment and of the need to avoid contact with almost all forms of hazardous materials and chemicals.

Safety
Neither type of high-temperature protective clothing affords protection from chemical hazards.

Chemical Protective Clothing The third type of PPE is designed to provide protection against chemical hazards. CPC encompasses a wide range of materials that can be subdivided into categories based on the exposure potential. The NFPA standards divide CPC into the following three types:

1. *Vapor Protective Suits for Hazardous Chemical Emergencies*—NFPA 1991
2. *Liquid Splash Suits for Hazardous Chemical Emergencies*—NFPA 1992
3. *Support Function Protective Garments for Hazardous Chemical Operations*—NFPA 1993

Note
NFPA standards provide guidance on various types of CPC.

While each of the other two types of PPE provides protection from the physical hazards associated with fire or flame, CPC alone does not provide heat or fire protection and in fact can have the opposite effect because the material that the CPC is made from can accelerate a fire. Before the specific issues associated with CPC are discussed, it is important to identify a method to provide limited protection from a combination of fire and chemical exposure. This is accomplished through a type of flash protection.

Note
The use of CPC can create greater fire concerns because the materials used in the clothing often contain plastics.

flash protection
a flash suit is worn over or incorporated into CPC

flash fire
the ignition of flammable vapors accumulated in a given area

Flash protection is most often achieved by wearing a flash suit over the CPC selected. Flash suits are intended to protect the user only for a few seconds from a **flash fire**, providing enough time to rapidly escape the area. Flash fires occur when the vapors or gases accumulate and are suddenly ignited. To provide protection, flash suits are typically composed of a flame-resistant material such as those used in firefighter protective clothing. This material is capable of withstanding heat for a short duration, but will not withstand sustained exposure to heat or fire.

Two types of flash suits are commonly used. Some systems incorporate a fire-resistive over-suit that snaps over another chemically resistive suit. Newer systems incorporate the flash suit as part of the CPC ensemble and require no additional assembly. In effect, some of these newer suits are a single suit with both chemical and flash resistance. Figure 7-2 shows a flash suit worn over a chemical-resistant suit.

Note
A flash protection suit can be used when the potential for both a fire and hazardous materials exposure is present.

Safety
A flash protection suit provides protection from a flash fire only. It will not withstand sustained heat without failure.

Figure 7-2
An example of flash protection worn over other chemical-resistive suits.

The selection of PPE to protect against chemical hazards is more complicated than selecting either type of fire protective PPE. When a situation involves chemical exposure, other questions must be answered, given the potential for a hazardous material to be harmful through more than one route of entry. To help determine this, we now move to the next question in our discussion to ascertain the level of protection that is needed.

Which of the Four Levels of Protection Is Required for the Health Hazard(s) Present?

levels of protection
groupings of CPC and respirators to create a systematic approach in selection of PPE appropriate to given situations that may be encountered

Selection of the appropriate type of protection is simplified through one of the four levels of protection. Each of the four **levels of protection** specify equipment that provides protection from a specific group of hazards. This equipment includes CPC with some form of respiratory protection because inhalation of harmful materials is also a potential problem for those exposed to hazardous materials. Both OSHA and the EPA divide CPC into four levels to assist personnel working with various hazards to select the appropriate level of protection. The levels of protection are Level D, Level C, Level B, and Level A. Each level is defined in a number of OSHA documents, including one of the appendixes in the HAZWOPER regulation. To work safely with hazardous materials, you need to know the hazards present and make a selection appropriate to that level. Once you have selected the level of protection, you will need more specific information to ensure that the specific garments are resistive to the hazards.

Level D protection
the lowest level of protection, which includes PPE without a respirator

Level D Considered the lowest level of protection in this system, **Level D protection** is defined as a normal work uniform. The definition is somewhat misleading in that the term "work uniform" seems to imply that there is no chemical protection afforded to those working with this level of protection. However, this is incorrect because Level D protection can include a wide variety of chemical-resistant clothing such as suits, gloves, and boots. Additionally, other protective equipment can be worn, including a face shield, hard hat, and earplugs. Because Level D covers such a wide range of equipment, perhaps the easiest way to understand what Level D includes is to identify what is not covered, and that is simply any type of respirator or respiratory protection equipment.

■ Note
Level D protection is determined by the type of respiratory protection provided—none at this level.

While the lack of a respirator may seem to limit the usefulness of Level D protection, it is likely the most commonly used of the four levels. With Level D, a range of other types of PPE can be worn to protect the worker from exposure to hazardous substances other than in the air. Natural or mechanical ventilation systems are used to ensure that the airborne concentration is safe, and yet there are countless other materials from which the worker needs to be protected. Level D protection allows the use of the PPE that would be required to provide protection from skin or eye contact with hazardous substances, making it the proper choice when the airborne concentration is below the PEL. Figure 7-3 shows a worker in a typical Level D ensemble.

■ Note
Level D is protective for most types of splash or contact exposure.

Figure 7-3
An employee using Level D protection while working with hazardous materials.

Note that the type of clothing may change with the task, but as long as a respirator is not used, the level is defined as Level D protection. The following list provides examples of some common types of PPE used in Level D protection:

- Safety glasses or goggles
- Coveralls
- Chemical-resistant safety shoes or boots with a steel toe and shank
- Chemical-resistant booties over work shoes or boots, if necessary
- Gloves (these may vary with the hazard encountered)
- Hard hat
- Face shield

> **■ Note**
> Level D protection can include all types of PPE other than respiratory protection.

Level C protection
PPE with an air-purifying respirator

Level C The type of PPE worn with **Level C protection** can be the same as for Level D protection. What differentiates this from any other level is that Level C requires the use of a NIOSH-approved air-purifying respirator. The criteria for the use of an APR are:

1. The atmosphere contains a known hazard above the PEL.
2. The level of contamination is below the IDLH value for the material.
3. There is a minimum of 19.5% oxygen in the air.
4. There is an approved APR cartridge available for the hazards present.

Wearing the respirator would provide protection from airborne hazards, and at this level, almost any type of PPE could provide the other types of protection necessary. Often, the type of clothing worn with Level C would meet the standards specified in NFPA 1993, *Standard for Support Function Protective Garments for Hazardous Chemical Operations*. Figure 7-4 shows a worker in Level C protection. Other protective equipment can be used, including the following:

- NIOSH-approved full-face or half-face APR
- Chemical-resistant clothing compatible with the material being used or present. This could range from a simple splash apron to a coverall or jumpsuit.
- Chemical-resistant outer gloves with chemical-resistant inner gloves
- Chemical-resistant safety shoes or boots with steel toe and shank
- Hard hat
- Face shield or safety goggles if half-face APR is used

> **■ Note**
> This level of protection is denoted by the use of an air-purifying respirator—not the clothing worn with it.

Level B protection
PPE with the inclusion of a SAR

Level B Whenever the situation encountered includes a significant respiratory hazard requiring the use of a supplied-air respirator (SAR), **Level B protection** is used. Technically, the type of CPC worn can be identical to that worn with either Level C or even Level D because it is the SAR that defines the level. It is the atmospheric hazard and not the body contact hazard that determines the need for this level of protection.

Figure 7-4
A worker using Level C protection.

non-encapsulating suit covers the body only and leaves the breathing-air system outside the suit

encapsulating suit Level B suit that encapsulates the breathing-air system under the CPC

The CPC used with this level comes in two variations depending on how the respiratory protective equipment is worn. In most cases, the suit is placed on the user and the respiratory protection equipment such as the SCBA tank and mask are placed over the top of the CPC. This is termed the **non-encapsulating suit** because the air system is not encapsulated or covered by the CPC. In other cases, the Level B CPC suit is zipped over the entire body and encapsulates or covers the breathing-air system. This is termed an **encapsulating suit** because most of the body is covered/encapsulated by the suit with the exception of the gloves, which must be added to the suit separately.

> **■ Note**
> Gloves are generally not a part of a Level B non-encapsulating suit.

Level B is probably the most commonly selected level for emergency response situations because it provides the highest degree of respiratory protection, the SAR, and allows for the use of a wide range of acceptable suits, boots, and gloves. Level B attire should adhere to the NFPA 1992 standard for *Liquid Splash Suits for Hazardous Chemical Emergencies* and can include a range of other types of PPE depending on the circumstances encountered. Examples of some commonly used types of PPE for this level include:

- NIOSH-approved positive-pressure SAR. This is most often a self-contained breathing apparatus (SCBA), although an air-line system may also be used.
- Chemical-resistant clothing compatible with the material being used or present. This clothing may consist of long-sleeved chemical-resistive

overalls, a hooded chemical splash suit, non-encapsulating chemical suits, or encapsulating chemical-resistive suits. Figure 7-5 shows a worker in Level B protection with an SCBA.
- Chemical-resistant outer gloves with chemical-resistant inner gloves.
- Chemical-resistant safety boots with steel toe and shank. Disposable over-booties are recommended when contamination levels are expected to be high.
- Recommended equipment includes a hard hat and fire-resistive coveralls underneath the CPC.

Note
Level B is used in emergency response operations because it provides a high degree of respiratory protection.

Safety
Level B protection is the lowest level that can be used for entering an area with an unknown airborne hazard.

Level A protection
the highest level of protection; PPE includes a fully encapsulating, gastight suit

Level A The highest level of protection utilized when dealing with possible exposure to a hazardous substance is **Level A protection**. The level is sometimes referred to as "gastight" and "fully encapsulating" because it eliminates all types of exposure of the worker to the hazardous substance, including airborne and body contact. Standards for construction of these garments can be found in the NFPA 1991 standard for *Vapor Protective Suits for Hazardous Chemical Emergencies*. The suit defines Level A because the suit is what adds to the protection. Level B affords the same level of respiratory protection as Level A because Level A requires a SAR. However, at Level A, the user is zipped inside a fully encapsulating, gastight suit that has the air supply inside the suit either as an SCBA unit worn underneath the suit or as an air-line unit with an air line passing through sealed openings in the suit. An example of workers in Level A suits is shown in Figure 7-6.

Figure 7-5
A heavy-equipment operator using Level B protection with an SCBA.

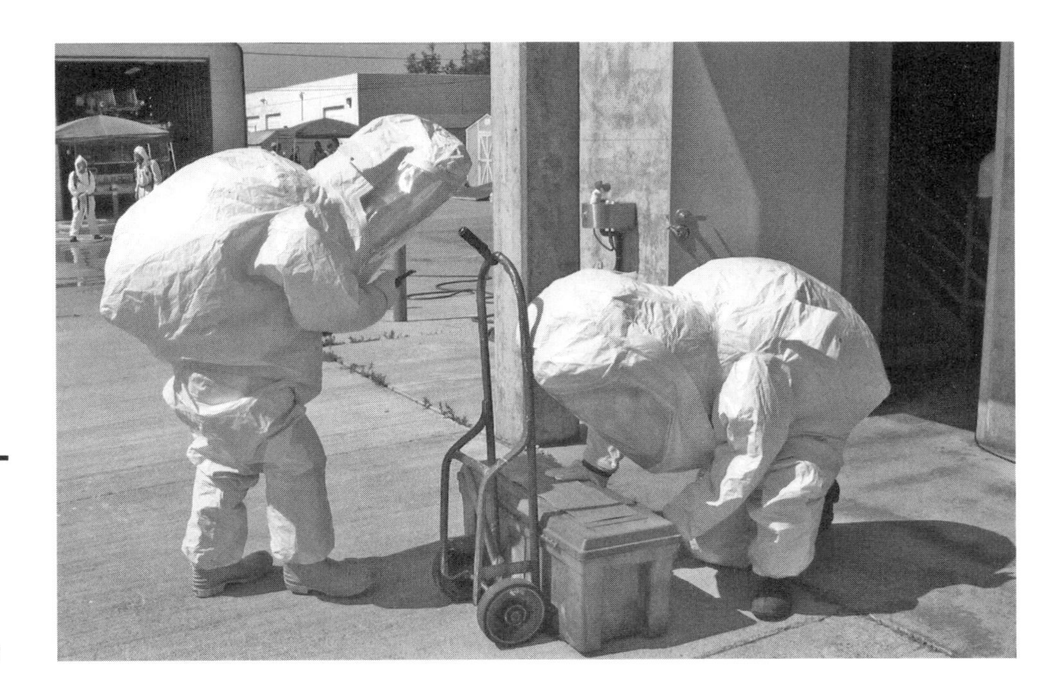

Figure 7-6 *Workers using Level A suits for protection. (Courtesy of San Ramon Fire Protection District.)*

Safety
Level A protection is the highest level of protection and eliminates the potential for any skin exposure.

As can be seen from Figure 7-6, the Level A suit is a one-piece unit with attached outer gloves and attached booties that are usually covered by heavy-duty outer boots. What is not so clear from the picture is that the suit is designed with double-sealed seams, high-tech zippers that seal the openings, and pressure relief valves that allow air to exit the suit when pressure builds inside. The result of these seals is that air exhaled from the SAR remains inside the suit itself, causing pressure to build. With the building pressure from the air, the suit inflates much like popcorn fills a bag as it pops. The resulting positive pressure in the suit works with the zippers and seams to keep materials from entering the suit and contacting the user. The additional bulk created by the expanding suit is also potentially hazardous because it can get snagged on items in the work area.

Safety
Level A suits create a positive pressure inside due to the exhaled air that builds inside the suit.

Note
Level A suits have one-way relief valves that open to allow overpressure to escape from the suit.

fully encapsulating suit completely covers all of the body inside the suit; no body area is ever exposed

Because the Level A suit can look like an encapsulating Level B suit, it is possible to get the two confused. The key in differentiating the two is to note that Level A suit is **fully encapsulating**, meaning that all parts of the suit are a one-piece unit including the gloves. You can determine if a suit is a fully encapsulating Level A suit or an encapsulating Level B suit by noting if the gloves are part of the suit or whether they need to be added. The encapsulating Level B suit does not have gloves attached to the arms of the suit and also does not have one-way valves built into the suit. Table 7-1 identifies some of the major differences between an encapsulating Level B suit and a fully encapsulating, gastight Level A suit.

Level A suits are more standard than other levels of protection because the fully encapsulating, gastight protection is what defines the level. Most Level A protection includes the following items, which can be seen in the donning sequence shown in Figure 7-7.

- *NIOSH-approved positive-pressure SAR.* This is most often an SCBA, although a supplied-air system with an escape pack could be used.

Table 7-1 *The differences between a Level B encapsulating suit and a Level A suit.*

Level B **Encapsulating**	**Level A** **Fully encapsulating, gastight**
No gloves attached to the suit	Gloves attached to the suit
Booties may or may not be attached	Booties always attached to the suit
Pressure is relieved by openings in the hood of the suit. The openings are generally covered to prevent materials from entering the suit.	One-way relief valves are built into the suit to allow pressure to exhaust as needed.
Zippers are protected with adhesive flaps to provide additional protection.	Zippers are heavy duty and are most often covered with a flap of material that is sealed tightly with materials such as Velcro.

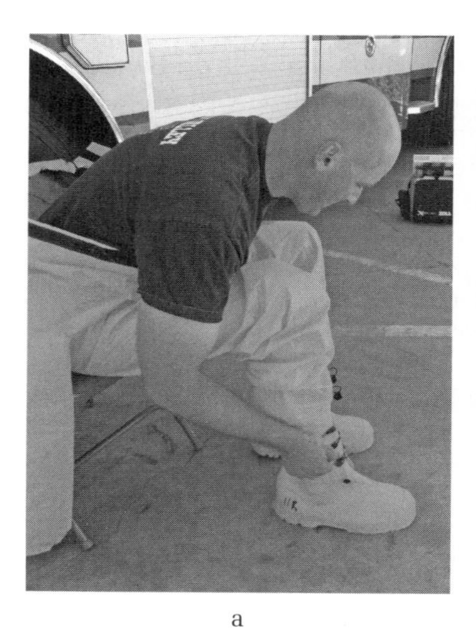

a

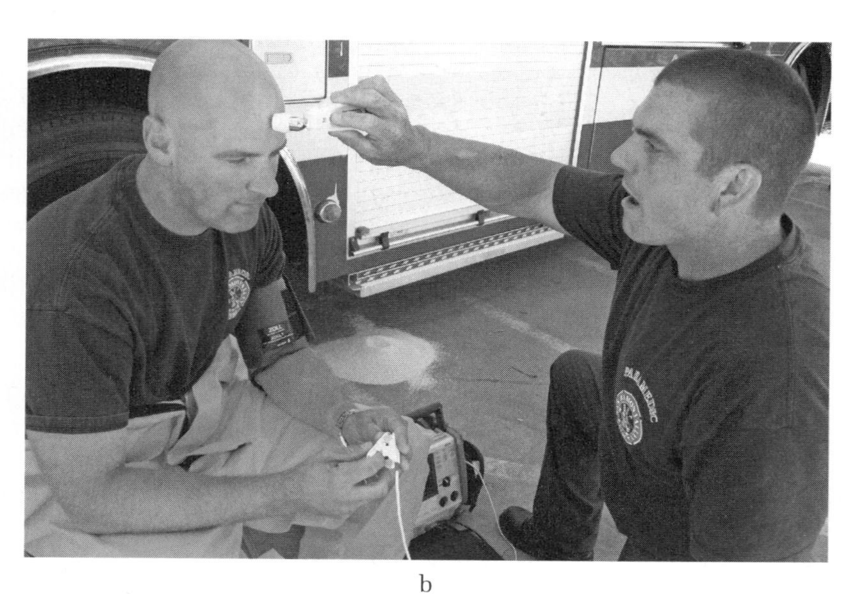

b

Figure 7-7
(a) Donning suit and boots; (b) medical monitoring done prior to final donning of the suit; (c) donning inner and outer gloves; (d) suit is overlapped or bloused over the outer glove to create additional barrier; (e) tape is used to seal openings between the suit, gloves, and boots; (f) tape is used to seal the opening between the suit and the mask; and (g) SCBA is placed on the team member and the donning is complete. (Courtesy of San Ramon Fire Protection District.)

(Continued)

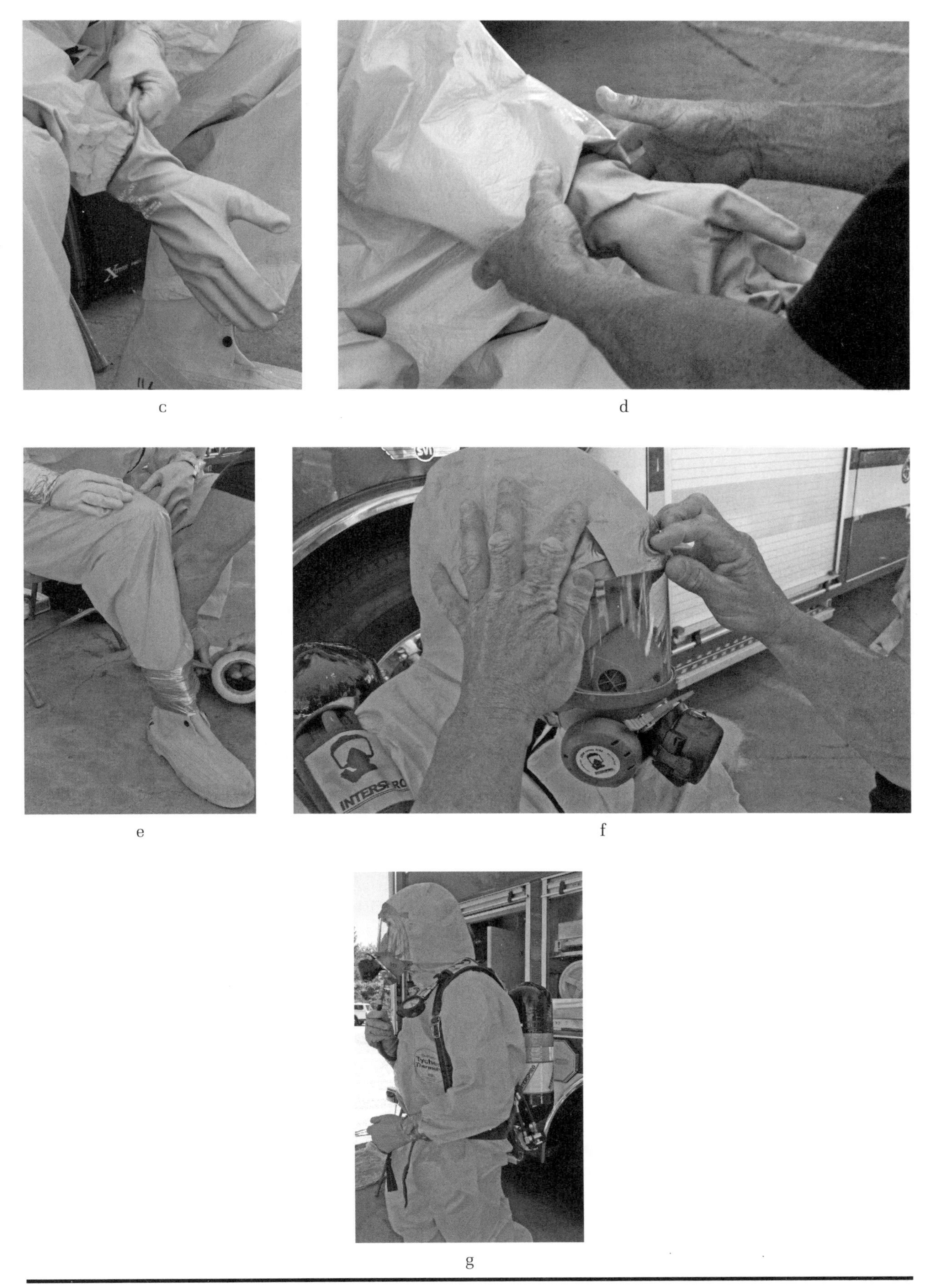

Figure 7-7 (*Continued*)

- *Vapor protective suits.* These are totally encapsulating, gastight garments that completely enclose the user inside the suit with a suitable air supply.
- *Chemical-resistant glove system.* Some manufacturers employ a three-glove system: a cotton underglove covered by a layer of chemical-resistant fabric, then a flame-retardant glove when the threat of fire is present. The entire system should pass the construction standards as listed in NFPA 1991 standard for *Vapor Protective Suits for Hazardous Chemical Emergencies.*
- *Chemical-resistant safety boots with steel toe and shank.* These are worn over the booty that is part of the Level A suit. At times additional over-booties are worn when high levels of contamination are anticipated.
- *Recommended equipment* such as a hard hat, fire-retardant coveralls underneath the CPC, cotton or fire-retardant long underwear, and cooling vests as indicated.

Levels of Protection Selection Criteria and Considerations

Making a decision regarding the appropriate level of protection can be complicated because many people do not have a full understanding of the additional protection that each of the levels provide, nor the understanding that the higher levels of protection can introduce hazards that often can exceed the potential exposure. While each of the levels of protection differs in the degree of respiratory protection provided, the actual type of CPC that can be worn among three of the four levels is largely the same.

Note
The type of CPC selected for Levels D, C, and B can be the same.

To understand this concept, consider that a person working in a chemical-resistive hooded jumpsuit and gloves without a respirator would be wearing Level D protection. The considerations for the selection of this level are consistent with the decision process used for respirator selection in Chapter 6. If the level of airborne hazard is below the PEL for the materials expected, there are no thermal considerations, and there is adequate oxygen in the air, the selection of Level D is most appropriate. Adding more respiratory protection would not increase the safety of the worker because there is no hazard from which protection is required. In fact, wearing a respirator in a situation where one is not needed can decrease peripheral vision, make communication more difficult, increase the effort it takes to breathe through some type of cartridge or air system, increase the heat stress on the person whose face is covered, and add time and cost to the project.

Safety
Using a higher level of protection can introduce hazards if the lower levels of protection are adequate.

However, if the level of respiratory hazard increased to the degree that the worker wearing the CPC in the example required the use of an APR, the level of protection would increase to Level C. Level C protection is worn when the decision to use an APR is made. The clothing for this level can be the same as with Level D or even Level B. To determine that Level C is the most appropriate level for a particular situation is simply to ascertain through the evaluation of the area that a respiratory hazard is present and that the criteria for wearing an APR have been met. Once that decision is made, Level C becomes the safest level of protection for those particular circumstances. At Level B, a SAR system would be worn. However, this additional protection may not be needed because it has been determined that Level C with an APR would provide enough protection for the worker to safely perform the task. The change to Level B with a SAR introduces some significant disadvantages, including obtaining the additional equipment, the limitations on the amount of time that SCBA equipment can provide, the limitations by the additional weight and bulk that restricts access into areas, and the additional required maintenance and service. If Level C with an APR provides the required protection, more protection through Level B with a SAR will add disadvantages that could be greater than the advantages that might be achieved.

Safety
Hazards may be added as the levels of protection increase.

However, if the decision process leads to the choice of a SAR, the level of protection selected can be either Level B or Level A because both levels require some type of SAR. While it was relatively easy to ascertain that Level D was appropriate by confirming that there were no respiratory hazards, or to select Level C as the most appropriate level because an APR would provide the required protection, the decision to use Level B or Level A in a given situation can be difficult since it contains a degree of subjectivity. The difficulty in the selection process occurs because the distinction between Level B and Level A protection is not as clear as with the other two levels. Both Level B and Level A provide exactly the same level of respiratory protection, the highest type provided, a SAR system.

Note
Both Level B and Level A provide the same type of respiratory protection and provide the same degree of respiratory protection.

The additional protection afforded by a Level A suit comes at a high price, and that price is not simply the cost of the suit, which can be several thousands of dollars. The real price is in the additional drawbacks that the user encounters when wearing Level A protection. As we have studied, Level A protection fully encapsulates the user in an environment that can prove hostile through the significant increase in heat stress, the reduced ability to communicate, the loss of vision caused by the fogging of the face-piece from the moisture within the suit, the additional bulk of a Level A suit filled with air, and the additional time and assistance that are required for a Level A suit. These factors are serious disadvantages when using Level A protection. If the benefits of wearing Level A protection do not offset the disadvantages, then Level A may actually be less safe than Level B in some instances.

So what are the differences between the two highest levels of protection? We know it is not the additional respiratory protection afforded because each uses a SAR. The real difference is in the ability of a Level A suit to ensure that no skin contact is possible for the times when an exposure of the skin to the hazard could be rapidly fatal. But consider that someone wearing a well-taped Level B ensemble or even using an encapsulating Level B suit is protected from most skin contact and only the most extreme exposures to significantly hazardous materials would be of concern. These factors must be considered when selecting between the two levels. The decision process may lead to second-guessing, and after using either level of protection encourages some people to play the "Monday morning quarterback game."

> **Note**
> Protecting the worker from any skin exposure is the real advantage to Level A protection.

Following are some guidelines that can be used to select the appropriate level for a job. They are based on the premise that the safest level is the lowest level required to safely do the job. The use of more protection when it is not required only adds problems and can introduce unnecessary drawbacks, as has been discussed. When there are questions regarding what level to choose, always default to the higher of the levels for that particular circumstance. But when information is available that leads to the selection of a lower level of protection than first thought, have confidence in the decision process and in the equipment and protection afforded.

Select Level D when all of the following conditions are met:

- The atmosphere contains no known hazards.
- Work functions preclude a potential for an airborne chemical exposure.
- A degree of basic safety has been met at incidents or work sites. This base level of protection may allow for a quick upgrade into higher levels of protection through the inclusion of a respirator should it be required.

Select Level C when all of the following conditions are met:

- All criteria for using an APR are met. This includes:
  - The materials in the atmosphere are known and there is an approved cartridge available.
  - The levels of contamination are below IDLH levels.
  - Oxygen levels are above 19.5%.
  - No thermal extremes are present.
- Any atmospheric contaminants, splashes, or direct contact with the substances have been adequately blocked by wearing appropriate non-gastight CPC.

Select Level B when all of the following conditions are met:

- All criteria for using an APR cannot be met. This includes:
  - Oxygen levels could be below 19.5%.
  - Thermal extremes can be present.
  - There is no approved cartridge available.
  - The material may not be known.
  - IDLH levels of contamination may be present.

- There is a high inhalation hazard from a material that can be adequately blocked by appropriate non-gastight CPC.
- Spills of unknown origin are present but adequate ventilation can be assured to dilute concentrations to below IDLH levels for skin contact.

Select Level A when all of the following conditions are met:

- The area being entered is confined or poorly ventilated and the materials are known to present a serious threat of immediate skin destruction.
- The area being entered is confined or poorly ventilated areas and the materials are known to present a high potential for splash, immersion, or exposure to a material that could produce serious IDLH conditions due to a contact exposure.
- The situation involves a material that is completely unknown and that could have the potential for serious skin contact effects. While Level B is the minimum level of protection for dealing with unknown materials in some situations, some response agencies require Level A for emergency situations involving unknown materials.

Note

Select the lowest level of protection that will protect you from the hazards present.

Which Specific Clothing Types Will Provide Protection Against the Hazard(s)?

Once the appropriate level of protection required for a particular situation has been determined, the last consideration is to select the exact type of CPC that will be worn and the specific fabric that the clothing will be made from. This is required because each of the levels of protection uses some type of CPC based on the hazardous materials present and many types of CPC can be used with each level. CPC could be an apron draped over the front of the person to protect from splashes, gloves that extend up the arms, or a one-piece hooded jumpsuit that covers the entire body and has fewer openings for splashes to get into and expose the worker. After the specific type of clothing has been selected, the specific fabric or material that the apron, gloves, or jumpsuit is made of must be considered. Even Level A, which uses a standard type of suit with no openings at all, requires that each suit selected be evaluated as to whether it can resist a particular hazard or not. The following two examples illustrate some key points in the CPC selection process.

Safety

Selection of the specific fabric or material that the CPC is made of is critical to protection from exposure.

Put gasoline into a Styrofoam cup and the gasoline will immediately destroy the cup, resulting in the release of the gasoline. Gasoline dissolves or degrades the material that the cup is made of. Figure 7-8 shows the results of this process. In this case, the bottom of the cup dissolved when it contacted the gasoline. Now consider what would happen if the CPC selected was also degraded by the material present. Without some form of testing, every type of suit fabric could be subject to similar results.

Figure 7-8
An example of the degradation of a cup exposed to gasoline.

Filling a standard latex balloon with helium results in the balloon floating to the ceiling. But the balloon falls to the ground after only a few short hours as the helium works its way out of the balloon through the latex. No material is totally impervious to all other substances, and helium gas simply worked its way through the pores until it was gone and the balloon sank. In much the same manner, hazardous substances can migrate through various types of chemically resistive fabrics, exposing the worker wearing the fabric.

So the final selection process for the appropriate type of PPE includes a discussion of three principles. These include penetrations, degradation, and permeation.

> **Safety**
> Criteria for the selection of the specific type of CPC requires consideration of penetrations, degradation, and permeation.

penetrations
openings in the PPE

Penetrations are simply the openings in the PPE. The openings can also be areas where the clothing does not completely cover the exposed worker, or they can be created by the zippers and seams in the CPC. In the example of the apron, other than directly in front, many openings allow the user to be exposed to a splash hazard. Such a choice would not be appropriate with materials that pose a high contact hazard. In such cases, the more protective hooded jumpsuit is chosen because it has fewer openings for exposure. Depending on the type of suit and the manner of construction, penetrations can be significant and allow for the passage of materials through the openings onto the body of the user.

Considerations of the openings or penetrations is an important point to evaluate in selecting appropriate types of protection whether that will be worn with Level D, Level C, or Level B protection. There are countless choices that need to be made because protective clothing ranges from the apron with gloves to the encapsulating Level B suits. Simply defaulting to the clothing that covers more of the body may not be the solution since we know that adding more protection comes at the cost of additional heat stress from wearing the suit and the limitations in mobility. If the cost is necessary to offset the hazards expected, then these types of protective clothing should be worn.

To reduce the number and types of natural openings in specific types of CPC, various types of tape are used to seal areas where the arms of the suit contact the gloves, or where the boots fit over the top of the leg of the suit. Chemical-resistive tapes are available for this purpose, although standard duct tape is also popular. The tape is wrapped over the outer gloves or boots to form a tight seal with the sleeves or legs. Tape is also used to seal the areas where the respirator mask contacts the hood of the jumpsuit. The front of the zippers on the suit as well as any openings that would allow the hazardous materials to penetrate and expose the user of the CPC to the hazards present can also be sealed with tape. Refer back to Figure 7-7 to see how the tape is used to seal the arms, legs, and mask of the user, significantly reducing the potential for any type of exposure through the openings in the suit. In every case where the type of protective clothing is to be selected, consideration should be given to openings and whether sealing the openings with tape will afford the appropriate level of protection required for the circumstances faced.

Note

The use of various types of tapes helps reduce the potential for the materials that can penetrate CPC.

Penetration can also occur through tears or abrasions in the outer covering of the suit or gloves. Unlike abrasion-resistive materials such as leather, CPC generally provides limited protection against sharp objects. It is important that the user always pay attention to the environment where they are working to avoid contact with jagged objects because a torn suit will expose the user to significant hazards. In order to avoid a PPE failure due to penetration, it is important to follow some basic rules for protecting the suit from failure. These include:

- Never kneel or crawl in the suit. The action of grinding knees, elbows, or hands on the surface unduly abrades PPE and may lead to failure.
- Use heavy outer gloves over chemical protective gloves when moving drums or working with sharp objects such as glass, jagged metal, valve stems, or piping. The outer gloves can provide protection from the abrasion and be discarded after use.
- Ensure all zippers are fully closed and covered with a protective flap before entering contaminated areas. If tape is used to seal openings, inspect the taping to verify that no openings or gaps exist.
- Completely inspect any piece of PPE for holes, tears, thin areas, poor seaming, or any other defect that may compromise the garment. Even new equipment can have flaws that can lead to failure. It is always better to inspect twice than have the garment fail once.
- PPE does not last forever. Respect the listed shelf life of the item and always observe the recommended storage practices from the manufacturer. Old and outdated equipment can easily crack or peel, opening a route for penetration.

Safety

Extreme care is required when wearing CPC to avoid damaging the equipment and allowing penetrations!

The remaining two factors, degradation and permeation, that must be considered when selecting the specific CPC involve the chemical resistance of the material from which the CPC is made. Recall the examples of the Styrofoam cup and the helium-filled balloon. Neither of these were able to stand up to the material placed inside them and ultimately each failed to hold the substance. Like these examples, CPC has issues that must be considered when you select the specific types of clothing for use. CPC is made up of a plastic and/or elastomeric material overlaid on a paper or fabric substrate. Many of the names of these materials are registered trademarks of the individual manufacturer and the exact chemical compositions are trade secrets. Typically each type of material used in the construction of CPC has specific resistance to certain types of materials. The fabric or material provides a limited amount of protection based on a number of factors. Selecting the proper material requires understanding how hazardous substances react with the fabrics and materials that make up CPC.

Because there is no such thing as CPC that is chemical proof or "impermeable," the specific type of fabric or material selected should be the one that has the highest chemical resistance. Some fabrics and materials are not resistive at all and simply degrade or break down when they are exposed to a particular hazardous material. **Degradation** is a very rapid process and generally causes the immediate failure of the plasticized outer coating of the garment. Essentially, the fabric dissolves or otherwise fails due to contact with the hazardous material, leaving behind the stringy fibers of the paper substrate. If this were to happen, the person wearing the CPC would have cause for considerable concern because there is a serious loss of the integrity of the garment and therefore little or no protection left. Degradation is best illustrated by the earlier discussed example of the gasoline and the Styrofoam cup. Almost instantly, the gasoline begins to "eat" the Styrofoam and totally degrades whatever parts of the cup it touches. This occurs because the two substances have similarities in their chemical makeup and materials that are similar often behave according to the "like dissolves like" concept. Simply stated, similar chemical compounds have a tendency to be soluble in one another.

degradation
the rapid failure of the CPC caused by the hazardous substance dissolving the CPC fabric

Note
Degradation often occurs when materials that are similar contact each other.

If the CPC is not degraded by contact with the hazardous materials, the final concept in the discussion is that of permeation. **Permeation** is a process in which the hazardous material gradually migrates through the fabric without causing direct damage to the fabric itself. Permeation can occur as the result of exposure to either a liquid or vapor and involves a three-step process. First, the hazardous substance contacts the outer surface of the CPC. Then the substance enters it much like water soaking into a sponge. A slow migration of the substance is caused when the material diffuses through the fabric of the CPC. **Diffusion** is simply the process of migration from higher to lower concentrations. In the case of the fabric, the material is more concentrated on the outside and slowly migrates through to the less concentrated areas. Finally, the substance can complete its journey through the fabric and passes through to the other side where it contacts the person using the CPC. An illustration of this slow process is shown in Figure 7-9.

permeation
the gradual migration of a material through the fabric by diffusion

diffusion
the movement of a material through the fabric of the CPC from higher to lower concentrations

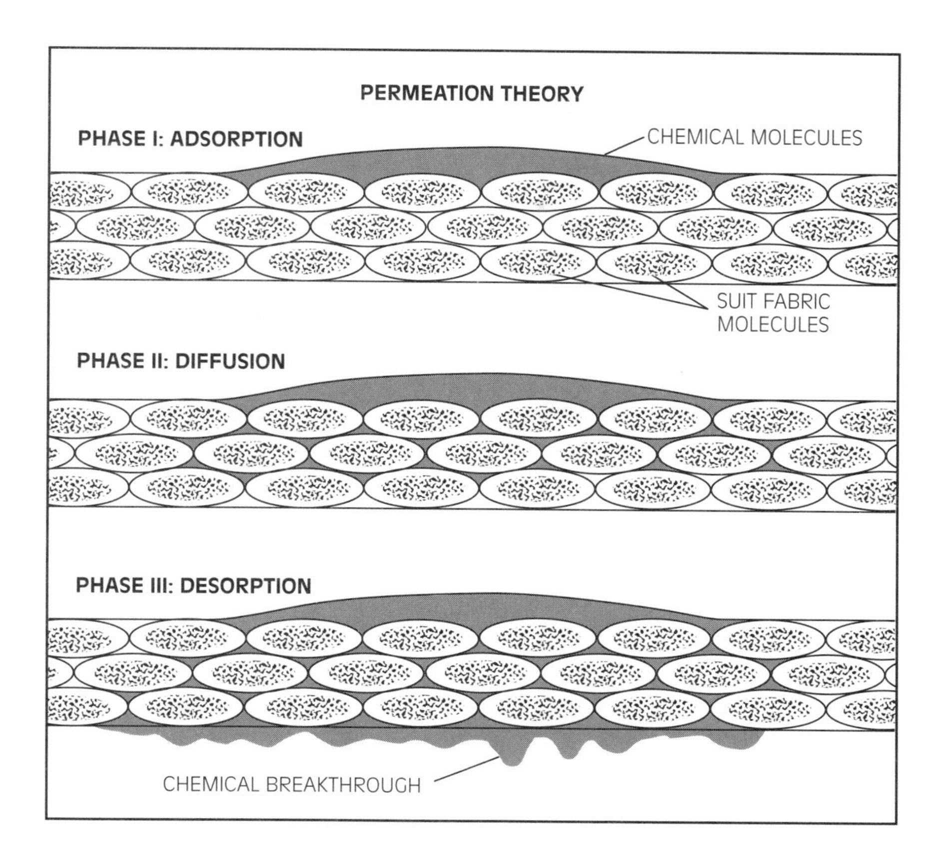

Figure 7-9
An example of CPC permeation.

> **■ Note**
> Standardized testing is needed to determine the ability of a particular fabric to withstand exposure to various hazardous materials.

To help in the selection of CPC, specific tests are used to determine the appropriateness of the CPC with various hazardous materials. The tests expose the CPC fabric to the hazardous materials in laboratory conditions that ensure consistent test results. Following the exposure, one test procedure requires that the fabric be examined for obvious signs of degradation, swelling, or weight changes. This has been the traditional method for generating the chemical resistance tables that are included in many PPE brochures. The test largely indicates the degrading effects of the hazardous material on the fabric on an acute basis. However, this test falls short of determining the permeability of the fabric to the material because permeation can occur with little or no visible or physical effect on the fabric.

breakthrough time
the time between contact with a hazardous material on the outside of the CPC and when it can be detected on the inside

To determine the permeability of a given fabric to specific materials, a test is conducted to measure the time between the contact with the test agent and the detection of the agent inside the CPC. This is known as **breakthrough time** and is a key factor in selecting CPC because it indicates the relative resistance of the fabric. In the test, a standard testing protocol is used that was developed by ASTM International, a group formerly known as the American Society for Testing and Materials. The testing protocol requires the fabric used to make the CPC be exposed to the hazardous substance in a standard test cell. A diagram of the ASTM F739 test cell is shown in Figure 7-10.

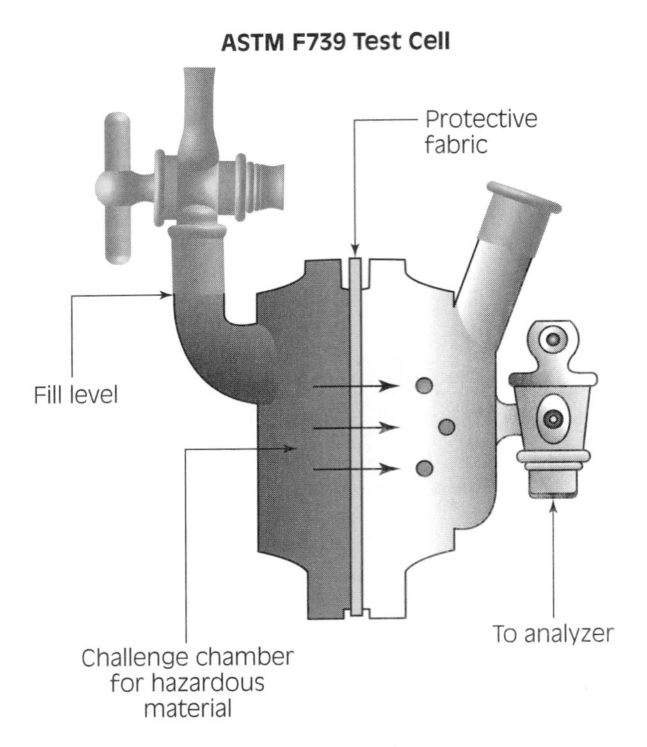

Figure 7-10
A diagram of the standard ASTM F739 test cell.

> **■ Note**
> A standard ASTM test is required to determine breakthrough time for CPC.

compatibility chart
document supplied by the manufacturer of the CPC that provide data on degradation and permeation

In the test, the fabric is evaluated to determine the specific breakthrough time for the particular fabric against a range of test agents. The results are then listed for the user and available on a **compatibility chart** that can be requested when the CPC is purchased. Compatibility charts contain the data regarding how a particular suit fabric or type of CPC will hold up against a certain type of chemical exposure. They are very useful in determining suit selection because they contain the testing data obtained from exposing various types of fabrics against a battery of test chemicals. While the format for reporting can vary, most chart guides show the results to help determine its appropriateness against both permeation and degradation as they relate to the exposure of the chemical to the selected fabric. Most guides are also color coded for quick reference as to chemical resistance and can be easily read and understood. An example of a part of a compatibility chart is shown in Figure 7-11. An exercise using the data from the chart is given at the end of this chapter and will help in the proper understanding of the charts and their use.

> **■ Note**
> Compatibility charts should be consulted any time PPE is to be used.

permeation rate
the speed at which hazardous materials seep into CPC

In addition to the breakthrough time, the compatibility chart also identifies the permeation rate. **Permeation rate** is often expressed in terms of the amount of a chemical that passes through a given area of clothing per a given unit of time. Common units are micrograms of material, per square centimeter of clothing, per minute of time. Thus, the total amount of chemical permeating an article of

Chemical Name	CAS No.	Silver Shield			Viton			Butyl			Chemsoft			Nitrile			Natural Rubber		
		D	BT	PR	D	BT	PR	D	BT	PR	D	BT	PR	D	BT	PR	D	BT	PR
p-Dioxane	123-91-1	I/D	I/D	I/D	P	23 min	26.8	E	>20 hrs	N/D	I/D	I/D	I/D	P	28 min	77.1	I/D	I/D	I/D
Perchloric Acid (70%)	7601-90-3	I/D	I/D	I/D	I/D	I/D	I/D	I/D	I/D	I/D	E	>8 hrs	N/D	E	>8 hrs	N/D	I/D	I/D	I/D
Perchloroethylene	127-18-4	E	>8 hrs	N/D	E	>17 hrs	N/D	P	I/D	I/D	F	1 hr	3.8	F	1.3 hrs	5.5	I/D	I/D	I/D
Perchloromethane	56-23-5	E	>8 hrs	N/D	E	>13 hrs	N/D	I/D	I/D	I/D	F	1.3 hrs	3.45	F	3.4 hrs	5	I/D	I/D	I/D
Phenol (85% in water)	108-95-2	E	>8 hrs	N/D	E	>15 hrs	N/D	E	>20 hrs	N/D	I/D	I/D	I/D	P	39 min	>1500	F	2.2 hrs	4.64
Phenylamine	62-53-3	E	>8 hrs	N/D	P	6 min	18.7	E	>8 hrs	N/D	I/D	I/D	I/D	F	1.1 hrs	45	I/D	I/D	I/D
Phosphoric Acid (85%)	7664-38-2	I/D	I/D	I/D	I/D	I/D	I/D	I/D	I/D	I/D	E	>8 hrs	N/D	E	>8 hrs	N/D	E	>8 hrs	N/D
Pimelic Ketone	108-94-1	E	>8 hrs	N/D	P	29 min	86.3	E	>16 hrs	N/D	I/D	I/D	I/D	I/D	I/D	I/D	F	2.1 hrs	0.07
2-Propanone	67-64-1	E	>8 hrs	N/D	P	2 min	383	E	>8 hrs	N/D	P	1 min	42.3	P	3 min	291	P	10 min	12.2
Propyl Acetate	109-60-4	E	>8 hrs	N/D	P	I/D	I/D	G	2.7 hrs	2.86	I/D	I/D	I/D	P	17 min	72.5	I/D	I/D	I/D
Propyl Alcohol	71-23-8	I/D	I/D	I/D	I/D	I/D	I/D	I/D	I/D	I/D	G	3.8 hrs	0.35	E	4.4 hrs	1.1	I/D	I/D	I/D
Propylene Oxide	75-56-9	I/D	I/D	I/D	P	1 min	1790	F	2.2 hrs	7	I/D	I/D	I/D	P	<6 min	>3.9	I/D	I/D	I/D
p-tert-Butyltoluene	98-51-1	E	>8 hrs	N/D	E	>8 hrs	N/D	F	1.78 hrs	8	I/D	I/D	I/D	P	I/D	I/D	I/D	I/D	I/D
Pyridine	110-86-1	I/D	I/D	I/D	P	38 min	74	E	>8 hrs	N/D	I/D	I/D	I/D	P	I/D	I/D	I/D	I/D	I/D
Sodium Hydroxide 50%	1310-73-2	E	>8 hrs	N/D	E	>8 hrs	N/D	E	>8 hrs	N/D	E	>8 hrs	N/D	E	>8 hrs	N/D	E	>8 hrs	N/D
Stoddard Solvent	8052-41-3	E	>8 hrs	N/D	I/D	I/D	I/D	I/D	I/D	I/D	E	>8 hrs	N/D	E	>6 hrs	N/D	I/D	I/D	I/D
Styrene	100-42-5	E	>6 hrs	N/D	E	>6 hrs	N/D	F	35 Mins	0.19	P	16 min	39	P	11 min	>3.35	I/D	I/D	I/D
Sulfuric Acid (50%)	7664-93-9	E	>6 hrs	N/D	E	I/D	I/D	E	I/D	I/D	G	>8 hrs	N/D	G	>6 hrs	N/D	G	>6 hrs	N/D
Sulfuric Acid (93%)	7664-93-9	E	>8 hrs	N/D	E	>8 hrs	N/D	E	>8 hrs	N/D	P	2 min	N/D	F	1.9 hrs	11.4	G	5.1 hrs	N/D
Tetrachloroethylene	127-18-4	E	>8 hrs	N/D	E	>17 hrs	N/D	P	I/D	I/D	F	1 hr	3.8	F	1.3 hrs	5.5	I/D	I/D	I/D
Tetrachloromethane	56-23-5	E	>8 hrs	N/D	E	>13 hrs	N/D	I/D	I/D	I/D	F	1.3 hrs	3.45	F	3.4 hrs	5	I/D	I/D	I/D
Tetrahydrofuran	109-99-9	E	>8 hrs	N/D	P	0 min	327	F	27 min	112	P	I/D	I/D	P	0 min	167	P	5 min	360
Thioglycolic Acid	68-11-1	I/D	I/D	I/D	E	>8 hrs	N/D	E	>8 hrs	N/D	I/D	I/D	I/D	I/D	I/D	I/D	I/D	I/D	I/D
Toluene	108-88-3	E	>8 hrs	N/D	E	>16 hrs	N/D	P	6 min	511	P	I/D	I/D	P	11 min	68.1	P	3 min	32.2
Toluene Diisocyanate	584-84-9	E	>8 hrs	N/D	I/D	I/D	I/D	E	I/D	I/D	F	1 hr	2.52	G	I/D	I/D	I/D	I/D	I/D
1,1,1-Trichloroethane	71-55-6	E	>8 hrs	N/D	E	>15 hrs	N/D	P	I/D	I/D	I/D	I/D	I/D	F	37 min	76.4	I/D	I/D	I/D
Trichloroethylene	79-01-6	E	>8 hrs	N/D	E	7.4 hrs	0.24	P	14 min	550	I/D	I/D	I/D	P	4 min	283	P	<5 min	894
Trichloromethane	67-66-3	E	>8 hrs	N/D	E	9.5 hrs	0.46	I/D	I/D	I/D	I/D	I/D	I/D	P	4 min	352	I/D	I/D	I/D
Triethanolamine	102-71-6	I/D	I/D	I/D	I/D	I/D	I/D	E	>8 hrs	N/D	E	>8 hrs	N/D	I/D	I/D	I/D	E	>8 hrs	N/D
Triethylamine	121-44-8	I/D	I/D	I/D	E	>8 hrs	N/D	P	I/D	I/D	E	5.8 hrs	0.18	E	>8 hrs	N/D	I/D	I/D	I/D
Vinegar Naphtha	141-78-6	E	>8 hrs	N/D	P	I/D	I/D	E	7.6 hrs	3.4	I/D	I/D	I/D	P	8 min	145	I/D	I/D	I/D
Vinylstyrene	1321-74-0	E	>8 hrs	N/D	E	>17 hrs	N/D	F	2.2 hrs	238	I/D	I/D	I/D	P	I/D	I/D	I/D	I/D	I/D
Xylene	1330-20-7	E	>8 hrs	N/D	E	>8 hrs	N/D	P	I/D	I/D	P	I/D	I/D	P	21 min	18.5	I/D	I/D	I/D

D = Degradation
BT = Breakthrough Time
PR = Permeation Rate

E = Excellent
G = Good
F = Fair
P = Poor

N/D = None Detected
I/D = Insufficient Data

Good for total immersion
Good for accidental splash protection and intermittent contact
Only use with extreme caution. Glove will fail with only short exposure

Figure 7-11
An example of a compatibility chart. (Courtesy of North Safety Products.)

clothing increases as the area exposed to the chemical is increased and also as the duration of exposure is lengthened. For a given chemical/material pair, the permeation rate decreases as the material thickness is increased. This leads to the conclusion that the thicker the fabric, the slower the permeation rate, resulting in greater time required before breakthrough occurs.

> **■ Note**
> Thicker types of CPC generally provide more protection than thinner types because the permeation time is greater.

Permeation rate is also a direct function of the solubility of the chemical in the PPE material. Solubility is the amount of chemical that can be absorbed by a given amount of PPE material. This is often expressed as grams of liquid per gram of suit material. This absorption may be accompanied by swelling or other physical changes in the suit fabric. Immersion testing used to determine solubility is an effective means for evaluating this property of PPE.

While compatibility charts provide much useful data, they do have some serious limitations. These include the following:

- Testing of fabrics is done at a standard temperature. The actual temperature where the CPC is used could be different. Temperature variances can increase or decrease the permeation rate and change the breakthrough time.
- During the testing, the fabric does not move. Using that fabric to make a glove will cause that fabric to stretch or constrict depending on where it is located. As the fingers close, the fabric on the outside stretches and thins. This can change the permeation rate as the material gets thinner by the movement of the fabric.
- With continual movement of the fabric, the chemical can be *pumped* into it through the loosening and tightening of the fabric material. This is not reflected in the data provided in the chart.
- Pushing or pressing on the fabric can change the actual protection times because of the increased pressure. Consider the palm side of a glove that is exposed to both the hazardous material and the physical action of gripping something. The gripping action can push or squeeze the substance into the fabric and cause the reported data to become invalid.
- Testing is done using one hazardous substance at a time. If multiple materials are present, the data could be less valid for that particular situation.
- Similar fabrics from different manufacturers can produce different results. This is due to the fact that while CPC is made up of the same fabric, the method of manufacturing may be different. This can result in increased or decreased thickness of the fabric, facts that can alter the data. Never rely on data from one manufacturer and assume that it works for all others. Even if the fabric is the same, the results can be significantly different.
- Compatibility charts do not always take into account how the fabric is used to create a given type of CPC. The manufacturer may leave large seams or use different types of closures. Remember that penetration issues are not covered on a typical compatibility chart.

> **! Safety**
> Compatibility charts generally do not provide data for CPC use at differing temperatures or when the fabric moves.

Safety
Pressure on CPC can change the time during which the CPC can provide protection.

Safety
Data provided on compatibility charts is not necessarily representative of actual field conditions.

Reuse of CPC

The issues associated with the reuse of CPC continue to be a subject of significant debate. Compatibility charts give data that is limited by factors such as those cited above as well as the fact that the testing done by the manufacturers is for a limited time. The typical test stops at 480 minutes, or one 8-hour workday. Data showing no breakthrough during this period does not imply that the test results can be extrapolated further. After the time has elapsed, the testing stops and any use of the equipment beyond that time is done without any testing data. Couple this with the concept of permeation that says that once a chemical has begun to diffuse into a suit material, it will continue to diffuse even after the chemical on the outside surface is removed. This is because a concentration gradient has been established with the material, and there is a natural tendency for the chemical to move toward areas of lower concentration. This movement will likely progress at a slower rate if the exposure stops and decontamination of the outer layer of fabric occurs, but to some degree, it will continue. This phenomenon has significant implications relative to the reuse of PPE. For example, a possible field scenario is as follows:

1. A worker uses a chemical-resistive glove that is rated to provide protection from the specific material to which it will be exposed for greater than 480 minutes.
2. The worker uses the glove for 2 hours during the day. Breakthrough does not occur during the time that the glove is worn because the glove has low permeability relative to the material to which it is exposed.
3. Prior to removal, the worker carefully washes the glove to remove surface contamination and places it in an area where no additional contamination will occur.
4. The next day, the worker reuses the glove. By that time, some fraction of the absorbed chemical might have reached the inside surface of the glove due to continued diffusion.

Note
The topic of reuse of CPC is subject to debate.

Safety
Reuse of CPC is an issue that needs to be carefully reviewed and should become part of the organization's standard operating procedures.

This problem is further complicated by the concept that once a piece of protective equipment is reused in the same contaminated environment, the subsequent exposure could move more rapidly through the already established permeation pathway. Protective clothing decontamination and reuse remain controversial and unresolved issues at this time. Most authorities agree that every effort should be made to avoid direct exposure to hazardous substances. Boots and gloves are most easily contaminated and require a conscious effort to keep clean. Often, surface contamination can be removed by scrubbing with soap and water. In other cases, especially with highly viscous liquids, surface decontamination may be practically impossible, and the PPE should be discarded. Many types of chemical protective equipment are considered disposable and are generally removed from service after a single use. It is a fact that once the chemical is absorbed into the fabric, some of the material could continue to diffuse through the fabric. For highly resistant clothing, the amount of chemical reaching the inside may be insignificant. However, for moderately performing fabrics, significant amounts of chemical may reach the inside, resulting in exposure. Remember that PPE is designed to protect the wearer in the event of an unplanned exposure. In other words, your goal should be to leave the work area as clean as possible. In addition to chemical resistance, the duration of the exposure and the surface area exposed affect the amount of chemical that may reach the inside surface. The reuse of PPE that has been contaminated is not advisable.

Regardless of whether the equipment is reused or not, other considerations must be taken into account that relate to the topic of wearing PPE. These include the following:

- While many organizations, including response agencies, use duct tape to reduce penetrations in CPC, duct tape is not classified as chemical resistant and care should be taken when working with it. Duct tape used to seal PPE should be at least two layers thick.
- Stitched seams of clothing may be highly penetrable by chemicals if not overlaid with tape or sealed with a coating. It is advisable to cover the areas where materials can get through CPC.
- Lot-to-lot variations in the manufacturing of PPE may impact the effectiveness of the PPE. Thickness may vary from point to point on the clothing item. Depending on the manufacturing process, the finger crotch area of the glove is particularly susceptible to thin coverage. Inspection of the equipment prior to use can identify defects before they expose the user to the hazards.
- Pinholes may exist in fabrics because of the manufacturing process. Again, inspection of the PPE is necessary prior to using the equipment.
- Garment closures differ significantly from manufacturer to manufacturer and within one manufacturer's product line. Attention should be paid to button and zipper areas and the number of fabric overlaps in these areas.
- Do not wear jewelry, watches, badges, nametags, earrings, or any other items not necessary for protection while wearing PPE. Not only can these items tear or abrade the suit, they may also become contaminated and end up in some landfill instead of your immediate possession. For example, gold is very permeable and can be easily contaminated.
- Facial hair other than closely trimmed mustaches may impede a proper seal around a face mask. While the topic is still a subject of debate in some areas, OSHA clearly states that no facial hair is allowed if it contacts the sealing surface of the respirator.
- The degree of protection provided by an item of protective clothing is also a function of the application it is to be used in. Factors such as abrasion,

puncture and tear resistance, reaction to perspiration, and crumpling of the fabric should be considered. Temperature and, to some extent, humidity influence the performance of some types of PPE. It is important to recognize that protective clothing can be cumbersome and restrictive, and thereby hasten the onset of worker fatigue. A result is that the period of safe and effective worker activity may be reduced.

- Clothing performance is determined by the specific type, the method of formulation of the plastic or elastomer, and the clothing manufacturing process. Materials classified as nitrile rubber can differ significantly in their composition, thickness, and ultimately chemical resistance. Testing the specific brand and product is the only means of identifying the superior equipment for a particular application.

Numerous factors must be taken into account when using PPE.

HEALTH CONSIDERATIONS AND CPC

While wearing PPE helps to control exposure to specific hazards, wearing it also exposes the user to a number of potential hazards. As the layers of protection are added and the levels of protection increase, the body is exposed to added stressors. These are the health conditions that can occur when working with any type of PPE. We already know from the discussion on respiratory protection in Chapter 6 that medical clearances are required for persons who wear respiratory protection. So the addition of other forms of PPE also increases the possibility of health hazards and in some cases require the user to receive medical surveillance as noted in Chapter 1. Consider that working with the protection comes at a cost. This cost includes psychological stress and physiological stress.

Wearing PPE produces both psychological and physiological stress.

Psychological stressors become more pronounced as the layers of protective equipment and levels of protection increase. Again, more protection is not always better because it comes at a price. Getting zipped into a Level A suit can be quite dramatic and has been known to cause great concern to many folks getting into the suit. Claustrophobia and a sudden fit of breathing difficulty can be observed on a regular basis. Consider the fact that most people never think about their breathing at all. It is quiet and without effort. But wearing a respirator, especially a SAR, makes breathing more of an effort, so the user becomes conscious of breathing since they also hear each time they breathe in and out. Few people appreciate the mental toughness that is required to work in extremely uncomfortable, hot, and restrictive equipment while being exposed to dangerous chemicals. The grim reality is that anyone should recognize that if they need to have a protective suit on to protect them from injury or death, they are engaged in a serious endeavor. In addition to this stark realization of potential danger, anyone in almost any level of protection will get hot, sweaty, and cranky. It is important to recognize that wearing high levels of PPE takes some mental discipline to help stay calm and focused. Only with this can the user control the rate of breathing and maintain a workable level of stress. This discipline increases with higher levels of learning about the equipment coupled with repeated practice using the various types of PPE available.

Safety
Psychological factors can lead to physiological ones, resulting in rapid breathing and an accelerated heart rate.

evaporation
the process of drying when liquids are exposed to air

In addition to the mental stress wearing a PPE can cause, a great many physical effects can result from using PPE. The physical effects occur as a result of the body becoming covered by protective materials that limit exposure with the outside environment. While the environment may be hostile due to contamination levels and require covering the body to prevent exposure, it is the covering that not only keeps the hazards out, but also keeps some hazards in. This relates to the body's ability to cool itself, which is a function of **evaporation**. Evaporation is the process where liquids on our skin dry when exposed to air. This drying process is one of the key ways that the body regulates the heat that results from work that is being performed. To be effective, there must be adequate supplies of water, usually from the production of sweat, coupled with the exposure to dry air. The dryness of the air allows the sweat to evaporate and subsequently cools the skin. Either a lack of sweat or a lack of exposure to the air will stop this process and allow heat to build up. Clearly, wearing protective clothing will have an impact on this process and result in heat stress.

Safety
Wearing PPE will reduce the body's ability to cool itself and result in heat stress.

heat stress
conditions such as heat cramps, heat exhaustion, and heat stroke caused by excess heat in the body

Heat stress can manifest itself in several common conditions ranging from minimal to literally life-threatening. It is critical to any study on the use of PPE that the types of heat stress be thoroughly understood so that efforts can be made to reduce the potential for the effects to occur, to recognize symptoms indicating the presence of a form of heat stress, and to treat persons experiencing one or more forms of heat stress. The most commonly encountered forms of heat stress include heat cramps, heat exhaustion, and heat stroke. Each of these is the result of an increase in body temperature coupled with a loss of the ability of the body to regulate the amount and buildup of heat inside itself. In effect, these are the result of a disruption in the normal process of evaporation.

Note
The most common types of heat stress include heat cramps, heat exhaustion, and heat stroke.

Preventing heat stress becomes more difficult as layers of protective equipment are added. In addition to the equipment insulating the body from the outside environment and reducing any chance for the sweat to evaporate, the layers of equipment take additional time to put on and take off, and add weight. Again, the importance of using only the type and amount of protection that is required to protect you from the hazard is of utmost importance. In addition to limiting what protection is used, the following considerations protect workers using high levels of PPE from the effects of heat:

- Do not close up the PPE any sooner than is necessary. Sometimes one person gets ready faster than the other team members. Coordinate the dressing of everyone wearing PPE to ensure that they all are ready to close up at the same time.
- Dress in shady areas with adequate ventilation when possible. This will help ensure maximum cooling prior to closing the equipment.
- Avoid caffeinated beverages prior to extended use of PPE. Caffeine causes a range of conditions including restriction of blood flow that can aggravate heat stress, particularly cramping, and increases the need to urinate more frequently. This can lead to further fluid loss.
- Ensure that everyone using PPE is adequately hydrated before they use the equipment. Drinking moderate amounts of water or an electrolyte-enhanced drink such as many sports drinks is the best way to accomplish this.
- Provide rehydration of personnel wearing PPE at regular intervals. The amount and frequency of rehydration will depend on a number of factors, including the personal condition of each user, the type and level of PPE worn, the amount of physical exertion involved, and even the psychological stressors present.
- Consider placing cooling devices underneath or inside the PPE. These may include cascade-fed air systems attached directly to the suit, chilled-air systems, vests with ice packs, or actual circulating water systems. While the type of cooling system selected varies in its advantages and disadvantages, all of them add weight, make the use of PPE a bit more complicated, and take longer to put on and take off. Figure 7-12 shows a cooling pack used under PPE. No one type of system has been accepted as superior, but wearing one can have a psychological impact on the wearer.

Safety
Prevention of heat stress is the best form of protection from heat.

Heat Cramps

Heat cramps are one of the more common and least dangerous forms of heat stress. Heat cramps are intermittent, painful contractions of skeletal muscles because of fluids shifting inside the body, resulting in a loss of both water and sodium (salt). The consequences of this are cramping and tightening of the muscles, most commonly in the arms, legs, fingers, toes, or abdomen. Sweating brings sodium and other electrolytes to the surface of the skin where evaporation would ordinarily cool the person and reduce the sweat needed to cool off. With the addition of layers of PPE, the sweating does not result in cooling because the sweat does not evaporate under the PPE. This leads to further sweat production with a subsequent loss of fluids, sodium, and other electrolytes. As a result of this, the muscles in the body begin to tighten, which causes the characteristic muscle cramping. Working in hot environments can result in the loss of up to two liters of fluids per hour with substantial sodium loss. While most cases of heat cramps are not particularly dangerous, the sudden tightening of a major leg muscle or a spasm in the fingers could cause a temporary inability to move that could

Figure 7-12
An example of a cooling vest worn under PPE.

become hazardous if it occurs at an inopportune time. The key to working with this and other types of heat stress is to prevent the occurrence of heat cramps by ensuring appropriate hydration.

> **Safety**
> Loss of fluids can result in heat cramps that could create serious complications and could be dangerous.

Common symptoms of heat cramps include the following:

- Cramping sensations or spasms in the hands, feet, arms, legs, or abdomen
- A normal body temperature
- A rapid heartbeat
- A normal mental status

If symptoms occur, treatment of heat cramps includes:

- Removing the person from the hot environment and removal of the outside PPE.

- Rehydration with fluids. Water is acceptable but not the best source if cramping is severe. Because the cramping is due to electrolyte loss, electrolytes most often need to be replaced. Sports drinks containing balanced electrolytes and sodium are often a better substitute, but the helpful effects may be slow. Generally, fluids that are not overly cold are better absorbed by the body, so avoid iced drinks when possible. If the condition of the person does not improve, medical treatment should be sought. Part of the therapy for heat stress includes an intravenous (IV) line supplying fluids and electrolytes such as normal saline (0.9% concentration) or Lactated Ringers. Lactated Ringers is an **isotonic solution** containing potassium chloride, sodium chloride, calcium chloride, and sodium lactate in water. Isotonic solutions contain approximately the same levels of salt and other electrolytes that are normally found in the body and are easily absorbed into the system. IV therapy is an advanced life support skill accomplished by paramedics and hospital personnel.
- Seeking medical attention if cramping does not resolve.

isotonic solution
solution that contains similar amounts of material commonly found in the body

Heat Exhaustion

Heat exhaustion, also called heat prostration, is another commonly encountered heat-related condition. Unlike cramping, which occurs at specific areas of the body, heat exhaustion is more systemic in its effects. As with cramping, the symptoms of heat exhaustion are also created by excessive dehydration or fluid loss due to sweating. The mechanism by which it occurs is related to the loss of fluid without any significant reduction in the heat inside the body. This happens when the sweating takes place under the PPE, which does not allow the sweat to evaporate because there is limited contact with dry air. As the heat rises, sweating becomes more pronounced, resulting in higher levels of sweat production. As this process continues, fluids begin to shift inside the body to provide the sweat glands with more liquids to form sweat. Urine output shuts down, and fluid diverts from other parts of the body. When this occurs, some of the liquid parts of the blood start to give up fluid as well, resulting in a decrease in the blood volume as the liquid portion of the blood drops even further. In effect, a shock-like condition is now underway inside the body and classic signs and symptoms of shock occur.

A person suffering from heat exhaustion may exhibit the following signs and symptoms:

- Profuse sweating
- Flushed face
- Anxiety
- Headache
- Dizziness
- Transient loss of consciousness or fainting in severe cases
- Low urine output
- Accelerated heart and respiratory rate
- Declining blood pressure or **orthostatic hypotension**

orthostatic hypotension
the low blood pressure that is found when you stand up

Orthostatic hypotension is simply a low pressure that results when the vital signs are taken when the person is lying down versus sitting or standing. The condition is caused when the body is able to compensate for fluid loss and maintain normal vital signs such as pulse rate, respirations, and blood pressure when lying down but not when the legs are lower than the heart such as when the person is sitting upright. When sitting, blood can pool in the lower extremities and result in lower blood pressure and higher pulse rates. To ascertain if this condition exists, the vital signs are taken and recorded with the person lying down. The vital signs are then repeated with the person sitting down or standing. When sitting, the legs should be hanging below the torso as if properly sitting in a chair. If the systolic blood pressure (top number) drops by more than 20–30 points or the pulse rate increases by more than 20 beats per minute, the person is considered to have a positive orthostatic change in vital signs. In the setting of suspected fluid loss, this could indicate a more serious medical condition that could require aggressive fluid replacement including IV therapy. In cases of substantial postural hypotension, the person will not be able to tolerate standing for any length of time. When taking orthostatic vitals with the patient in the standing or sitting position, make sure to pay attention as the sudden decline in blood pressure could result in a feeling of dizziness or even fainting momentarily. If this occurs, immediately allow the person to lie flat until fluids have been replaced and the orthostatic changes do not occur.

Treatment of persons with heat exhaustion symptoms includes:

- Removing the person from the hot environment
- Removing outer clothing to allow sweat to evaporate
- Providing air movement by fans or other means if available
- Applying cool compresses may be applied to the armpits, back of the knees, and/or back of the neck
- Rehydration with fluids. Water is an acceptable method of replacing the lost fluids because the body still has the ability to produce sweat. If water is administered, it should not be ice cold as this can slow the absorption into the system. If the condition of the person does not improve, medical treatment should be provided because IV fluids may be required as with heat cramps. Note that monitoring the urine production is an effective means of preventing this condition. If urine production is low or if the urine color turns dark, it indicates a rising lack of fluid within the body. Regular replacement of fluids should result in regular urine production that is normal colored.

Heat Stroke

Heat stroke is the type of heat stress that constitutes a true life-threatening emergency condition because it occurs when the body's temperature regulation mechanism is not able to control the heat in the body. Heat regulation by the body is done through the hypothalamus, a gland inside the brain that monitors and regulates body functions and ultimately the body temperature. When the regulatory mechanism is not working, the person will suffer severe and uncontrolled **hyperthermia**. The result is the inability of the body to cool itself, rising internal body temperatures, brain damage, and ultimately death if not resolved. The fatality rate for this condition can be as high as 20% and, while rare, it is a condition that needs to be carefully monitored.

hyperthermia
a condition where excess heat builds inside the body

Heat stroke can occur by itself or in combination with unresolved heat exhaustion that could occur first. It is sometimes associated with taking certain types of medications or with specific medical conditions. A hazardous materials medical surveillance program should screen workers for this condition because the consequences can be extreme. When it occurs, the core temperature of the body can reach 105°F, well above the normal temperature limit. Because the body is unable to cool itself, sweating may not be present. The lack of sweating is one visible sign of this condition and should be carefully monitored since it can lead to further heating. Without the ability to regulate the heat, the brain quickly starts to show the effects of the rising heat, which further exacerbate the problems because the person may continue to work in the hazardous environment with a rapidly declining mental state. Other medical conditions within the body also begin to occur, resulting in a deadly combination of symptoms if not quickly resolved.

A person suffering from heat stroke may be exhibiting the following signs and symptoms:

- Hot, red, and often dry skin. The noticeable absence of sweat is something that should attract attention to the possibility of this condition.
- A declining mental state with symptoms of heat exhaustion such as anxiety, confusion, light-headedness, or delirium
- Loss of consciousness or coma in extreme cases
- Seizure activity related to the high brain temperatures
- A core temperature above 105°F
- Declining vital signs such as rapid pulse rates, rapid respiratory rate, and ultimately low blood pressure

Treatment of heat stroke includes the following:

- Remove the person from the hot environment.
- Remove outer clothing and begin applying water to the exterior of the body. Because sweat may not be present, the application of water can begin evaporation and cooling. Water should be cool but not cold because shivering could result in the production of more body heat.
- Fan the person, if possible, to allow for rapid cooling.
- Seek medical attention immediately. An ambulance will be needed to transport the victim to get prompt medical care.

Safety

Persons suspected of suffering from heat stroke must receive immediate medical attention.

The presence of heat stroke symptoms requires immediate medical intervention. Minor cases of heat cramps or heat exhaustion may be managed at the site with fluid and electrolyte administration, rest, and time to allow the body to recover. Once urine output has returned and the patient is fully oriented to the conditions, it may be appropriate to allow him to return to work. The brain is often one of the best indicators of what is going on inside the body relative to heat. Because it suffers the effects of heat stress early in the process, it can lead us to know whether the person is capable of going back to work. If there is ever a doubt, keep the person from wearing PPE until a medical clearance is obtained; otherwise it would lead to increased heat stress.

SUMMARY

- PPE stands for personal protective equipment. This includes but is not limited to CPC, fire-resistive clothing, boots, gloves, splash aprons, face shields, and goggles.
- Because of the range of hazards and the fact that no one type of PPE that is protective from all hazards, the selection of the proper PPE starts with an in-depth assessment of the hazards present and the work to be done.
- There are three major types of PPE: structural firefighting clothing, high-temperature protective clothing, and CPC. Only specific types of CPC will provide protection from chemical hazards.
- PPE is the final line of defense against hazardous materials and requires that the items be carefully selected to ensure that protection is afforded.
- There is no one type of CPC that is chemical proof. It is important to remember that one suit does not cover all hazards.
- PPE is not worn to enable or encourage you to intentionally contact the chemicals you are working with. It is worn to protect you in the event of an *unplanned* exposure.
- CPC is divided into three subcategories by the NFPA. They are vapor protective suits, liquid splash suits, and support function protective garments. The following standards are relevant to PPE.
  - NFPA 1991 *Vapor Protective Suits for Hazardous Chemical Emergencies*
  - NFPA 1992 *Liquid Splash Suits for Hazardous Chemical Emergencies*
  - NFPA 1993 *Support Function Protective Garments for Hazardous Chemical Operations*
- In order to more effectively choose an appropriate level of protection, ask yourself these questions:
  - Have I positively identified the product and all the hazards associated with it?
  - Have I chosen the type and level of protection appropriate for the hazard?
  - Have I selected the specific fabric or material that is appropriate to the hazards present?
- Level D protection is composed of a normal work uniform and is considered the lowest level of protection available. Level D is utilized when there is no danger of exposure to the worker from any airborne materials, or when the chemicals used do not pose a hazard by inhalation.
- The following list illustrates some common items found in Level D protection, although specific types vary with the hazards present.
  - Safety glasses or goggles
  - Face shield
  - Coveralls
  - Chemical-resistant safety shoes or boots with a steel toe and shank
  - Chemical-resistant boots or booties over work shoes or boots, if necessary
  - Gloves (these may vary with the hazard encountered)
  - Hard hat
- Level C protection includes a NIOSH-approved APR and is used when conditions are appropriate to the APR. A Level C ensemble includes a range of equipment based on the hazards present, including the following:
  - NIOSH-approved half-face or full-face APR
  - Chemical-resistant clothing compatible with the material being used or present. This could range from a simple splash apron to a one-piece coverall or jumpsuit.
  - Chemical-resistant outer gloves with chemical-resistant inner gloves
  - Chemical-resistant safety shoes or boots with a steel toe and shank
  - Hard hat
  - Face shield or safety goggles if half-face respirator is used

- Level B protection is used whenever a chemical has a high level of inhalation hazard present. It is different from Level D and Level C in that it includes a SAR.
- Level B attire may adhere to the NFPA 1992 standard for *Liquid Splash Suits for Hazardous Chemical Emergencies* and can include the following equipment:
  - NIOSH-approved positive-pressure supplied-air respirator. Usually a self-contained breathing apparatus, although an air-line system may also be used
  - Chemical-resistant clothing compatible with the material being used or present. This clothing may consist of long-sleeved chemical-resistive overalls, a hooded chemical-resistive splash suit, and non-encapsulating or encapsulating chemical-resistive suits.
  - Chemical-resistant outer gloves with chemical-resistant inner gloves
  - Chemical-resistant safety boots with a steel toe and shank. Disposable over-booties are always recommended when contamination levels are expected to be high.
- Level A protection is the highest level of protection and is required when the hazardous substance that is being used or which has been released presents a significant threat by inhalation and skin contact. It uses the highest level of respiratory protection worn inside a fully encapsulating, gastight suit. It includes the following items:
  - NIOSH-approved positive-pressure supplied-air respirator. Usually a self-contained breathing apparatus or supplied-air system with escape cylinder
  - Vapor protective suits. These are totally encapsulating, gastight garments that completely enclose the user inside the suit with a suitable air supply.
  - Chemical-resistant glove system
  - Chemical-resistant safety boots with a steel toe and shank. These are worn over the Level A suit.
  - Recommended equipment includes a hard hat, fire-retardant coveralls underneath the CPC, cotton or fire-retardant long underwear, and cooling vests as indicated.
- Select Level D when all of the following conditions are met:
  - The atmosphere contains no known hazards.
  - Work functions preclude a potential for an airborne chemical exposure.
  - A degree of basic safety has been met at incidents or work sites. This base level of protection may allow for a quick upgrade into higher levels of protection through the inclusion of a respirator should it be required.
- Select Level C when all of the following conditions are met:
  - All criteria for an APR are met. This includes:
    - Oxygen levels are above 19.5%.
    - No thermal extremes are present.
    - An approved cartridge is available for a known material.
    - The levels of contamination are below IDLH levels.
    - Any atmospheric contaminants, splashes, or direct contact with the substances can be adequately blocked by appropriate non-gastight CPC.
- Select Level B when all of the following conditions are met:
  - All criteria for an APR cannot be met. This includes:
    - Oxygen levels could be below 19.5%.
    - Thermal extremes can be present.
    - No approved cartridge is available.
    - The material may not be known.
    - IDLH levels of contamination may be present.
    - A high inhalation hazard is present from a material that can be adequately blocked by appropriate non-gastight CPC.
    - Assessment of spills of unknown origin when adequate ventilation can be assured to dilute concentrations to below IDLH levels for skin contact.

- Select Level A when all of the following conditions are met:
  - All criteria for the use of an APR cannot be met. This includes:
  - Entering confined areas or poorly ventilated areas where the materials present are known to present a serious threat of immediate skin destruction.
  - Entering confined areas or poorly ventilated areas where the materials present are known to present a high potential for splash, immersion, or exposure that could produce serious IDLH conditions due to a dermal exposure.
  - Dealing with a situation involving a material that is completely unknown and that could have the potential for serious skin contact effects.
- While Level B is the minimum level of protection for dealing with unknown materials in some situations, some response agencies require the use of Level A for emergency situations involving these materials.
- If the additional benefits of wearing Level A protection do not offset the disadvantages, then Level A may actually be less safe than Level B in some instances.
- Reuse of CPC is an issue that needs to be carefully reviewed and should become part of the organization's standard operating procedures.
- Hazardous substances can work their way into CPC through the processes of penetrations, degradation, and permeation.
- Penetrations are openings in the CPC, which include zippers, seams, tears, and other openings where the body is exposed.
- Degradation is the destruction of the CPC fabric caused by exposure to materials that break down the CPC fabric.
- Permeation is the slow movement of a chemical from the outside to the inside of the particular type of PPE.
- Commonly encountered forms of heat stress include heat cramps, heat exhaustion, and heat stroke.
- Heat cramps and heat exhaustion are common in warmer climates and work environments, especially when workers are using PPE. Both conditions are not usually life-threatening and are the result of profuse sweating with subsequent loss of fluids and electrolytes.
- Heat stroke constitutes a potentially life-threatening emergency that occurs when the body's temperature regulation mechanism is lost, resulting in lack of sweating and increasing internal body temperature. Immediate medical attention is required for a person suffering from heat stroke.

REVIEW QUESTIONS

1. Briefly describe the three major types of PPE and their application to hazardous situations.
2. List the four questions that need to be answered in order to select the appropriate type of PPE that is required for a particular situation.
3. List the four levels of protection and describe the selection criteria for each level.
4. Identify equipment used in each of the four levels of protection.
5. List the differences between an encapsulating suit and a fully encapsulating gastight suit.
6. Describe how penetrations, degradation, and permeation allow hazardous materials to enter PPE.
7. List the three types of heat stress, signs and symptoms of each, and treatment of a worker suffering from each of the conditions.
8. List four ways to prevent heat stress.

ACTIVITY

COMPATIBILITY EXERCISE

Use the chemical compatibility chart for gloves in Figure 7-11 and obtain an MSDS for the materials presented in the questions.

- Step 1: Read the following scenarios and research the listed chemical or material.
- Step 2: Identify the properties of the material involved.
- Step 3: Review the format of the compatibility chart in Figure 7-11.
- Step 4: Review the information on the chart to determine the type of glove that provides the highest resistance and the one with the least resistance to the material involved.
- Step 5: Answer the questions regarding each scenario presented.

1. *Carbon Tetrachloride Spill.* The response team has been called to the scene of a carbon tetrachloride spill in the laboratory. Upon arrival, you find approximately five gallons of liquid on the level tile floor. The student who was pouring the chemical has a minor splash exposure to the right forearm. The liquid has a sweetish odor and the room has poor ventilation.

Make sure to address at least the following points:

a. What level of protection would you choose to deal with the problem?
b. What is the recommended glove?
c. Are any glove materials not recommended?
d. Is respiratory protection advisable for this situation?
e. How would you handle the exposure of the student? What are the signs and symptoms of overexposure?

2. *Odor Investigation.* The response team has been called to investigate an odor coming from the storage room. Upon arrival you look through the glass door and see a four pack of containers lying on the floor. You see a red label on the side of the box that reads "Flammable" and the word "Toluene" on the side of one of the spilled bottles. You have been assigned to research the chemical and recommend a glove for the entry team.

Make sure to address at least the following points:

a. What is the recommended glove?
b. Are any glove materials not recommended?
c. Is respiratory protection advisable for this situation? If so, what type of respiratory protection would you recommend?

3. *Unknown Spill.* The response team has been called to investigate an odor emitting from the cargo area of a large delivery truck. The truck is parked at the rear of the building by the loading ramp. The driver tells you that he started feeling dizzy when he was unloading some packages. Other people have also stated that they could smell something in the air. The driver is a little shaken but tells you the boxes he was unloading were marked "sulfuric acid" on the side. You are assigned to help determine the appropriate PPE for this operation.

Make sure to address at least the following points:

a. What is the recommended glove?
b. Are any glove materials not recommended?
c. Is respiratory protection advisable for this situation? If so, what type of respiratory protection would you recommend?
d. Where would you find information on the signs and symptoms of overexposure to this material?
e. Would an APR be acceptable if the airborne vapor concentration was below the IDLH for sulfuric acid? Why or why not?

PRINCIPLES OF DECONTAMINATION

Learning Objectives

Upon completion of this chapter, you should be able to:

- Define decontamination.
- List the two purposes of personnel decontamination.
- Describe when and how emergency decontamination procedures should be used.
- Describe the various methods of decontamination.
- List seven factors that determine the extent of decontamination.
- Identify the major factors in selecting a decontamination site.
- Describe the process of setting up a decontamination corridor.
- Describe the location of a decontamination corridor and the importance of Control Zones.
- List the seven steps of Level B decontamination.
- Describe the responsibilities of the Decontamination Team Leader.

CASE STUDY

Sometimes emergency responders accidentally expand a chemical release beyond its original site. This usually occurs when responders' emotions become too involved in the emergency response, causing responders to become narrowly focused on seeking medical care for the injured without properly considering the risks involved in relocating them. This is the case when responders decide to move chemically contaminated victims from a disaster site to another location without first decontaminating the victims. In fact, the emergency departments of many large hospitals have been closed to facilitate the decontamination of chemically contaminated injured.

Relocating contaminated victims may seem like a good idea, but it can create serious problems beyond the original emergency site. One of the most notorious of these incidents, which has become known as the "toxic lady" situation, occurred as the result of a series of very unusual circumstances. The story is summarized below. It illustrates how chemical hazards can affect more than the original victim and makes the important point that exposed personnel and victims should be decontaminated before they are relocated to a hospital.

The toxic lady appeared in 1994 in Riverside, California, when a 31-year-old woman was brought to a hospital in severe distress from cardiac and respiratory problems. During the resuscitation efforts, emergency personnel reported a foul odor. This odor caused some of them to become sick and pass out. Consequently, the entire emergency department was sealed shut, and all other patients were evacuated. The county emergency responders eventually were able to decontaminate the emergency department, and the cause of the incident is still unknown.

This case was extreme, so it may differ slightly from other cases where contaminated persons were delivered to an emergency department. Nevertheless, it illustrates critical concepts about decontamination, which are discussed in this chapter. Failing to implement these principles makes it likely that the emergency release simply will be relocated to the emergency department and that the victims will suffer prolonged exposure until they can be properly decontaminated.

INTRODUCTION

Decontamination is one of the most important aspects of working with hazardous materials, whether you are a worker or an emergency responder. The process of decontamination can be as simple as washing your hands or taking a shower at work before going home, or as complex as going through an elaborate system of pools, hoses, brushes, and other personnel. Regardless, decontamination reduces the potential for workers' exposure to hazardous substances and prevents the spread of contaminants outside the work area. The steps and techniques involved in decontamination vary from situation to situation, but the concepts should be universally applied to situations where decontamination is required.

Safety
Decontamination procedures are important to ensuring the safety of everyone who works with hazardous materials.

The variation of decontamination processes can be illustrated by activities that occur outside the workplace. Taking a shower, for example, is a process that decontaminates you by removing chemical hazards. Washing your hands also is a method of decontamination. The method chosen, the type of soaps used, and the time expended in each of these activities determine the level of decontamination. Unlike these examples of home decontamination, however, the consequences of not doing a complete job of decontamination can be considerably more severe when you are working with hazardous materials. In such cases, we need to ensure that decontamination is thorough and systematic so as to reduce the risks of working with these materials. Taking a systematic approach allows the decontamination process to be complete and thorough, no matter what materials are involved. We must be thorough because many contaminants are not visible to the naked eye or to our other senses. Even if the presence of a contaminant is only suspected, the rule is to provide for thorough decontamination. Figure 8-1 shows an example of a relatively complex decontamination process.

Note
The process of decontamination can vary, depending on a number of factors, but it must be systematic.

Safety
It is vital to assume that all personnel who work with a hazardous material are contaminated and require appropriate decontamination.

Figure 8-1
An example of extensive decontamination activities being practiced by students. (Courtesy of San Ramon Valley Fire Protection District.)

THE WHAT, WHY, HOW, AND WHERE OF DECONTAMINATION

The concepts of decontamination can be explained by reviewing what decontamination is, why it is done, how the process is completed, and the location where it is performed. Once these concepts are understood, their application to a variety of situations becomes clear.

What Is Decontamination?

Numerous definitions of decontamination exist, encompassing people and property. Because our study involves personnel safety, we will use the definition in the Hazardous Waste Operations and Emergency Response (HAZWOPER) regulation. In the regulation, Occupational Safety and Health Administration (OSHA) defines decontamination as the systematic removal of "hazardous substances from employees and their equipment to the extent necessary to preclude the occurrence of foreseeable adverse health effects." Expanding on this, we can state that decontamination is the systematic removal of contaminants from personnel and equipment because it is critical that the process be done in a standardized and complete manner. Other definitions include decontamination of the area where hazardous materials are found, such as decontamination of the soil around an underground storage tank that may have leaked. However, personnel decontamination is the focus of our study.

Additionally, note that the process of decontamination often is referred to as **decon**.

decon
the shortened term for decontamination

Why Is Decontamination Necessary?

The purpose of personnel decon is twofold. First, and primary in establishing a reason for decon programs, is to provide a system of protection for personnel who may have been exposed to a hazardous substance. This is accomplished by the systematic, and as complete as necessary, removal of contamination from the person and their personal protective equipment (PPE) in an effort to reduce the potential for harm if the person has been exposed. Recall from Chapter 7 that all PPE is subject to failure, and the longer that a contaminant is in contact with the PPE, the greater the likelihood that an exposure can occur.

> **Safety**
> The primary "why" of decon is to remove contamination from personnel in an effort to reduce their potential for exposure to the hazardous materials present.

> **Safety**
> The longer the amount of time before decon, the greater the potential for exposure to that material.

The second purpose of decontamination also is related to personnel protection, but of personnel who may not have entered the hazardous area and who may not be aware of being exposed. In other words, the second purpose of decontamination programs is to ensure that the contamination does not spread from the release area to other areas. Effectively controlling contaminants through the decontamination of personnel and equipment limits the spread of that contamination to other personnel who were not originally exposed, to other equipment, and to the environment outside the area of original contamination.

> **Safety**
> The secondary "why" of decon is to prevent the spread of contaminants from the area of release.

The Three Methods of Decontamination How decontamination is performed is a topic of considerable debate. No one way is accepted by everyone involved in decontamination practices because there are so many variables in the process and there are multiple ways to accomplish the task. The procedures used in one decon program may vary widely from the practices used in others. For example, many organizations, including the military, have developed procedures that are used consistently regardless of the circumstances and of factors that might cause other organizations to use different methods. Depending on the material(s) involved, the location of the material(s), the availability of water and other equipment, and local procedures, the process varies considerably. Water is one of the most widely used materials, but it may not be effective for washing off thick, heavy oils or caked-on powders. As long as the principles of decontamination are followed, however, many different practices may be acceptable.

> **Note**
> Decon practices can vary as long as the principles of decontamination are followed.

Although the specific activities vary, most decontamination programs use one or more of three generally accepted methods. These methods are dilution, absorption, and chemical degradation. Each method varies in how it accomplishes decontamination, and each method may be more or less appropriate, depending on the circumstances.

> **Note**
> The three generally accepted methods of decontamination are dilution, absorption, and chemical degradation.

dilution
using water to flush contaminants from personnel or equipment

Dilution probably is the most common method of decontamination and involves the use of water to flush contaminants from personnel, protective clothing, and equipment. It is the most common method because water is readily available and able to dilute many hazardous materials by itself or in combination with soaps and other materials. However, using water can present issues that need to be addressed, including the following:

- Water is not effective with a hazardous material that is not soluble in water. In such cases, detergents or other materials may need to be added to water to help break down the contaminant. Oil is an example of this.
- The water used to wash items must be properly contained and disposed of because it probably is considered to be hazardous waste itself as it now contains hazardous materials.
- Water may not effectively wash off powders. Large amounts of powder can be trapped inside the outer surface of a wet material, much like pockets of dry ingredients in pancake batter that is not properly mixed. When

large amounts of powder are present on clothing or equipment, a brush may need to be used to remove the majority of the powder before water is used to wash away the rest.

- Some materials react with water and generate heat. If that happens, large amounts of water are required to negate the effects of the heat.
- Large amounts of water are required for highly concentrated materials. Applying water to hazardous materials only reduces the concentration of the material, but it does not necessarily change the strength of the material. For example, if you can attempt to dilute sulfuric acid with water to change the pH, but a huge amount of water will be required to bring the pH close to neutral. Figure 8-2 shows an example of using water to dilute a substance during a decon operation.

Note

Dilution is the most common method of decontamination and uses water to flush contaminants from people or equipment.

Safety

Water should not automatically be used for decontamination because it can present other problems.

absorption
a decon process where the contaminant is picked up by a neutral material

Absorption, the second method of decontamination, uses absorbents or absorbent materials to soak up a contaminant. This method is effective for decontaminating equipment or property, but it is far less effective for decontaminating

Figure 8-2
Decon operations using water to dilute the materials involved. (Courtesy of San Ramon Valley Fire Protection District.)

personnel. Absorbent materials include soil, clean dry sand, cat litter, pads and socks made of absorbent fabric, and commercially available products, such as expanded clay.

An advantage of the absorption technique is its ability to minimize the surface area of a liquid spill. Additionally, many absorbents are inexpensive, easy to use, and readily available. The major disadvantage of this method is that, as we have said, it is not practical for the primary focus of decontamination—use on personnel. When an absorption material is used to decontaminate an area, its use is limited to flat surfaces on which the contaminant has not soaked into the ground, such as paved areas. Keep in mind that the hazardous materials confined to an absorbent remain chemically unchanged and retain their original health and safety hazards. Finally, the use of absorbents usually creates a larger amount of contaminated material that, again, has to be properly disposed of. For example, absorbent spill pads are very efficient at picking up oils, but the dirty pads must be properly collected for proper disposal at a later time (Figure 8-3).

■ **Note**

Absorption is the form of decontamination most useful for picking up spilled materials.

degradation
a form of decon that involves altering the chemical hazards present

Much like the degradation of chemical protective fabrics, chemical **degradation** in the decontamination process alters the chemical structure of a hazardous material. Degradation reduces a hazard material in a controlled fashion by exposing the chemical to other materials that negate the hazardous substance's primary properties. Some commonly used chemicals for degradation are sodium hypochlorite (household bleach), sodium hydroxide as a saturated

Figure 8-3
An example of absorbents that can be used to decontaminate sites. (Courtesy of San Ramon Valley Fire Protection District.)

solution (household drain cleaner), sodium carbonate slurry (washing soda), calcium oxide slurry (hydrated lime), liquid household detergents (such as dishwashing liquids), and ethyl alcohol.

It is essential to know the specific type of contaminant to select the appropriate chemical to use for degradation. For example, a common practice is neutralizing corrosives by using a mild base to neutralize a strong acid. However, neutralization and many other types of degradation results in an exothermic reaction—in other words, heat is generated by the reaction of the materials. For this reason, degradation is not commonly used to decontaminate personnel or their PPE because heat may damage personnel and PPE. Remember, the goal of chemical degradation is to render the hazardous material less harmful than it was prior to decontamination. Degradation is most often used to make the final disposal of the contaminants less hazardous.

■ Note
Chemical degradation breaks down a contaminant's hazardous properties and is most useful on equipment.

! Safety
Chemical degradation should not be used to decontaminate personnel or PPE because the heat produced by the process can damage personnel and PPE.

How Much Decontamination Is Required?

Before any decontamination can be done, we must determine how much decontamination is needed. For example, showering is a form of decontamination, but we do not use the same method each time we shower. Some showers are quick ones to wash off a little sweat. Other showers are more complete and include washing our hair or dissolving various greases or oils on our skin. Additionally, the amount of showering (decontamination) is dependent on what we are going to do next. If we are going to a social event, the shower may be lengthy and involved. If we are going to work, where we will quickly get dirty again, the process may be far less involved. These same considerations are part of the process to determine the required amount of chemical decontamination. Minor exposure to low-risk hazardous substances may require a relatively simple decontamination process. More complex incidents involving significant potential for exposure to highly hazardous substances will require more decontamination using considerably more equipment, personnel, and time.

Seven factors must be considered when determining how involved the decontamination process should be. These factors raise obvious questions that will help you identify the amount, type, and complexity of the decon process. The seven factors involved in the decision process are:

1. Type of contaminant
2. Amount of contaminant
3. Level of protection
4. Work function
5. Location of contaminant
6. Reason for leaving the site
7. Type of protective clothing worn

> **■ Note**
> The required amount and type of decontamination is determined on the basis of seven factors.

Some organizations do not follow any type of decontamination decision process because they already have established a standard protocol to follow when decontamination is required. However, simply following an existing protocol may not lead to the best results—remember that more is not always better. Like the thought process involved in selecting PPE, when determining what level of decon is needed, it is best to use only what is required to safely perform the required tasks, because taking more time than needed and using more equipment than required probably will expose more personnel to the hazards for longer periods of time. Instead, try to keep all of your steps, processes, equipment, and personnel at the minimum levels required to do the job properly. Decontamination is far too costly a process to involve unneeded time and equipment. Also, the more complex the process, the greater the room for error. However, *never* take shortcuts when performing decontamination. Assume that everyone and everything that is in an area where contaminants are present is contaminated and must be appropriately decontaminated to reduce individuals' potential exposure and to prevent the spread of contamination from the area.

The first factor in determining how much and what type of decontamination process is required is identifying which hazardous material(s) are present. Knowing the specific types of contaminants helps you identify the materials' hazards and properties. Highly toxic materials present a high hazard and increase the risk of exposure if appropriate decontamination is not conducted. Materials that do not pose a significant hazard may not require significant amounts of decontamination. The material's properties can guide you toward the appropriate type and amount of decon. Knowing whether the material is soluble in water, for example, is critical if water is to be used to flush the contaminant. If the material is not water soluble, decontamination needs to include soaps or other materials to break the material down. Also, if a material is volatile, it may release more vapors and rapidly evaporate. This could create additional hazards for those involved in the process, and also reduce the amount of material that ultimately will need to be removed. The more that we know about a material, the more confidently we can select the appropriate type and amount of decontamination that is required.

> **! Safety**
> Identifying the materials present is critical to determining the type and amount of decontamination required.

> **■ Note**
> Once a material is known, its properties must be evaluated to determine appropriate decontamination.

The second factor in determining the necessary type and amount of decontamination is to ascertain the amount of contamination present. The amount of contaminant cannot always be determined; in these cases, preparing for significant exposure is the appropriate course of action. If the amount of contaminant is measurable and is small, however, the amount of decontamination required likely can be reduced. Clearly, the potential for serious contamination increases with

entry team
the personnel who enter an area where contamination is present; they can be part of a cleanup group or an emergency response team

greater amounts of material. If a material is well-contained and not spreading, however, exposures to it probably will not result in serious contamination. But, if equipment or the persons who enter a site as part of the **entry team** are heavily covered with a contaminant, then more thorough decon is required. Decon team members must consider that any hazardous material on protective clothing may penetrate, degrade, or permeate the clothing and harm the user. Therefore, the decontamination process must be organized, rapid, and thorough.

> **■ Note**
> The amount of contaminant present is an important factor in determining the amount and type of decontamination required.

The question of how much contaminant is present affects how many people require decontamination and which of those who enter a site will be decontaminated first. This question spurs considerable debate among those who work in the various industries where hazardous materials decontamination is conducted. The HAZWOPER regulation under 29 CFR, Part 1910.120 requires the use of the "buddy system," so decontamination teams must determine which of the people who enter a contaminated area will be decontaminated first.

> **■ Note**
> Decontamination teams need a method for determining which member of the entry team will be decontaminated first.

The answer to the question may seem clear, but the debate still rages. On one side of the debate are those who insist that the least contaminated person leaving the site should be decontaminated first. Their logic is that the decontamination of this team member results in less contamination being brought into the decontamination area because they surmise that decontamination concentrates contaminants into the pools or other areas where other personnel may then be exposed. This logic is flawed, however, given the very premise of decontamination and an understanding of the protection afforded by various parts of PPE systems. Recall that the longer that a material remains in contact with the PPE, the greater the potential for exposure of the person wearing it. Obviously, the one with the greatest amount of contamination has the greatest exposure potential.

> **■ Note**
> Some believe that the cleanest entry team member should go through the decon process first.

The contrary approach is that the person leaving the site with the most visible or greatest potential for contamination should be decontaminated first. Doing this reduces the potential for exposure to the contaminants for the person with the highest potential for exposure given the amount of contaminant present. Recall that the primary definition of and reason for decontamination is to reduce the potential for exposure to exposed entry team members who are more contaminated than the others. The greater amount of contaminant on clothing presents the highest potential that the material could enter the PPE through one of the routes discussed in Chapter 7. Recall that hazardous materials can enter through openings or penetrations, degrade the fabric through contact, or permeate gradually through the fabric.

The longer a substance is in contact with PPE, especially in higher amounts, the higher the potential for the user to be exposed to the substance. For this reason, it makes the most sense to have the most contaminated team member, or the team member with the most potential for contamination, to be decontaminated first.

> **■ Note**
> Decontaminating the dirtiest member of the entry team first is more in line with the purpose of decontamination than decontaminating the least dirty member of the team first.

The logic of decontaminating the least contaminated first is flawed in another way, as well. The argument against decontaminating the most contaminated team member first also assumes that the most contaminated person brings the most contamination into the decontamination area. Further, it assumes that because the water used to remove the material usually is captured in pools or other containers, it may become contaminated. If this happens, the argument goes, subsequent members of the entry team are forced to stand in water that has been significantly contaminated. Even if this occurs, however, the water exposes only the boots of team members who follow the first decontaminated member. Given that boots are one of the most protective parts of the PPE ensemble, the potential for a user to be exposed in this way is extremely low and should not be a consideration.

> **■ Note**
> The potential for exposing personnel to the hazards of the decontamination water is minimal.

Although the rationale of decontaminating the most contaminated person first seems most defensible, other factors may enter into the decision about who is decontaminated first. One of the most critical factors to consider is the amount of available air in self-contained breathing apparatus (SCBA) units. If one team member has significantly less air available for breathing, that factor may determine who goes first. Other issues, such as potential suit failure or damage, or the medical condition of an entry team member, also may override the premise that the most contaminated person will be decontaminated first. Regardless of the method used to determine who is decontaminated when, each team or organization must develop protocols that should be understood and followed by everyone working at a site.

> **■ Note**
> Factors such as low air, damaged PPE, or medical issues also affect the decision as to who goes first through the decontamination process.

The third factor in determining the amount of decontamination is the level of protection worn by entry team members. This factor relates to the first two factors because materials that pose higher hazards and are present in larger quantities usually require that the entry team wear an upgraded level of protection. Generally speaking, higher levels of protection require higher amounts of decontamination because the existing hazards are a major consideration in determining which decontamination program is used. This factor also affects the level of protection selected for the decontamination team.

There are many variations of standard practices, but most hazardous material organizations require that decontamination team members use a level of protection one level below that chosen for the entry team. This practice is not universal but it is common, because it presumes that the hazards inside the contamination area are higher than the resulting exposure of the entry team members. In other words, the entry team will not be as contaminated as the site given that the entry team members have standard procedures that minimize their exposure to the hazards. This practice results in some contamination being brought out of the contaminated area, but not at as high a level as exists inside the area. Other organizations require that decontamination team members use the same level of protection as the entry team or some variation of that practice. Regardless of what is decided about levels of PPE, all personnel working at the site should be fully apprised of what is required and follow the standard protocols selected.

> **■ Note**
> Decontamination team members often wear protection one level below the level worn by the entry team.

The fourth factor in our decontamination decision process is the work functions of the entry team. Regardless of the other factors involved, the potential for exposure to a hazardous substance is a key indicator of the potential for harm to entry team members. If the entry team is going into the contaminated area on a fact-finding mission to determine the extent or specific types of the materials involved, they are less likely to be contaminated by the materials. Team members often enter the area and maintain considerable distance from the hazardous substances, and therefore have limited exposure to the substances. A decontamination operation still should be set up based on the expected exposure because there is always the potential for contamination. However, knowing the job function of the entry team assists the decontamination team in setting up the appropriate decontamination operations.

Knowing which materials may be present at a site must be factored into this aspect of our decision process. For example, a spill on a floor of a viscous liquid that has limited volatility has the potential for only limited contact with the entry team if the team uses good judgment and proper entry procedures. However, if a team enters an area contaminated by a leak from a pressurized piping system (such as ammonia), the likelihood that the team members will be exposed is much higher regardless of the team's work activities.

> **■ Note**
> The work activities conducted by the entry team affect the amount and type of decontamination conducted.

The fifth factor in our decision process is the physical location of a contaminant on potentially contaminated personnel. From our prior discussions of toxicity and PPE, we know that some areas of the body are more vulnerable to exposure than others. If the upper portion of the bodies of the entry team members is exposed to the contaminants, the exposure may be more harmful because the material will be closer to the team member's respiratory system and face. As we know, these areas pose the greatest risk of exposure from inhalation or from rapid absorption through mucous membranes in the face or eyes. These areas of the body also are more vulnerable to a material entering PPE through its openings because more openings occur in the top half of protective ensembles. For this reason, the decontamination process must be more complex when there is potential exposure to the head and chest areas.

However, the chest and head are not the most commonly contaminated areas of the body in most operations. The most commonly exposed areas are the hands and feet because contamination on the ground exposes the feet to a material's hazards and working with the hands results in their contamination. If the only exposure potential is to the feet, the amount of decontamination can be significantly reduced because, as we have said, boots provide high levels of protection, and exposure to the critical parts of our bodies would be minimal.

Note
The two areas most commonly exposed to high levels of contaminants are the hands and feet.

Safety
Potential exposure to the upper portions of the body presents the highest risk of exposure and requires the most extensive decontamination operations.

The reason for leaving the contaminated area is the sixth factor that determines the need for and extent of decontamination operations. For example, if an entry team member leaves the area of a spill to get a new air cylinder, to change an air-purifying respirator cartridge, or to drop off or pick up new equipment, less decontamination is required for this aspect of the operation because the contaminants do not leave the decontamination area. As we will see in the discussion of where decontamination activities take place, a special area of the site is designated to become contaminated by those who leave the site. This area is a buffer zone between the contaminated and noncontaminated areas. Conducting decontamination operations in this buffer zone contaminates it, which means that the area then requires its own decontamination. However, when entry team members enter the buffer zone for a bottle change, little decontamination is required because they remain in their suits and do not take the contaminant out of this area. In fact, once team members get a new air cylinder or tool, they usually reenter the contamination area to continue work. Figure 8-4 shows a worker who has been partially decontaminated and is getting a new air cylinder prior to returning to the Hot Zone.

Note
Less decontamination is needed when entry team members leave the contaminated area only to enter the buffer zone then reenter the contaminated area.

The last factor in determining the type and amount of decontamination involves the specific type of PPE worn by the entry team. Because many types of chemical protective clothing (CPC) are disposable and are taken off in the decontamination area where they can be placed into bags or other sealed containers, they require minimal decontamination. For example, disposable Tyvek® suits are inexpensive and designed to be used only once. When someone using this garment requires decontamination, less decontamination and time are needed because in this case, the purpose of the decontamination operations is to remove enough of the contaminant from the disposable clothing to ensure that entry team

Figure 8-4
Team members changing air cylinders in the decon area.

members do not become contaminated as they remove the garments. Why waste time and effort in decontaminating a suit that will be placed into a container for later disposal? The possibility that the suits will carry contaminants out of the area after being placed into a sealed container for disposal is extremely limited.

> **■ Note**
> Disposable PPE reduces the amount of decontamination required.

Conversely, some PPE needs very high levels of decontamination—for example, air packs used in Level B operations, where they are not enclosed in an encapsulating suit, and reusable suits. In these cases, a primary decontamination operation is set up to get enough of the contaminant off the equipment to allow the entry team member to remove the items without risking exposure. The items are then placed into containers, such as plastic bags or bins, where they are later decontaminated in a thorough manner before being reused. Thorough decon can take place either at the site of the original decontamination operation or at a location where appropriately protected decontamination workers can conduct the additional cleaning.

> **■ Note**
> Some PPE and equipment may need to be contained for later decontamination.

All seven factors must be taken into account before a decontamination operation is established. Each organization should establish protocols for determining the type and amount of decontamination required and for deciding which factors determine the order in which the entry team is decontaminated. Once the process is under way, it is difficult to change, so decontamination programs must be properly planned and implemented. For this reason, decontamination operations often are defined in advance in the site safety plan required for cleanup operations or by the decontamination officer in the case of an emergency response. Chapter 11 introduces the various positions and responsibilities that are used to ensure that all activities, especially those as critical as decontamination, are properly managed.

Note
The site safety plan or decontamination officer often decides which decontamination operations are required.

But what happens when a situation involves unknown materials, and information about decontamination operations is not readily available? How is the appropriate type and amount of decontamination determined when the material, its properties, and even the amount that might be present are not known? In such cases, the person in charge of the operation must plan for the worst-case scenario based on the available information. In many of these cases, the entry team may enter with higher levels of protection than turns out to be required, and the decontamination operations will be based on this level of protection. The level of protection and the decontamination operation may be scaled back once the scope of the problems is better understood. But in every case, taking precautions to provide higher protection in cases of unknown hazards is the appropriate course of action.

Safety
When dealing with unknown materials, plan decontamination operations based on the worst-case scenario. Never cut corners when you do not know what you are dealing with.

work zones
areas designated for specific operations

Hot Zone
the area where the potential for hazards exists

Warm Zone
the area just outside the Hot Zone, where decon operations take place

Cold Zone
the area where no contamination or hazards are present

Where Are Decontamination Operations Conducted?

Once we have determined the type and amount of decontamination that is required, we must determine where the operations will take place. Again, as with many other aspects of decontamination, a number of variables must be considered when making this decision, and choosing a location for decontamination operations often requires a degree of judgment. Regardless of this, the location can be changed as the operation or incident evolves.

Before we discuss the specifics of selecting a location, it is necessary to understand the concept of **work zones**. Work zones are simply areas within a response or cleanup operation where contamination is or is not present. Standard terms identify these areas: **Hot Zone**, **Warm Zone**, and **Cold Zone**. These terms are relatively standard within the hazardous materials industry, but may be interchanged with other terms. Table 8-1 lists some of these variations.

Regardless of which terms are used, each organization must ensure that all personnel working at a scene understand which name denotes each zone and what the term means. This understanding is critical to the safety of all personnel working at the site because mistakenly referring to one area when you actually

Table 8-1 *Names used to denote zones.*

Zone	Also Known As	Alternative Term
Hot zone	Exclusionary zone	Red zone
Warm zone	Contamination reduction zone	Yellow or orange zone
Cold zone	Support zone	Green zone

mean another can result in serious consequences. This will be clear when you understand what happens in each zone.

The Hot Zone is the area where the potential for contamination exists and is generally where a hazardous material is present. Establishing this area takes considerable knowledge and is necessary for both emergency and nonemergency operations. The specific issues surrounding, factors affecting, and the methods for establishing the boundaries of a Hot Zone are discussed in Chapter 12. For this discussion, we need to understand only basic concepts of this zone and decontamination operations. They include:

- Anyone entering the Hot Zone is at risk for exposure to hazardous materials and their properties.
- Once the Hot Zone is established, only properly trained personnel are allowed to enter it. The required level of training is determined by the situation and can include many HAZWOPER training levels.
- Personnel entering the Hot Zone must be protected with appropriate levels and types of PPE.
- Personnel entering the Hot Zone must have a mission or purpose for entering it. The entry team's work functions help determine the amount and type of decontamination operations that are required.
- All personnel leaving the Hot Zone are contaminated to some degree and require decontamination prior to entering the Cold Zone.
- The purpose of the Hot Zone is to clearly indicate the hazardous area within an operation and to help keep personnel from entering that area.

Once the Hot Zone is established, any area outside the Hot Zone is defined as the Cold Zone. Establishing the Hot Zone's boundaries effectively creates the boundaries for the Cold Zone. This area is technically free of contamination, and personnel working in this zone are not exposed in any way to the hazards in the Hot Zone. Supportive operations take place in the Cold Zone and can be conducted without specialized PPE.

The Warm Zone is created outside the Hot Zone in the area of the Cold Zone immediately adjacent to the Hot Zone. The Warm Zone is a buffer zone that becomes contaminated as the operation evolves, because it provides an area for decontamination operations. Like the Hot Zone, the Warm Zone must be clearly identified, and its boundaries must be visible to all personnel working at the site. Factors in Warm Zone operations include the following:

- Initial setup of decontamination operations within the Warm Zone can be done with limited PPE because the area is initially free of contamination. Level D is the level of protection required for setup of this area.
- The area should be as close to the Hot Zone as possible to prevent contaminants from spreading beyond the Hot Zone.
- The area should be upwind of the incident to allow any airborne contaminants to be carried back into the Hot Zone and away from the Cold Zone.

Shifting winds and migrating gas clouds should be taken into consideration before the Warm Zone is established. The wind direction often can be predicted by either anticipating commonly encountered prevailing winds or contacting a local weather information source, such as an airport or the Coast Guard.

- The area should be set up in a location that allows liquids to run back into the Hot Zone and away from the Cold Zone. The ground slope and grade must be considered when determining this.
- The area should be set up in a location close to a water supply because water often is required for decontamination operations. Water can be obtained from fixed systems, such as hose connections or fire hydrants, or from portable systems such as fire engines.
- The Warm Zone should be in an area that allows access to support activities in the Cold Zone. This can include access to roads and areas where medical assistance is available.
- Once the Warm Zone is established, personnel working in the Warm Zone should wear PPE at no less than one level below that worn by the entry team.
- The size of the Warm Zone is determined by the type and amount of decontamination that is required for the specific operations.
- The location of the Warm Zone can be adjusted as wind, weather, or other conditions change.

■ Note

Decontamination areas should be located upwind and uphill of the Hot Zone but downwind and downhill of the Cold Zone.

Figure 8-5 illustrates Hot, Warm, and Cold Zones.

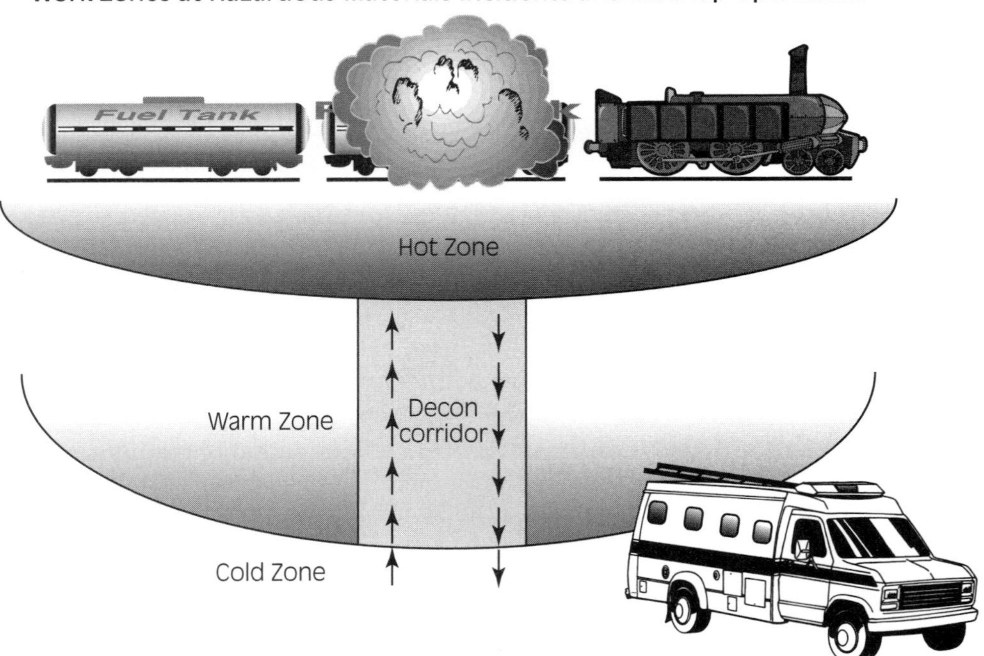

Figure 8-5 *Diagram showing the relationship of the Hot, Warm, and Cold Zones.*

How Are Decontamination Operations Conducted?

Knowing why decontamination operations are required, the factors involved in determining the appropriate type and amount of decontamination, and the location of the activities leads us to the final aspects of the process. Now, we must determine which steps and operations need to be conducted. How best to do this, like several other topics discussed in this book, is subject to debate. A number of variables, such as an operation's location, influence which steps are included in a decon operation. For example, a spill along a roadway presents different challenges than one that occurs in a fixed facility with access to water supply, lighting, and safety showers and eyewash stations. Many factors must be considered when you are determining how decontamination will take place.

Because so many variables affect this decision process, our discussion focuses on how to conduct decontamination operations. The examples we use outline a system that is useful in most types of decontamination programs. Remember that the level of decon required by a given situation can range widely, from simple hand washing to extensive, complex operations. Regardless of the level required, the basic concepts of decontamination must be followed.

The first issue in conducting a decon operation is to prevent the spread of contaminants. Recall that one of the reasons for decontamination operations is to ensure that contaminants are controlled and not allowed to leave the Hot Zone. To accomplish this, many decon operations start by setting up a **decontamination corridor** over a water-resistant tarp or plastic sheet. The decontamination corridor is the designated area where operations are conducted and is set up to ensure that runoff does not leave the area and enter the Cold Zone. Recall too that the decontamination area, including the decontamination corridor, must be uphill of the Hot Zone to allow runoff to flow back into the already contaminated area. Using a water-resistant tarp is even more important when decon is conducted on dirt or other surfaces where runoff can enter the environment. Once the tarp or plastic sheet is set, cones or other visible identifiers must be used to clearly indicate the decontamination area's boundaries. These cones also can help keep the tarp on the ground in windy conditions.

decontamination corridor
the area in the Warm Zone where decon operations are conducted

■ Note

Water-resistant tarps often are used to help contain splashes that occur during the decon process.

After the specific area for the decon corridor is established and identified with cones and tarp, the next step typically involves placing containers on top of the tarp to enable personnel to be washed. Some organizations use plastic tubs for this purpose, but disposable swimming pools are often the best choice because they are inexpensive and can be easily cleaned or disposed. Inflatable pools often are used because they require little storage space. Depending on the amount of decontamination required, one or more pools can be used to help trap the water and contaminants generated by the decontamination process. The example used later in this chapter involves a two-pool operation, which is widely used in the industry.

Once the pools or tubs are placed on the tarp, water hoses and nozzles are needed. These often are brought from the Cold Zone into the decontamination corridor for use. Some organizations use standard squeeze nozzles to control the flow of water, but flower-watering wands are a better alternative because they can move large quantities of water at lower pressures than hoses can. The lower pressures prevent splashing from occurring as water is sprayed onto the entry team. Flower-watering wands also have longer handles, which allow the decontamination team member to flow water over the top of the entry team member while maintaining some distance from the entry team member. This distance helps reduce the potential for contamination to transfer from the entry personnel to the decontamination personnel. Figure 8-6 shows a practice decontamination operation in which the decon workers are using long-handled brushes and shower wands.

Depending on the decontamination operations, brushes and detergent solutions may also be required. Five-gallon plastic buckets with short-handled brushes often are used in each decon pools. The buckets may contain detergent solutions containing Dawn dishwashing soap, which is well-known for its ability to break down oily materials. The brushes are used to apply the solution to the clothing of an entry team member, and the flower-watering wands are used to rinse it off.

Decontamination corridors may contain other equipment, depending on the operations being conducted. Soft car-wash sponges, for example, often are used in each pool to wash respirator masks. Entry team members are given a sponge soaked in the detergent solution to wash the masks' face pieces. The sponges are pliable enough to get into all areas of the masks and do not scratch the masks' material, as brushes might. Other equipment can be added as local protocols dictate.

After the decontamination corridor and equipment are set up, the decon team prepares itself for the entry team to use the area. *The decontamination process must be fully established prior to entry.* If entry team members enter the Hot Zone before the decon operations are set up, they take unnecessary risks if something forces them to leave the area, such as suit failure or other unplanned events. In all cases, the decontamination team must be in place, properly protected, and ready to perform decontamination operations as soon as they are required.

Figure 8-6
An example of a decon operation being practiced.

Safety
The Hot Zone should be entered only after the decontamination corridor is set up and staffed.

The Hot Zone can be entered through numerous routes, but most organizations require that all personnel enter the Hot Zone through the decontamination corridor. This helps establish accountability for personnel in the Hot Zone because the decontamination team can see the entry team members as they walk the corridor and check the entry team's PPE to ensure that it is completely sealed. This practice also establishes the corridor as a point of reference for the entry team members as they enter the Hot Zone. Someone, often the designated Decontamination Officer, will be assigned to monitor access into the Hot Zone. This individual is responsible for monitoring the access point into the Hot Zone, coordinating information among the various zones, and establishing and maintaining a log or list of who is in the Hot Zone. When the incident is terminated, this log, or documentation, can be filed with the incident report.

The number of people required to conduct a decontamination operation varies with the scope of the process. A rule of thumb is to have two decon team members for each entry team member. Typical two-pool setups use three to four decon personnel. Two of these decon team members are responsible for scrubbing or otherwise physically removing the contamination from the entry team member with brushes. A third person can be placed in between the two pools to control the hose operations. This person works in both pools and is often referred to as the *rinser*. The fourth person in the operation is sometimes referred to as the *bagger*. The bagger is the last person in the decontamination corridor and is responsible for helping personnel remove their PPE following decon. Often, the PPE is placed into plastic bags for later decontamination or disposal, hence the term *bagger*. Regardless of the terminology used, all decon personnel must have a specific assignment. If the process is more complex or involves injured victims, additional personnel may be required. Figure 8-7 shows a properly staffed decon corridor in which three of the four decon team members are working with two entry team personnel.

Figure 8-7
Entry team members being decontaminated in the decon corridor.

> **■ Note**
> A good rule of thumb is that there should be two decon team members for every person entering the Hot Zone.

tool drop
the designated area just inside the Hot Zone and adjacent to the decon corridor where equipment and tools are kept pending later use, disposal, or decontamination

Once the entry team is ready to leave the site and enter the decontamination corridor, it becomes the responsibility of the decontamination team members. As entry team members approach the edge of the Hot Zone, they often stop and leave their equipment just inside the Hot Zone in an area called the **tool drop**. The tool drop is a designated area where equipment and tools can be kept pending later use, disposal, or decontamination.

Because the primary purpose of decontamination is to reduce the potential for exposure to the entry team members, the process must begin as quickly as possible. Equipment and other items can wait. Appropriate warning signs or labels should be placed near or attached to the containers awaiting disposal. Security, support, and lighting may be required if the contaminated materials remain onsite overnight. Equipment that has been exposed to hazardous materials, such as fire hoses, air packs, and vehicles, should be completely decontaminated, inspected, and tested before being placed back in service.

Once entry team members are inside the decontamination corridor, the decontamination team directs them in what they need to do. Generally, one entry team member enters the first pool, depending on the criteria established for that operation. That person starts the process while a second person stays just outside the corridor. As the process continues, the decontamination team members tell the people being decontaminated what steps they must take to complete the process. For example, stepping from one pool into another requires lifting one foot, having the bottom of that boot rinsed and scrubbed, and then stepping into the next pool. This may sound obvious, but entry team members usually are hot, tired, and have only limited communication because of their PPE, so it is critical that decon team members coordinate the process. Decontamination team members may need to speak loudly or use nonverbal communication, such as helping to lift an entry team member's foot with the shower wand or a brush, or tapping entry team members when they need to move into the next pool.

Throughout the process, everyone must work to prevent exposure of the decontamination team members to the contamination that might be present on the entry team members. For example, decontamination team members must avoid any direct contact with entry team personnel and should instead use their brushes or wands to contact entry team members.

> **! Safety**
> The decontamination team should avoid direct contact with contaminated equipment and personnel as much as possible.

Another recommended practice is keeping all decontamination team members on one side of the decontamination corridor to reduce the possibility of cross-contamination from overspray. It is much easier for the entry team personnel to turn around while standing in the pool than to have decontamination team members walk to the other side of the pool. This reduces the potential for exposure and maintains better control over the areas of contamination within the decontamination corridor. Figure 8-8 shows an entry team member moving between the pools in the decontamination corridor. In the picture, the

Figure 8-8
Entry and decon teams coordinating the decon process.

decontamination team member is using a brush to direct the movements of the entry team member and avoiding making physical contact with the entry team member.

SIX-STEP LEVEL B DECONTAMINATION

There is no single method of conducting decontamination, but the following six-step process for decontaminating personnel in Level B protective clothing is typical. Some organizations may use different equipment or techniques, but the basic concepts are the same.

Step 1 Equipment drop
Step 2 Gross decontamination
Step 3 Secondary decontamination
Step 4 Breathing apparatus removal or bottle change as needed
Step 5 Suit and boot removal
Step 6 Field shower and redress

The first step in this process is the tool drop. Because the tool drop is actually in the Hot Zone, nothing needs to be done with the tools or equipment at this point other than ensuring that they are placed into the appropriate containers.

The tool drop controls the tools' location, makes them easy to recover, and prevents the contamination on them from spreading.

The second step is *gross decontamination*. As Figure 8-8 shows, during this step, an entry team member is directed to step into the first wash-and-rinse pool. These pools should be large enough to contain all runoff from the person being decontaminated. Decon team members should avoid overspraying, spreading runoff, or filling the pools. Brushes or sponges are used to scrub all PPE that has been exposed. Remember that decontamination is a systematic process, so scrubbing should be done systematically. Generally, the entry team member is completely rinsed with water, and then the sponge is used to remove contaminants from the mask area. Next, the front portion of the person is scrubbed using the brushes, followed by a second rinse. This process should start at the head and work down to allow the contaminants to be flushed into the pools. Following this, the person is directed to turn around, and the process is repeated on the back.

■ Note
It is important to get as much of a contaminant off entry team members as soon as possible.

■ Note
Decontamination is best done from the head down to wash materials away from the vulnerable parts of the body.

As soon as the first entry team member has been through gross decontamination in the first pool, he steps into the second pool, where the process is repeated. This third step is called *secondary decontamination*. In the second pool, there should be less contamination and additional time can be taken to ensure that all contaminants are removed from the entry team member's PPE. As the first entry team member moves into the second pool, the second entry team member enters the first pool and begins gross decontamination. Before entry team members can move from the first pool to the second pool, however, their boots must be washed and rinsed. This includes the bottoms or soles of the boots because these areas are often significantly contaminated. Once one boot is washed, the person places that clean boot into the next pool. Obviously, the pools must be placed close enough to allow entry team members to step from one to the other without having to step onto the tarp. Personnel who are wearing an SCBA may feel off balance as they move through the decontamination corridor, so the decontamination team may need to assist entry team members from one pool to another. Poles can be placed into cones and positioned near each pool to provide a crutch for the entry team members to use as they move from pool to pool or when they hold their feet up for decontamination. Again, it is the responsibility of the decontamination team members to direct the actions of the entry team.

The fourth step is removing SCBA units or changing their cylinders. This can be done safely once gross and secondary decontamination are completed. Depending on the type and configuration of the equipment and PPE involved, various methods can be used to remove SCBA. Generally, the air pack assembly is removed first, and the mask is allowed to stay on the entry team member's face. Leaving the mask in place provides the wearer with some protection from splash exposure. If a cylinder change is required, the cylinder can be removed from the pack without removing the pack from the entry team member's back. The specific steps in this operation must be clearly identified, understood, and practiced by all persons involved in the operation prior to the operations being conducted.

Step five is removing suits and boots. Figure 8-9 shows a decon team member at the end of the process directing an entry team person to step into a large plastic bag that has been rolled down to accommodate the PPE. This allows the decontamination team bagger to remove protective clothing systematically and place it into the bag for later disposition. The best method of removing clothing is to begin at the top and work down, allowing all the items to drop into the bag. Much like peeling a banana, the suit is best peeled or rolled off the person wearing the PPE and turned inside out as the process continues. Even though the suit and other equipment have been through the decontamination process, this technique provides another layer of protection for the entry team members, who are now exposed to splashes or contact with the used PPE or equipment. As the suit is rolled down toward the feet, entry team members can sit on a stool and continue removing their suits right into the plastic bag. They can then step out of their boots and swing their legs to the side, out of the bag. Once they step out of the bag, entry team members can remove the breathing apparatus mask and inner gloves, placing them in the bag with the suit or in another designated container.

> **■ Note**
> Peeling clothes off entry team members helps prevent contamination from getting inside the PPE.

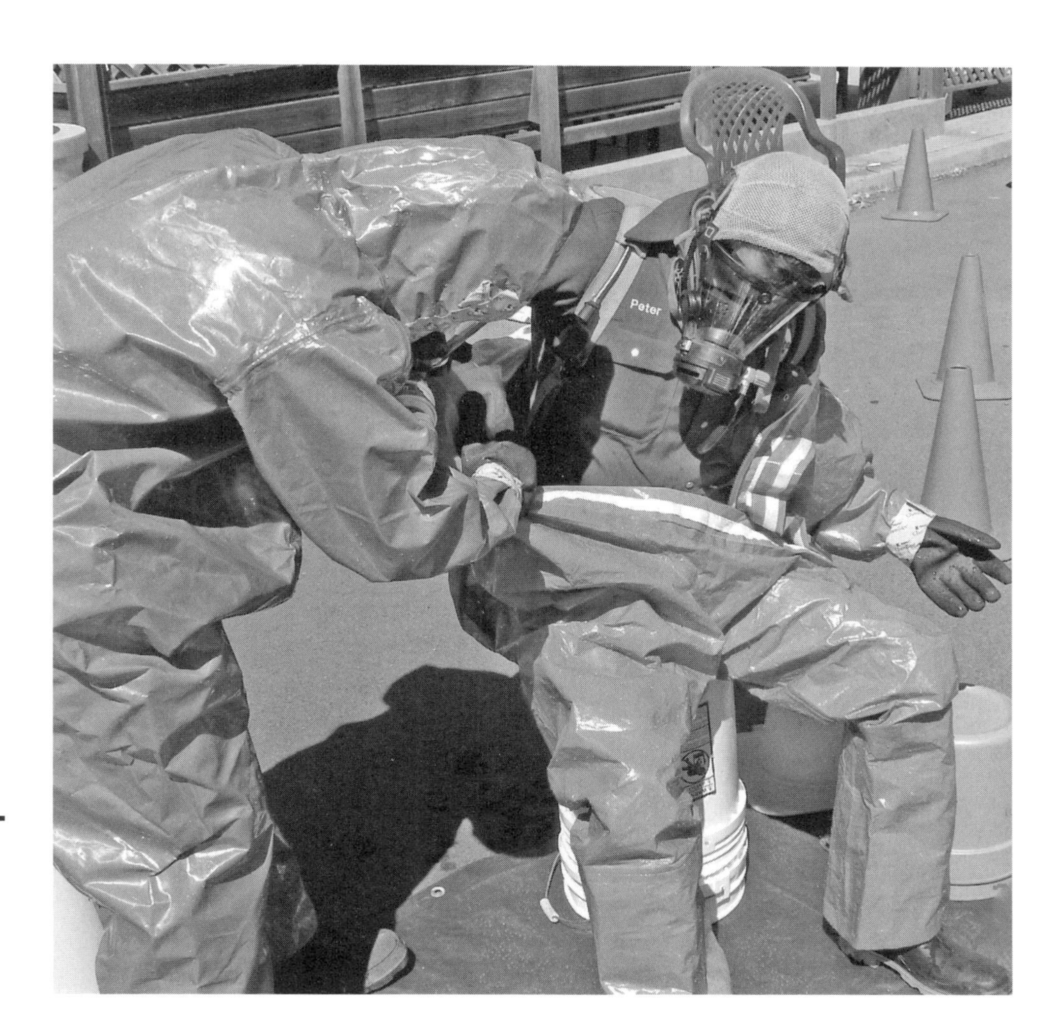

Figure 8-9
Decon team members peeling the entry team member out of his PPE.

The final step in the decon process, personal decontamination in a shower, is completed in the Cold Zone. Because of the complexity of this activity, and depending on the process, location, and circumstances, this step may occur away from the contaminated area, such as in a change room on the site, or offsite, such as back at the fire station. Regardless of where it takes place, all entry team members should shower using clean, warm water and a lot of soap following decon operations.

Putting these steps together and working as a cohesive team takes practice. The process takes considerable time to set up, and even more time is required to get contaminated personnel through the process safely and thoroughly. You will not perform decontamination procedures perfectly the first few times you do them, which is why decontamination drills and exercises must be practiced as much as possible before they are required for a real incident. No matter how often you practice, however, some stages will always go faster than others. For example, Figure 8-9 shows personnel waiting to undergo the final step in the process. This occurs often, because the steps to remove clothing and equipment take the most time, so while other decontamination team members, such as rinsers and scrubbers, can do their tasks relatively rapidly, baggers have the most work to do and their job requires the most time. If you are undergoing decon, be patient and understanding that the bagger has the most demanding job.

Finally, once entry team members have been decontaminated, the decontamination team itself requires some decontamination based on its potential exposure. As with the entry team members, the decon person with the most potential for exposure should be decontaminated first. In most cases, this is the person first in the system, who was closest to the Hot Zone. Other decontamination team members follow suit based on their position in the process. The bagger should be the least contaminated team member and should be the last to go through decontamination. When performing this step, it is best to stay at the assigned position and work out from the Hot Zone to the Cold Zone. The first person in line should start at the first pool and work toward the second, and so on. The rinser in the second position can start in the second pool and work out to the Cold Zone. When everything is done correctly, the decontamination corridor has performed its task of preventing contaminants from spreading and has protected the most important asset of any job—you.

Note

When the entry team has been decontaminated, the decon team starts its own decontamination.

EMERGENCY DECONTAMINATION PROCEDURES

Sometimes, entry teams encounter unforeseen problems or victims in the Hot Zone and need to leave immediately. If this occurs, all team members and victims must go through emergency decontamination. The most common reason for a team member to leave the Hot Zone before the containment or cleanup task is completed is because the member's SCBA is running low on air. Different respiratory rates and the varying levels of physical exertion demanded by the job result in some entry team members running low on air before others. In situations where two entry team members are in the Hot Zone, the team members must leave at the same time, and the person with the least amount of air must go through decon first.

Remember that if you must leave the Hot Zone to deal with an unforeseen problem or emergency, *do not panic.* SCBA systems are designed to give approximately 5 minutes' warning before their air supply is exhausted. This is more than enough time to perform a thorough decon procedure, and the decon team should reassure entry team members that this is the case.

There are also situations that put standard practices to the test and require a rapid decontamination operation, such as when someone is acutely exposed to a hazardous substance in an emergency response situation. For example, the failure of a piping system, splashes from operations, or accident situations can expose personnel to hazardous substances and require emergency decontamination. When an acute exposure occurs, time is critical, because as we know, the longer the exposure, the greater the likelihood for harm. This is why OSHA regulations mandate that safety showers and eyewash stations be placed within a 10-second walk of an area containing certain types of a hazardous material.

Safety showers are good tools in emergency decon activities.

Because there are so many variables in emergency decontamination practices, and the specific activities are similar to those of standard decontamination practices, it is better to discuss the general concepts of emergency decon rather than its specific tactics. The following list summarizes the principles of emergency decontamination.

- Avoid all direct contact with the exposed individual. This may require that the victim stay in the newly formed Hot Zone, the area of exposure, until the proper PPE can be obtained.
- Evaluate the circumstances as rapidly but thoroughly as possible. Knowing which material(s) is involved can shorten this process, but take time to identify the hazards before taking action.
- Set up a full technical decontamination corridor. This can be done prior to initiating rescue and decontamination activities. The exposed person can be verbally directed out of the most contaminated area into another area or into a safety shower without actual contact with them.
- Use tools such as broom handles to guide contaminated persons to safe areas while helping to ensure that no direct contact takes place between them and unexposed personnel. For example, the broom handle can be placed gently against the chest of someone who has chemicals in their eyes and cannot see. The person holding the handle can direct the exposed person to a shower or eyewash station to begin emergency decontamination.
- Begin decontamination at the scene, because acute exposure with symptoms requires it. If the exposed person is not decontaminated at the scene, the contamination may be relocated to the back of a waiting ambulance or a local hospital emergency room.
- Remove the victim's exposed clothing as soon as possible to stop the contamination.
- Ensure privacy for the victim as soon as possible. If possible, set up screens, such as plastic sheeting or blankets, to avoid exposing the victim to undue attention. Remove all noninvolved persons from the area. If the victim's clothing is removed, have a clean Tyvek suit or rescue blanket for him to step into following decontamination.
- Medical care must take place *after* decontamination—otherwise, emergency medical personnel may be exposed to the contaminants.

Safety
Never expose medical personnel to contaminated victims. Decon must be done prior to most medical intervention.

Regardless of the process or system used, emergency decontamination activities present considerable challenges even for the most experienced personnel. Practicing the procedures, however, decreases the time required to put operations into effect in an actual emergency. Finally, remember to provide as much information as possible about the contaminants to emergency medical and hospital personnel. When possible, a material safety data sheet (MSDS) for the material should accompany the victim to the medical center so that physicians will have information on the proper treatment options.

Disaster Decontamination—Mass Decon

Large-scale disasters require mass emergency decontamination. The decontamination of large numbers of personnel or victims may be necessitated by several scenarios, including:

- A chemical release resulting from an industrial accident near a population center
- A major transportation accident releasing hazardous materials over a populated area
- An act of terrorism involving the intentional release of standard industrial chemicals or substances
- An act of terrorism involving the intentional release of weapons of mass destruction, including radiological materials that create a "dirty bomb"

Such incidents create tremendous confusion and chaos. In these situations, even if the hazardous material is not a weapon of mass destruction, it becomes a weapon of mass *disruption*, because the region's systems are quickly overwhelmed. For example, hospital systems may be disrupted when many people present at the emergency departments of local hospitals. This already overwhelming situation will be even made worse if contaminated people appear at medical facilities without receiving proper decontamination in the field.

To prevent this problem, decontamination should be conducted outside hospital emergency departments whenever possible. Emergency response personnel may be called to an emergency department to set up large decontamination stations where multiple victims can be decontaminated prior to being allowed to enter the emergency department. Given the extreme difficulty of relocating critically ill or injured persons, it is often necessary to station security staff at hospital entrances to prevent contaminated personnel from entering the facility.

The process of decontamination is necessary even when other medical conditions are present. Also remember that the longer a victim is exposed to a material, the greater the potential health effect of the material on the victim. To facilitate mass decon, the U.S. Department of Homeland Security provides local agencies, including hospitals, with grants that can be used to purchase large tents with decontamination stations. These tents can be rapidly set up outside hospitals or at other locations where large numbers of victims require decontamination. Additionally, OSHA has developed standard procedures for hospitals to assist them in preparing for such events. These procedures define the receiving hospital as the *First Receiver* and require that personnel involved in decontamination activities be appropriately trained to the first responder operations (FRO) level to ensure their safety. A copy of these procedures is available on the OSHA Web site.

SUMMARY

- Decontamination is the systematic process of removing contaminants from people and equipment.
- Decontamination is performed to protect personnel, the environment, and other resources.
- Dilution, absorption, and chemical degradation are three acceptable decon techniques. The most common of these techniques is dilution, and water is the most commonly used material.
- The decontamination corridor is located in the Warm Zone.
- The Warm Zone is sometimes referred to as the Contamination Reduction Zone, Yellow Zone, or Orange Zone.
- Outdoor decontamination sites should be located uphill, upwind, and upstream of the Hot Zone.
- The Hot Zone is sometimes referred to as the Exclusionary Zone or the Red Zone and is the area where a release has occurred or where hazards are present.
- The Cold Zone, sometimes referred to as the Support Zone or Green Zone, is an area where no contamination is present and from where site operations are coordinated.
- Decontamination areas should have access to water and emergency medical support.
- The area selected for decon may change based on wind and weather conditions.
- A Decontamination Officer or Team Leader should be appointed and is responsible for operations at the decontamination site.
- Seven factors determine the extent of the decontamination activities. These include:
  1. The type of contaminant present
  2. The amount of contaminant present or released
  3. The level of protection worn by the workers
  4. The work function performed
  5. The location of contaminant on the personnel
  6. The reason for leaving the site
  7. The type of protective clothing worn
- Emergency decon is the rapid removal of contaminants from those who are exposed and should be done as soon as exposure is noted.
- Six steps commonly found in a technical decontamination operation include:
  1. The equipment and tool drop
  2. Gross decontamination of the personnel by washing and rinsing
  3. Secondary decontamination by washing and rinsing
  4. Tank change (if required) or SCBA/respirator removal
  5. PPE removal
  6. Field shower and redress
- Emergency decontamination of mass casualties requires rapid intervention to prevent relocating the disaster to a hospital or other medical facility.
- Hospitals may be the first receivers of chemically contaminated personnel and may be required to have FRO training.

REVIEW QUESTIONS

1. What is decontamination (decon)? Give two examples that may apply to you at work or at home.
2. List the reasons that decontamination operations are necessary.
3. Identify the work zones used in decon and their purposes. List other names used to describe each zone.
4. List the seven factors used to determine the amount and type of decon required.
5. Why must decon operations be in place before personnel enter the Hot Zone?
6. Describe some of the issues involved in emergency decon operations.
7. List the steps in a two-pool decon operation.
8. Describe the rationale for determining which entry team member is decontaminated first.

ACTIVITY

The following scenarios will help you apply what you have learned about decontamination practices. Read each scenario and prepare a decontamination plan for the operation described. Include the following elements in your plans:

- The level of protection assigned to the decon team members
- The number of decon workers required to complete the operation that you establish
- Factors involved with the location of the decontamination area
- Equipment that will be used to perform the established decontamination operations
- Any special issues that will be faced by the decon team
- Draw a diagram of the decon area as it relates to these operations

1. A crew must clean up an abandoned waste site containing twenty drums of unknown chemicals. The drums are 55-gallon polyethylene drums without labels. Your first assignment is to set up staging areas within the Hot Zone to store the drums and assess their condition. The entry team members will be required to wear Level B protection with SCBA units. The area is located on a slight grade flowing downwind. The wind direction is from the east. Some of the drums may be slowly leaking and may already have leaked material. Small amounts of vapor are occasionally visible around the drums.
2. You are called to an emergency situation in which a worker has been exposed to a chemical used at the site. The material is powdered sodium hydroxide, which is contained in drums. One of the drums fell from an elevated location and has contaminated the worker, causing him to become disoriented. A safety shower is nearby, and other response personnel are coming to assist.

ASSOCIATED PHYSICAL HAZARDS

Learning Objectives

Upon completion of this chapter, you should be able to:

- Identify the OSHA regulations related to working in confined spaces.
- Describe confined space operations including:
  - Characteristics of confined spaces and Permit-Required Confined Spaces
  - Hazards associated with confined spaces—particularly atmospheric hazards
  - Management of confined space hazards
  - The hazards posed by the release of dangerous materials and energy within a confined space.
- Identify the different types of hazardous atmospheres including oxygen-deficient and oxygen-enriched atmospheres, environments with flammable vapors, and potentially toxic atmospheres.
- Describe how to monitor the atmosphere to detect hazards.
- Identify the respirators and other equipment used to reduce exposure to hazards.
- Describe the basic emergency procedures for confined space operations.
- List the types of hazardous energy sources that can be present at work sites.
- List the steps used in controlling hazardous energy sources.
- Describe noise hazards and methods to control them.
- Identify the types of controls used to reduce the risks involved in elevated work.

CASE STUDY

Two electricians working at a site in Southern California were found dead inside an underground concrete vault. The vault was located in a newly built water treatment facility and had never been used to store any hazardous material prior to this incident. The facility was in the last phases of construction, and the electricians, one of whom was an apprentice, were completing the building's electrical hookups.

The details of this incident can only be surmised because no one actually observed the incident, but it is believed that the apprentice electrician went down a ladder into the vault to retrieve an electrical cord attached to a sump pump. Knowing that the vault was completely clean and had not contained any known hazardous materials, the electrician anticipated no hazards and took no precautions. Shortly after descending into the vault, the young man was overcome by the vault's oxygen-deficient atmosphere. It is believed that the journeyman electrician who was the job supervisor saw his coworker collapse inside the vault and immediately went down the ladder to help. The atmosphere also affected the would-be rescuer, who lost consciousness while descending the ladder and fell, unconscious, on top of his coworker. The men's bodies were later discovered and retrieved by a local fire department rescue team.

As a result of this incident, several lawsuits were filed by the electricians' families resulting in considerable litigation and award payments to compensate the families for the loss of their fathers and husbands.

The greater tragedy occurred, however, when it was determined during the lawsuits that the two men died needlessly because they failed simply to recognize the existing hazards and follow safety rules that should have been in place. What is believed to have happened was that both men entered the vault without understanding the situation's hazards. Both men may have looked down the ladder before entering the space and concluded that it looked safe enough. The vault contained no outward indications of problem: no visible odors, no stains on the floor or walls, and no hazardous equipment. The vault had never been used for anything because the facility was not in operation. Additionally, there were no posted signs or warnings indicating the presence of a hazard. Yet, one existed, one that ultimately claimed two lives.

The lessons learned were obvious after the fact. Hazard recognition and implementation of basic controls could have easily prevented these deaths. Even a warning sign at the top of the stairs would likely have caused at least one of the men to question the vault's safety. The space they entered was what OSHA defines as a *Permit-Required Confined Space*. Such areas are easily recognized by anyone who has even a basic understanding of how physical hazards can cause serious injury or death.

INTRODUCTION

In addition to the effects caused by hazardous materials themselves, a range of other common hazards are encountered at sites where hazardous materials are present, whether in the workplace or at emergencies involving them. These are **physical hazards** and are associated with a site and its operations. The term describes the non-health-related issues of hazards commonly encountered in many types of site operations. Physical hazards can cause significant harm and even death, as illustrated by the case study. These dangers include electrical hazards, trenches or other unstable workplaces, elevated working locations, confined spaces, and

physical hazards
non-health-related dangers present at many sites where hazardous materials are found

exposure to other environmental issues. Because these elements can be quite hazardous on their own, OSHA directed Hazardous Waste Operations and Emergency Response (HAZWOPER) training programs to include several of these topics as part of the required curriculum. The range of these hazards is far too broad to discuss here, so we focus on a few of the most widely encountered problems found at work sites. These include confined space operations, control of hazardous energy, noise exposure, and working at elevated locations. In addition to these dangers, you should always consider the potential for slips, trips, and falls; or additional threats that complicate your chemical response or cleanup operation.

■ Note

OSHA mandates that HAZWOPER training cover commonly encountered hazards in addition to those posed by hazardous materials.

CONFINED SPACE OPERATIONS

Confined spaces are one of the most commonly encountered hazards in many hazardous materials operations, from storage to response. These dangers are so prevalent that OSHA specifically mandates that confined space activities be a component of HAZWOPER in HAZWOPER training programs. Although the areas classified as confined spaces vary, common practices must be followed when working in and around them.

■ Note

Training on confined space operations is specifically mandated by the HAZWOPER regulation.

confined space
an area large enough to enter, not designed for occupancy, and hard to get in or out of

Permit-Required Confined Space (PRCS)
a confined space that contains potential safety hazards

The specific OSHA regulation for most types of workers involved in confined space operations is in 29 CFR, Part 1910.146. In it, OSHA defines two classes of confined spaces. The first is simply called **confined space**, and the second is called a **Permit-Required Confined Space (PRCS)**. The regulation defines a confined space as any space that meets all of the following criteria:

- The area is large enough and so configured that an employee can bodily enter the area and perform work.
- The area is not designed for continuous human occupancy.
- The area has limited or restricted means of entry or exit.

■ Note

OSHA has specific confined space regulations in 29 CFR, Part 1910.146.

Common confined spaces include storage tanks, storage and process vessels, silos, storage bins, hoppers, vaults, trenches, and pits. Often, however, determining whether a particular area is classified as a confined space is not easy. For example, a shallow trench for underground pipes is questionable. The following list provides generally accepted guidelines to help you determine whether an area should be considered a confined space:

- Small spaces, such as those found inside walls or above ceilings, that only part of the body can enter generally do not qualify as a confined space, because the OSHA regulation that employees must be able to "bodily enter" a space.

- Trenches or pits can qualify as a confined space if they are big enough to bodily enter. A reasonable standard for this is that the trench is approximately 4 ft deep, although shallower areas can be defined as confined spaces if they are covered.
- Rooms or parts of buildings that are designed for occupancy by people do not ordinarily meet the requirements for being a confined space. Usually, areas where personnel are expected to work are easily entered by doors and do not meet the definition.
- Doors that allow access into an area do not qualify as limiting entry or exit and do not meet the standard criteria for having limited ingress and egress.
- Stairs leading to upper or lower areas do not generally restrict entry or exit if they meet the standard criteria for easy entry and exit.
- Openings that require employees to crawl into an area qualify as limiting entry or exit (Figure 9-1).
- Entry or exit by way of ladders is generally accepted as limiting entry and exit and meets the criteria for limited ingress or egress.

■ Note
Classifying an area as a confined space can be complicated.

■ Note
Common confined spaces include pits, tanks, vaults, and storage vessels.

The key indicator in this determination process often is the ability to quickly exit an area. OSHA places considerable importance on the ability of an employee to quickly and safely exit all areas where they work. Any limitations on this action are reason for OSHA to examine and regulate. It is safest to define questionable

Figure 9-1
An example of a confined space common where hazardous materials are found.

spaces as confined spaces. Once a space has been determined to be confined, it should be determined whether it is a Permit-Required Confined Space.

> **Safety**
> An area with restricted exits should be carefully studied to determine whether it is a confined space.

OSHA defines a Permit-Required Confined Space as any properly defined *confined space* that also includes any of the following hazards:

- Contains or has a potential to contain a hazardous atmosphere. Hazardous atmospheres are those that expose employees to the risk of death, incapacitation, impairment of ability to self-rescue (i.e., to escape unaided from a permit space), injury or acute illness from either a flammable atmosphere or an oxygen-deficient or -enriched atmosphere, or one that has levels of exposure above the permissible exposure limit (PEL).
- Contains a material that has the potential for engulfing an entrant, such as grain silos or areas where loose solids are present.
- Has an internal configuration such that an entrant could be trapped or asphyxiated by inwardly converging walls or by a floor that slopes downward and tapers to a smaller cross section. These configurations can be found inside piping systems and a variety of industrial processes.
- Contains what OSHA terms "any other recognized serious safety or health hazard." Because of the wide range of hazards encountered in the various operations that involve confined spaces, it is impossible for OSHA to list all such hazards. It is clear, however, that those working in the confined space must evaluate all the hazards, as well as those that are not chemical in nature. This could include a range of issues, such as mechanical equipment that may start up, heated environments, or elevated areas from which workers might fall.

The presence of any of these conditions in a confined space is enough to classify the space or the area as a PRCS and subject it to the additional OSHA-mandated requirements for operations conducted in them. This is because these areas are potentially harmful to workers. And because this potential for harm exists in a space with restricted exits, OSHA imposes considerable requirements for all activities that take place within them. If there is the potential for a work area to contain a PRCS, OSHA mandates that the following occur:

- Assess the hazards within each confined space and determine whether the spaces are PRCS.
- Identify every PRCS in an appropriate manner to alert workers of the presence of a hazard that requires protection and procedures beyond those normally used. This can be done with the standard label or with any other equally effective system. Figure 9-2 shows an example of a sign that is typically used to alert workers to the presence of PRCS.
- Identify the types of hazards inside each space and determine how to manage them. Written procedures for entering these areas, including the use of an employer-issued permit, are required. The permit serves as a safety checklist and documents the hazards within the PRCS, the appropriate procedures to be used when working in it, the measures taken to eliminate the hazards using engineering systems, and the names of personnel authorized to work in the area. The specific items required to be included on a permit are found in the OSHA regulation. An example of a permit is found in Figure 9-3.

Figure 9-2
An example of a sign indicating a Permit-Required Confined Space.

CONFINED SPACE PERMIT

NO CONFINED SPACE MAY BE ENTERED UNTIL THIS PERMIT HAS BEEN COMPLETED.

SECTION 1 - CONFINED SPACE ENTRY PRECAUTIONS

DATE (GOOD FOR 24 CONSECUTIVE HRS MAX) | LOCATION

TIME STARTED | DURATION

VESSEL

WORK TO BE DONE

ENTRY PRECAUTIONS/ HAZARDS

	YES	NO	N/A
1. HAVE ALL VALVES BEEN LOCKED?	___	___	___
2. HAVE ALL LINES BEEN BROKEN, BLANKED OFF, OR OTHERWISE ISOLATED?	___	___	
3. HAVE ALL SOURCES OF INERT GASSES, SUCH AS NITROGEN, BEEN ISOLATED INCLUDING THOSE THAT FEED INSTRUMENTATION?	___	___	___
4. HAVE ALL SOURCES OF EXTERNAL HEAT BEEN LOCKED IN THE OFF MODE?	___	___	___
5. HAS THE CONFINED SPACE BEEN PURGED?	___	___	
6. HAS THE CONFINED SPACE BEEN CLEANED OR WASHED?	___	___	
7. HAS THE CONFINED SPACE BEEN VENTILATED? IF YES, ☐ FORCED ☐ NATURAL (CHECK ONE)	___	___	
8. HAVE ALL THE SWITCHES BEEN LOCKED OUT, TAGGED, AND, TESTED?	___	___	___
9. HAS THE AGITATOR BEEN IMMOBILIZED?	___	___	___
10. HAVE ALL SURROUNDING CONDITIONS BEEN INSPECTED SO AS TO PREVENT NEARBY HAZARDS FROM AFFECTING THE WORK?	___	___	

PPE / COMMENTS / SPECIAL PRECAUTIONS

I CERTIFY TO THE ACCURACY OF THE RESPONSES IN SECTION 1 AND THAT ALL OF THE SAFETY PRECAUTIONS LISTED ABOVE HAVE BEEN CONSIDERED.

TITLE SIGNED

SECTION - 2 AIR MONITORING

1. PERCENT OXYGEN (MUST BE GREATER THAN 19.5% AND LESS THAN 23.5%)

TIME	PERCENT OXYGEN	INITIALS	TIME	PERCENT OXYGEN	INITIALS
	%			%	
	%			%	
	%			%	
	%			%	

2. COMBUSTIBLE GAS TEST (MUST BE LESS THAN 2% OF LEL)

TIME	PERCENT OF LEL*	INITIALS	TIME	PERCENT OF LEL*	INITIALS
	%			%	
	%			%	
	%			%	
	%			%	

*LEL - LOWER EXPLOSIVE LIMIT

REVISED 11/26/2002

Figure 9-3
An example of a permit used for Permit-Required Confined Space.

- Post the permit in the area of the PRCS. Each person entering the space should read the permit and understand the hazards of and requirements for entering the PRCS.
- Ensure that a team of trained personnel is available to conduct rescue operations should they be required. The rescue team can stand by at the entrance to the confined space and be available on site during the entry operations, or, in some cases, it can be off-site.
- Train employees in safe operations regarding confined spaces.

! Safety
A confined space that poses potential hazard must be designated a Permit-Required Confined Space.

The following six subsections discuss the requirements for PRCS as they pose a much higher potential for hazard than simple confined spaces. To avoid injury or potential exposure to hazards, workers should assume that all confined spaces may present hazards and take the necessary precautions, which are described in the following sections. OSHA may not require these procedures to be used for non-Permit-Required Confined Spaces, but their use significantly increases the level of safety during any operations in confined spaces by identifying potential hazards that might otherwise be overlooked.

! Safety
Although not required by OSHA, the use of a permit system for entry into all confined spaces provides workers with a higher degree of safety.

Atmospheric Hazards of Confined Space Operations

The hazards that make a confined space into a permit-required space encompass two distinct groups, atmospheric hazards and physical hazards. Each group poses a number of potential hazards. Atmospheric hazards are some of the most common encountered in confined space operations and are the leading cause of death of those working in a confined space. Atmospheric hazards include the following:

- Oxygen-deficient atmospheres
- Oxygen-enriched atmospheres
- Flammable atmospheres
- Toxic or poisonous atmospheres

■ Note
Atmospheric hazards are common in confined spaces.

Perhaps the most common problem associated with working inside a confined space is the potential for a low level of oxygen to exist in the space. As we know from previous chapters, OSHA classifies an oxygen level below 19.5% by volume as oxygen deficient. We also know that in most cases, there are often no noticeable indications of this condition because most atmospheric gases do not have an odor. Monitoring the atmosphere is the only way to determine the lack of atmospheric oxygen. People exposed to low levels of oxygen do not realize that they are starting a process that often results in a fatality or even multiple fatalities, as this chapter's case study illustrated.

OSHA sets the safe level of oxygen at 19.5% because exposure to this level does not result in symptoms. Not until the level of oxygen drops below 17% do people typically exhibit symptoms. At this lower level, people can begin to feel confused, making it difficult for them to recognize and respond to the potential hazard that the lower level of oxygen creates. In fact, exposed people may become significantly confused and lose the ability to judge whether they should get out of the area. As we will see in our discussion of the requirements for workers inside a PRCS, OSHA mandates that an unexposed worker, whose job is to monitor the behavior of those working inside, be stationed outside a PRCS. This helps prevent the catastrophic consequences that occur as oxygen levels drop even more. Continued exposure to an oxygen-deficient atmosphere can lead to asphyxiation, which is the leading cause of death for workers in confined space operations.

Safety
Asphyxiation is the most common cause of death for workers in confined spaces.

Note
Symptoms of oxygen deficiency occur when the level of oxygen in the air drops below 17%.

A number of factors can lead to the decrease of oxygen levels in a confined space. For example, something as simple as the lack of ventilation within a space can reduce the available oxygen level. Lack of ventilation coupled with various other factors often results in oxygen levels becoming extremely low. Some of the common factors that lead to oxygen-deficient atmospheres include the following:

- *Displacement by other gases.* The introduction of other gases or vapors into the space may displace oxygen. Because most vapors or gases are heavier than air, they tend to settle in low areas when ventilation is limited. When this occurs, they can displace the normal atmosphere containing the required amount of oxygen and cause other gases or vapors to fill the space.
- *Bacterial action.* The decomposition of organic matter, such as sewage, leaves, grass, and wood, can use available oxygen and make the levels within a space deficient. Again, limited ventilation in the space means that oxygen cannot be replaced as it is used up in this process.
- *Oxidation.* Whenever metals oxidize (rust), they absorb atmospheric oxygen. While this process is not a huge cause of oxygen deficiency, it is worth noting, especially when a space is significantly confined and not subject to ventilation.
- *Combustion.* Burning, welding, or cutting operations inside a confined space use considerable amounts of oxygen. If only limited ventilation exists, oxygen is used up by the operations and replaced with other combustion gas by-products, such as carbon dioxide. These combustion gases also can displace the normal atmosphere and further complicate this situation, even when some ventilation is present.
- *Absorption.* Some materials, such as wet-activated carbon, absorb oxygen from the atmosphere. Therefore, before an area is entered, it is critical to evaluate a space and determine which materials it contains and whether they have this capability.

- *Painting operations.* As paint dries, it absorbs oxygen and releases other vapors that can displace oxygen in a space. Because many of the operations that take place within confined spaces involve painting, this common hazard must be considered.

> **Safety**
> Oxygen can be used up by a number of factors in a confined space.

Although OSHA sets the oxygen safety level at 19.5%, an atmosphere that contains 19.5% oxygen is not automatically safe. Remember that every atmosphere contains 100% of something. That *something* is normally composed of a fixed percentage of materials that include the following:

- 78% nitrogen
- 20.9% oxygen
- 1% carbon dioxide and other trace gases

An atmosphere that contains 19.5% oxygen instead of the normal 20.9% has had nearly 1.5% of the oxygen replaced with another material. Unless we know specifically what that material is, we may be working in an atmosphere that contains a harmful substance. Because of this, we must remember that the safest atmosphere for us to work in is one that is most similar to normal air with the normal amounts of constituents. Do not be fooled into thinking that an atmosphere is safe unless you know what is in it as well as what is not.

> **Safety**
> The safest atmosphere within a confined space is one that is closest in composition to normal air.

> **Safety**
> An atmosphere that contains enough oxygen to satisfy OSHA criteria still may not be safe.

oxygen-enriched atmosphere contains greater than 23.5% oxygen by volume

On the other end of the oxygen scale is an atmospheric condition known as oxygen enriched. An **oxygen-enriched atmosphere** contains greater than 23.5% oxygen by volume. This condition is one that is not easily created, but its presence can present considerable hazards to anyone working in a space. Oxygen enrichment usually is created when oxygen leaks from a tank or hoses containing it. It is never a good idea to take a cylinder containing oxygen or any other gas into a confined space because even a small leak can quickly become hazardous if it is not immediately detected and dealt with. And while the presence of more oxygen inside the space does not present health concerns to workers, high levels of oxygen can cause serious physical hazards, such as fires and other chemical reactions.

Recall that oxygen is an oxidizer, so it enhances the speed at which chemical reactions occur. If there are materials inside a space that can react with the high concentrations of oxygen, high levels of it can pose a serious threat to those working in the area. This is especially true when welding, cutting, or other hot work operations are being conducted. High levels of oxygen can rapidly accelerate the burning rate, resulting in extreme fire conditions. A historic example of this occurred early in the Apollo space project, when three astronauts burned to death inside the Apollo 1 capsule, which was supplied with an atmosphere high in oxygen. A small fire started inside the capsule and spread so quickly that the astronauts were unable to exit the capsule. Given the limitations for a quick exit from confined spaces, these spaces must be evaluated for the presence of high levels of oxygen.

Safety
Never allow a gas cylinder to be taken into a confined space because even a small leak can quickly become hazardous.

Safety
An oxygen-enriched atmosphere poses a serious potential for fire.

The next atmospheric hazard that can be present in a confined space, normal levels of oxygen, also relates to fire. Oxygen enrichment can create extremely hazardous fire conditions, but even normal atmospheres contain enough oxygen to sustain a fire. Any fire that occurs in a confined space presents the possibility of serious injury or even death, given the limitations of exiting the space. As we know from our discussion of flammability, a fire can occur when the right combination of oxygen, heat, and fuel react. The amount of oxygen present in normal atmospheres is more than enough for this process. In fact, any amount greater than 15% is all that is required for fire to start. So, controlling the potential for fire by diluting atmospheric oxygen is not practical. Preventing fires by eliminating ignition and heat sources also is not always possible or practical. Open flame from welding operations or even hot metal can be sources of ignition. However, the presence and amount of fuel or combustible materials can be controlled; it is a necessary element of safety programs for confined space operations.

Note
Fire is possible if the amount of oxygen in the atmosphere is greater than 15%.

Ignitable atmospheres in confined spaces are created by the presence of one of two elements: gases or vapors, or particulates or dusts. Examples of gases or vapors that commonly are encountered in confined spaces and present a hazard for fire include methane and hydrogen sulfide, which can be created by decomposing materials; hydrogen or acetylene used in welding operations; carbon monoxide, which is created by various conditions; and vapor from the evaporation of flammable liquids, such as alcohol, gasoline, or other solvents. Because most of these gases or vapors are not readily visible and many have little or no odor, their presence can go undetected.

stratification
the process by which gases and vapors settle into layers within an area based on their weight

Additionally, gases and vapors can stratify in various areas of a confined space. **Stratification** is the process in which gases or vapors settle into layers in the air in a space based on their density or weight. (Recall the discussion of vapor density and molecular weight from Chapter 4.) Stratification is commonly encountered in confined spaces because these spaces usually have limited air movement. Also, all vapors from flammable or combustible liquids are heavier than air, which makes them even more subject to stratification than other gases or vapors. When the concentration of vapors in a space reaches 10% of the minimum amount needed to produce an ignitable mixture in air (the lower explosive limit or LEL), the atmosphere is classified as flammable. Such areas should not be entered because their potential for fire is significant. In such cases, the space should be opened and ventilated to reduce the concentration of gases or vapors present.

Safety
Flammable atmospheres can be created whenever a space contains large amounts of flammable/combustible vapors, gases, dusts, or other particulates.

Safety
A flammable atmosphere is one that contains at least 10% of the lower explosive limit (LEL).

Particulates or dusts in confined spaces are actually a suspension (a cloud or mist) of combustible particles or liquids contained in the atmosphere. Examples of combustible particulates include flour, pulverized coal, charcoal, aerosols containing flammable or combustible liquids, and many finely divided metals. When these materials are suspended in air, they can be easily ignited if a flame or other ignition source is present. Therefore, areas that may contain these materials must be monitored to prevent ignition. Generally, if there are enough particulates to ignite, their airborne concentration is thick enough to be seen and obscure vision at a distance of 5 ft or greater. Some substances, however, may pose a danger at concentrations lower than the 5-ft standard. Be especially cautious in situations where this may be the case, because combined gases and particulates can act together to create an even more combustible atmosphere than either element can create alone.

Safety
Dust that obscures visibility to less than 5 ft must be considered ignitable.

The final atmospheric hazard common in confined space operations is the potential for other than oxygen-deficient/enriched or flammable materials to accumulate. Because confined spaces generally have poor air circulation, and because most toxic gases or vapors are heavier than air, these materials can settle into specific areas within a space. Some areas in a space may be free of these substances, but other areas may contain high concentrations of them. Therefore, when confined in a space, even small amounts of harmful materials can present a serious hazard to those who must work in the space.

Safety
Gases' ability to stratify can create different atmospheres within the same confined space.

A toxic or poisonous atmosphere contains excessive levels of toxic gases, vapors, mists, or dusts in concentrations that can cause illness or injury. Hazardous levels generally are defined as being above the exposure limits of the materials. Generally, the permissible exposure limit (PEL) is used to set the level at which an atmosphere becomes hazardous. Amounts higher than this level expose workers to hazardous situations and must be avoided.

Safety
Levels of contaminants above the permissible exposure limit (PEL) in a confined space should be hazardous and must be avoided by workers.

Depending on the nature of a confined space, many possible sources of toxins may exist in it, including the following:

- Bacterial action of materials in a confined space can produce toxic or flammable materials.
- Products or chemicals that were or may have been stored inside a confined space can emit harmful vapors. This can occur even when the storage vessel is empty because the materials may have permeated the vessel or be trapped in pockets inside it. Water blasting or steam cleaning a dry vessel can draw absorbed toxins from its walls and floors.
- Substances (e.g., cleaners, solvents, and paints) brought into a confined space may produce vapors or react with materials in the space to produce harmful materials.
- Work being performed, such as cleaning, welding, sandblasting, and painting can release toxins or other hazardous materials.
- Areas next to a confined space may release toxins, and these materials may migrate into the confined space as a result of vapor density or other factors.

Note
Atmospheric contamination can be created by a number of common conditions.

These are common causes of toxic atmosphere production in general. There are two very specific materials that are commonly found within confined spaces, hydrogen sulfide (H_2S) and carbon monoxide (CO). These materials should be well understood because they are so commonly encountered in a range of operations.

Note
Hydrogen sulfide and carbon monoxide are commonly encountered atmospheric contaminants.

Hydrogen sulfide (H_2S) is a by-product of decomposing organic matter and is produced by a variety of industrial processes. It is found in many vessels where organic matter is present and is prevalent in storm drains, sewers, manholes, and ditches containing vegetation. The gas itself is poisonous, flammable, and extremely irritating to the eyes, nose, and lungs at relatively low concentrations. Its characteristic odor often is described as being similar to that of rotten eggs. The ability to detect hydrogen sulfide by its odor, even at low levels, causes many people to become complacent when working in areas where it is present. In fact, the odor is so pronounced and distinctive that many people believe they do not need a detector to identify the material's presence. This may be one of hydrogen sulfide's biggest hazards, because its odor is detectable *only* when the levels are low. When the level of H_2S is high enough to cause harm, the material may not be detectable by its odor because it causes olfactory paralysis. A person whose sense of smell is temporarily paralyzed can not smell H_2S or other harmful materials. Additionally, if H_2S is inhaled at this level, its paralytic effects can spread to the respiratory system, causing respiratory arrest. Other symptoms of overexposure include eye irritation with tearing and blurred vision, irritation of the nose and the respiratory system, and pulmonary edema (fluid in the lungs). Rapid respiratory arrest can occur if exposures exceed 1000 ppm for even a short time. In addition to its toxicity, H_2S is extremely flammable, with a wide flammable range between 4% and 46%.

Safety
H_2S cannot be detected by smell even at concentrations below the level at which it is harmful. Never rely on its odor to determine whether it poses a hazard.

Safety
H_2S is very flammable and has a wide flammable range.

Carbon monoxide (CO) is another material commonly encountered in confined spaces. It is often present where flame or fire also is present, such as when open flames or welding occur, or where combustion takes place, as in the internal combustion engines of generators or power tools. Detecting this material is even more difficult than detecting hydrogen sulfide, because unlike H_2S, carbon monoxide gas is colorless, odorless, tasteless, and nonirritating. Once inhaled, CO interferes with the blood's ability to transport oxygen and leads to rapid asphyxiation at the cellular level. The mechanism for this is important to understand because this type of asphyxiation can occur even when adequate levels of oxygen are present.

Safety
Detecting CO is difficult because it has no odor, color, or taste.

hemoglobin
the part of the red blood cell that transports oxygen to the body's cells

When CO is in the atmosphere, it is brought into the lungs as we breathe and travels to the deepest levels of the respiratory tract. Once inside the alveoli, CO rapidly crosses into the blood, where it is picked up by red blood cells in the capillaries. **Hemoglobin** is the part of the red blood cell that is the predominant oxygen transport system of the body. In most cases, the oxygen in the air binds to the hemoglobin and is transported to the cells in the body for use. Unfortunately, hemoglobin has a much greater affinity for CO than for oxygen—this affinity is estimated to be more than 240 times that for oxygen. Therefore, if CO exists in the atmosphere and is inhaled, hemoglobin becomes locked up with it rather than with oxygen. The CO held in the hemoglobin remains there for a considerable period of time depending on a number of factors, reducing the blood's oxygen-carrying capacity. This results in a condition where the unoxygenated blood circulates around the body, depriving the cells of the fuel they need for survival. The brain cells are some of the first that show symptoms resulting in headache, confusion, and mild dizziness. Exposure to concentrations higher than 400 ppm can result in extreme dizziness, nausea, vomiting, and loss of consciousness. Once concentrations reach 2000 ppm, immediate unconsciousness and death may occur.

Safety
Hemoglobin binds with CO 240 times more than it does with oxygen.

Safety
Carbon monoxide can cause asphyxiation even when adequate oxygen levels are present.

Because of the serious consequences of exposure to CO, the behavior of personnel who may be exposed to it must be monitored for symptoms. Rapidly removing and intervening with victims of CO exposure can literally save lives. Extreme exposures are sometimes treated with hyperbaric chambers. These chambers, which often are associated with diving accidents, use high atmospheric pressures and high levels of oxygen to reduce the levels of CO in the blood by forcing it out of the red blood cells. The levels of CO within the blood can be carefully monitored to determine when the CO is gone and the oxygen levels have returned to normal. This process takes considerably less time inside a hyperbaric chamber than under normal conditions.

Note
Hyperbaric chambers can be used to treat CO toxicity.

backdraft explosion an explosion caused when air is suddenly introduced into a confined area that contains high levels of heat and combustible gases

Finally, and also like H_2S, CO also is flammable, with a flammable range of 12.5–74%. This property makes CO the most likely cause of **backdraft explosions**. A backdraft is the phenomenon that takes place when fires occur inside sealed spaces. When the CO builds to high levels and the oxygen level is reduced, the fire smolders until someone opens a door or a window. When this happens, the rush of air fans the smoldering fire, and CO's flammable range is realized, resulting in an explosion.

Safety
Like hydrogen sulfide, carbon monoxide is extremely flammable, with a flammable range of 12.5–74%.

The key to safely working in a confined space with a potentially hazardous atmosphere is to ensure that the space is properly ventilated. When outside air is drawn into a confined space, it dilutes harmful levels of contaminants (whether they cause oxygen deficiency, are ignitable, or are toxic) and creates a more normal atmosphere. Diluting toxic atmospheres with normal air provides adequate oxygen, reduces the concentration of ignitable vapors to levels below ignitable levels, even with an ignition source present, and dilutes levels of contamination below the PEL. Appropriate ventilation helps provide workers in confined spaces with the degree of protection required for safe operations by ensuring that no pockets of concentrated vapors or gases exist within the space. Increasing the air movement also lessens the potential for heat stress that is common when people are working in areas with limited air movement.

Safety
Confined spaces have a high potential for atmospheric contamination.

Physical Hazards of Confined Space Operations

In addition to the atmospheric hazards associated with confined spaces, these spaces commonly pose physical hazards. These hazards are called *physical* because they have more to do with the physical makeup or conditions of a space and less to do with the atmosphere in a space. The most common types of physical hazards include the following:

- Engulfment
- Entrapment resulting from the design of the space

- Falls, including those from elevated locations
- Exposure to hazardous energy
- Heat stress
- High noise levels

> **Safety**
> A number of other physical hazards are commonly encountered in confined space operations.

Most of these hazards strike with little or no warning and can prevent those inside the space from being able to rapidly evacuate. Accident studies show that most people get into trouble because they do not recognize the hazards, underestimate the hazards, or simply ignore the potential for a hazard to be present. Yet these types of hazards can often lead to situations where death is the result.

> **Safety**
> Many accidents involving physical hazards occur due to the lack of recognition that the hazard is present.

engulfment
a condition in which loose materials are released in a space and bury or crush people in the space

Engulfment is one of the most common physical hazards. It is so common, in fact, that the OSHA regulation identifies it as one of the major physical hazards encountered in confined spaces. Engulfment occurs when a person entering a space is buried or crushed by loose material in the space, such as grains, sand, sawdust, cement, or water. These materials may be contained in one part of a confined space, and workers may enter the area thinking that they can avoid them. If these materials are disturbed, however, they can move with considerable force and momentum, trapping victims. Once trapped under the material, victims suffocate from the pressure of the material on their bodies. The engulfing material may also be hot or corrosive, leading to other injuries if the engulfment itself does not kill the victims.

> **Note**
> Suffocation is a common result of engulfment in a confined space.

The very design of some confined spaces poses a hazard to workers. Some spaces have inwardly converging walls that taper down to a restricted area. Others contain sloping floors that narrow down, creating smaller areas within the space. In these situations, workers may become trapped in the narrow or shortened area. Even if the space does not pose other hazards, becoming trapped can lead to serious injury.

> **Safety**
> The configuration of a space can trap workers.

An often-overlooked physical hazard of working in confined spaces is the risk of slips, trips, and falls. Many confined spaces have elevated work locations that expose workers to falls, either from the elevations themselves or from the ladders or scaffolding used to access them. The potential for slipping while ascending or descending ladders is increased if floors are wet or contain debris.

Another risk associated with elevated locations is the potential for objects to fall onto workers from above. For example, tools or other equipment may be brought into a space from an opening above the workers inside the space. In such situations, tools or equipment must be handled with the utmost care because it may not be possible for workers inside the confined area to move out of the way if something is dropped. Obviously, falling objects present significant hazards in these situations. It is important that all personnel be aware of these types of hazards and take precautions to avoid an accident including the use of a hard hat when the potential for injury is present.

The potential for exposure to energy sources is another commonly encountered situation in confined spaces. One type of energy source is electrical energy. The electrical equipment already in a space and the electrical tools or devices taken into the space by workers both pose the potential for worker electrocution. Electrocution is one of the leading causes of death and injury within confined spaces.

In addition to electrical energy, numerous other energy sources are often found in confined spaces, including steam, pressure, hydraulic, pneumatic, and thermal. Later in this chapter, we discuss the importance of controlling energy sources by using *lockout/tagout* programs. Before workers are allowed to enter a confined space, it is absolutely required that all forms of energy be controlled and the potential for harm eliminated. If energy is unexpectedly released while workers are inside a confined space, the results can be catastrophic because exiting the space is difficult.

Safety

Controlling energy sources within confined spaces is essential to ensuring the safety of everyone working inside the space.

Heat stress is another hazard associated with working in a confined space. The potential for heat stress is far greater in a confined space than in other types of work given the frequent use of personal protective equipment (PPE) and the limited air movement in such spaces. From Chapter 7, we know that heat stress is more pronounced when evaporation is limited, and clearly, the potential for these conditions is high in confined space activities. Care must be taken to ensure that personnel are monitored for the symptoms of heat stress and provided with appropriate rehydration through the course of their work.

Safety

The potential for heat stress is magnified inside confined spaces as a result of limited air movement and the frequent use of PPE in these spaces.

The final hazard associated with work in confined spaces is exposure to high levels of noise. As we discuss later in this chapter, exposure to high levels of noise leads to a number of associated issues, including elevated stress levels among workers and their inability to hear warnings, including alarms. The configurations of these spaces, the fact that many of them have poor acoustical properties, and the use of equipment in or near them amplify noise. These factors make this hazard something that must be considered for anyone working in a confined space.

Controlling the Risk of Confined Space Activities

Because of the high risk often associated with confined space operations, OSHA regulations mandate that specific practices be employed to help reduce this risk by making work performed in these spaces safer. As we have emphasized throughout this text, it is important that the controls follow a prescribed order. It is essential to use these controls in confined space work, to enable workers to safely enter a space and to make the work safe.

As in other hazardous situations, controlling and reducing the risks posed by a specific operation begins with engineering controls. Usually, the most effective engineering control available is the use of ventilation systems. Whenever possible, spaces should be ventilated with fresh outside air to help make the atmosphere and conditions in the space similar to those outside of it. Ventilation methods can vary considerably, depending on the equipment and conditions encountered, but some common guidelines and practices include the following:

- The ventilation system, sometimes called a *blower*, should provide enough airflow to fill a space with fresh air, but not be so forceful that it stirs up dust and other materials in the space. Disturbing materials in the space can increase the potential for eye or facial exposure.
- Ventilation should be done in conjunction with air monitoring. (Air monitoring techniques are discussed later in this chapter.)
- The ventilation system should employ a hose that channels air into the space and distributes it into a safe area. (Figure 9-4 shows a confined space ventilation operation.) The dropped end of the hose should hang vertically in the space to ensure proper air movement. It should be positioned not more than 1 ft below the ceiling and not more than 2 ft

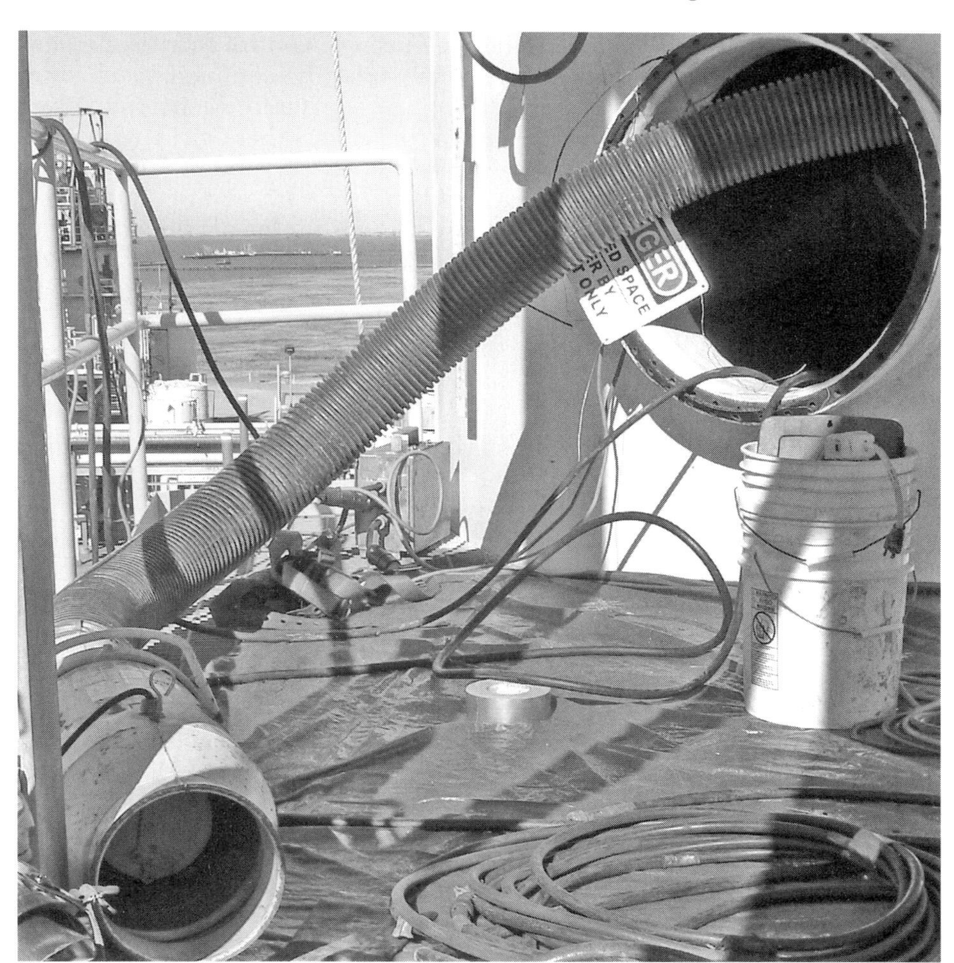

Figure 9-4
An example of a confined space ventilation operation.

above the floor of the space. This is required to ensure maximum air movement within the space.

- When using a blower to inject fresh air into a space, it is important to make sure that the blower intake is located away from vehicle exhaust or other airborne contaminants. This helps avoid introducing harmful airborne contaminants into a space. If a gas-powered blower is used, ensure that the exhaust is not immediately sucked into the blower air intake. Regardless of the blower selected, the air intake must be monitored to ensure that only clean outside air is drawn into a space.
- The blower and the air intake should be at a safe distance from the entry point into the confined space. This ensures that air is not simply recycled from within the space, that the equipment's noise does not impede communication, and that the equipment does not impede workers' ingress or egress from the space. When large lengths of hose are used, it takes longer for air to be injected into the space.

It is critical that the air brought into a confined space be free of contaminants.

Using blowers is the primary engineering control for confined space operations, but some situations may require additional equipment and other methods. For example, energy control systems, such as locking out hazardous energy from a space, are another engineering control that is often employed. As we discuss later in this chapter, the use of the confined space entry permit prescribes the other measures that should be taken for each confined space operation.

Administrative controls are another OSHA-required control, and they afford additional protection for workers in confined space operations. Administrative controls include a standardized work permit that details the hazards and methods of controlling those hazards for the confined space activity, requires specific amounts and types of worker training for confined space operations, and requires special equipment and standard operating procedures whenever confined space operations are taking place.

The required work permit is developed and implemented by the employer, who is responsible for the confined space operations inside a Permit-Required Confined Space. The permit serves as a safety checklist to help identify the conditions within a space and the steps necessary to eliminate the hazards posed by a space and control the risk of entry. The permit shown in Figure 9-3 is an example of one type of permit, but there can be a number of variations used that meet the OSHA requirements. Regardless of which format is used, the OSHA regulation requires that all permits contain specific information about the confined space and operations they cover. This information includes:

- The location of the confined space to be entered. This ensures that the space and the area authorized for entry are properly identified.
- The purpose of the entry. This ensures that the hazards associated with the operations are considered in the evaluation of the hazards in the space. Remember that in some cases, the work being performed introduces hazards or increases the existing hazards.
- The identification or listing of the authorized entrants (in other words, the workers who are allowed to enter the space) and of the attendants who monitor the entrants' activities. The actual names of the entrants are not required to be on the permit, but OSHA states that a system of tracking who is in a space while they are there must be used. This can be done in

a number of ways, including having workers sign the permit document, leaving their identification cards with the attendant, or by some other equally effective means. Regardless of the method chosen, the permit must indicate how the entrants are tracked by the attendant.

- The name of the entry supervisor. This is the person who has reviewed the conditions in the space and authorized the work to take place. (The duties of the entry supervisor are discussed later in this chapter.)
- The hazards in the space. This includes the hazards posed by the space itself, as well as any created by the work.
- The measures used to isolate the space and to eliminate or control its hazards. These can include the listing of lockout procedures, the use of ventilation systems, or a requirement to wash the space prior to entry.
- The acceptable conditions for entry.
- The initial and periodic atmospheric testing results. These are compared with the list of acceptable entry conditions. If the current test results do not match the acceptable entry conditions, the permit specifies what other steps are necessary to ensure the safety of workers in the space. (Testing is discussed in greater detail later in this chapter.)
- The type, location, and means to summon a rescue team if an emergency develops during the confined space operations. (A more comprehensive discussion of rescue personnel is found later in this chapter.)
- The method for communication between the attendant and the entrants. This helps to ensure that workers inside a space maintain contact with the attendant who is monitoring their activities. Often, communication is direct, such as simply visual or verbal contact between the two parties. In other cases, radios or video systems may need to be employed. The permit lists all methods of communication required for the operations to take place.
- The types of equipment required for workers in the space. This can include the required PPE, rescue equipment, lighting, ventilation systems, and other items required to safely perform the tasks authorized by the permit.
- Any additional permits required to be completed to ensure safe operations in the confined space. This provides a checklist of other hazards that may be present and includes completion of a hot work permit whenever activities involving hot work are performed as part of the confined space operations.

Note

OSHA requires that confined space entry permits include specific information.

Safety

Properly completing a confined space entry permit is the key to working safely in a confined space.

The entry permit is a key component of the steps taken to eliminate or control the risk associated with the confined space operations. Completion of the permit by the entry supervisor helps ensure that the appropriate steps are taken to ensure workers' safety and that the results of this are documented. This includes the need to adequately monitor the condition of the atmosphere in a space and to document these findings on the permit so that workers have information about the existing hazards and conditions. OSHA provides specific rules regarding atmospheric monitoring, and these rules list the specific steps in the monitoring process.

Atmospheric testing must be completed before a space is entered. Whenever possible, the initial air sampling should be done before the space is opened. It may be possible to insert the monitor probe into a sample hole for the space. If this is not possible, the next best procedure is to gently crack the opening just enough to insert the probe (Figure 9-5).

Once the space is open, its atmosphere must be tested not only at the entrance but also well within the space, where workers will conduct the operations. This can be done before and after the ventilation system is in operation to obtain information regarding the quality of the air at various levels in a space and the effectiveness of the ventilation system. All levels in the space must be tested, including the lowest level, workers' breathing zone, the level above their breathing zone, and just below the ceiling. This accounts for various vapor densities, temperature inversions, and other gas stratification.

OSHA also requires that when the atmosphere in a PRCS is tested, the level of oxygen in the space be established first. This is done for a number of reasons. While many types of instruments monitor several gases simultaneously, some operations still use multiple instruments to evaluate the hazards within a PRCS. It is important to know if oxygen is present in the correct amount prior to conducting other tests. Another is that low oxygen levels in a space may provide false readings for other tests that are going to be conducted. For example, testing for a concentration of flammable vapors is performed with the assumption that the atmosphere contains normal levels of oxygen. If the oxygen level is below 10%, however, this can result in an artificially low combustible vapor level readout. Therefore, the oxygen level should be established first, then tests can be conducted to determine the presence of flammable or combustible vapors and then of toxins.

Figure 9-5 *Atmospheric monitoring being conducted from outside a confined space to determine the hazards before entry is allowed.*

Safety
OSHA requires that the oxygen level in a confined space be tested first.

After the pre-entry testing is completed, OSHA requires that testing be repeated at intervals frequent enough to ensure that acceptable entry conditions are maintained. There is considerable confusion and a host of misinformation regarding the frequency of air testing, but the regulations do *not* specify continual testing of an atmosphere while personnel are working in a confined space. Many people mistakenly believe that continual testing is required, but such testing often is unnecessary and can damage equipment exposed to high levels of contamination in a space. OSHA does specify that when the space to be entered cannot be isolated from other spaces, the space and areas involved should be continually monitored to ensure that work in another area does not present hazards where employees are working. However, even in these conditions, the monitoring may not include continual testing of the air because that may provide little information and could cause the monitor to become saturated from being constantly exposed to the high levels. Monitor saturation can lead to inaccurate readings and reduce the level of safety otherwise afforded by continual air monitoring.

Safety
Air monitoring in confined spaces must be done at regular intervals to ensure that acceptable conditions for entry are maintained.

To illustrate this important point, take a hypothetical case in which workers in a confined space are using supplied-air respirators and a high level of PPE. Let's say that this level of protection is required by the entry permit, which was completed by the entry supervisor based on the pre-entry air testing. Once personnel are working in a space with the appropriate types of protection, continuously monitoring for levels of oxygen and contaminants does nothing to promote safety and could, in fact, distract personnel from observing other hazardous conditions. Even if unsafe levels of contamination or low oxygen levels were observed on the monitor, nothing more needs to be done to protect the workers because they are already appropriately protected.

More useful requirements for the frequency of air monitoring are as follows:

- Initial testing of the atmosphere of a confined space to be entered should be done prior to entry.
- Testing should be conducted in the order prescribed by OSHA: oxygen, LEL levels, and potential air contaminants.
- Testing should be done at all levels in a space where employees will be working.
- Testing should be repeated at intervals to ensure that acceptable entry conditions are maintained.
- Testing should be repeated as conditions change, such as when workers leave the space for breaks, etc.
- Testing should be repeated if employees request further testing.
- Testing air in non-isolated spaces should be done at more frequent intervals to ensure that work in other areas does not change the conditions in the confined space that is being entered.

Safety
More frequent testing of the air may not always promote higher levels of safety.

Another administrative control of the risk in confined space operations is ensuring that only properly trained personnel are involved in the operations. The groups of workers, the duties they perform, and the training required is clearly spelled out in the OSHA regulation. Personnel who work in confined space operations fall into one of four classifications: the entrant, the attendant, the supervisor, or rescue personnel. The duties of each type of worker are outlined in the regulation, and one person can fill more than one role depending on the circumstances. As a general rule, entry into a confined space requires that at least two properly trained personnel be at the confined space during the operation. Additional personnel may be standing by at other locations to assist in rescue operations should they be required. Some states have more restrictive requirements for confined space operations than others, so it is wise to review the specific state OSHA regulations that may also apply. Following is some information on the responsibilities of each of the three groups of workers involved in normal confined space operations. (A more detailed discussion of the rescue operations is found later in the chapter.)

Note
OSHA assigns specific responsibilities to personnel involved in confined space entry operations.

entrant
the title of the person who enters a space and performs work

Entrant is the term used to describe an employee who enters a space and performs work. Entrants run the highest risk of exposure to the hazards in the space because their activities take place inside it. The hazards of entering a space begin once entrants introduce any part of their bodies into the space. For this reason, OSHA states that a confined space entry occurs when any part of the body of the entrant breaks the plane of the Permit-Required Confined Space. This means that even inserting a hand or an arm into a space is technically defined by OSHA as entering the space. Because of this, OSHA requires that personnel who enter a space must be appropriately trained in the hazards they may face and operations they must perform.

To satisfy OSHA regulations, entrants must be trained in the following areas:

- Knowing the hazards and potential hazards within a given space
- Knowing the signs and symptoms of exposure to the hazards and the consequences of exposure to the hazards found in confined spaces
- Being able to properly use all of the PPE necessary for safe entry into a space
- Maintaining consistent communication with the attendant outside the space. Communication between the entrant and the attendant can be accomplished in a number of ways, and the entrant must be familiar with as many of these as is appropriate.
- Alerting the attendant to any observed hazard or condition not allowed by the permit
- Instantly obeying any order to evacuate the permit space

■ **Note**
Confined space entrants must receive specific training on the hazards of the work and ways to protect themselves.

confined space attendant the person positioned outside a confined space who monitors the conditions in and outside the space to ensure the safety of the entrant

A **confined space attendant** *attends* to the confined space entrant. The attendant is required to stay outside the confined space and monitor the activities that take place both in and outside the space. The attendant may be called the *hole watch* because the entrance to a space often is often a manhole, and the attendant watches the hole and what is below or behind it. Workers may alternate between being an entrant and being an attendant. This is relatively easy because many of the training requirements for the attendant position mirror those of the entrant (Figure 9-6).

To satisfy OSHA regulations, attendants must be trained in the following areas:

- Being familiar with the hazards in permit spaces and recognizing the presence of hazards in a space
- Monitoring the entrant's behavior closely for signs and symptoms of exposure, or other health problems
- Keeping an account of the workers in a space and allowing only authorized entrants to enter the space
- Evaluating entrants to ensure that they comply with all requirements in the confined space entry permit
- Maintaining constant communication with entrants
- Protecting entrants from external hazards, such as falling objects or other operations that can introduce hazards into a space
- Not leaving the entrance unless relieved by another qualified attendant
- Performing non-entry rescue techniques as needed. (These are discussed later in our study.)
- Not entering a space to perform rescues and being able to instantly contact the rescue team

Figure 9-6
A confined space entry being monitored by an attendant.

■ **Note**
Confined space attendants must be trained to perform duties that ensure entrants' safety.

entry supervisor
the person responsible for ensuring that the confined space operation is conducted safely; the person responsible often completes the confined space entry permit

The **entry supervisor** (an employer, a foreman, or a crew chief) is responsible for the overall safety of the project operations in a confined space. The supervisor must be trained to perform the following duties:

- Knowing the hazards that may be faced during entry, including understanding the mode of action of chemicals, the signs or symptoms of overexposure, and the consequences of exposure
- Verifying, by checking that the appropriate entries have been made on the permit, that all tests specified by the permit have been completed before endorsing the permit, and allowing entry to begin
- Verifying that rescue services are available and that the means for summoning them are operable
- Determining that acceptable entry conditions are maintained and that entry operations remain consistent with the terms of the entry permit whenever responsibility for a permit space entry operation is transferred, and at intervals dictated by the hazards and the operations performed in a space
- Authorizing entry into the confined space
- Terminating the entry and canceling the permit as required by this section
- Removing unauthorized individuals who enter or who attempt to enter the permit space during entry operations

■ **Note**
The entry supervisor has many duties related to the overall safety of the confined space operations.

Note that entry supervisors may also serve as attendants or authorized entrants, as long as they are trained and equipped as required for each role. The duties of entry supervisors may be passed from one individual to another during the course of an entry.

■ **Note**
Personnel working in confined space operations may be trained to serve in several positions.

■ **Note**
One person may serve as both the attendant and the entry supervisor, or as the entrant and the entry supervisor at the same time.

Various types of equipment are used and required for use by OSHA during confined space work. This equipment is another form of administratively controlling the risk to workers in confined spaces. In addition to the equipment required by the permit, such as PPE for protection, radios for communicating, and air monitoring equipment for testing the atmosphere, another type of equipment is required by OSHA to assist in rapid retrieval of anyone in a confined space should an emergency occur. Many organizations are not aware of the requirement for such systems, but the OSHA regulation is clear:

> *To facilitate non-entry rescue, retrieval systems or methods shall be used whenever an authorized entrant enters a permit space, unless the retrieval equipment would increase the overall risk of entry or would not contribute to the rescue of the entrant. Retrieval systems shall meet the following requirements. Each authorized entrant shall use a chest or full body harness, with a retrieval line attached at the center of the entrant's back near shoulder level, above the entrant's head, or at another point which the employer can establish presents a profile small enough for the successful removal of the entrant. Wristlets may be used in lieu of the chest or full body harness if the employer can demonstrate that the use of a chest or full body harness is infeasible or creates a greater hazard and that the use of wristlets is the safest and most effective alternative. The other end of the retrieval line shall be attached to a mechanical device or fixed point outside the permit space in such a manner that rescue can begin as soon as the rescuer becomes aware that rescue is necessary. A mechanical device shall be available to retrieve personnel from vertical type permit spaces more than 5 feet (1.52 m) deep.*

Safety
Personnel who enter a confined space must wear a retrieval system, unless the equipment increases the hazard or does not assist with the rescue.

This equipment enables the rescue of workers who cannot get out of a space if an emergency occurs. Again, many who work in confined space operations are either not aware of this requirement or choose to ignore it, but the logic of the requirement is clear: If someone is incapacitated and unable to exit a space on their own (i.e., is unable to perform **self-rescue**), then the use of retrieval equipment aids non-entry rescue operations. **Non-entry rescue** is simply the use of the retrieval equipment by the attendant. Without this equipment, an incapacitated entrant must wait to be rescued until confined space rescue personnel can arrive. This delay may prove fatal in many atmospheric situations because rescue operations can be complex and time-consuming, as is discussed later in this chapter. An example of a retrieval system is shown in Figure 9-7.

self-rescue
an operation in which entrants leave a confined space on their own power when an emergency occurs

non-entry rescue
an operation in which the attendant uses the retrieval equipment worn by entrants to pull them out of a confined space

Safety
It is critical that retrieval equipment be worn to enable rapid extraction of an incapacitated entrant.

Emergency Operations and Personnel

The easiest emergency to deal with is the one that never occurs. Complying with a properly completed confined space entry permit helps prevent emergencies, which is the larger goal of confined space safety programs. With these programs and permits effectively employed, it is possible to prevent nearly all emergency situations. Despite proper planning, however, emergencies can and do occur, and OSHA mandates that they be prepared for to protect personnel involved in confined space operations.

As we just discussed, the method for removing personnel from these spaces, which often have limited exits, may be complex or time-consuming. Entrants

Figure 9-7
An example of a retrieval system used during a confined space entry.

should be made aware of the need for and conditions of evacuation at the earliest moment. This facilitates their ability to self-rescue. When entrants are overcome and unable to escape a space, non-entry rescue techniques can be employed. Lastly, when these are not successful in extracting entrants from a space, properly trained confined space rescue team members may need to enter the space and extract anyone who is unable to escape. Numerous conditions can lead to entrants' inability to get out of a space on their own or to the attendant's inability to perform non-entry rescue. Examples of some conditions that warrant the rapid removal of personnel from inside confined spaces include the following:

- The oxygen level drops below 19.5% when the entrant is not equipped with a supplied-air respirator.
- The oxygen level rises above 23.5%, regardless of the respirator being worn.
- The level of combustible gases rises above 10% of the LEL.
- Concentrations of toxins rise above the permissible limits, and respirators are not being worn.
- Conditions outside a space change to the extent that they might endanger the entrants.
- An entrant begins to exhibit signs or other behavior indicating exposure to chemicals.
- Conditions that are not included on the entry permit arise and are observed.
- The entrant suffers a medical emergency or is the victim of trauma within a confined space.

Safety
Compliance with the requirements of entry permits is a key method of preventing an emergency.

Safety
Rapid action is required whenever a situation in a confined space is encountered because exiting the space may be difficult.

Safety
Any condition that arises and is not addressed on the entry permit is cause for immediate evacuation of a space.

At the first sign of changing conditions that could lead to any of these situations, it is the responsibility of the attendant to begin the appropriate emergency response. At the first sign of trouble, an alarm should be sounded to alert others to the problems and to get additional help. The conditions in a space can change rapidly, so swift, aggressive action is important. *If there is any possibility of danger to the entrants, remove them immediately.* Remember, "when in doubt, get them out!"

Following the sounding of the alarm, the attendant must quickly assess the ability of the entrants to self-rescue. Sometimes a combination of self-rescue and non-entry rescue is employed. The attendant may begin to pull on the retrieval system worn by an entrant to help them get out of the space. Keep in mind that attendants must NEVER enter the space to help in the rescue operations unless they are part of the rescue team and another qualified attendant is present. Most personnel who die in confined space operations are well-intentioned would-be rescuers who incorrectly believe that they can quickly enter a space and help the entrant out. This reasoning is clearly flawed as more than 60% of those who die are trying to effect a rescue without understanding that they are more likely to suffer the effects of exposure to the conditions in the space than to succeed in rescuing the original entrants. The most common cause of death in confined space operations is simple lack of oxygen and asphyxiation. In fact, would-be rescuers in such cases have an even higher potential for suffering harm than the entrants. The rescue activity creates a heightened rate of breathing and a greater need for oxygen as a result of adrenaline being produced in the body, which puts them at heightened physiological risk. In some cases, the original victim has been removed alive, while other rescue personnel died as a result of their exposure to the conditions in a space.

Safety
Most deaths in confined spaces involve well-intentioned would-be rescuers. *Never* enter a confined space to effect a rescue unless appropriate procedures are followed.

Safety
Self-rescue is the most efficient method of getting an entrant out of the confined space.

If self-rescue and non-entry rescue procedures are unsuccessful in removing entrants, sounding the alarm alerts the rescue team to the emergency. Because the entrants may require additional help getting out of the space in an emergency, OSHA requires that a team of trained personnel be available for rescue whenever a PRCS is entered. Although federal regulations allow the use of rescue teams that are not on site during the confined space entry—for example, local fire departments—that team's ability to respond quickly, to be properly trained, and be equipped to conduct the rescue must be confirmed by the entry supervisor before entry is made. Some state-plan states require on-site rescue teams for all operations. Even if allowed, many fire department rescue teams may not be able to respond in a timely manner to extricate an entrant from an oxygen-deficient space within the 4 minutes before brain damage can occur. If a company relies on an outside group to perform these hazardous duties, it is very important that the group's qualifications, including their response time, be thoroughly evaluated. Many fire departments do not have personnel who are qualified to do this work as it involves specific training (which is detailed in Section [k] of the regulation). Following is the list of HAZWOPER requirements regarding rescue team training and preparation:

- Unlike other confined space training, rescue team members must be retrained annually.
- Rescue team members must be trained in providing emergency medical care, including CPR and first aid.
- Rescue team members must practice actual rescue operations each year as part of their training in spaces that are similar to the types that the team may encounter in the field.
- Entering a hazardous environment to conduct rescue operations may trigger other OSHA-mandated training requirements, such as for respiratory protection, use of PPE, and even certification at the hazardous materials technician level when emergency response includes potential exposure to hazardous materials.

Safety
Only properly trained and certified personnel are allowed to conduct rescue operations that involve entering a confined space to rescue an entrant.

Safety
Confined space rescue operations can be technically difficult, require considerable training, and must be practiced annually.

Once rescue personnel arrive at the scene, they must be supplied with information about several factors, including:

- The number of entrants working in the space
- The condition of the entrants, which may be unknown at the time of rescue
- The types of hazards present in the space
- The cause of the emergency and the circumstances surrounding it

Confined Space Entry Permit Compliance Checklist

Confined space operations can be complex given all the requirements that are involved. Despite this, such operations are relatively common and can be done with a high degree of safety. The key to achieving this degree of safety is ensuring that the confined space entry permit is accurately completed prior to any activity. The permit serves as the all-important safety checklist. Following is a list of the ways in which a permit can be used to identify and control the hazards posed by a confined space. These steps help ensure a higher degree of safety for everyone involved in a confined space entry operation.

- Read the permit and note the hazards. Do they align with what you know about the space?
- Note the safety equipment required. Does it address all the hazards mentioned? Is the equipment available and have you inspected it prior to use?
- Look at the list of required tools. Are they available and have they been inspected?
- Check the list of authorized entrants and attendants. Do they have the required training?
- Go through the permit procedures for isolating the space and controlling energy sources. Make sure that all the required steps have been completed.
- Review the data from the atmospheric tests that were conducted. Were the instruments calibrated and zeroed? Have all the potential atmospheric contaminants been identified and measured? Were readings taken throughout the space?
- Has the space been properly purged and ventilated with a supply of fresh air? Will the ventilation system continue to provide adequate air movement in all areas of the space? Is the ventilation system clear of the opening to the space and not subject to drawing other contaminated air into the space?
- Does the entrant have the required retrieval system in place? Is there an appropriate rescue team available for response to the space in a timely manner?

Contractors Entering Permit Required Confined Spaces

A final word about confined space operations involves entry of the space by a contractor. This is a very common practice because many outside craftspeople may be required to perform work in a space, including welders, mechanics who may be needed to repair pumps and ancillary equipment at the site, or inspectors who need to confirm the status of the vessel or space. Whenever an outside contractor is authorized to enter a space, OSHA mandates that the host employer, the owner of the confined space, ensure the contractors are protected during the time they perform their work. A list of required and recommended practices for contractors working in confined spaces of another employer follows:

- The contractor must be informed that the work location contains a Permit-Required Confined Space and that entry is allowed only through compliance with the site written confined space program.
- The contractor must be told of any possible hazards that could be present in the space.
- The contractor must be informed of the precautions that have been taken to protect employees from the hazards associated with the confined space.
- The contractor must be informed that the site safety official must be notified prior to entry operations and at the completion of the confined space work.

- Upon notification from the contractor that the work has been completed, the site safety official or his designee must debrief the contractor regarding any hazards confronted or created in the permit space during entry operations.
- Any contractor who will perform work in a confined space must show proof to the site safety official or entry supervisor that the contractor's employees have completed training in confined space operations and are authorized by their employer to work in such spaces.
- Contractors who use specific equipment or PPE should have documentation that they are qualified to use the equipment through training records and other forms of documentation.

> **Safety**
> OSHA requires that contractors working in a confined space be properly informed of all hazards and debriefed when their work is complete.

CONTROL OF HAZARDOUS ENERGY (LOCKOUT/TAGOUT)

Uncontrolled energy sources account for a large number of worker deaths nationwide. What makes this particularly tragic is that almost all these deaths could have been prevented through the proper application of a range of OSHA regulations. If these workers had received effective training in the proper procedures for working with energy sources and had been aware of the potential dangers that energy posed, most of them would have been alive today. Lives are lost and injuries occur when employees fail to recognize energy's hazards and do not even take such simple precautions as turning off an energy source before working or ensuring that energy sources are not turned on while other employees are working on a system. No one seems to expect anything to happen under these dangerous circumstances, yet their consequences are extreme. The most common mistakes made when working with hazardous energy that lead to injury or death include:

- Failure to stop equipment prior to conducting work. Working on operating equipment has enormous potential for disaster.
- Failure to disconnect power sources to prevent equipment from accidentally restarting. Even if a machine is turned off, if the power source is still connected, accidentally hitting switches or buttons may unexpectedly energize the equipment.
- Failure to dissipate residual power. This includes failure to relieve stored pressure from pipes opened in the course of work and failure to dissipate stored electrical energy from capacitors, etc.
- Failure to clear work areas prior to restarting machinery. This can cause clothing or gloves to snag in gears or conveyors, paths of conduction to be created by metallic objects contacting electrical wires, or shards of metal or glass to become projectiles when machines are activated.

> **Safety**
> Energy sources must be controlled to avoid unexpected equipment startup that may expose employees to hazards.

Because many forms of energy are present at work sites where HAZWOPER-regulated projects take place, OSHA requires that personnel be trained on the site hazards and that procedures to protect employees be documented in the site health and safety plan.

Energy comes in many forms and from a variety of sources. The most commonly encountered energy sources found at work sites include electrical, mechanical, pneumatic, chemical, thermal, hydraulic, and radioactive processes. Other than hand tools, almost all types of equipment used at a work site contain some kind of energy that can present a hazard. Additionally, a site itself may contain a range of hazardous energy sources in the form of electrical panels, pipes that contain chemicals or steam, vessels that contain powered equipment, pressurized systems that could run through an area, and others. Figure 9-8 shows an example of a site with numerous potential sources of energy.

kinetic energy
the energy used to make items move

Energy exists in two forms, both of which are capable of causing harm or even death if not properly controlled. The first type of energy is **kinetic energy**. This energy is usually the easiest to identify because it is the type most often associated with the term *energy*. Kinetic energy is the actual energy required for a process to take place, for an item to move. It can be seen in the work or motion that something does. For example, a drill requires electricity in order for its motor to turn. Likewise, a certain amount of pressure may be required to push liquids through a pipeline. Each of these is an example of kinetic energy. Often, simply looking at a process can indicate that energy may be present, for example, in the form of the noise of an operation, pressure readings in gauges, and observation of items in motion.

Note
Energy can present itself in two forms: kinetic and potential.

potential energy
stored energy or energy that has not been released and still exists in a system

The second form of energy is **potential energy**. Potential energy is different from kinetic energy in that it is not the main source of energy that causes work or movement to take place, but rather is the energy that is left in a system after the work or movement has occurred. This energy is found in hot equipment,

Figure 9-8
An example of a site with multiple types of potential energy.

depressed springs, electrical capacitors that hold some electricity after a system is de-energized, and pressure that remains in a pipe after the compressor is shut off. For example, if a coiled spring is allowed to uncoil, it releases its potential energy by converting it to kinetic energy in the form of movement. Because potential energy often is not the primary energy source present in an object, and may be stored within the process and becomes active only upon release, its sources may be harder to identify. In order for any energy control program to be effective, it must include procedures to identify and control both types of energy.

Because of energy's high potential for harm, a specific OSHA regulation, 29 CFR, Part 1910.147, addresses controlling energy sources in the workplace. Like many OSHA regulations, this one is complex and warrants a full reading if it applies to your operation. The scope of this regulation is to prevent the unexpected startup of machinery or release of energy when employees may be exposed to the equipment or energy source and potentially injured. The regulation follows the control steps that we have discussed throughout this book.

lockout/tagout (LOTO) programs
a series of steps to control energy in the workplace

Energy is controlled by a lockout/tagout (LOTO) process, although this term only partially describes the steps required. **Lockout/tagout programs** are simply a means of eliminating existing energy by using a standard system and then alerting others to the existing hazards. As we will see, this process combines engineering and administrative controls to deal with energy that may be present in a system.

> **■ Note**
> Lockout/tagout programs are a common method of controlling various types of energy.

OSHA requires lockout/tagout procedures to be implemented whenever employees are exposed to an energy source that, if suddenly released, poses a danger to the exposed employees. A common example of this is when an employee must bypass safety guards on equipment to perform maintenance. In some cases, the employee may need to actually enter the equipment or area to do the work. In such cases, confined space entry requirements also may apply. The rules for LOTO do not apply in all cases, such as when a cord-and-plug device for minor hand tools is used. The key to knowing when to implement a LOTO program is simply to ascertain whether energy may be released suddenly, unexpectedly, or at an inopportune time.

Lockout refers to physically controlling energy sources with a variety of equipment and a lock. The lock is placed so that it locks, or controls, the switch or other means of activation of the energy system, such as on an electrical circuit panel after the electricity to a given circuit is turned off. Simply turning off switches is not enough protection for employees who must work on a system that they believe to be turned off. Time after time, these switches have been accidentally bumped or turned on, causing power to be restored to the circuit and injuring or even killing the exposed worker. Using locks protects against this potential harm.

Lockout systems and equipment vary considerably from site to site. For this reason, OSHA requires that site personnel always be involved with contractors who may work on a system that requires energy control. Within any facility, the equipment used in the lockout program should be standardized. But each worker should have their own keys to the locks on any equipment used. Sharing locks and keys defeats the purpose of this control system. If multiple employees are working on the same system, multiple locks should be used, one applied by each person who may be exposed to the energy source. Figure 9-9 shows examples of energy control when several people are working on a system.

a

b

Figure 9-9
Examples of a lockout system in which multiple employees are working on a single system.

Once a system has been locked, the next step in the process is for each worker to place a tag on the system stating the reason for the energy control measures. A tag is placed on the system by each worker, indicating that the system must not be activated. In effect, the tag becomes a form of administrative control to alert others to the hazards posed by the system's energy, which might not be obvious. Figure 9-10 shows an example of a tag placed onto a system that has been locked out. Because locks and tags are an important part of the system mandated by LOTO, there are a number of requirements for their use, including:

- Lockout and tagout devices shall be standardized within the facility in at least one of the following criteria: color, shape, or size, and additionally, in the case of tagout devices, print and format shall be standardized.
- Lockout and tagout devices shall be singularly identified; shall be the only devices(s) used for controlling energy; shall not be used for other purposes.
- Lockout and tagout devices shall be capable of withstanding the environment to which they are exposed for the maximum period of time that exposure is expected.
- Tagout devices shall be constructed and printed so that exposure to weather conditions or wet and damp locations will not cause the tag to deteriorate or the message on the tag to become illegible. Tags shall not deteriorate when used in corrosive environments, such as areas where acid and alkali chemicals are handled and stored.

Figure 9-10
An example of a tag used to identify locks and the need for energy control.

- Lockout devices shall be substantial enough to prevent removal without the use of excessive force or unusual techniques, such as with the use of bolt cutters or other metal cutting tools.
- Tagout devices, including their means of attachment, shall be substantial enough to prevent inadvertent or accidental removal. Tagout device attachment means shall be of a nonreusable type, attachable by hand, self-locking, and nonreleasable with a minimum unlocking strength of no less than 50 pounds and having the general design and basic characteristics of being at least equivalent to a one-piece, all environment-tolerant nylon cable tie.
- Lockout and tag-out devices shall indicate the identity of the employee applying the device(s).
- Tagout devices shall warn against hazardous conditions if the machine or equipment is energized and shall include a legend such as the following: Do Not Start. Do Not Open. Do Not Close. Do Not Energize. Do Not Operate.

Lockout/tagout procedures apply to a range of activities, including cleanup and emergency activities involving hazardous materials. If flammable vapors are present and ignition sources need to be secured, it may be necessary to completely control machinery or other processes. Rather than leaving personnel in harm's way, lockout/tagout systems accomplish the goal of safety and reduce workers' exposure to potential hazards. OSHA mandates that specific steps be taken to control the various energy sources, including:

1. *Prepare for shutdown*: This step requires knowledge of the system and the identification of the type(s) of energy present. As part of this step, it is important to identify how much energy is present and how it is best controlled. Essentially, this means completely evaluating the system before shutting it down. You may be required to inform other persons or divisions that might be affected by the shutdown, so they are aware of the circumstances. It may be necessary to consult manufacturer's notes or specifications for complex equipment.
2. *Shut down the system*: The system should be turned off by regular means. This may include the normal on-off switches or closing valves.
3. *Secure the main power source for the equipment*: This may include isolating electrical subpanels or major valve junctions. Do not remove fuses—they are too easily replaced. Review all potential sources of energy, including backup supplies, and remember to consider multiple sources. Also, determine whether your own work may energize or start the machinery.
4. *Lockout/tagout the system*: Use a standardized system that is known and understood by all. Ensure that all lockout devices function properly. Valve covers should be properly sized, and locks for circuit breakers should be appropriate or adjusted using lock adapters. It is acceptable to use multiple locks with multiple workers, but if more than one person is working on a system, each worker should place a lock on the energy-isolation device.
5. *Release all stored or potential energy in the system*: This phase includes releasing energy stored in springs, compressed air or gaseous systems, electrical capacitors, etc. Essentially, it is necessary to *bleed off* whatever energy exists after shutdown. To control sources of static electricity, you may need to install ground wires or another control method.
6. *Verify that everything has been done*: This is a final check—walk through the procedure to ensure that all potential energy sources are identified and secured. It is better to check twice than to be wrong once. Try to operate the

normal switches, control valves, etc. in an attempt to activate the system. This helps ensure that the machinery will not operate as you are working on it. Use meters, temperature gauges, or pressure gauges to help you verify that all harmful energy is removed prior to working on the system.

Once these steps are completed, work can be done on the system or other activities involving the system can be performed. However, to ensure the safety of everyone working on a system, all steps in the process must be completed and verified.

When work on a system is completed and it is safe to return power to the equipment, the lockout/tagout devices may be removed. Before power is restored, however, guards should be replaced, tools should be removed from the machine, junction boxes should be covered, valves should be replaced, and a final safety walk-through should be performed. This ensures that the equipment is indeed ready to be returned to normal operating status. Because lockout/tagouts are individualized systems, if you put a lock on, you are the only one who should remove it. Once locks are removed and the system is operational, notify anyone who may have been affected by the shutdown that it is completed. Such announcements formally notify everyone that the systems are operating normally and that no restrictions apply to using appropriate power sources. Controlling the energy sources that could be inside a confined space involves lockout/tagout procedures. This process is part of isolating a confined space from the hazards prior to entering the space. Many accidents are caused when hazardous materials or energies enter a permit space and endanger or contact the entrants. The procedures for isolating a confined space vary but generally include the following elements:

- Locking out/tagging out all electrical sources
- Bleeding hydraulic/pneumatic power energy in the system
- Disconnecting all mechanical links that can activate machinery or equipment within a confined space
- Securing all mechanical moving parts within the permit space
- Alerting others in the area of the work to be done

Safety

Confined space entry permits specify measures that must be taken to control energy sources in or connected to a confined space. These procedures must be followed to ensure the safety of everyone working in the space.

NOISE HAZARDS

Noise exposure can cause serious hearing damage and is a commonly encountered hazard in a number of hazardous materials operations. High levels of noise can be created by the equipment that is frequently used in hazardous materials cleanup and emergency response activities, it is present in the process equipment used in many chemical plants, and it is often present in confined space operations, where noise is concentrated by the space's walls. Noise may be intermittent or continuous, but in either case, it poses a range of hazards to workers exposed to high levels of it. The effects of noise exposure are not limited to hearing loss. In fact, exposure to high levels of noise can result in any of the following effects or conditions:

- Instructions cannot be heard clearly, resulting in exposure to other worksite hazards.

- Communication among workers in a loud area is limited, resulting in confusion and potential danger.
- Alarms indicating hazards cannot be heard through excessive noise.
- The noise may distract those exposed to it.
- Prolonged noise exposure leads to higher incidences of fatigue.
- Exposure to high levels of noise increases stress in workers.
- Chronic exposure to noise can lead to permanent hearing loss.

As with other workplace hazards, OSHA regulations mandate that employers control the levels of noise exposure through a combination of control measures. The specific regulation related to general noise exposure is in 29 CFR, Part 1910.95. The regulation's premise is that there is a safe level of noise exposure just as there are safe levels of exposure to other workplace hazards, such as chemicals. The regulation sets an **action level** for workplace noise exposure at 85 dB (decibels) for 8 hours. Above that exposure level, hearing protection may need to be used. Slightly above the action level, at 90 dB, is the PEL for noise. The action level is used to alert employers of the need to protect workers. The PEL sets the safe level for that 8-hour exposure.

action level
the amount of exposure that requires an employer to implement a program to protect workers

■ Note
The action level for noise is 85 dB averaged over an 8-hour work period.

■ Note
The PEL for noise is 90 dB averaged over an 8-hour work period.

The hazard posed by noise depends on several factors, not just the intensity of the sound. These include the length of exposure, the force of the sound waves (loudness), and the frequency (pitch) of the sound. Specific components of noise hazards and hearing conservation programs can be found in the regulation, which is comprehensive and lists such items as hearing conservation programs, audiometric testing programs, and audiometric testing criteria. Note that just because noise exists at a work site does not mean that it exceeds the accepted OSHA levels—many variables determine the effect of noise on someone. When levels of noise exceed the action level, specific testing of personnel and the areas where the noise is present must be performed to provide adequate data from which to base hearing conservation programs. The bottom line is that if a potential noise hazard exists, it must be dealt with using the same regard for safety as any other workplace hazard. Usually, the hazard can be eliminated by using engineering controls such as placing mufflers on equipment, using sound-dampening foams, or isolating noisy equipment from the work area.

ELEVATED WORK

Elevated work safety is not a topic, but no discussion on hazardous materials is complete without information on this subject. *Elevated work* is working at a level at which falling to a lower level is possible. Usually, elevated work involves climbing to a work platform or scaffolding; however, falling into trenches or excavations also is a potential hazard and should be considered when a space is being evaluated for elevated work hazards.

A number of levels or heights are in the range of the OSHA regulations that define elevated work, depending on the type of industry involved. The most common level used to alert workers to the hazard of falling is 6 ft. Essentially, if workers can fall from a distance of 6 ft or more, OSHA mandates that they must be protected from falling by using a combination of engineering, administrative, and PPE controls.

A range of engineering controls are used to protect workers who are exposed to falls in excess of 6 ft, including installing railings, either permanent or temporary, at elevated locations. OSHA specifically regulates the types and construction of approved railing systems to protect employees from falling. These regulations specify the railings' height, number, and weight limitations, and even the materials from which they can be constructed. Temporary railings may be installed at a construction site to protect workers from falling from a building. Another example of common engineering controls for fall protection is scaffolding. Scaffolding is a temporary railing system used in a number of operations involving hazardous materials. Figure 9-11 shows an example of scaffolding constructed at a chemical plant to allow workers to access high levels in the facility to perform their work.

Another example of engineering controls that protect workers at heights is specialized equipment that elevates personnel to locations where they work inside an engineered system of platforms and rails. This equipment includes human lifts, scissor lifts, and elevated cages raised by forklifts. One of these pieces of equipment is shown in Figure 9-12. As you can see, personnel

Figure 9-11
An example of a scaffolding system erected to protect workers from falling.

Figure 9-12
An aerial lift that can protect personnel while they work at elevated locations.

working inside this equipment are protected by railings similar to those on scaffolding.

Administrative controls for reducing the risk associated with working at elevated locations include a variety of measures, mostly involving signage or other visual indicators. The most common administrative controls are signs indicate areas of exposure and barricade tape that alerts workers to the edges of elevated locations. Although not as effective as railings and scaffolding in reducing the actual fall hazard, such items are required to be used whenever engineering controls cannot be fully effective.

The final system used to prevent injury from falls is fall protection PPE. Like other types of PPE, workers must be trained in the use of this equipment for it to be effective. The use of this equipment does not prevent falls, it simply reduces the injuries caused by falling. The equipment has three components: a body harness that attaches to the worker, a lanyard that connects the body harness, and an anchor point. When used together, these items provide a high level of fall protection and reduce the potential for serious injury.

Protection from fall hazards is best accomplished by using all three forms of control. Railings and other engineering controls are effective in preventing most falls if they are properly installed and maintained. Administrative controls, however, such as rules, warning signs, and standard operating procedures, also are essential to alerting workers to hazards that might be hidden. Finally, the proper use of fall protection PPE limits the damage from falls.

SUMMARY

- A confined space is defined as any space that meets all of the following criteria:
  - ☐ The area is large enough and so configured that an employee can bodily enter the area and perform work.
  - ☐ The area is not designed for continuous human occupancy.
  - ☐ The area has limited or restricted means of entry or exit.

- OSHA defines a Permit-Required Confined Space as a *confined space* that has any of the following hazards:
  - Contains or has a potential to contain a hazardous atmosphere. Hazardous atmospheres are those that expose employees to the risk of death, incapacitation, impairment of ability to self-rescue (i.e., to escape unaided from a permit space), or injury or acute illness from a flammable atmosphere; an oxygen-deficient or -enriched atmosphere; or an atmosphere that has exposure levels above the PEL.
  - Contains a material that has the potential for engulfing an entrant.
  - Has an internal configuration such that an entrant could be trapped or asphyxiated by inwardly converging walls or by a floor that slopes down and tapers to a smaller cross section.
  - Contains what OSHA defines as "any other recognized serious safety or health hazard."
- If there is the potential for a work area to contain a Permit-Required Confined Space (PRCS), OSHA mandates that specific rules and operations be in place to ensure that workers are safe when involved in confined space operations.
- Atmospheric hazards are some of the most common hazards encountered in confined space operations and are the leading cause of death of workers in confined spaces.
- Physical hazards have more to do with the physical makeup or conditions in a space and less to do with the atmospheres in a space. The most common physical hazards include:
  - Engulfment
  - Entrapment due to the design of the space
  - Falls including those from elevated locations
  - Exposure to hazardous energy
  - Heat stress
  - High noise levels
- Ventilation systems are an effective form of engineering control in a confined space. When used correctly, these systems provide a high level of safety for workers in confined spaces.
- Administrative controls for confined spaces include using a standardized work permit that details the hazards and methods of control for the confined space activity, the requirement for specific numbers and types of training for workers involved in confined space operations, and the requirement to utilize special equipment and procedures for confined space operations.
- OSHA requires that the atmosphere in confined spaces be monitored in a specific order, with oxygen levels tested first. OSHA also requires that testing be done prior to entry and at regular intervals to ensure that acceptable entry conditions are maintained.
- Entrants are workers who enter a space to perform work.
- OSHA states that a confined space entry occurs when any part of the body of the entrant breaks the plane of the Permit-Required Confined Space.
- A confined space attendant attends to the confined space entrant. The attendant is required to stay outside the confined space and monitor the activities that take place within and outside the space. This is sometimes called a hole watch.
- The entry supervisor is responsible for ensuring that confined space operations are conducted safely.
- Personnel who enter a confined space must wear a retrieval system to facilitate non-entry rescue unless the equipment increases the hazard or does not assist with rescue.
- The preferred method of rescue from any confined space is the entrant's self-rescue; non-entry rescue is the second choice. If neither of these methods is successful, properly trained confined space rescue team members may need to enter the space and rescue anyone unable to escape.
- Energy comes in a range of forms and from a variety of sources, and the most common energy sources at work sites include electrical, mechanical, pneumatic, chemical, thermal, hydraulic, and radioactive processes.

- Lockout/tagout programs are a means of eliminating energy by using a standard system and then alerting others to the existing hazards.
- OSHA's regulation on noise exposure is in 29 CFR Part 1910.95. The regulation sets an action level for workplace noise exposure as 85 dB for 8 hours.
- Elevated work is working at a level where it is possible to fall to a lower level. The most widely used level to alert workers to the hazards associated with falling is 6 ft.

REVIEW QUESTIONS

1. List the criteria used to define a confined space.
2. A confined space becomes a Permit-Required Confined Space when it contains a hazard. List the types of hazards that cause a confined space to become a Permit-Required Confined Space and give examples of each.
3. What is the most common cause of death in confined spaces, and which group of employees is most likely to become a victim of it?
4. List four possible reasons for a confined space to contain low oxygen levels.
5. When the atmosphere of a confined space is monitored, the monitor must consider stratification. What is stratification and how can it be addressed?
6. List four responsibilities of the confined space attendant.
7. List four responsibilities of the confined space entrant.
8. Describe the three methods of rescue from a confined space.
9. List five items that are required to be on a confined space entry permit.
10. Failure to control energy sources can result in death. List five types of energy that must be controlled as part of a lockout/tagout program.
11. List the steps required to be used in energy control programs.
12. Identify three ways to protect a worker from falling.

ACTIVITY

Your team is involved in a project that involves entry into a large vessel that previously contained sulfuric acid, which has been drained. The vessel has several levels that need to be inspected by your crew. The vessel is entered through a small manhole approximately 18 in. in diameter. This limited size does not allow the entrant personnel to wear SCBA units. Some residual sulfuric acid is found in the vessel and may present a hazard to those entering the area. You have been selected as the project safety officer. As part of your responsibilities, address the following pre-entry safety issues.

1. List as many hazards as possible that may be present for those entering the vessel, and list which safety systems need to be developed to protect the workers.
2. How do you determine the level of protection required for those entering the vessel, and what type of PPE is required?
3. How do you address the need to access several levels of the vessel? How will you address the need to provide appropriate respiratory protection if personnel cannot enter the vessel wearing an SCBA unit? How will the need for respiratory protection be affected by the need to access several levels of the vessel?
4. Develop a written work plan for the entry and include the following components:
   - What type of air monitoring is necessary to ensure safe entry?
   - How many people are needed to safely work on the project? Provide a list of the required positions.
   - How should the need to provide for rapid entry be handled?
   - If heat stress is an issue, what precautions are in place to deal with it?

AIR AND ENVIRONMENTAL MONITORING

Learning Objectives

Upon completion of this chapter, you should be able to:

- Identify the reasons for performing air monitoring.
- Describe the principles of when and how to operate air monitoring equipment, including radiation detectors, oxygen meters, combustible gas indicators, photoionization detectors, flame ionization detectors, and colorimetric tube systems.
- Describe the proper use of air monitoring equipment in confined spaces.
- Describe the conditions for use of air monitoring equipment in outside areas.
- Identify the effects of local weather on monitoring operations.
- Describe environmental sampling that may be required in certain circumstances.

CASE STUDY

Following the hurricane season of 2005, the Gulf Coast region was a major focus of not only the public's attention, but also various safety and environmental organizations. One group that paid particular attention to the area was the Environmental Protection Agency (EPA). One of their concerns was about the air quality of the region. The area was home to many chemical plants, industrial complexes, and petrochemical refineries. The potential for severe air pollution was amplified by the decay of demolished buildings, and particulates from the fires that followed Rita and Katrina. This highly industrialized region suffered massive flooding, and reports of a "toxic soup" filled headlines for several days. With that toxic soup came a host of air quality issues that needed to be assessed so that rescue and recovery operations could begin.

To locate the areas of potential concern, the EPA used some of the most sophisticated monitoring equipment available. In addition to multiple ground-level monitors, they also used a remote-controlled aircraft, Airborne Spectral Imagery of Environmental Contaminants Technology (ASPECT), to locate chemical spills and determine the level of danger those materials posed. This information further directed them to the ground-monitoring activities, which also spanned a large area. Once the EPA pinpointed the locations of various chemical concentrations, they dispatched ground teams that used a handheld instrument to get more specific information.

Monitoring air and environmental conditions at both site and emergency response operations is a critical component of developing safety systems to protect employees, emergency responders, and the public. The EPA was able to assess the Gulf Coast region using monitoring techniques that are not available to many of us, but determining what is in the air, whether it is hazardous, and whether it exceeds the permissible exposure limit (PEL) or other limits such as the ERPG for public exposure, is what many of us who work with hazardous materials do on a regular basis. When the need to determine whether contamination exists in the air or the environment, some sort of monitoring must be done.

INTRODUCTION

Monitoring the area and atmosphere where there is a potential for exposure to hazardous materials is an essential component of any safety system. It is also an important element of any Hazardous Waste Operations and Emergency Response (HAZWOPER) training program. Whether the operation involves working with hazardous materials at a chemical handling facility, entering a confined space where gases may be concentrated or oxygen deficiency may exist, or dealing with the release of a hazardous substance at an accident site, knowing what is in the air and in what concentrations is critical. Without a working knowledge of what is there and in what quantity, determining the appropriate actions, including selecting PPE, becomes a guessing game.

Personnel who work in any area of the hazardous materials industry must be aware of the basic concepts of air and environmental monitoring whether or not their work will involve the actual use of the equipment. Knowledge of the concepts of air monitoring will alert those who work in the industry to both the advantages of such equipment as well as the limitations. This chapter provides an overview of the types of equipment commonly used to determine levels of air contamination and what that equipment will and will not do. Workers or responders whose jobs involve using this equipment will need additional training on the specific equipment, because the equipment is made by a several manufacturers and can vary. While similar in their overall capabilities, each specific type of equipment will have different methods of use, maintenance, and other specific operational features that need to be understood by those using the equipment.

■ Note
Monitoring is a required component in a number of HAZWOPER programs.

■ Note
Personnel should be familiar with the types and limitations of air monitoring programs.

AIR AND ATMOSPHERIC MONITORING FUNDAMENTALS

To be able to detect the presence and determine the concentration of hazardous materials in any atmosphere, we must first understand the behavior of these materials. From our previous discussions on respiratory protection and gases, we know that airborne contaminants can take one of several forms. The form they take is largely determined by their basic state of matter—solid materials, liquids, or gases. Each form has the ability to create airborne contamination based on a number of factors. The hazards associated with the presence of these materials may not be realized without appropriate air monitoring.

■ Note
Airborne contamination can come from any form of matter—solids, liquids, or gases.

The first form of potential airborne contamination is that created by solid materials. Solid materials are often overlooked as materials capable of presenting airborne hazards; however, many types of solid materials can pose a hazard under some conditions, once they are released as a particulate. This usually involves the conversion of the solid to smaller particles through processes such as grinding, sanding, or sawing. Additionally, the application of heat can start to break down solid material thermally, releasing smoke and vapors. Other solids may be volatile enough to release their own vapors. Regardless of how it occurs, it is important to consider the possibility that solids can present airborne hazards.

■ Note
Although less hazardous than other forms of matter, solid materials can be broken down by physical or thermal processes and present airborne hazards.

Liquids also are capable of creating airborne contamination through the process of evaporation, emitting vapors into the atmosphere in concentrations that pose a hazard. A liquid's ability to release vapors depends on a number of factors, which were discussed in previous chapters. These factors include the liquid's volatility and other physical properties, such as its boiling point and flash point. The application of heat also is a major factor in the process of vapor production. Interaction between liquid and the other materials can cause an exothermic reaction that releases heat. Again, regardless of how it happens, the potential for a liquid to present an airborne hazard must always be considered.

Highly volatile liquids can present serious hazards similar to those posed by gases.

Safety
Heat and chemical interactions can produce more airborne contamination.

Finally, gases are the most common materials associated with airborne contamination. By their nature, gases tend to be more migratory than other materials and are more subject to diffusion in the area of the release. Gases can be influenced by minute air movements and tend to move around more than other types of materials. This movement continues depending on a number of factors, but eventually gases diffuse in the container or area of release as their pressure and concentration are dispersed. Depending on a number of factors, including a gas's density or weight, the air movement created by wind or ventilation, and the presence of thermal extremes, gases are generally more mobile than other materials and the process of their movement is more extreme. These three factors must be considered because they can influence a gas's movement in the area of its release.

Safety
Gases can present a high potential for airborne contamination because of their migratory nature.

Safety
Wind/ventilation, thermal extremes, and high/low vapor densities can modify the speed at which a gas diffuses into an area.

Regardless of which form of a material creates a hazard, airborne contaminants must be detected and measured before their hazards can be fully appreciated and addressed. As we know, a material may present a hazard even if it is not visible to the naked eye and does not have an odor or other warning property. The smallest and least visible materials are often those most capable of movement and of entering the depths of the lungs, where they can cross the membranes separating the alveoli and blood. Additionally, some materials can adversely affect us at extremely low levels or in small amounts. Arsine is a good example of a gas that can cause harm at very low levels—it has an Immediately Dangerous to Life or Health (IDLH) level of only 3 parts per million (ppm). When dealing with highly hazardous materials, it is vital that the level of contaminants be known so that appropriate protective measures can be implemented. One should never assume that air is clean even if it shows no clear signs of contamination or has no noticeable odor. If the potential exists for a material to be present, we must ascertain whether the material is present and if so, at what level.

Safety
An area can be determined to be free from airborne contamination only if monitoring has confirmed this.

Safety
Never rely on sense indicators such as visible vapors or odors to determine whether an airborne hazard is present.

Before we begin discussing air monitoring, a word of caution is appropriate. Some organizations believe that air monitoring should be conducted until everything is known about an area's atmosphere—to the extent of attempting to detect materials whose presence is significantly unlikely. This is much like insisting that continuous monitoring of a confined space is the best method to protect employees working in the space, even though they are already wearing the highest level of protection against the worst-case scenario. A degree of good judgment must be exercised by anyone responsible for monitoring operations because monitoring activities take time to conduct. During emergency response operations when time is critical, it is not always possible or advisable to determine all of the hazards in a given area. When time is of the essence, a basic evaluation of the atmosphere through a standard series of tests using a multi-gas meter may be in order. In such cases, responders entering the environment may default to using higher levels of protection rather than taking time to conduct repeated tests to determine the presence and levels of materials. As we will learn in a later chapter, the Occupational Safety and Health Administration (OSHA) mandates that self-contained breathing apparatus (SCBA) units be used primarily until appropriate air monitoring can confirm that they are not required. Wearing an SCBA unit largely eliminates the risk of respiratory exposure to all types of airborne contaminants. Therefore, it is wiser not to waste time confirming and reconfirming levels of contamination when SCBAs provide the required amount of protection.

Safety
Do not monitor for the sake of monitoring. Monitoring activities should have a purpose and should affect the work plan.

The basic rule for how much and what types of monitoring operations are required begins with the site evaluation. The potential for airborne contamination is determined on the basis of an overall evaluation of the scene and the potential for specific materials or types of materials to be present. This is done using many of the techniques described in Chapter 5, which covered identification systems. Depending on the location and type of incident, monitoring operations may be extensive or limited. Using indicators such as container profiles, shipping papers, placards on vehicles, the presence of NFPA identification systems, and even the location within the site, the types of materials that need to be monitored may be easily ascertained. Without knowing what is likely to be present, a proper monitoring program cannot be designated.

Safety
The presence of identification systems can help determine whether and what type of monitoring is required.

Once materials are identified and their properties and hazards are determined, a monitoring program can be established to provide additional information that may alter the response activities or levels of protection required. The chosen monitoring program should include the capacity to repeat testing at various intervals depending on the circumstances. If a material may be present but the initial monitoring shows no level of contamination, the monitoring program should provide for repeated testing based on a number of factors. These can include the potential that a release of the material may occur, the impact that a release may have on other operations, and whether the level of protection worn by potentially exposed responders is adequate if a release occurs. Monitoring operations should provide valuable information. It should not use valuable time and resources simply to provide information that will not alter the activities taking place.

Air monitoring may be an ongoing process during specific types of operations.

Once the monitoring plan is set up on the basis of the types of materials that are potentially present and for which levels need to be determined, the next step is to properly identify the types of equipment necessary to conduct the monitoring activities. Before we discuss this equipment, however, we need to emphasize that *all monitoring operations must use correctly calibrated and maintained equipment.* Depending on the type of instrument, the manufacturer, the frequency of use, and a range of other factors, calibration may be required over a range of frequencies. Many types of instruments are required to undergo **field calibration** at some level before any use. Much like the requirement for conducting positive and negative pressure tests on respirator masks prior to each use, some instruments must have a quick field check even if they regularly receive more extensive calibration at an instrument shop. Other instruments may not require such steps, but even when that is the case, all instruments and equipment used in monitoring operations must be properly maintained and tested in accordance with the manufacturer's recommendations and requirements. Doing otherwise violates OSHA requirements and can place the personnel who rely on the data at risk. Instrument calibration and maintenance are discussed in more detail later in this chapter.

field calibration
the process of quickly checking the monitoring equipment prior to use in the field

Selecting the proper type of equipment to conduct monitoring is not easy. While there are instruments that can detect a wide range of materials, these instruments are generally not available for the average responder, are not generally portable, and are extremely costly. Such equipment could be present at national laboratories but is not available to most response agencies or sites. Most of the equipment used in the field performs only one or two functions and identifies the levels of only around six materials at a time. This chapter provides information on the more commonly used types of equipment and how they can be incorporated into a monitoring program. Keep in mind, however, that this technology is continually changing and that new equipment is produced regularly. Nonetheless, many standard types of equipment and their methods of operation have not changed and probably will be used for a considerable time.

Safety
No one instrument can indicate all the hazards present. Multiple instruments and tests are generally required to obtain this information.

THE SEQUENCE OF AIR MONITORING

Most monitoring operations are conducted using handheld, direct-reading instruments. The equipment is generally small and portable so that it can be carried by personnel into the area that needs to be evaluated. The choice of a specific type of equipment is up to the individual organization, but it is recommended that whichever instrument is used, it should be rugged and able to withstand a range of conditions, easy to operate and read under field conditions, intrinsically safe (not a spark hazard), and able to produce reliable and useful results. Personnel who use the specific equipment must be properly trained in its operation and care. It is never a good practice to allow personnel to train with this equipment "on the job" for such critical operations. Personnel should be confident in their ability to correctly use monitoring equipment and to determine the levels of contamination present, because this is essential to the safety of workers who rely on the test results.

Some instruments can simultaneously monitor for several different materials, but no one meter can detect and measure all materials that may be present. To properly determine the atmospheric conditions of a given space, tank, cylinder, or area, several types of monitors usually are required. Given this, a suggested sequence of testing should be followed when air monitoring is being conducted. The sequence is as follows:

1. If the site assessment identifies a potential for a radioactive material to be present—although this occurs extremely rarely—the first test should monitor the atmosphere for the source of that material. This test usually is not required, as we have suggested, and provides little additional information that can benefit the operation. Keep in mind that some of these radiation detection monitors may not be safe to use when flammable atmospheres are present; so some judgment may need to be used to ascertain this prior to entry.
2. If there is no indication that radioactive material may be present, the first test generally monitors the levels of oxygen. As we know from our discussion of confined space activities, OSHA requires that this should be the first test when individual tests are required because the level of oxygen present in an area may affect other test results.
3. The next test generally performed determines the presence of flammable or combustible atmospheres. A combustible gas indicator (CGI) is one of the most common types of instruments used to detect the presence of combustible gases or vapors and is discussed later in this chapter.
4. Organic gases and vapors should be detected next. This is done using a flame ionization detector (FID). This device may include a gas chromatograph (GC) that is part of the unit or located in a remote area from which the data are transmitted.
5. Organic and some inorganic gases and vapors should be monitored for next, using an ultraviolet photoionization detector (PID). Like using an FID, using PIDs may be considered optional.
6. Finally, the presence of specific gases may be sensed by using a system of colorimetric indicator tubes.

Note that this standard testing sequence does not preclude multiple tests being conducted at the same time. In some cases, the personnel conducting the testing may carry equipment capable of detecting multiple hazards at the same time. Additionally, the entry team may tape strips of wetted pH paper to their suits, or it may have a monitoring instrument that is capable of detecting specific air contaminants that it expects might be present. As they enter the area, any visible color changes can be immediately noted and the appropriate actions taken.

THE USE OF AIR MONITORING EQUIPMENT

While the specific types of equipment used will likely have their own procedures for operation, there are certain general concepts that can be seen for each type of instrument. The following information will provide some general guidance on the use of the equipment, but always consult with the specific information available from the manufacturer of the equipment.

Radiation Detectors

Radiation detectors are more commonly known as Geiger counters. While several types of these detectors can be used based on the type of radiation present, many units are designed to detect and measure ionizing radiation. Some devices detect only one type of particulate radiation, such as alpha or beta particulates, while others can also detect gamma radiation. While these units differ in terms of manufacturer and type, most of them work by using an inert gas-filled tube (usually containing helium, neon, or argon with halogens added), called a Geiger–Müller tube, that briefly conducts electricity when a particle or photon of radiation temporarily makes the gas conductive. The tube then puts out a current pulse, which is displayed by a needle and audible click. These instruments read the actual amount or radiation that is present in any given area on the basis of the amount per a given time period (per hour). So a reading from such an instrument will generally indicate that a given amount of radiation is emitted per hour (Figure 10-1).

A radiation dosimeter is a smaller device than a radiation detector. It is usually worn clipped to one's clothing, and it measures the cumulative dose of radiation received by people while they are working. This information is important because it is the actual dose of radiation received by an individual that causes problems. Two popular types of dosimeters are electronic and film badge

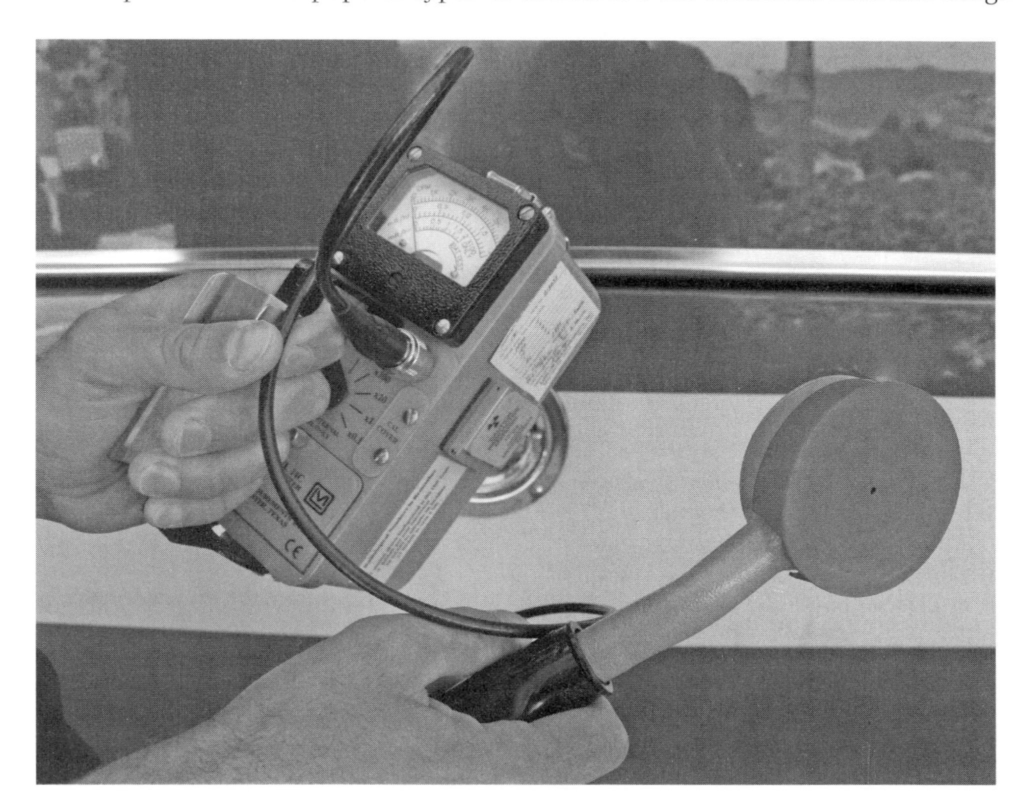

Figure 10-1
A radiation detection device. (Courtesy of the San Ramon Valley Fire Protection District.)

dosimeters. The electronic dosimeter utilizes a magnifying lens as a low-power microscope, the results of which are readable through an illuminating lens. Such units generally are reusable. An alternative is the film badge dosimeter, a plastic badge containing a small piece of photographic film. Radiation exposure gradually exposes the film, which is periodically removed and developed. The amount of exposure can be measured by comparing the developed film's optical density to a cumulative dosage measurement. Once used, the film badge can be disposed of just like other types of film.

Because it is critical to know whether radiation is present in a given situation, radiation detectors must be used in areas where the potential for radiation exposure exists. The growing concern about radiation since September 11, 2001, has renewed interest in this area, and new equipment is gradually replacing equipment that has not changed in a number of years.

Oxygen Meters

The first monitor commonly used in hazardous materials site operations and most emergency response incidents is the oxygen meter. As we know, the presence of oxygen at specific levels is critical to ensure our safety. If the level is lower or higher than the normal range, problems can occur. Most oxygen meters work by allowing oxygen to diffuse into a detector cell and then measuring a chemical reaction that establishes a current between two electrolyte cells. Oxygen meters are often used to determine which type of respiratory protection is required for a given situation because a low level of oxygen in an area requires the use of a supplied-air respirator. Too much oxygen, on the other hand, increases the potential for fire in a given area.

Note that the presence of oxygen in a given area is assumed when other gases are being testing for. The measuring of flammable vapors in an area is based on the percentage of that vapor in normal atmosphere, one that contains approximately 21% oxygen. If that level is higher or lower, the information relative to flammable vapors will not be as valuable. In fact, many types of oxygen meters warn users of inaccurate readiness when levels of oxygen are low. The presence of high levels of oxygen, however, may enhance the ability to a given amount of vapors to ignite.

As seen in Figure 10-2, oxygen meters are relatively simple instruments that have a visual readout and an audible warning that alarms when the oxygen levels drop below 19.5% or rise above 23.5%. Like all instruments, they require calibration and **zeroing** before being used. This should be done in a normal oxygen level atmosphere before the instrument is exposed to a test atmosphere. The accuracy of an oxygen indicator also may be affected by changes in altitude or barometric pressure or the presence of certain gases, such as ozone or carbon dioxide, which can damage the detector cell. Many of these units, such as the one shown, also are capable of detecting gases in addition to oxygen.

zeroing
bringing a monitoring device to a baseline or zero readout on gauges or digital readouts

> **Safety**
> If the level of oxygen is outside the normal limits, other atmospheric testing results may be affected.

Combustible Gas Indicators

Another instrument commonly used to detect gases and vapors is a combustible gas indicator. The market for these instruments is large, and so is the variety of instruments available. Many CGIs are incorporated into instruments that also conduct other testing, such as multi-gas detectors that have a built-in oxygen sensor. Others may also be able to detect specific gases, such as carbon monoxide,

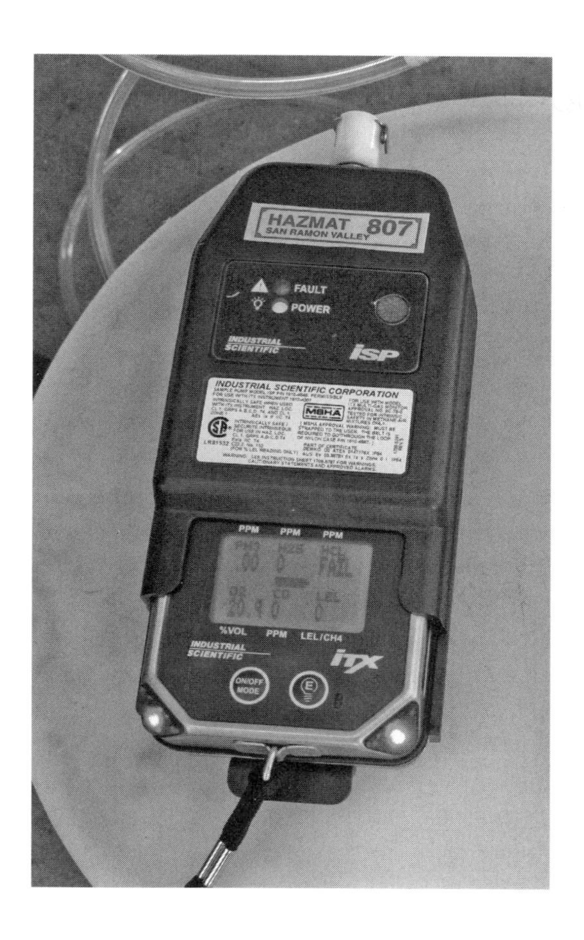

Figure 10-2
An oxygen meter used in emergency response and site operations. (Courtesy of the San Ramon Valley Fire Protection District.)

methane, or hydrogen sulfide. Regardless of whether CGI units detect or measure other gases, their primary purpose is to detect the presence of flammable or combustible gases and vapors. These units, however, may be able to determine only whether such a material is present, not which specific material is present.

Many CGIs use a common process for determining whether materials are present that may present a flammability concern. Most units contain a fan that draws a sample of the test atmosphere and allows it to flow over a heated catalytic filament in the instrument. If a flammable or combustible gas or vapor is present, electrical resistance is created on one of the filaments connected to a circuit, causing a measurable imbalance. This technology, which is used in many other types of circuitry, often uses a *Wheatstone bridge*. Remember to *never* allow the intake port of a CGI to come in contact with any liquid. Drawing liquid into the unit compromises and damages the device, resulting in the need for extensive (and expensive) repairs.

Typical CGIs, such as the one shown in Figure 10-3, are designed to give visual and audible alarms when the units detect flammable or combustible vapors at a specific level. The level chosen for this alarm set point is usually 10% of the lower explosive limit (LEL). This allows for a large safety factor despite the large number of variables that can affect the detection of combustible gas or vapor. These variables include oxygen levels that may not be within the normal range, the calibration gas (such as hexane or methane) used to set the device's alarm set point, variations in temperature or humidity, and the presence of other materials in the area that may confuse the meter. The alarm set point is low (as we have said, at 10% of the LEL for the test gas that was used to calibrate the instrument) to allow enough time for personnel to leave the area before a serious contaminant level is reached. If the instrument is a multi-gas detector, it may also be able to alert you to which gas or vapor is present because many of these units also

Figure 10-3
A combustible gas meter used in emergency response and site operations.

detect the presence of carbon monoxide, hydrogen sulfide, or other common materials. Like all monitors, remember that CGIs require proper calibration and zeroing before each use and periodic recalibration with the test gas.

> **Safety**
> Remember *never* to allow the intake port of a CGI to come in contact with any liquid, including water.

> **Note**
> CGIs do not usually indicate which specific material is present, only that a hazardous atmosphere exists.

> **Note**
> The alarm set point for most CGIs is 10% of the LEL.

Photoionization Detectors

ionization potential (IP) a unique measure of the energy required to displace the electrons in a compound by using an ultraviolet light

Photoionization detectors are designed to detect relatively low concentrations of contaminants, usually in the range of 0.1–2000 ppm. These devices efficiently detect aromatic hydrocarbons such as benzene, toluene, xylene, and vinyl chloride. They work by drawing a sample of the test atmosphere into a chamber with an ultraviolet (UV) lamp. The UV light breaks down the sample by displacing electrons and then measures the energy required to do so. This is a material's **ionization potential (IP)**, and each chemical compound has a unique IP. Therefore, it is possible to determine which gas is present in an atmosphere by reading its IP on the PID and comparing it with known levels of gases. One of the most commonly used reference sources for this purpose is the *NIOSH Pocket Guide to Hazardous Materials*, which lists the IP for every compound. Note that PIDs contain a variety of UV lamps with different electron volt (eV) capabilities. Because IP is measured in electron volts, and because the contaminant gas must have an IP less than the eV capacity of the UV lamp, the correct lamp must be used to determine which gas is present. (Although many models now have built-in correction or correlation factors.) These devices must be properly calibrated and accurately **spanned** before being used to ensure that they can properly determine materials' IPs. This is best done by consulting the manufacturer's data tables. Table 10-1 lists some IPs.

spanned
to have an approximate calibration for a certain number of chemical compounds

Figure 10-4 shows a typical PID. Many of these units are small and handheld, allowing them to be carried easily into an area.

An important note about PIDs is that although they work well with small concentrations, high concentrations may cause false low readings. Other variables that can affect the performance of these instruments include water vapor (humidity), non-ionizing gases, diesel exhaust, smoke, soil, and a dirty UV lamp. Also, while a PID can detect many ionizing materials at low levels, they do not detect everything, including some common materials such as methane. Because of this, it is recommended that they be used in conjunction with the other instruments.

Flame Ionization Detectors

Flame ionization detectors detect many organic gases and vapors by using charged particles or ions to detect materials in the air and a hydrogen flame to burn organic materials in the air. As the test contaminant is burned, positively charged ions are produced, and a current is generated. This current is then measured on a scale relative to a calibrant gas. Because many of the capabilities of an FID are similar to those of a PID, the high cost and maintenance of these types of instruments keep them from use by many except for very specialized hazardous materials teams. When they are used, however, they have an advantage over PIDs—since FIDs in survey mode can read 0–1000 ppm or up to 10,000 ppm. Many models of FIDs also are configured to work as a GC, but personnel require further training and experience to be able to correctly interpret the data when the device is used in this way. Figure 10-5 shows one model of an FID. Other models, including those that function solely as a GC, may be more portable than this one.

Gas Chromatographs

If a PID or an FID incorporates a GC, then the device can be set in one of two modes. In the survey mode, the instrument functions simply as a detector, determining whether a given material is present. In the GC mode, the unit operates as both a detector and an instrument that determines the amount of material present. As we have said, however, proper use of the device in this latter mode requires extensive training, and obtaining correct readings from it requires near ideal conditions.

Table 10-1 *Examples of IPs for common materials.*

Chemical	IP (eV)	Chemical	IP (eV)
Acetone	9.69	Ethylene oxide	10.56
Ammonia	10.18	Hydrogen peroxide	10.54
Carbon dioxide	13.77	Toluene	8.82
Chlorine	11.48	Vinyl chloride	9.99

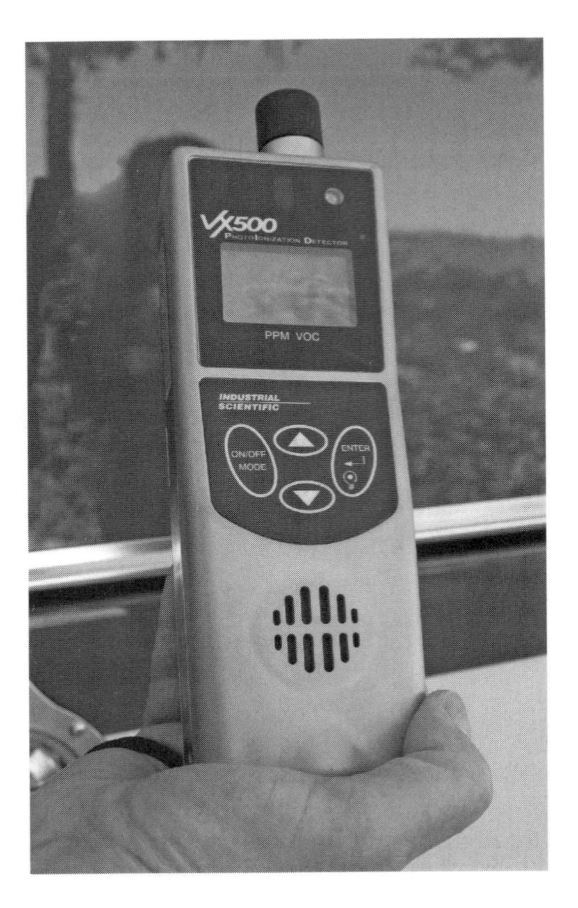

Figure 10-4
An example of a photoionization detector. (Courtesy of the San Ramon Valley Fire Protection District.)

As a GC, an instrument is able to separate various materials in a complex sample. GCs draw a test sample into a tube that contains a medium that adsorbs the sample to determine the variation in bonding strength. Depending on the strength of the bond, the sample is retained in the tube and eventually passed out the other end. This retention time is measured and compared with known properties of gases. Under ideal conditions of temperature, medium, and flow rate, GCs can determine the quantity of a gas present based on retention time. As we have said, GCs often are built into PIDs and FIDs. It is possible to purchase separate GCs for field use, but the cost is significantly high, making them too expensive for most organizations.

Colorimetric Tubes

One of the most common and oldest systems used to detect the presence of specific hazardous materials is the colorimetric tube. This system uses very simple technology to detect specific gases in an atmosphere and indicate the approximate amounts of those gases. Like other air monitoring devices, a colorimetric tube draws a sample of atmosphere into it. The tube is filled with a material that changes color in response to which gas is present. Although colorimetric tubes pose some limitations and can be time-consuming to use, they are far easier to use and maintain than most other air monitoring instruments, including PIDs and FIDs. Figure 10-6 shows an example of a colorimetric tube system.

Colorimetric tubes work well for identifying which gases are present in an area, but they can only approximate the quantity of those gases. In fact, test results from these tubes, which read in parts per million, may vary from the amount actually present by a factor of ±25%. Figure 10-7 shows the types of

Figure 10-5
An example of a flame ionization device.

Figure 10-6
A colorimetric tube system. (Courtesy of the San Ramon Valley Fire Protection District.)

Figure 10-7 *An example of colorimetric tubes and a pump. (Courtesy of the San Ramon Valley Fire Protection District.)*

tubes used in this system. As air is drawn through the tube by a pump, the color of the media inside the tube changes based on the amount of material present, just as it does to identify which gas is present. Note that the number of times the hand pump is activated determines the amount of sample that is pulled into the tube. However, as a sample enters the tube, any concentration reading should be taken as somewhat limited due to slight variations in color, irregular changes in color, or "bleeding" effects in color change. This is why colorimetric tubes are used primarily to determine that a particular gas is present in *some range* of concentration but not its *actual* concentration.

It is not necessary to purchase a tube for every type of contaminant. This is not only extremely expensive, but also unwieldy. Manufacturers such as Drager and Sensidyne offer qualitative tubes for acids, bases, organic amines, unsaturated hydrocarbons, halogenated hydrocarbons, and aromatic hydrocarbons. Polytest tubes that detect multiple gases, such as ammonia, hydrogen chloride, hydrogen sulfide, chlorine, etc., also are available.

> **■ Note**
> Colorimetric tube systems work well for identifying a specific gas, but they can only approximate the quantity of those gases. In fact, test results from these tubes, which read in ppm, may vary from the amount actually present by a factor of ±25%.

Various factors may affect the use of colorimetric tubes. Changes or extremes in temperature, high or low humidity, barometric pressure, sunlight, and other interfering gases may alter the tubes' reading. Also, the tubes have a shelf life, and if that shelf life has expired, the tubes cannot be used. Therefore, the most reliable method of using these tubes is to use three tubes. The first can perform monitoring, the second can continue monitoring if the first tube becomes saturated, and the third can be a blank to determine subtle color changes. This three-tube technique helps determine whether any cross-sensitivities occur when two gases are similar and are picked up by the tube, creating false or confusing readings. Each box of tubes includes a sheet of directions that describe any cross-sensitivities that may exist.

> **■ Note**
> Changes or extremes in temperature, high or low humidity, barometric pressure, sunlight, and other interfering gases may alter the reading in colorimetric tubes.

Other Detection Devices

Other monitors may be used in some situations to determine unknown contaminants in an atmosphere. For example, infrared vapor analyzers use an infrared absorption filter to detect and measure gases and vapors in the air. These instruments contain a microprocessor that controls a spectrometer. As the device's pump draws a sample in, the infrared energy is absorbed and its wavelength is read. These analyzers are recommended for detecting a range of substances, such as carbon monoxide, carbon dioxide, nitrous oxide, halothane, and fumigants, if their absorption band is in the infrared region of 2.5–14.5 μm. If not, the infrared analyzer may provide only semi-specific results for sampling some gases and vapors because of interference by other chemicals with similar absorption rates.

Additionally, there are specialized monitors for materials such as lead, mercury, ozone, and solvent vapors. These are used when a specific contaminant is suspected or known to be present and verification is required.

Another type of air monitoring equipment is pH paper. Using such a low-tech device may sound rudimentary, but it can be essential. If corrosive vapors are known or suspected to be present, a simple procedure can protect sensitive equipment against them. Simply attach a piece of pH paper wetted with deionized water to an extension device as simple as a broom handle and put it into the test atmosphere. If the paper detects the presence of a strong corrosive gas (pH less than 2 or greater than 12), then steps should be taken to reduce the amount of gas or neutralize the gas as much as possible. Exercise caution when using pH paper, however, because even when using an extension, you may still be contaminated by the atmosphere. Ventilating the area or using some other engineering control can help reduce the level of contaminant, as can neutralizing techniques.

GENERAL PROCEDURES FOR AIR MONITORING

As with any piece of equipment, you must fully understand how air monitoring equipment works before you use it in the field. Becoming familiar with this equipment includes reading the manufacturer's guidelines, practicing with personnel experienced in its use, and performing basic drills before actually using it.

Waste site workers and emergency responders also should be familiar with the procedures for calibrating and zeroing monitors before they are used. Usually, this requires energizing the monitor in a clean atmosphere before entering a test atmosphere. A primary reason for equipment failure is that battery levels are not checked before the equipment is used. So, ensure all monitors are properly charged and ready to go before being used. Never go into the field assuming that all equipment is properly operating. The operation personnel should check all equipment before it is needed to make sure that it is in good working order. An ongoing maintenance program, including a log that documents all work and tests performed on equipment, is an excellent way to ensure that air monitoring equipment can be counted on when the need for it arises.

> **Note**
> One of the most common reasons for monitoring equipment to fail is low battery levels.

When testing for contaminants in the atmosphere, remember that materials may migrate depending on a number of factors, including a material's vapor density. That is, some vapors will rise and others will settle and possibly collect in low areas of the test atmosphere. This is why it is important to test several levels within the area where suspected materials could be present. Always test at high levels within the area as well as low. In confined spaces or other areas where stratification could occur, this is even more critical.

Before workers bodily enter an area or a confined space, an extension device should be used to insert the probe head of a monitoring instrument into the area. When using any monitor, with or without an extension, you must allow enough time for the machine to draw a useable sample into the test chamber. Some instruments, for example, have pumps that can draw in samples from more than 100 ft, but the process takes time. It may take a few moments for the sample to reach the detection device and be analyzed. In other words, once you have begun taking a sample, do not keep walking into the area until you have a reading. Stop, wait a few moments, and then proceed. Continue this process until the entire area has been tested.

If you are dealing with gases in a confined space or any enclosed area, determine the most logical place in which the gases may accumulate and in which pockets of more concentrated materials may be present. As we know, some gases stratify, allowing them to occupy different layers depending on the vapor density or atmosphere's temperature. Other gases, such as hydrogen, acetylene, and ethylene, usually accumulate near the ceiling. Most hydrocarbon vapors and flammable gases, such as propane and butane, are heavier than air and tend to follow the path of least resistance by settling in the lowest area possible in an area or on the ground. This is why it is important to check areas such as sumps, subfloors, and trenches.

hydrophobic filters those that swell when they contact liquids

When checking for gases and vapors that may hover over liquid spills, take care not to put a probe head into any standing liquid. Even if a monitor has a **hydrophobic filter**, which resists water, drawing any liquid into the instrument can alter readings and cause severe internal damage. Figure 10-8 shows an illustration of a hydrophobic filter inserted into the pickup tube of an instrument. If this filter contacts any liquid, it quickly swells and stops the airflow into the instrument, affording some level of protection for the unit and the sensors. But the best method of preventing liquids from entering monitoring instruments is to maintain control of the pickup tube.

> **Safety**
> Do not walk into an area without first monitoring it or getting a reading.

> **Note**
> If dealing with gases in a confined space or any enclosed area, determine the most logical place in which the gases may accumulate.

When monitoring atmospheres in confined spaces, it is safest to attempt the initial monitoring before the space is physically entered. It may be necessary to use an extension device to extend the monitor's probe into the space because the air quality at the entrance may be different than in the pockets of air inside. When monitoring confined spaces that do not have outside ventilation, you must ensure

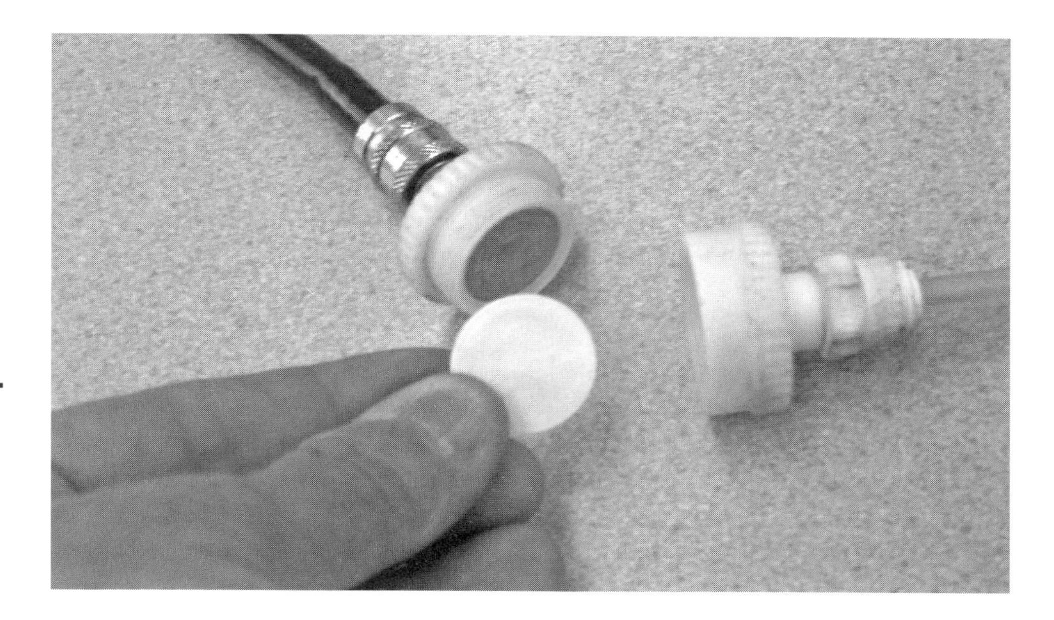

Figure 10-8
A hydrophobic filter placed into a pickup tube prevents liquids from getting to the instrument.

that all corners of the space are tested as well as any area in which a suspect gas may pocket or accumulate. Once the initial readings are obtained, it may be necessary to repeat the monitoring several times during the workday, including any time after workers have left the space and are re-entering it, when there is any change in temperature or barometric pressure, if potential gases are produced or disturbed, or if any other condition exists that may warrant atmosphere monitoring. Monitors often are kept energized from the beginning until the end of the task.

Safety
In confined spaces, initial monitoring is most safely performed before a space is physically entered.

When monitoring outside areas, it is best to approach the area from uphill and upwind if possible. This reduces potential exposure by the personnel conducting the monitoring operations but requires that a close approach to the area be made because vapors may be influenced by the wind and terrain. Extensions also are useful in these cases. Remember to allow a reasonable amount of sampling time to enable the monitor to cycle what it is sampling.

When working outside in open areas, the use of an oxygen monitor is not usually required unless the gas discharge is large enough to displace the normal atmospheric oxygen level and is continuing to discharge. In most outdoor settings, any displacement of the normal levels of oxygen is minimal. If, however, a visible cloud of contaminant is present and relatively stable conditions are in effect (i.e., no significant wind), then monitoring for oxygen deficiency may be required unless personnel are already using a supplied-air respirator. Remember too that most gases and vapors are heavier than air and follow the path of least resistance into low areas such as ravines, gullies, drainage ditches, and trenches.

Even after you approach an outdoor contamination area from uphill and upwind, it will be necessary to enter it to detect which gas or vapor is present and in what amounts. In these situations, proper personal protective equipment (PPE) should be used. Contaminant detection procedures should include looking for the source or origin point of the gas, such as a compromised cylinder, and then checking at various points downwind to determine the extent of the contamination's spread. Monitoring also should be done on the cross-axis of the discharge area to determine the degree of dispersion. Finally, monitoring should be done upwind of the source to determine whether there is cross-contamination and to verify the contamination's source.

THE EFFECTS OF LOCAL WEATHER ON MONITORING

Air monitoring procedures and the movement of gases and vapors can be significantly altered by changes in the weather. Changes in temperature can increase vapor production (volatilization) and possibly lower vapor density, which would keep gases near the ground instead of allowing them to dissipate into the atmosphere. An increase in temperature also may affect the stability or reactivity of normally stable materials. A change in temperature may also cause a change in **relative humidity**. As the humidity level increases, so too may the rate of vaporization of water-based soluble materials.

relative humidity
the ratio of the actual water vapor pressure to the saturation of vapor pressure

The most significant weather event that may alter a contamination situation is wind. In conditions of low wind, clouds of gases and vapors are slow to dissipate, and vapor concentration remains high in the area of the release. In these conditions, the weather is considered stable and downwind dispersion is limited. When conditions are more dynamic and involve high winds or variation in the wind direction, atmospheric contaminants tend to spread further and faster. The good news is that as they disperse, their concentration downwind generally decreases. An increase in wind also can make dusts and particulate-bound contaminants more likely to become airborne and to spread. These factors, as well as expected daily changes, called **diurnal effects**, must be taken into account for any task involving air monitoring. Diurnal effects include changes in barometric pressure, wind direction, and temperature.

diurnal effects
daily weather changes such as barometric pressure, wind direction, temperature, fog, or rain

Safety
Air monitoring procedures and the movement of gases and vapors can be significantly altered by changes in the weather.

Note
The most significant weather event that may alter the situation is wind.

When dealing with a release of a hazardous material, whether in an emergency setting or during cleanup operations, it is very important to know how the local weather will affect the operations. In many situations, outside agencies such as the National Weather Service or local airports can provide weather forecasts. Many areas have standard weather patterns that vary during the day. A prime example of weather variation is **land breezes** and **shore breezes**. Because of the differences between the cooling ratios of land and water, coastal areas often are subject to changes in wind speed and direction during the day. Site operations may take this into account as the standard patterns that are usually present. In some cases, however, such as those involving highly hazardous and migratory materials, it may be necessary to obtain more direct information from an on-site source. To assist with this, some agencies have on-site weather reading-equipment that can be affixed to a vehicle at the site. Figure 10-9 shows a remote weather station being used during a hazardous materials operation. This system is sophisticated—it uses remote monitoring that sends the data to a computer located in the HazMat vehicle. However, weather-monitoring equipment does not have to be complex. A simple wind sock, such as the one shown in Figure 10-10, or any type of flag or banner attached to a pole can determine the approximate wind speed and direction. Similarly, simple barometers and thermometers can be kept ready for use.

land breezes/shore breezes
those that occur when a land mass cools faster than an adjacent body of water and creates an offshore or onshore breeze

a

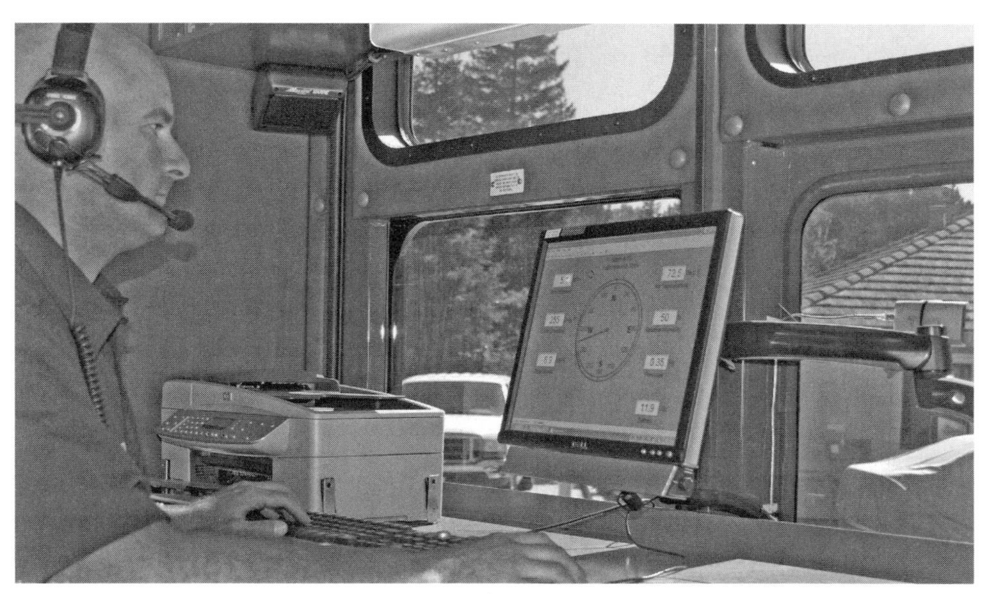

b

Figure 10-9
An example of a remote weather station system used during hazardous materials operations. (Courtesy of the San Ramon Valley Fire Protection District.)

Figure 10-10
A simple wind sock can help determine wind speed and direction.

ENVIRONMENTAL MONITORING

As we have just discussed in detail, monitoring an atmosphere for the presence of harmful contaminants is critical to ensuring the safety of anyone in the area. It also may be important, however, to monitor for or evaluate other materials that might be present in the environment. This is often the case when an unknown material has spilled onto the ground or area near a release site. The material may be a solid or a liquid that can also present an atmospheric hazard, but the material must be identified so that the appropriate level of cleanup operations can be conducted. Such is the case following incidents caused by materials being accidentally or intentionally released or discarded into the environment, perhaps even years previously. Regardless of how such materials arrive at a site, the area must be evaluated to determine which material is present, at what level, and whether it poses a hazard and requires abatement by a cleanup firm. Examples of actual incidents that required abatement are the cleanup activities throughout the United States Gulf Coast region following the major storms in 2005; the cleanup of buildings contaminated with anthrax following the September 11, 2001, terror incidents; and countless sites throughout the country where hazardous materials have been illegally dumped. When such incidents occur, not only does the atmosphere pose a contamination hazard, but also the environment and the materials in it. All such instances require that samples of the material be taken and analyzed so that the appropriate cleanup operations can be set in motion.

The operations involved in environmental analysis are a major topic of study in themselves and require considerably more training and information than can be provided in this text. This is partly because soil analysis usually is not like air and atmospheric monitoring, in which the types of materials present and their concentration often can be quickly determined. An area that has been exposed to a potential biological material cannot be easily analyzed because most of environmental monitoring operations include the complex analysis of soil or other samples. In most cases, samples are taken from a site to a remote biological laboratory where the organism is allowed to grow, which often takes days. Only in the most sophisticated systems can field verification be accurately performed. Therefore, it is much more difficult to determine what an unknown solid or liquid material is and whether it is hazardous in its present form. The good news is that the steps carried out by field personnel during environmental monitoring are standard and involve a few basic operations that we can discuss here.

The process of obtaining samples of materials at a release site begins with the application of most of the concepts that already have been presented in this text. The site is evaluated, it is determined which containers may be there, and any signs or labels indicating what might be present are assessed. These and a host of other steps are taken to acquire as much information as possible about the materials and the hazards they pose so as to provide the proper protection for workers or responders who must obtain samples from the area. Appropriate site control, PPE, and decontamination operations are set up and implemented prior to entry. In addition, the appropriate sample containers must be selected to preserve the materials. This is important for a number of reasons. For example, the material may be part of a criminal investigation, and the samples may become evidence. In addition to preserving material samples, a **chain of custody** must be established for them. The chain of custody documents the who, where, how, and what of a sample and includes appropriately labeling samples and identifying who picked them up, where they were acquired, how they were preserved, and who received them. This chain continues to the laboratory, where the lab workers who conduct the analysis must sign a form indicating that they received the sample and documenting the condition of that sample. In some cases, the person who takes the original sample seals and initials the sample container. In these situations, the Laboratory Technicians confirm that the sample container's seal was intact when the material arrived at the laboratory for the analysis. An example of a vacuum container used to collect air samples at a release site is shown in Figure 10-11. Such equipment collects and preserves the material for further study and use.

chain of custody
documentation of the who, where, how, and what of a sample; includes appropriately labeling the samples, identifying who picked them up, where they were acquired, how they were preserved, and who received them

Figure 10-11
An example of a vacuum container used to collect air samples at a release site. (Courtesy of the San Ramon Valley Fire Protection District.)

The techniques used to collect samples vary depending on a number of factors, including what the material is, where it is located, its relative hazard, and a host of other factors. Similarly, the tools to obtain samples vary and include scoops that aid in the collection of solid materials, the vacuum cylinders shown in Figure 10-11, and specially designed pickup tubes that suck liquid out of an area or a drum. An example of a pickup tube used to collect samples from drums is shown in Figure 10-12. This device often is referred to as a *drum thief* because it can collect small samples from inside a drum without damaging or affecting the material. Note that in the figure, the material is made of plastic that does not normally react with most types of materials, making it safe to use with a range of substances often found in drums. In addition to the tools just mentioned, various swabs are used to wipe an area; these swabs are then analyzed to determine what is on them.

haz-cat
short for hazard categorizing; field analysis that can be performed by appropriately trained personnel working at a site

Once a material is collected, rudimentary field analysis can be performed on it by appropriately trained personnel working at the site. Field analysis often is called **haz-cat**, a term that is short for *hazard categorizing* of materials. Using a standard haz-cat methodology, Field Technicians run part of the sampled material through a range of field tests to help determine which class of material the substance may be. This process usually does not definitively analyze a substance or determine its concentration, but it is a valuable tool during the early phases of a response or cleanup operation, because it at least provides some initial information about the material at a site. Variations of the standard haz-cat system also are used. An example of a flow chart of a standard system is shown in Figure 10-13. Figure 10-14 shows an example of a haz-cat kit that contains all the agents used to perform sample analysis.

Figure 10-12
An example of a sampling tool used to extract materials from inside a drum.

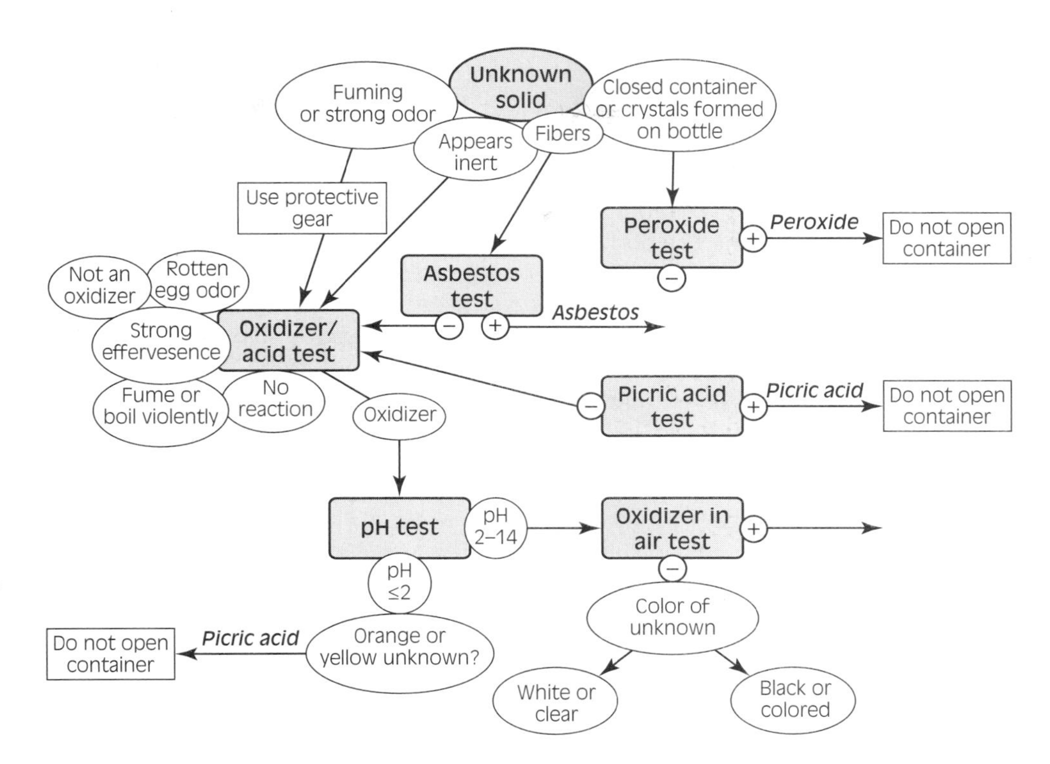

Figure 10-13
An example of a flowchart of a standard haz-cat system.

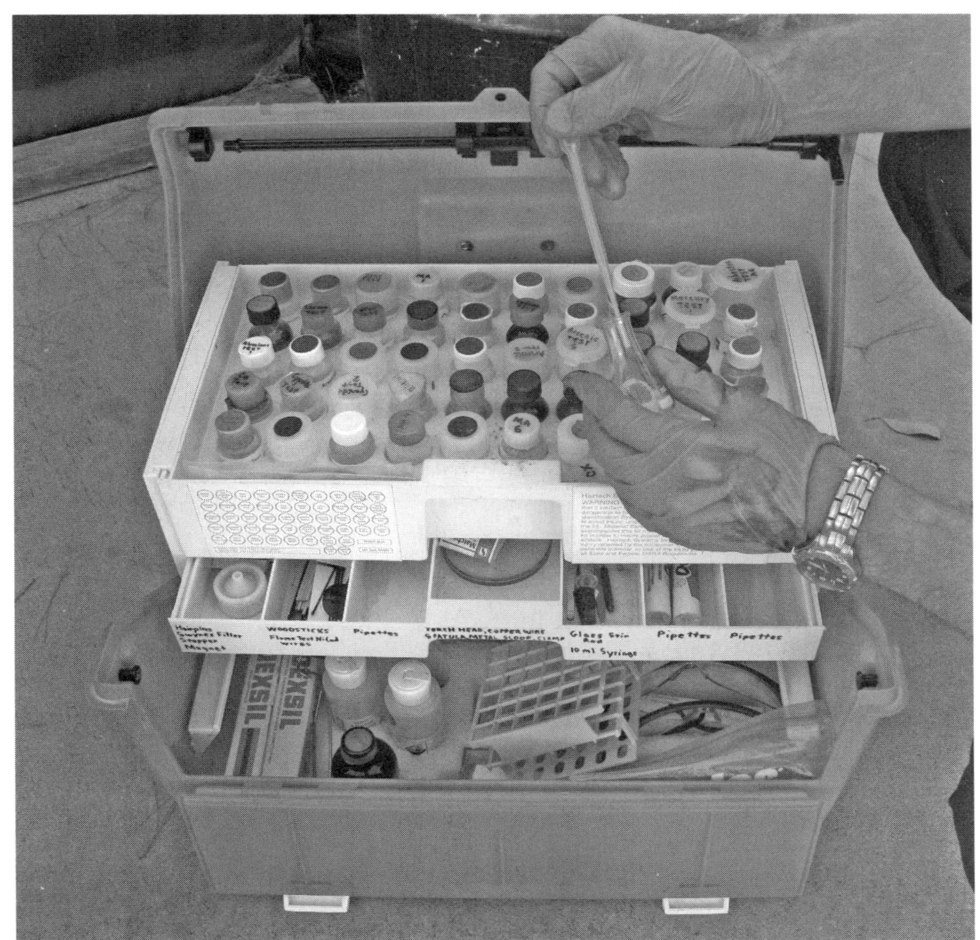

Figure 10-14
An example of a haz-cat kit used to help identify the specific hazards of unknown substances.

THE FUTURE OF ATMOSPHERIC AND ENVIRONMENTAL MONITORING

The events of September 11, 2001, served as a wake-up call for the hazardous materials industry. Its events interjected new realities into our world, including the concepts that not only accidental, but also intentional releases of hazardous materials could, and very probably would, occur. September 11 taught us that our own systems and materials could be used against us to cause considerable damage to and profoundly affect the world that we live and work in. From that, we realized that we needed to review our methods of security and storage, and even our response systems to ensure that we can continue to provide the level of safety we previously provided. As a result, storage and emergency response programs began to incorporate safety practices based on the possibility of terrorists attempting to access and use materials already in the United States.

Another result of these events and realizations was renewed interest in atmospheric and environmental monitoring. Unfortunately, however, we still are in the infancy stages of making concrete changes to monitoring practices. In effect, a lot of activity is taking place and new technologies are being developed. The Department of Homeland Security has awarded countless dollars to studies to develop new technologies, and national laboratories, such as Lawrence Livermore National Laboratory, are working to develop systems and equipment that will significantly improve our ability to detect and analyze materials in the air and environment. But today, several years after this process began, we still largely rely on the systems and technologies used prior to 9/11. This makes it even more important that people who deal with hazardous materials understand the concepts presented in this book.

Nonetheless, gradually, the equipment is changing and new technologies are being developed. Air monitoring equipment is getting smaller and can respond more quickly and provide more complete data. Hand-sized units are being developed by more manufacturers, allowing personnel to be less burdened by carrying larger, bulkier items. But many teams who use the new equipment complain that the smaller units are easier to lose in the chaos of response efforts or are not as easy to read when one wears the PPE required for hazardous entries.

Another change in monitoring equipment is that many units are able to connect or be combined with other equipment and therefore provide more complete data more rapidly. For example, many multi-gas monitors also incorporate a PID, which allows for a wider spectrum of analysis. Some units can now transmit data from a scene to a remote site for more complete analysis. In such cases, the speed of analysis is improved considerably.

However, the changes in the technology come at a cost, and that cost is often too high for most sites or hazardous materials response organizations. As we noted in the case study at the beginning of this chapter, only highly specialized teams with considerable funding usually can purchase the new technology because much of it has an initial cost of more than $100,000 and requires considerable ongoing maintenance. Such equipment often has a limited shelf life, and its sensor technology is so sophisticated that the sensors must be replaced even if the unit is not used. The bottom line is that the equipment that has been used in this industry for many years may likely continue to be used.

Fortunately, the equipment currently in use is sufficient to manage what we face today, even in the event of an intentional release of a hazardous material. As we have said, one of the lessons learned from September 11 was that our own infrastructure can be used against us. It is far easier to release a cylinder of chlorine gas from a railcar than it is to create the exotic chemical weapon systems sometimes described by the media. Chlorine with an IDLH of 10 ppm, for example, carries a significant potential for harm when released into a populated area. If this were to occur, the air monitoring equipment we already use can effectively detect chlorine. So, we must ensure that we know how to use the equipment we already have to detect the hazardous materials that we use on a daily basis. As monitoring technology continues to evolve, the equipment will become less complex, more available, less costly, and more capable of quickly and accurately providing us with the information that we need to be safe in our response and site operations.

SUMMARY

- Airborne contamination can come from any form of matter—solids, liquids, or gases.
- Gases can present a high potential for airborne contamination because of their migratory nature. Depending on a number of factors, including a gas's density or weight, the air movement created by wind or ventilation, and the presence of thermal extremes, gases generally are more mobile than other materials, and their process of movement is more extreme.
- Airborne contaminants must be detected and measured for their hazards to be fully appreciated and addressed. A material may present a hazard even though it is not visible to the naked eye and does not have any odor or other warning property.
- To properly determine the atmospheric conditions of a given space, tank, cylinder, or area, various types of monitors usually are required. Given this, a suggested sequence of testing should be followed for conducting air monitoring. This sequence is as follows:
  - If the site assessment identifies a potential for a radioactive material to be present, the first test should monitor the atmosphere for the source of that material.
  - If there is no indication that radioactive material may be present, the first test generally monitors the levels of oxygen.
  - Testing for the presence of flammable or combustible atmospheres generally is the next test performed.
  - A flame ionization detector (FID) is used to detect organic gases and vapors. This device may include a gas chromatograph (GC), but in many situations, using this device is optional.
  - An ultraviolet photoionization detector (PID) is used to detect organic and some inorganic gases and vapors. Like an FID, the use of this device is optional in many situations.
  - Finally, a system of colorimetric indicator tubes is used to identify specific gases.
- Combustible gas indicators usually determine the *presence* of a combustible gas but do not indicate which gas is present. These monitors should be calibrated to alarm at 10% concentration of the lower explosive limit (LEL).
- PIDs must be accurate adjusted, spanned, and read to determine the ionization potential (IP) of each material.
- Many FIDs have a built-in GC, but extensive training is required on these devices before they can be properly used and their results correctly interpreted.
- Colorimetric tubes efficiently verify the presence of many gases but should not be counted on to accurately determine the amounts (ppm) of gases.
- If corrosive gases are known or suspected to be present, then testing the atmosphere with a wet strip of pH paper is recommended before any activities are conducted.
- The initial monitoring of confined spaces should be done without personnel entering the space. Attach extension devices to the probe head and wands, if necessary.

- Keep in mind that air monitoring procedures and the movement of gases and vapors can be significantly altered by changes in the weather, particularly wind movement.
- Some hazardous material incidents and releases create a contamination hazard in the environment. These situations require that samples of the material be taken and analyzed so that the appropriate cleanup operations can take place.
- The techniques and tools used to collect samples vary depending on a number of factors, including what the material is, where it is located, its relative hazard, and a host of others.
- The haz-cat process is a valuable tool that enables a Field Technician to run a part of the sampled material through a range of field tests to determine which class of material a substance may be.
- Many national laboratories are developing new technology, but most workers and emergency responders who deal with hazardous materials still rely on the monitoring systems and technologies used before 2001.

REVIEW QUESTIONS

1. How do solid materials present inhalation hazards?
2. What monitoring technique can be used if the presence of a corrosive gas is suspected?
3. What is the proper sequence of testing when conducting air monitoring?
4. You must monitor for the presence and concentration of aromatic hydrocarbons, such as benzene. What monitoring equipment do you choose and why?
5. Describe why testing the atmosphere in a confined space is especially critical.
6. What factors should you take into account when conducting air monitoring in an outside area?

ACTIVITY

You are part of a site cleanup operation whose purpose is to empty and clean an underground storage tank. You have been given the responsibility of developing a monitoring plan to assess the atmosphere in the tank. When you open the tank's access plate, you observe that the tank's interior is heavily rusted and contains what appears to be about 6 in. of a water-like material on the bottom. You have access to any monitoring devices that you require. Develop a monitoring plan that identifies the types of instruments that you will use, the rationale for selecting the instruments, and how the monitoring will be conducted.

SITE CONTROL, SUPERVISION, AND INCIDENT MANAGEMENT

Learning Objectives

Upon completion of this chapter, you should be able to:

- Identify the importance of safety management systems when working with hazardous materials.
- Describe why the Incident Command System (ICS) is beneficial to both site operations and emergency response involving hazardous materials.
- Describe other emergency management systems related to the ICS, including the National Incident Management System (NIMS), and identify similarities and differences between the two systems.
- Describe the characteristics of the ICS, including its common terminology, modular format, unity of command, and span of control.
- Identify the job functions of the following command positions: Incident Commander (IC), Safety Officer, Information Officer, and Liaison Officer.
- Identify the General Staff Sections and list their responsibilities.
- Identify supervisory functions that may exist beneath the General Staff functions, such as security, decontamination team leader, entry team leader, technical reference, etc.
- Describe the need for a Safety Officer on all hazardous materials incidents.

CASE STUDY

On June 18, 2007, a large fire in a sofa superstore claimed the lives of nine employees of the City of Charleston (S.C.) Fire Department. Following an exhaustive investigation, the South Carolina Department of Labor, Licensing, and Regulation, Office of Occupational Safety and Health, issued citations against the city fire department in part for failing to provide a system of command that provided for the overall safety of the emergency personnel and their activities. In total, four violations were issued against the fire department, which was charged with three areas: not implementing an appropriate Incident Command System (ICS), failing to develop plans and procedures for safe operations, and failing to provide the necessary protective equipment.

According to the OSHA report, once the blaze started, the fire and smoke spread rapidly, and the firefighters became lost and separated from their hoses. The firefighters' air supply quickly was exhausted in the suppression efforts, and when they could not find their way out, they died of smoke inhalation. While contributing factors were listed in the report, including numerous violations by the property owner, this incident underscores the importance of maintaining control in all incidents where there is a potential for personnel to be injured or killed as a result of their efforts.

Loss of control is a condition that often leads to injuries or death. The loss of control in routine operations presents serious problems, but dire consequences can occur when control is lost in situations involving hazardous materials. Control simply must be maintained in such operations—the alternative is the loss of life and property. The South Carolina incident was tragic, but its tragedy is amplified because it is not an isolated incident. Reports issued after the evaluation of response efforts to major and minor incidents show repeatedly how loss of control compromises the safety of the responders and inhibits the response efforts. The chaos that has been reported following major incidents such as the shootings at Virginia Tech, the attacks on the World Trade Center, and response to hurricanes in the Gulf Coast can be partially attributed to a loss of control. Surprisingly, the majority of injuries to response personnel do not occur during major crises, but during smaller, mundane emergencies. In these cases, a strong system of control must be in place. Hazardous materials response and site operations demand that we control the actions that take place or risk the inevitable results.

INTRODUCTION

Any operation involving hazardous substances requires some degree of effective management to ensure that appropriate actions are taken and employees are not exposed to the hazardous substances. This is important whether the activity involves the general handling of hazardous materials at a work site or cleaning up hazardous wastes at a release site. It is even more critical when the activities involve emergency response to a release of a hazardous substance. Without a standardized and easily implemented system of management, the risk associated with response activities significantly increases. This is complicated by the potential for losing track of resources, including personnel, and for endangering personnel by unnecessarily exposing them to the situation's hazards.

! Safety
Operations involving hazardous materials require an effective system of management to ensure the safety of those working at the site.

Public safety agencies across the United States have long recognized this and have developed a system of emergency management to help coordinate response efforts. The system, which is generally known as the standardized *Incident Command System (ICS)*, has undergone a number of refinements since its introduction in the 1970s. From its original development, it has expanded to provide a universal framework upon which all types of emergency and even nonemergency activities can be coordinated and controlled. The ICS is so well suited to the management of hazardous materials incidents that in the Hazardous Waste Operations and Emergency Response (HAZWOPER) regulation, OSHA mandates its use in the management of emergency response activities involving releases of hazardous materials. While ICS is required for emergency response activities, it is also extremely valuable as a management tool for all site operations where hazardous materials are present. For those who supervise waste site cleanup operations as Waste Site Supervisors, this system provides an effective means of managing the site operations in an effort to provide a high degree of safety for everyone involved. For this reason, it is recommended that the system be used for all hazardous materials operations, including site cleanup, handling, and processing of hazardous materials, and all operations involving hazardous wastes.

■ Note
The standardized system used to manage both emergency and nonemergency operations involving hazardous materials is called the Incident Command System (ICS).

! Safety
OSHA mandates that the ICS be used to manage emergency response operations involving hazardous materials.

Although the ICS is mandated and used throughout the United States, variations of the system have been developed that allow this very effective system to be applied to even larger incidents than those originally in the scope of the system. These variations include the National Incident Management System (NIMS) and several state versions of NIMS, called the Standardized Emergency Management System (SEMS). This chapter reviews the ICS and NIMS and in the context of waste site operations and emergency response activities.

■ Note
Variations of the ICS include the National Incident Management System (NIMS) and the Standardized Incident Management System (SEMS). NIMS and SEMS are based on the concepts present in the ICS.

OVERVIEW OF THE ICS/NIMS INCIDENT MANAGEMENT SYSTEMS

As we have said, ICS and NIMS are useful systems in operations involving hazardous materials activities. These systems have considerable internal flexibility and can grow or shrink to meet the differing needs presented by different operations. They can be adapted to small, relatively simple operations, or they can be expanded to effectively manage larger, more complex ones. ICS and NIMS share many common features that make them suitable for use in a range of operations. Despite this similarity, the systems have minor differences, which are discussed later in this chapter. (For example, because NIMS is a federal government program, it may not be applicable outside this country, while the basic concepts of the ICS are.) First, we discuss the fundamental concepts introduced by the ICS, which form the basis for both systems. Understanding these concepts creates the foundation for our discussion of how these systems apply to a range of activities involving hazardous materials and other emergency operations. These include:

- Fires of all types
- Hazardous materials releases at fixed sites as well as in air, rail, water, or ground transportation accidents
- Medical emergencies, including multicasualty incidents
- Multi-jurisdiction and multi-agency disaster response to both natural and man-made catastrophes
- Wide-area search and rescue missions
- Pest eradication programs
- Oil spill response and recovery incidents
- Single- and multi-agency law enforcement incidents
- Planned events such as celebrations, parades, and concerts
- Private sector emergency management programs

One advantage of the ICS is that it is very simple to understand and use. Although training on the system can seem complicated, experienced users understand that the system was developed around just a few basic concepts. Once you understand these concepts, using the system is fairly straightforward. The concepts are similar to many of the day-to-day management tools effective managers already use, which is why they are often good emergency managers. They only need to apply the basic management tools they already know to an emergency situation. Four essential concepts that form ICS and NIMS include the following:

1. Module structure
2. Unity of command
3. Effective span of control
4. Standardized language and communication—i.e., common terminology

> **Note**
> ICS and NIMS are simple systems that incorporate four basic concepts. Understanding these concepts is a key part of learning how to use the systems.

Modular Structure

The first element of the ICS and NIMS systems is their modular nature, meaning that they have a number of components that can be used individually or in groups. This modularity allows the systems to grow according to operational demands. These management systems can be thought of as large toolboxes in which many types of tools are found. Like any operation involving the use of tools, some need only a single tool, while others at times require several tools to complete the tasks. In addition, to effectively use these tools, personnel must be trained on how to use each of them and on which tool is appropriate for which job—every job does not involve the use of every tool in the toolbox. Simple tasks require fewer tools and more complex tasks require more tools. Using more tools than is needed may not be better than simply using the right amount of tools.

> **■ Note**
> ICS and NIMS are modular management systems that contain a variety of tools for managing an operation.

So, if an incident or operation is small, only one or two of the ICS's "tools" may be needed. As an operation grows in size or in complexity, or as hazards and risks increase, the person in charge of the operation can grow the structure of the system—in other words, can use more tools as needed. ICS's modularity is one of its most useful aspects. Senior officials in an organization, such as the Site Supervisor, the Fire Captain, or the Emergency Response Team Chief, may be responsible for assigning tasks and functions as dictated by the conditions presented. Ultimately, one person will be responsible for directing the overall activities that dictate how large the operation is going to be. That person is called the Incident Commander or the IC.

As an operation grows, shrinks, or changes in complexity, the IC is responsible for choosing which tools or modules of the system are used. ICs break the operations into manageable pieces or chunks, make assignments based on those pieces and tasks, and select which positions will be filled by other individuals at the site. Figure 11-1 shows an example of a standard ICS chart with many of

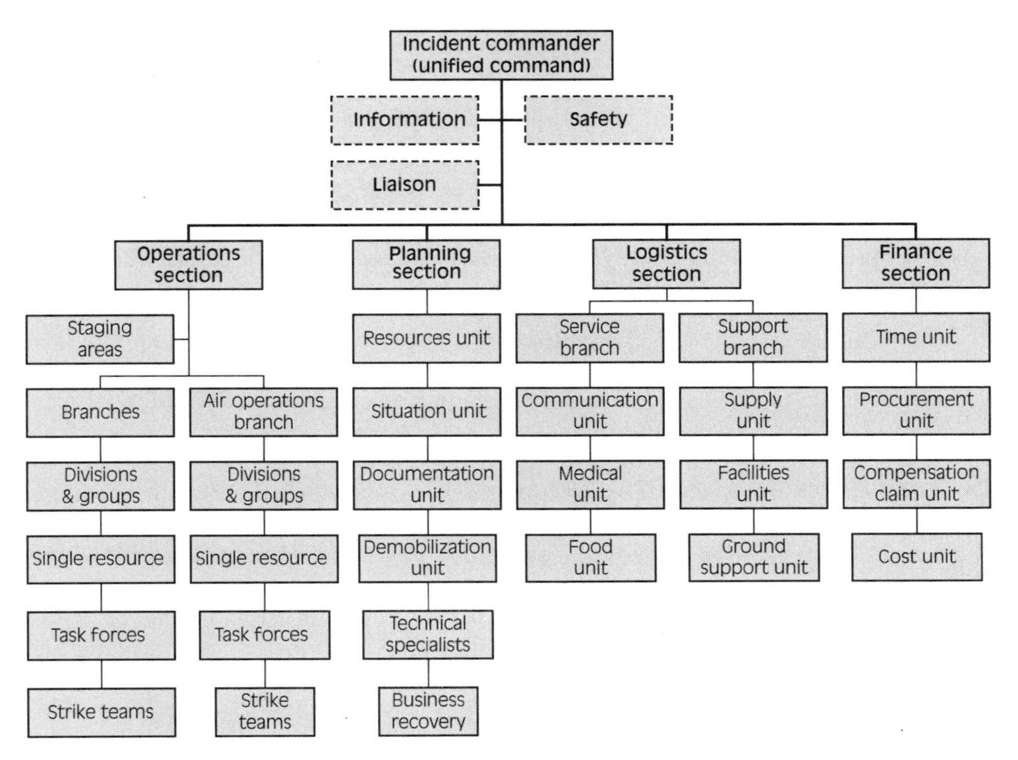

Figure 11-1
The standard ICS organizational chart.

the positions identified by the various boxes. As you can see, the system seems comprehensive—in fact, its comprehensiveness can be daunting if you do not understand that the system is modular. Like the example of not using every tool in the toolbox every time, when the operation is small or uncomplicated, it is not necessary to assign personnel to each task, in other words, to fill in all the boxes representing the different positions available. The IC has overall responsibility for setting up the system and determining which tools are used during an operation. However, the IC also is responsible for ensuring that all the activities required by an operation are carried out, regardless of whether a specific person has been assigned to a given task.

■ Note
When using the ICS or NIMS, select only the tools needed to safely conduct the operations being performed.

■ Note
The IC is responsible for setting up the organization required to conduct the site or the emergency response operations.

Deciding which of the position boxes to fill is not much different during an emergency than it is during any other complicated task. It often involves giving away or *delegating* key functions or areas of responsibility to others to accomplish the desired goals. If possible, all key functions or supervisory positions that are established should have some form of direct or indirect communication with the IC. In emergency operations, this usually requires using radio communications linked directly to the IC. If radio use is not possible, other ways to communicate must be developed. Often these involve visual monitoring by personnel working inside and outside the hazard areas, using paper forms, and other means for site operations where time is less critical. *The IC or the supervisor must be kept informed of progress or problems, so that they can facilitate the decision making process.*

■ Note
Delegating key functions and responsibilities is the responsibility of the IC.

As we have said, smaller incidents or operations may require a scaled-down use of the system, as examples later in this chapter illustrate. If an operation is smaller, do not use personnel simply to fill every position identified in the ICS because not every position may need to be filled. In the way that some tools have multiple purposes, one person may be able to accomplish multiple jobs without compromising his ability to perform each task. For example, a hammer can be used to both insert and remove nails depending on which side of it is used. So too can a good supervisor handle multiple tasks if the operations are simple and safe. As we have said, the beauty of the ICS is that it can grow as an incident grows and can be used to oversee and safely manage even the most complex operations, such as response to the terrorist attacks as on September 11, 2001, and in major disaster operations, such as response to Hurricane Katrina in 2005. As we learn more about ICS's components, we will see how best to apply the various modules to all types of activities.

ICS's modular system is key to the success of incident management because incidents or operations involving hazardous materials can vary tremendously. Some incidents are bigger, are more complex, involve more risk, and require a lot of people, all of whom must be coordinated and managed. Other operations may be relatively small and can be handled by a few personnel. As long as the system's fundamental structure is honored, the system's modular nature provides considerable flexibility.

Unity of Command

The second element of the ICS is *unity of command* or a Unified Command system. This element has two parts. The first part of this is that everyone in an operation is supervised by someone else at all times. In other words, everyone is supervised by someone who then is supervised by someone else, and so forth. This establishes what is called a *chain of command* among personnel working at a site. Everyone involved needs to be responsible to a single someone, and someone needs to be responsible for everyone at all times. This structure is especially important when hazardous materials are present and the operation's activities must be controlled to reduce personnel's exposure to those materials. The process of each person being supervised by only one other person at one time is vital to ensuring accountability. Accountability is an essential element of safety because without it, we run the risk of losing track of personnel in the chaos that often accompanies emergencies.

ICS's chain of command and modular nature make it expandable and allow it to include layers of supervision when the situation demands. Using the chain of command and appropriate levels of supervision helps ensure the safety of everyone working on an operation and prevents anyone remaining unassigned to a supervisor, which in turn prevents **freelancing**. Freelancing occurs when unassigned personnel self-initiate independent actions and work on their own. Freelancing is a common hazard of emergency response operations and one of the leading causes of injury and death, because freelance personnel often are unknown to the person in charge and are not part of the system used to ensure the safety of everyone working on an operation.

freelancing
the self-initiated actions taken by unassigned personnel during an operation without direction from the IC or the person in charge

The second part of unity of command also is related to the chain of command. It is that each operation must have only one person in charge overall, the person at the end of the chain. With limited exceptions that will be discussed later, most activities are managed by a single person, such as the IC for emergency operations, or the Site Supervisor for cleanup or other operations involving hazardous materials. The IC or the Site Supervisor is responsible for developing and implementing the single set of operational goals and plans. Without a final supervisor overseeing everything, an operation may have multiple goals and plans that contradict each other or unnecessarily drain the available resources. To prevent chaos from erupting, all participants in the operation must understand that they work toward the goals developed by the IC. Remember that the Site Supervisor in a nonemergency situation may fill the role of the IC even though the Site Supervisor may not be called by that title.

Manageable Span of Control

The third element of the ICS is called the *span of control*. It works on the premise that each of us can do only so many things efficiently at one time. Our brains can process only so much information, and our actions can be focused only on a limited number of things at once. This is true in everyday life, where we have multiple tasks facing us or several people who each want or need our time.

When working with hazardous materials and in emergency situations, however, it is critical that we retain control of the tasks we are responsible for and that we do not overextend ourselves. In an emergency, we must observe our limits and not stretch ourselves to the breaking point. Otherwise, we endanger ourselves and our co-workers.

span of control
the number of people that one supervisor can effectively supervise at any one time

Span of control is defined as the maximum number of people that one can effectively supervise at a given time. This concept has been widely studied, and despite situational variables, including the type of circumstances encountered, the expertise of the subordinates, and the skill of the supervisor, an effective span of control has been determined to include no more than seven people at one time. In emergency operations, despite the variables, the maximum span of control drops to five, based on the potential for hazards. Most authorities use the 5:1 span of control ratio as being optimum for emergency operations and other hazardous operations in which a high potential for exposure is possible.

Safety
In hazardous operations, an effective span of control is five personnel to one supervisor.

Using this concept in the field is easy. In keeping with the unity of command, personnel working on a given operation are required to be assigned to a supervisor. As a single supervisor becomes responsible for more than five subordinates, the supervisor begins to divide the personnel into groups based on criteria such as where they are working or the types of tasks involved. Recall that the ICS allows the use of various modules to help ensure the safety and efficiency of the system. In practice, this works as outlined in the following example.

A spill occurs at a work site. Personnel respond to help stop the spill and conduct the cleanup. The supervisor gathers the personnel and makes assignments. Initially, only three or four workers may be at the site. Each receives assignments. As more personnel arrive, the supervisor divides the response personnel into teams. Assume that there are twenty people now available for assignment at the incident. Now, the IC can assign one team leader with up to five subordinates to be responsible for getting the necessary equipment for disposing of the material once it is ready to be cleaned up. The IC now supervises one team leader who in turn directly supervises his team members. Another team leader may be assigned to assist with site control, keeping other employees outside the area of the spilled material. Assume that up to five people now work for this team leader. An additional team leader may be assigned up to five other workers to accomplish such tasks as getting the necessary personal protective equipment (PPE), beginning the process of stopping the release, and picking up the spilled material. Still another team may be set up to assist with the decontamination of the cleanup team. In this example, only five personnel report directly to the IC, each of whom is responsible for his team members and for maintaining the effective span of control.

As this example shows, using ICS's first three elements together enables you to establish a modular system that grows as the operations grow. The system has a unity of command because everyone is accountable to someone else and a single IC commands the activities. Finally, no one person in the system is responsible for overseeing more than their effective span of control, which further ensures the safety of everyone working on the operation. Additional examples will be discussed later as we describe small-picture and big-picture ICS operations.

Common Terminology

The last element of the ICS is its use of standardized language to communicate within the system. NIMS also uses standard language to communicate within it, and the two systems' languages are largely similar. This is important for whichever system is chosen because without a common communication system and language, numerous problems can occur. Imagine an operation where the participants spoke different languages. This may sound like an exaggeration, but all too often this is effectively the case in emergency situations in which a standard language is not used. This is particularly true when the public and private sectors work together to handle an incident at a private sector site because these sectors do not use similar languages. You probably would expect the result of such a situation to be chaos, and it often is.

To rectify this potentially catastrophic problem, the ICS uses a language composed of common terminology to describe key roles, functions, operations, or locations. The terminology is standard within the system and provides considerable assistance to everyone involved in an ICS operation. Because ICS can be used for all types of operations in both the public and private sectors, and in both emergency and nonemergency situations, the standard terminology is vital to convey important points relative to the operations being conducted. Depending on the type of operations and job responsibilities, an entire language may be used to describe the positions, responsibilities, and various parts of an operation. However, it is not necessary for all personnel to learn each and every term for every possible part of the system. Many people, for example, do not need to learn the terms for the job responsibilities of each position box on the ICS chart shown in Figure 11-1 because they would never need to use all that terminology. Most private sector operations do not usually involve activities, for instance, that require a *ground support unit* or a *demobilization unit*. Knowing something about more commonly used positions, however, is required for personnel who work in the hazardous materials industry. Following are some examples of standard ICS positions that have predefined roles and are commonly found in operations involving hazardous materials cleanup operations and emergency response activities. Personnel who may have other responsibilities outside those of hazardous materials operations may be required to learn other or additional terms related to their activities involving the ICS.

- Incident Commander
- Safety Officer
- Information Officer
- Liaison
- Hazardous Materials Group Supervisor
- Decontamination Team Leader
- Technical reference

A summary of the duties and the responsibilities for these positions follows:

Incident Commander or IC Incident Commander (IC) is the term given to the person who is ultimately in command or in charge of the operations taking place. This person manages the incident and is responsible for ensuring that the operations are carried out in the safest and most effective manner. Because this position is so important, we will discuss the duties and responsibilities of the IC in more detail later in this chapter. The responsibilities of the IC also can be applied to the Site Supervisor who is overseeing operations involving hazardous materials.

Safety Officer Many types of operations involving hazardous materials pose considerable risks that require a high degree of emphasis on the safety of everyone involved. While the ultimate responsibility for safety rests with the IC or the Site Supervisor, it is often valuable to have someone assigned to specifically monitor the operation's safety. For this reason, the ICS identifies a specific position to oversee the various safety systems in play and to address the operation's safety concerns. As we will discuss later, not only does OSHA require that the ICS be used to manage hazardous materials emergency response activities, but also it clearly requires that the Safety Officer position be addressed in all response operations.

A summary of the Safety Officer's major areas of responsibility and duties includes the following:

- Obtains a briefing from the IC, the Site Supervisor, and other key personnel as necessary
- Participates in the preparation of, aids in the implementation of, and oversees the safety concerns of the Incident Action Plan for emergency activities or for the Site Specific Health and Safety Plan ensuring that safety concerns are included in the plan
- Advises the IC or the Site Supervisor and other responsible person of deviations from the adopted plans as they relate to safety
- Has full authority to alter, suspend, or terminate any activity that may be judged to be unsafe
- Oversees the protection of site personnel from physical, environmental, and chemical hazards/exposures
- Ensures provision of required emergency medical services for assigned personnel and coordinates with others as necessary
- Ensures that medical related records for the involved personnel are maintained as required

Information Officer The position of Information Officer is sometimes referred to by the previously used term *Public Information Officer* or PIO. This position includes considerable responsibilities because of the range of potential public concern and of the issues involved with hazardous materials operations, whether they involve a major site cleanup operation or an emergency response. For example, a train derailment almost anywhere in the United States is quickly picked up by local news outlets, which are now on a 24-hour cycle and linked to national news organizations. Within minutes, large numbers of media personnel converge on the scene and demand information. The IC or the Site Supervisor probably is not going to have the time to talk to them and provide the information that they demand.

For this reason, particularly in emergency response operations, the IC usually will need to quickly fill the position of the Information Officer. That person will be the point of contact for the media or other organizations seeking information directly from personnel at the incident. Several agencies might respond to the media during a major emergency involving hazardous materials, but it is generally best to adopt a unified information program in which information is provided by a single source. Without such a program, incomplete, conflicting, or inaccurate information may be given out, and this information might affect the operations. In a typical ICS or NIMS operation, the Information Officer generally is responsible for the following activities:

- Obtains briefing from the IC or the Site Supervisor and other key personnel as directed
- Prepares or approves all statements or materials to be released
- Prioritizes the release of information

- Advises those in charge of any potential problem area that may occur related to information
- Designates an area to meet with the media and other involved organizations
- Meets with the media and others as needed to provide factual and timely information
- Ensures that all related records and reports are maintained
- May assist in required notifications under federal and state laws in coordination with the Liaison

Liaison Because of the potential for significant harm to people and the environment, operations involving hazardous materials and wastes are highly regulated. Oversight occurs from those whose interest may be protecting the air, the ground, the water, employees, the surrounding populations, and wildlife. So, a release or an accident involving hazardous materials often attracts a small army of representatives from the various agencies to the site. Furthermore, in the case of emergency activities, additional agencies, such as law enforcement and others that need to be involved but do not necessarily provide direct support, may need a place to coordinate their activities with those of the IC. The Liaison Officer is responsible for providing a point of contact for all agencies that respond to the incident or need to access the site. The Liaison Officer represents the IC or the Site Supervisor when dealing with outside agencies and maintains a flow of information among the interested parties. The Liaison Officer also may make the necessary legal notices to governmental bodies in the event of a hazardous material's release, or may keep those agencies informed of the progress of any mandated cleanup operations. In the event of a major disaster, the Liaison Officer from one site serves as the site contact to the governmental agencies who may have also established their own emergency operations center (EOC) to coordinate the response efforts throughout the jurisdiction. An EOC is a central location where major command and control activities are coordinated. It is generally a building or an area away from the actual emergency activities and is set up to help facilitate emergency operations during significantly large emergency operations, such as disaster response.

A good Liaison Officer keeps everyone "in the loop" and frees up the IC and the Site Supervisor to focus on their areas of responsibilities. Some of the typical duties of the Liaison Officer include:

- Obtains regular updates and briefings from the IC or the Site Supervisor
- Identifies the agencies that might be involved in a particular operation
- Contacts the agencies that require notification of specific types of hazardous materials events or operations
- Maintains a visible point of contact for the outside agencies
- Approves the information that is released to involved outside agencies.

Hazardous Materials Group Supervisor A position commonly filled in many operations involving hazardous materials is that of Hazardous Materials Group Supervisor. Hazardous materials emergencies can involve several major types of operations, including securing the scene, evacuating surrounding populations, fire control, emergency medical care, and actions related to the hazardous material involved. In such cases, someone must oversee the technical aspects of the hazardous materials activities and fill the position that oversees the hazardous materials operations. This position is called the Hazardous Materials (HazMat) Group Supervisor. In emergency response operations, this position often is filled by a knowledgeable Hazardous Materials Technician or Specialist who is responsible for managing all aspects of the hazardous materials component of the incident, including identifying the material(s), determining their properties and hazards, assisting in the formulation of tactics and strategy, and coordinating the activities of the decontamination group, technical references, and equipment procurement. In site operations, the Site Supervisor may assign a key member of the crew to oversee the specific operations involving the hazardous materials activities. The major responsibilities of the Hazardous Materials Group Supervisor are as follows:

- Obtains briefings from the IC, the Site Supervisor, and other personnel as required
- Coordinates the tactics and operations within the Hot Zone
- Coordinates information from the Safety Officer or the Site Specific Health and Safety Plan regarding accepted work practices and procedures
- Coordinates the team meeting with all involved personnel prior to entry into the Hot Zone
- Recommends and ensures that the PPE and tool selection for entry team members is appropriate to the tasks to be performed. (These recommendations also may be reviewed by technical reference personnel, the Safety Officer, and other entry personnel prior to use.)
- Maintains effective communications with the entry team

Decontamination Team Leader This position is used in complex hazardous materials operations to isolate this major chunk of the problem so that it can be dealt with more effectively. In emergency operations, the Decontamination Team Leader reports directly to the IC unless a Hazardous Materials Group Supervisor or Operations Chief has been designated. In site operations, this person reports to the Site Supervisor or the Hazardous Materials Group Supervisor. The Decontamination Team Leader is responsible for the operations of the decontamination team. These duties include the following:

- Obtains briefings from the IC, the Site Supervisor, or the Hazardous Materials Group Supervisor
- Establishes the decontamination corridor and equipment required to properly conduct the decontamination of site personnel
- Identifies contaminated people and equipment
- Supervises the operations of the decontamination element in the process of decontaminating people and equipment
- Maintains control of movement of people and equipment within the Decontamination Zone
- Maintains communications and coordinates operations with the Entry Team Leader
- Coordinates the transfer of contaminated patients requiring medical attention (after decontamination) to the medical group
- Coordinates handling, storage, and transfer of contaminates

Technical Specialists—Hazardous Materials Reference The Technical Specialist position often is valuable in a hazardous materials operation. This position is key to providing information needed to develop a plan of action in emergency response activities and in resolving issues that may arise in other site operations involving hazardous materials. This person reports to the IC or the Hazardous Materials

Group Supervisor if one is designated in emergency operations or to the Site Supervisor in site operations or cleanup activities. Technical Specialists provide technical information and assistance to the people who make key decisions regarding the operations. This information is obtained through reference sources such as computer databases, technical journals, and facility representatives. The Technical Specialist-Hazardous Materials Reference may provide product identification using hazardous categorization tests or any other means of identifying unknown materials. This information may be of value to the information section in the event that public information needs to be disseminated. Other activities that this position may be responsible for include the following:

- Obtains briefings from the IC, the Site Supervisor, or the Hazardous Materials Group Supervisor
- Provides technical support to the IC, the Site Supervisor, or the Hazardous Materials Group Supervisor
- Provides and interprets environmental monitoring information
- Provides analysis of hazardous material samples that may have been taken at the site
- Determines PPE compatibility to hazardous materials that may be present. (This can be done prior to site operations in the development of the Site Specific Health and Safety Plan for site operations.)
- Provides technical information management with public and private agencies, that is, Poison Control Center, Tox Center, CHEMTREC, State Department of Food and Agriculture, or the National Response Team
- Assists the IC or the Site Supervisor with projecting the potential environmental effects of the release

Command Staff ICS's standard terminology includes two groupings of personnel that are important to those engaged in emergency response operations. As we have seen, an IC may be called on to perform several tasks at once that are not directly related to the incident. For this reason, these personnel are grouped to help the IC manage an incident.

One of the groups is the *Command Staff* and is composed of the Safety Officer, Information Officer, and Liaison Officer. These three positions serve as extensions of the IC into their areas of responsibility. Their position on the ICS chart shows them indirectly connected to the IC, meaning they are not among the personnel who are directly supervised by the IC. Although the IC has the overall responsibility for each of the incident management functions and may perform all of them simultaneously in smaller operations, it is crucial to have key personnel who can fill these critical positions when an incident or the operations become more complex. Again, it is not necessary to fill every position in the ICS structure in every case. However, personnel in the organization who can help the IC when circumstances demand should be identified. In fact, the responsibilities of these positions are such that assistants often are brought in to help a designated officer. For example, the Information Officer may have numerous assistants to prepare and disseminate information through several channels. Likewise, the Safety Officer may have assistants who oversee the technical aspects of the hazardous materials operations.

General Staff The second personnel grouping is the *General Staff*. Its key positions, located just below the IC on the ICS chart, comprise four functions essential to conducting any operation. The positions are called Operations, Logistics, Planning/Intelligence, and Finance/Administration. Personnel who hold these positions generally are referred to as the Operations, Planning/Intelligence, Logistics, and Finance/Administration Section Chiefs, respectively. These positions are useful in emergency and nonemergency site operations involving major types and amounts of activity. The IC or the Site Supervisor may use these positions to manage complex operations and ensure that all aspects of them are effectively controlled by breaking them into these four key areas.

Each General Staff position may have one or more deputy positions as needed. The deputy's role is flexible. The deputy can work with the person in the primary position, work in a relief capacity, or be assigned specific tasks. Deputies should always be as qualified as the person for whom they work. In large operations, especially where multiple agencies or jurisdictions are involved, assigning deputies from other agencies can greatly increase interagency coordination.

Using standard position descriptions from both the ICS and the NIMS, the four General Staff areas are responsible for the following duties.

Operations Section Managing operations is the responsibility of the IC or the Site Supervisor, but the people in these positions may want to assign someone to oversee this function so that they can step back and look at the larger picture. When activating an Operations Section, the IC or the Site Supervisor assigns an individual as the Operations Section Chief. The Operations Section Chief develops and manages the Operations Section to accomplish the incident's or operation's objectives as defined by the IC or the Site Supervisor. Operations involve those activities conducted either to handle an emergency or to conduct site operations or cleanup. If the operations occur over a protracted period, two or more personnel may share this role over the course of a day; however, only one person at a time is assigned to the role. If the operations are complex and require further division into smaller and more manageable chunks, the Operations Section Chief may elect to assign one or more of the other positions mentioned earlier, such as HazMat Group Supervisor, Decontamination Team Leader, or others to facilitate the management of the operations and maintain an effective span of control. In other cases, the operations may involve multiple activities taking place in several areas. In such cases, the IC or the Site Supervisor may elect to break up the operations into standard pieces called divisions, groups, or branches. These terms simply describe the types of subdivisions used by the IC or the Operations Section Chief to maintain their effective control. An example of an operation in which these terms are used is shown in Figure 11-2.

Keep in mind that there is no one best way to organize any particular incident or operation. The organizational structure should develop to meet the functions required to safely and effectively conduct the operations required. The characteristics of each incident and the management needs of the IC or the Site Supervisor determine what organizational structure elements should be established. The incident organization may change to reflect the various phases of an incident.

Planning/Intelligence Section This division of the General Staff serves a key role in the formulation of the overall operations by assisting the IC or the Site Supervisor in critical areas. As its name implies, this section is responsible for gathering and evaluating data on the operations and the materials involved. Because hazardous materials operations can involve a variety of materials, it is critical to have someone responsible for determining what the materials will do, how they will react, and what specific tasks need to take place to safely manage the operations. In emergency response operations, the Planning Section, sometimes referred to simply as *Plans*, reviews the hazardous materials present, evaluates how factors such as weather and topography may affect the release, and recommends action to the IC, who then has the Operations Section carry out

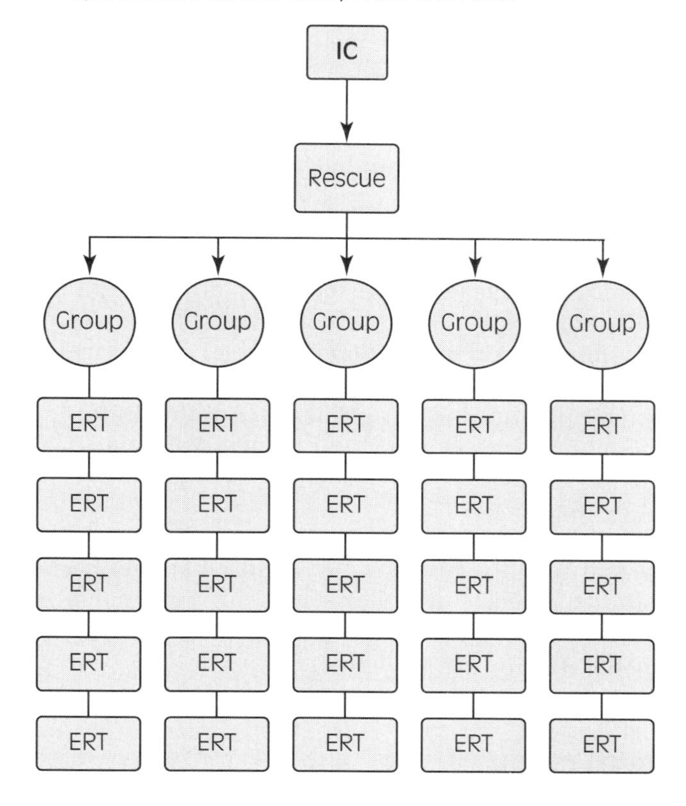

Figure 11-2 *An example of a large operation in which divisions are used to help manage the activities.*

the required tasks. In site operations, this section can monitor the progress of the work taking place, recommend changes to the overall Site Specific Health and Safety Plan, and provide other technical support to the Site Supervisor.

Briefly stated, the major activities of the planning/intelligence section are as follows:

- Collect, evaluate, and display information about the incident or the operations
- In emergency response, to develop Incident Action Plans for each operational period, conduct long-range planning, and develop plans for demobilization at the end of the incident
- Maintain resource status information on all equipment and personnel assigned to the incident
- Maintain incident documentation

Technical Specialists assigned to the operations most often are assigned to the Planning/Intelligence Section if one is established. Depending on their assignment, Technical Specialists may work in the Planning/Intelligence Section or be reassigned to other incident areas. Such specialists are invaluable in incidents involving hazardous materials or complex site operations.

Logistics Section Because hazardous materials operations are often complex and require a lot of equipment, PPE, and other materials, another key General Staff position frequently used by the IC or the Site Supervisor is a Logistics Section

Chief. The Logistics Section Chief is responsible for establishing an organizational structure that helps provide all the services and support needs of incident operations, including obtaining and maintaining essential personnel, facilities, equipment, and supplies. Like the other sections, a large number of personnel may be assigned to this section so that all the special equipment required by an operation can be quickly obtained. For example, major cleanup or emergency activities may require that the Logistics Section assign someone whose sole job is overseeing the PPE used by the personnel at the site because such equipment often is contaminated and needs to be frequently replaced. For other operations, the amount and types of equipment needed for decontaminating personnel or the site may require obtaining specialized materials. When the Logistics Section is activated, there are standard groupings on the ICS chart that are usually filled. These include communications and obtaining and managing internal and external resources, such as facilities, food, and other supplies.

Finance/Administration Section The final group in the General Staff is the Finance/Administration Section. The IC or the Site Supervisor determines whether an operation needs a Finance/Administration Section and designates an individual to oversee it. If the Finance/Administration Section is not established, the IC or the Site Supervisor needs to perform all the functions normally handled by this section.

This section is set up for any incident that requires on-site financial management and tracking of equipment and personnel costs. This is an important responsibility because many hazardous materials operations are subject to reimbursement or cost recovery by a governmental agency or the party responsible for the materials that are spilled.

THE ROLE OF THE INCIDENT COMMANDER OR SITE SUPERVISOR IN MANAGING ACTIVITIES

Managing an operation involving hazardous materials can be difficult even for the most experienced person. These situations are dangerous by their very nature, and mistakes can happen, making bad situations worse. Using the concepts of the ICS is critical to avoiding mistakes by providing a strong command and control program. In the case of emergency response activities, using such a system is mandated by the federal government. ICS's various tools provide an effective system of management can be established and maintained over the course of the operations. Especially during an emergency response, there may be periods that seem chaotic and when things are happening quickly. At the conclusion of an event, the actions of the IC or the Site Supervisor may be criticized, questioned, or second-guessed by many, including his own team members. Because of this, it is important that everyone who assumes the role of IC or Site Supervisor be knowledgeable enough to realize their own limitations and when to delegate assignments. Remember that the ICS is a toolbox and it contains all the tools required to do an effective job.

> **■ Note**
> All incidents include a certain amount of chaos and disorganization. It is the job of the IC or the Site Supervisor to manage this turbulence and direct the incident toward a safe resolution.

An IC or a Site Supervisor makes or approves countless decisions regarding an operation. We summarize several key areas of the IC/Site Supervisor's responsibility to help you see this role in the context of the "big picture." These responsibilities include the following:

- *The IC or the Site Supervisor is responsible for the health and safety of all responders working at a scene.*

  The IC or the Site Supervisor is the premier safety person on the scene at all times. He is responsible for his own safety first but must also ensure that everyone involved in the operations is accounted for at all times and that they are performing their function as safely as possible. In this capacity, it is the IC or the Site Supervisor's responsibility to make the situation better, not worse, by his involvement. To achieve this goal, it is important, particularly in emergency response operations, that emotions do not drive actions. Remember first and foremost that the IC or Site Supervisor did not cause the incident and is therefore not responsible for what has occurred prior to his involvement. Only after the IC or the Site Supervisor arrives and takes control of the operations is he 100% responsible for all that transpires and everyone operating under his direction. Even appointing a Safety Officer does not totally relieve the IC or the Site Supervisor of the duty of the overall scene safety.

> **Safety**
> The IC or the Site Supervisor is directly responsible for the safety of all personnel working at a site.

- *The IC or the Site Supervisor is responsible for deciding which operations are conducted.*

  "What are we going to do?" This simple question has tripped up many who assume a leadership role in a hazardous materials operation. It is an easy question to ask but sometimes a difficult one to answer. Addressing this question is made even more difficult when others question an IC's or Site Supervisor's decisions. Even when a plan has been developed, many critical decisions still need to be made by the person who is in charge of the overall operation. That person may feel like a duck swimming on the smooth surface of a lake. Above the water, the duck appears serene and to be moving almost without effort. Below the surface of the water, however, the duck may be paddling like crazy. The IC's or the Site Supervisor's decisiveness goes a long way toward gaining the attention and respect of those who are being supervised. Remember too that while the ICS allows for delegating responsibility for specific aspects of an operation, it is ultimately the responsibility of the IC or the Site Supervisor to ensure that the job gets done in a safe and an effective manner. Good information, sound planning, and involving others who can provide assistance is often the key to achieving this. Once a problem is fully understood and a plan implemented, the rest becomes an issue of time management and working toward the desired objective. The million-dollar question usually is deciding what the plan will be and how it will be carried out.

  Often, the IC or the Site Supervisor must determine the overall strategy for dealing with an incident and, in some cases, even for deciding which specific operations—which tactics—are required to implement that strategy. Strategy and tactics may sound similar but are quite different, and ICs/

Site Supervisors must understand the differences between them to effectively manage operations. Essentially, strategy is big-picture thinking—the overall goal—while tactics are the individual activities that carry out the strategy. Football teams provide a good example of the use of tactics and strategy. The entire team should understand the overall strategy of the game, which is that the offense is supposed to score points, while the defense prevents them from being scored. The coach or other key team members use individual plays as the tactics for accomplishing these goals.

> **■ Note**
> Strategy is big-picture thinking for an incident.

> **■ Note**
> Tactics are the individual activities taken to carry out the overall strategy.

It is the role of the IC or the Site Supervisor to set the strategic tone for operations and determine the tactics necessary to solve the problem. In many cases, strategy is developed using a team concept or previously established plans, but the final decision on implementation is left to the person in charge. A common pitfall here is the "analysis to paralysis" syndrome. This occurs when a group "what ifs" a situation to death instead of taking appropriate action. This is not to say that the team should rush into action without proper thinking, but developing and implementing a plan of action often takes too long.

- *The IC or the Site Supervisor is responsible for developing and implementing the battle plan for an operation.*

  The person placed in charge of the operations or activities is responsible for setting up the incident-management structure and organization that will be required as the operation evolves. As we have said, in smaller operations, the person in charge may perform all the functions that are needed. This person may oversee incident safety while also being available to answer questions from the media or outside agencies. The IC or the Site Supervisor also may be responsible for overseeing the operations, getting the required equipment, making the plan for the required activities, and monitoring the equipment and costs of the operations. In effect, in smaller operations, the IC or the Site Supervisor may undertake all the responsibilities of the Command Staff and the General Staff at the same time.

As the operations increase in size or in complexity, however, the IC or the Site Supervisor must delegate some job functions to others so that he can stay on task, provide for the safety of everyone working on the operations, and maintain an effective span of control. The IC or the Site Supervisor pulls the tools out of the ICS toolbox and ensures that the operation's overall objectives are achieved. In most cases, IC/Site Supervisors accomplish this without becoming involved in the operations themselves. IC/Site Supervisors should not close the valves, put on a suit, or reset alarm panels. Instead, they must think through and direct the operations and help control the activities of the responders who are their responsibility. It is critical that the IC or the Site Supervisor stay focused on managing the operations and taking steps to adjust the activities as necessary.

> **■ Note**
> The incident command structure is built from the bottom up and is used to organize the overwhelming number of jobs to be done at a scene.

ICS—The Smaller Picture

Now that we have described the ICS's main functions, we can break the system into smaller components and illustrate how it all works by using two examples—a small operation and a larger one. Smaller operations occur most frequently and include small spills of hazardous substances—those that present an emergency and those that do not—and day-to-day activities at sites of cleanup operations. In small operations, the person in charge uses less of the ICS because only a limited number of personnel need to be assigned and supervised. In such cases, the management system used may resemble the chart in Figure 11-3.

This system may be appropriate for a number of minor operations, such as a small spill of diesel fuel at a site. In this example, the spill occurs at the receiving area of the facility when the fuel tank of a delivery truck develops a slow leak. OSHA usually does not define such a small spill an emergency response operation because the material presents very little hazard and can be easily managed using standard methods. One of those methods involves using the ICS in the following way.

You arrive first at the scene with the other members of your site response group. You may activate the ICS shortly after arriving and pronounce that you are the person who will be in charge of the operations from this point forward. This is a critical first step because it establishes that a system is in place and prevents others from developing their own plan to handle the problem.

Once the ICS is set up, you may begin to direct the members of the response group to control the spill, shut off any ignition sources, and prepare for cleaning up the mess. Because this incident is very small, you elect not to appoint anyone to Command Staff positions and instead decide to fill each role concurrently with your role as the IC. Is this allowed? Absolutely. This flexibility is one of the advantages of the ICS.

> **■ Note**
> The ICS allows for considerable flexibility in developing an incident management structure.

In this example, the IC can oversee the safety at the incident. Because only a few people are involved in the response, the area involved is small, and the chemical is known, the IC can effectively fill the role of Safety Officer.

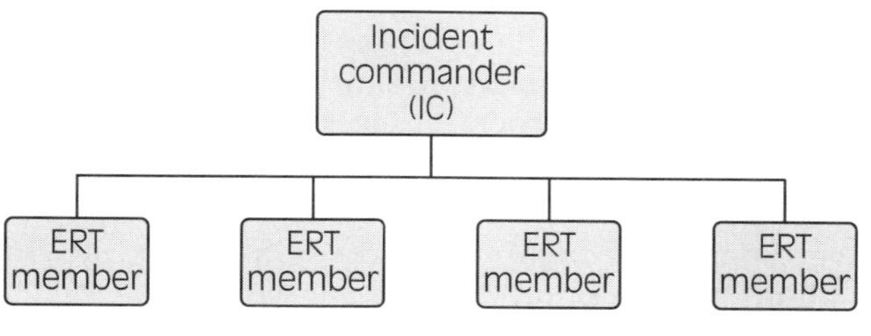

Figure 11-3
An example of an ICS organization for a small operation involving hazardous materials.

Also in this example, the IC elects to serve in the capacity of the Information Officer and Liaison Officer. As such, the IC deals with any media or information concerns should they arise, coordinates and meets with other agencies that may arrive, and ensures the proper disposal of the spilled material once it is cleaned up.

Finally, the IC in this case develops the plan of action, ensures that the required equipment is available, and oversees the activities of the small group of response personnel. Essentially, all of the General Staff responsibilities discussed earlier are being conducted by a single individual. If this can be done safely and effectively, it is the appropriate choice.

> **■ Note**
> Smaller incidents require smaller management structure. Delegate when necessary but do not give away decision-making authority just for the sake of creating a big management system.

ICS—The Big Picture

Not all operations involving hazardous materials are as simple as the one just described. Many emergency response operations require a complex system of management to oversee many types of complicated response activities. In such cases, multiple responders may need to conduct a range of operations, from evacuation to controlling site access. Additionally, if there are victims who require emergency medical care, that aspect of the operation must be managed. And finally, the actual response operations involving entry into the site where the spill occurred may require a very complex and sophisticated series of activities, including the proper selection of PPE, tactics for stopping the release, decontamination issues, and the proper disposal of the materials. A single person may not be able to do all these jobs on his own, so using a more complicated ICS management structure may be required.

Like emergency response operations, major cleanup operations may also require a more complex ICS. In such cases, the Site Supervisor may assign one or more of the Command Staff functions. A Liaison Officer may serve as the point of contact with the variety of regulators who may show up to monitor the operations. A Safety Officer may be assigned to be a mobile set of eyes and ears for the Site Supervisor and to assist in making decisions that affect the overall safety of the project staff and personnel. Additionally, a Logistics Section may be set up to monitor the types and amounts of equipment used. A Plans Section may be used to track the activities and monitor the project's progress. In larger, more complex operations, the ICS management structure may look like the ICS chart shown in Figure 11-4.

In addition to the regularly assigned positions on the ICS chart, another useful position is the staff function that documents the event. Documentation often is required during cleanup and emergency operations involving hazardous materials because these incidents often are subject to close scrutiny. When an event needs to be documented, a scribe is assigned to closely follow the IC or the Site Supervisor and document the information going through that job function, including pertinent conversations and why decisions were made. This information can be added to the report that documents the operation's activities and it can help response agencies procure reimbursement.

Larger operations require that the IC's or the Site Supervisor's presence be clearly identified. This indicates to everyone working at a site that a system of

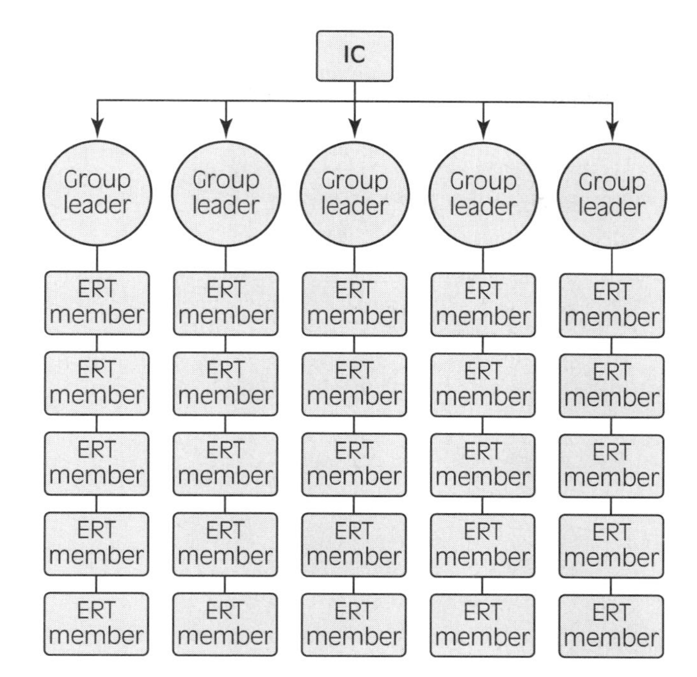

Figure 11-4
An example of an ICS management program for a more complex incident or operation.

Command Post
the single place where management operations are conducted

management is in place and that they can work through the appropriate structure to influence the plan of action by providing the IC with information or suggestions. In single-jurisdiction operations, this often is accomplished by having one area where management operations are conducted. This area is often called the **Command Post**, and personnel assuming key positions in the Command Post wear vests indicating their positions in the management system being used. Figure 11-5 shows an example of such a system. This system is quite effective, especially when standard ICS terms are used on the vests. The system alerts everyone on the scene that the ICS has been established and that someone is in charge. As responsibilities change, the persons wearing the vests simply exchange them for the one that designates their new role.

> **■ Note**
> The IC should be visible, relatively stationary, and organized. Because of this, it is recommended that the IC establish a command post at or near the incident.

Unified Command
a system of ICS management in which several jurisdictions assign a person to share key responsibilities, including that of IC

Large incidents may cross jurisdictional boundaries and involve multiple agencies, each of which has some degree of responsibility for the operations. Again, ICS's flexibility allows for this activity by using a **Unified Command**. In ICS, a Unified Command is a system of management in which several jurisdictions assign a person to share key responsibilities, including that of IC. While we have established that a single person usually serves as the IC in an emergency response, several agencies often may need to be involved in an operation. Rather than having multiple ICS programs that overlap, which might cause chaos, a single unified ICS can be established in which two or more persons share the responsibility for key ICS operations.

For example, two response team members from separate agencies may be assigned to collectively serve as the Safety Officer for an incident. This is a very effective system when outside emergency responders arrive at a site in response to an emergency at the site. Often, a fire department person and a site person

Figure 11-5
A command post where personnel are wearing vests indicating their positions.

may work together to fill the role of a Safety Officer since the site person may see things that are hazardous that the fire department person may not. Additionally, the IC position may be filled by more than one person such as the fire department chief officer and a management representative from the site. Together they work to achieve the goals and manage the operations in a safe and an effective manner.

NIMS

Before we conclude the discussion on controlling an operation by using the ICS, we must discuss the NIMS. As we know, this system is an extension of the basic ICS and is most useful when an emergency situation grows so large that it requires response from several levels of government. Although most emergency situations are handled locally, major incidents may require help from other jurisdictions, the state, and the federal government. NIMS was developed to enable responders from different jurisdictions and disciplines to better work together to respond to natural disasters and emergencies, including acts of terrorism. NIMS's benefits include a unified approach to incident management, standard command and management structures, and emphasis on preparedness, mutual aid, and resource management.

On February 28, 2003, the White House released Homeland Security Presidential Directive (HSPD)-5, 2003. The directive was intended to establish a standard management approach to major incidents by creating a "single, comprehensive national incident management system." The Secretary of Homeland Security is charged with administering this national system, the NIMS, as part of the National Response Plan. The National Response Plan is used to integrate federal government activities involving domestic prevention, preparedness, and response and recovery plans into a single, multidisciplinary, and multihazard plan. NIMS is used to enable federal, state, and local government agencies to work together during incidents that are complex or that involve multiple jurisdictions, such as the response efforts to the September 11 attacks and Hurricane Katrina. NIMS provides concepts, terminology, coordination systems, training, and the use of a Unified Command system that allows the interoperability and compatibility among responding agencies. In essence, it formalizes a standard response program using the ICS as its framework.

The NIMS Integration Center (NIC) was established by the Secretary of Homeland Security to provide "strategic direction for and oversight of the national incident management system (NIMS) . . . supporting both routine maintenance and the continuous refinement of the system and its components over the long term." The Center oversees all aspects of NIMS, including the development of compliance criteria and implementation activities at federal, state, and local levels. It provides guidance and support to jurisdictions and incident management and responder organizations as they adopt the system.

The Center is a multidisciplinary entity made up of federal stakeholders, and over time, it will include representatives of state, local, and tribal incident management and responder organizations. It is situated in the Department of Homeland Security's Federal Emergency Management Agency.

THE INCIDENT COMMANDER ROLE DURING EMERGENCY RESPONSE OPERATIONS

A number of rules must be followed during emergency response operations that involve hazardous materials. These rules are in HAZWOPER regulation section (q), which deals with emergency response, and identify specific actions that must be taken by the IC, who in some cases may be you. You should particularly familiarize yourself with section (q)(3), titled "Procedures for handling emergency response," which lists ten explicit rules detailing what OSHA requires in an emergency response. These ten rules are quoted in the following paragraphs. Note that the regulation refers to the IC as "the individual in charge of the ICS," and not specifically as the IC, so that is how that position is referred to in the following list. As the IC, it is critical that these actions be implemented as part of the response operations. Failing to do so constitutes a violation of the OSHA requirements for handling an emergency response. More importantly, failing to enact these rules keeps you from being as effective as you need to be because these rules are very good practices to implement, whether or not they are required. It is strongly recommended that you write down these rules in the form of an incident command checklist to ensure that they are considered in your response efforts. We can call these rules our Ten Commandments for hazardous materials response operations. They include the following:

1. *The senior emergency response official responding to an emergency shall become the individual in charge of a site-specific Incident Command System (ICS). All emergency responders and their communications shall be coordinated and controlled through the individual in charge of the ICS assisted by the senior official present for each employer.*

   The first of the rules is the statement that the ICS must be used to manage emergency situations. The specific ICS this rule refers to can be a subject of debate, in part because a variety of systems are used throughout the country. Also, the appendix that provides guidance on the implementation of the regulation refers to it in several ways, including "their company's Incident Command System," "an ICS," and most often as "the ICS." While the specific system referred to is debated, in the end OSHA mandates using a standardized system of management with clear rules. Given that a standardized version of ICS and NIMS is now used by most public safety agencies, it is recommended that this system be used.

2. *The individual in charge of the ICS shall identify, to the extent possible, all hazardous substances or conditions present and shall address as appropriate site analysis, use of engineering controls, maximum exposure limits, hazardous substance handling procedures, and use of any new technologies.*

   The second rule is a major one and requires considerable effort to fully implement. All too often response personnel are overly time-sensitive and rush into action when they confront an emergency. This rule states that the IC must make sure that a thorough hazard and risk assessment is conducted and that all materials present are properly identified before offensive actions take place. Recall from the discussion in Chapter 2 that the hazard and risk assessment is required for all operations that have a high risk of potential harm for operation personnel. This second rule also implies that the full extent of the problems must be determined because many hazardous materials can react with others creating situations and conditions that far exceed what might be expected from any one material. This is sometimes referred to as *chemical synergy.*

   A second implication of this rule is that many materials have multiple hazards and that the additional hazards may also be serious, which means that they must be considered when response actions are being determined. As responders, we often rely on identification systems that provide only limited information about a material. Recall that the Department of Transportation system of labeling and placarding generally lists only one of a substance's hazards—the one DOT considers the primary hazard. Often, however, a material may have other and even more hazardous properties, which could change with the introduction of other materials or conditions.

3. *Based on the hazardous substances and/or conditions present, the individual in charge of the ICS shall implement appropriate emergency operations, and assure that the personal protective equipment worn is appropriate for the hazards to be encountered. However, personal protective equipment shall meet, at a minimum, the criteria contained in 29 CFR 1910.156(e) when worn while performing fire fighting operations beyond the incipient stage for any incident.*

   The third rule encompasses a broad scope because it requires the identification of engineering controls at a site and the type of PPE needed. Using a Unified Command may be valuable in implementing this rule if the materials are released at a site where site personnel have been trained in emergency response operations and are familiar with the engineering systems and the PPE that is required. OSHA states emphatically that the PPE selected must be appropriate for the hazards that could be encountered. Because selecting the proper level and type of PPE requires a high degree of understanding of the potential hazards, the IC probably will need the assistance of technically trained personnel if there are no site personnel available, such as is the case in transportation incidents.

4. *Employees engaged in emergency response and exposed to hazardous substances presenting an inhalation hazard or potential inhalation hazard shall wear positive pressure self-contained breathing apparatus while engaged in emergency response, until such time that the individual in charge of the ICS determines through the use of air monitoring that a decreased level of respiratory protection will not result in hazardous exposures to employees.*

   This rule addresses protecting personnel from the highest of the hazards. All too often emergency response personnel are faced with a hidden, unseen inhalation hazard. Because many hazardous materials pose a significant inhalation hazard, OSHA demands that we start an operation using the highest level of respiratory protection. As part of that requirement, OSHA states that the use of respiratory protection must be maintained until appropriate air monitoring shows that the use of the equipment is not necessary, in other words, when the exposure level falls below OSHA's permissible exposure limits.

5. *The individual in charge of the ICS shall limit the number of emergency response personnel at the emergency site, in those areas of potential or actual exposure to incident or site hazards, to those who are actively performing emergency operations. However, operations in hazardous areas shall be performed using the buddy system in groups of two or more.*

   This is a critical rule for handling hazardous materials, although it is counter to the practice in many other types of emergency response. For example, the fire service routinely uses large numbers of personnel in combating fires or handling other emergencies. Hazardous substance spills, however, should be approached with more finesse. Handling a hazardous materials emergency has been described as being like a chess game: actions should be carefully calculated, players should be slow to react, and only the minimum number of pieces should be exposed to the risk of being taken out. In other words, the IC must ensure that only a minimum number of personnel are exposed to the hazard. This always includes a team approach, watching each other to further ensure safety.

6. *Back-up personnel shall be standing by with equipment ready to provide assistance or rescue. Qualified basic life support personnel, as a minimum, shall also be standing by with medical equipment and transportation capability.*

   Two major concerns are addressed by the regulation's sixth rule. The first is that a backup team of personnel be available to assist the entry team if needed. The backup team needs to have the equivalent level of protection as the entry team and must be immediately available to respond should they be needed to rescue the entry team. The second concern is that if a medical emergency occurs, appropriate personnel with adequate levels of training and resources are immediately available at the scene. In this rule, OSHA specifies provision for medical care and transport capability in the event of an emergency.

7. *The individual in charge of the ICS shall designate a safety officer, who is knowledgeable in the operations being implemented at the emergency response site, with specific responsibility to identify and evaluate hazards and to provide direction with respect to the safety of operations for the emergency at hand.*

   Appointing a safety official, known as the Safety Officer in the standard ICS, is significant given the broad range of risk associated with emergency response operations related to hazardous materials releases. As we know, the IC has a great deal of responsibility, so OSHA, whose primary mission is to promote safety, requires that additional personnel be assigned the specific role of overseeing the safety of an operation. In effect, this means that the IC is required to place safety at the same level of priority as the entire response operation. The result is that the Safety Officer is the only person other than the IC who has the level of authority to completely suspend response activities.

8. *When activities are judged by the safety officer to be an IDLH and/or to involve an imminent danger condition, the safety officer shall have the authority to alter, suspend, or terminate those activities. The safety official shall immediately inform the individual in charge of the ICS of any actions needed to be taken to correct these hazards at the emergency scene.* [*IDLH* refers to Immediately Dangerous to Life or Health.]

   As noted, the Safety Officer has the authority to completely suspend activities that are deemed to be unsafe. Working as part of the Command Staff in the ICS, the Safety Officer reports directly to the IC and advises the IC if he changes the plan that the IC has implemented.

9. *After emergency operations have terminated, the individual in charge of the ICS shall implement appropriate decontamination procedures.*

   Decontamination is one of the steps that must start at the beginning of a response and continue through the end. Until all personnel and equipment are adequately decontaminated, the potential for exposure exists. Only after decontamination is finished are the hazards completely contained.

10. *When deemed necessary for meeting the tasks at hand, approved self-contained compressed air breathing apparatus may be used with approved cylinders from other approved self-contained compressed air breathing apparatus provided that such cylinders are of the same capacity and pressure rating. All compressed air cylinders used with self-contained breathing apparatus shall meet U.S. Department of Transportation and National Institute for Occupational Safety and Health criteria.*

    This final rule establishes the criteria for selecting self-contained breathing apparatus (SCBA) cylinders. Generally, this selection is done well in advance of an incident and is not something that the IC needs to consider. In this last rule, OSHA simply restates its mandates regarding these important pieces of equipment.

How to Take Charge from the Beginning

Taking charge of an operation from the earliest moment is critical to effectively applying the concepts of the ICS and its rules. Strong and visible command sets the right tone for a response and is vital to the overall success of the effort. In the following paragraphs, we investigate initial steps for taking charge of an operation, but these are suggestions, not required actions. Various circumstances apply to any incident response. Use the following information as guidelines to help you formulate an initial plan of attack.

Sample Command Procedures for Emergency Response Operations A suggested procedure for those who may be called upon to respond to incidents and institute the ICS is as follows:

Step 1: Respond to the Incident

- Respond from an uphill, upwind direction, maintaining a safe initial distance from the problem. Note weather conditions, time of day, and other information that will need to be considered as you formulate a plan to manage the emergency.

Staging Area
a location near the incident where units assemble for later deployment once the plan of attack is set

- Assign a safe area where other responders can stage their equipment or themselves prior to being assigned. A **Staging Area** is a location near the incident where units assemble for later deployment once the plan of attack is set. Sending emergency response vehicles as they arrive to one Staging Area will provide you with some time to assemble your thoughts, gather additional information about the situation, and start sizing up the incident.
- Begin to think about the location of your command post. The area selected for it should be visible from the problem area but in a safe location so that the operations can be conducted without exposing command post personnel.

> **! Safety**
> Do not put yourself or others at risk or become part of the problem. Maintain a safe distance from the problem area until the full extent of the emergency is known.

size-up
a mental evaluation of the problem that takes into account the issues you will face

Step 2: Size Up the Situation A **size-up** is a mental evaluation of the problem that takes into account the issues you will face.

- Obtain a briefing from those involved in the incident prior to your arrival.
- Try to see as much of the incident as possible from a safe distance or get effective reconnaissance from team members.
- Think about assigning tasks based on your initial assessment. Start thinking about how to match up the number of jobs with the number of responders. Use these questions to help assign priorities.
    - *What are the facts of the situation?*
        - What is the time of day?
        - What is the weather, and will it affect the incident in any way?
        - Is there a chemical spill? If so, what is it and how much of it is there?
        - Is there a medical emergency? If so, what type?
        - Are people involved or trapped?
        - Is evacuation necessary?
        - Is something on fire?
    - *What are the probabilities/possibilities of the situation?*
        - Is the material moving from the area of release?
        - Where is the spill headed or likely to go?
        - Will this result in an exposure to others outside the area of the release?
        - Will the problem require building evacuation?
        - Is the weather likely to affect the situation?
    - *What is your current situation?*
        - Will immediate action solve any part of the problem?
          Example: Shutting off a valve from a remote location.
        - Are you in a safe location?

- *What resources are available?*
  - How many responders are responding?
  - What additional resources might be available if needed?
  - What is the time frame for getting additional resources to the scene?
  - Do other specialty response teams need to be activated?
  - Does site security or the police department need to be called for scene control?

Step 3: Assume Command and Direct the Incident At the earliest time possible, using your radio or another form of communication, advise your team of the following information:

- Provide your name or other radio designator.
- State that you are taking command of the incident.
- Verify the actual location of the emergency.
- Confirm in your own mind, and for the other responders, the type of emergency you are facing.
- Announce the location of the Staging Area and advise responding units to go there unless a plan of operation has been established.

Safety
Tell responders what you want them to do. Do not leave them waiting because they may initiate their own actions.

Safety
Do not let responders who are waiting for an assignment push you to make hasty decisions. Take time to think and formulate a plan of action before making assignments.

Following the initial transmission, further information can be given as needed, including:

- Any other vital information for the responders, including life hazards
- The location of the command post if pertinent
- What you are doing or planning to do, including calling outside response agencies

By this point in the incident, other team members will be arriving at the Staging Area and reporting in for assignments. As IC, you may deploy team members immediately or wait for a work force to assemble. Keep in mind that any offensive action should be carried out in pairs of team members.

Safety
Think of the incident as you would a game of chess—contemplate all moves prior to taking action and always ask yourself "what if?"

A number of checklists can help you with size-up. A common one uses the acronym D-E-C-I-D-E. Figure 11-6 shows the DECIDE model, which you can use

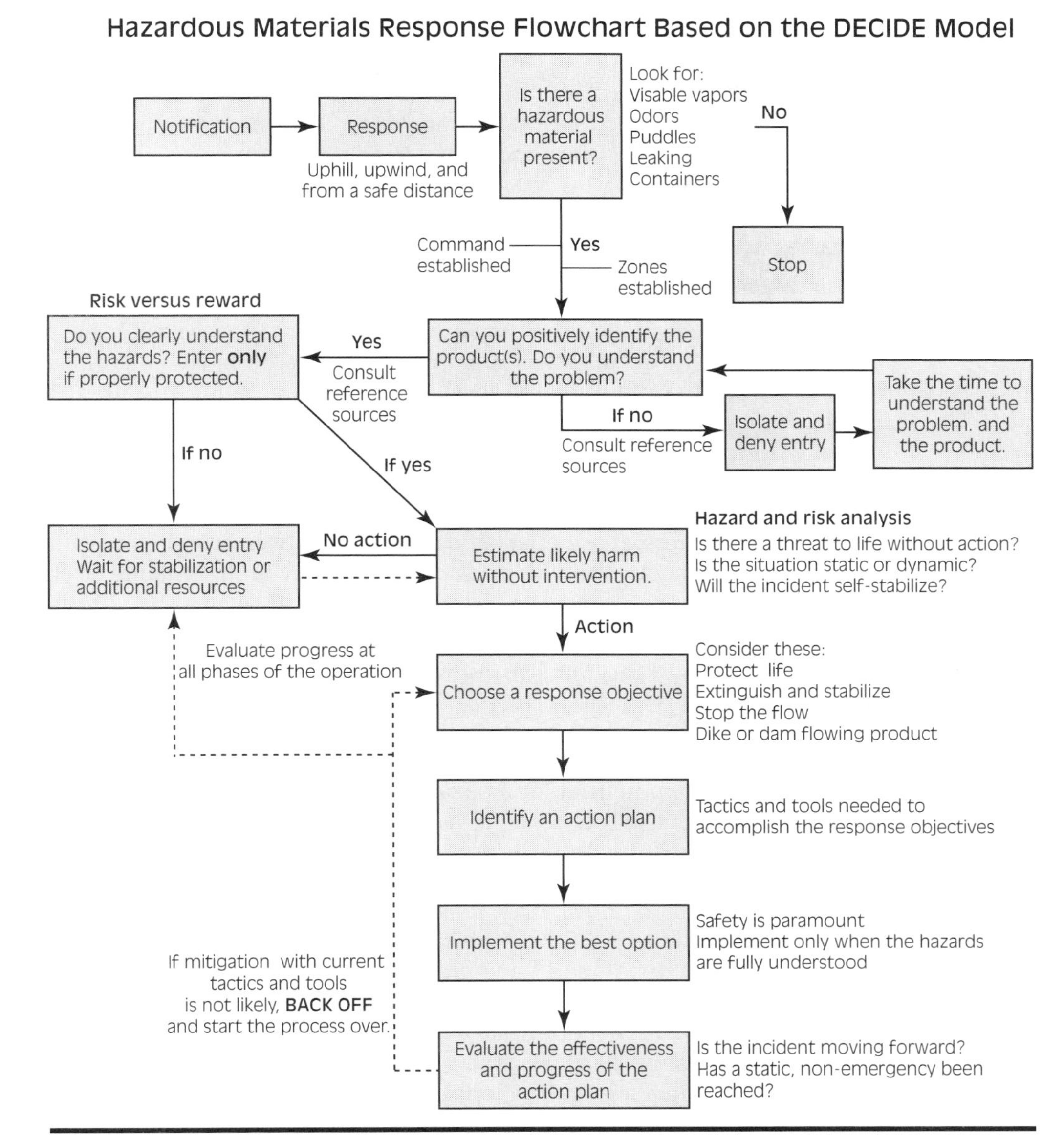

Figure 11-6 *The D-E-C-I-D-E model.*

to help you determine what needs to be done. This model contains the following questions:

- **D**etermine whether a hazardous material is present. Not all responses involve hazardous materials, but if such substances are present, the rules of HAZWOPER must apply.
- **E**stimate likely harm without intervention. This simply means asking yourself what would happen if you did nothing. In some cases, incidents can self-stabilize.
- **C**hoose the response objectives. List what needs to be done and who can do it.
- **I**dentify the priority for getting those tasks done. You cannot do everything at the same time, so prioritizing jobs is important.

- **D**o the first things on the list and keep moving through it until all the objectives are met.
- **E**valuate the progress and circumstances regularly during the response. Do not put your head down and just keep moving. Make sure that any course corrections are made in a timely manner.

SEVEN COMMON MISTAKES MADE BY INCIDENT COMMANDERS

Even the most effective and experienced hazardous materials ICs can make mistakes. Some of those mistakes are tragic, but their tragedy increases if we do not learn from them. Following is a list of seven common mistakes that a commander may make during an incident. This list has been compiled through interviews with numerous commanders who have made mistakes in their emergency response efforts. Some errors have been minor and inconsequential to the success or the failure of the incident. Others have had a huge and detrimental impact on the response. Regardless, these mistakes have already been bought and paid for, so to speak, so wise ICs will study them and avoid duplicating someone else's errors.

1. *Failing to plan.* It is important to think about what the real potential of an emergency at a site is, whether that site is the facility where you work each day or a site in your jurisdiction. Not planning for "the big one" or the worst-case scenario is mistake number one. If a situation can happen, you should plan for that eventuality. Incidents usually occur at the one place you do not think about and at the worst possible time. So, it is important to analyze the potential for a hazardous situation and expect that it will happen. Do not use the word "if" when preplanning for an emergency because this assumes that you do not actually believe anything bad will happen. Always preface preplanning with the phrase "when an incident occurs at . . . or during. . . . " This makes the threat a bit more real and helps you focus on the realistic possibilities. The emergency that is the easiest to handle is the one you have planned for and practiced countless times on paper or in drills. Anticipate the disaster potential for any site, whether it is a fixed location, a railroad system that runs through your jurisdiction, or a major transportation system. Whether it occurs during training or in real life, experience is an invaluable benefit to managing an incident.
2. *Failing to take charge of an incident.* Setting up command and calling yourself the IC does not constitute taking charge of an incident. The IC must be a leader and someone who the team is comfortable following. Weak leadership encourages weak performance and generates confusion, which results in undesirable outcomes. Be strong and decisive and instill confidence in the people you are leading.
3. *Failing to understand the nature of the incident.* As the IC, it is paramount that you fully understand the nature of the problem at hand. It is one thing to know that the toxic gas sensor is going off but the astute IC focuses on finding out why the alarm is occurring. This very basic concept is often overlooked.
4. *Failing to base decisions on fact rather than emotion.* Emergencies are great for getting everyone pumped up and ready to act. They are also great for making people behave in the most illogical and irrational fashion imaginable. Time after time, commanders give orders or take action based on nothing more than emotion. Avoid getting caught up in the moment—instead, be the one who is thinking clearly, unaffected by the nature of the event.

As an emergency responder, it is your job to bring order out of chaos, and this is best accomplished by calm, cool thinking based on good information. Again, take the time to get as much information as possible about the situation and make your decisions based on fact, not emotion.

5. *Failing to anticipate changes.* A good IC formulates a plan of action and implements it anticipating that it will work like it was conceived. A great IC conceives the same plan but also figures out alternative plans as the original plan is being implemented. School yourself to think 10–30 minutes ahead of the incident and always be considering alternative courses of action. Even if the initial plan looks like it is working, consider different approaches and strategies—just in case. This is beneficial in the long run because eventually you will be faced with a plan that is not working. Do not resign yourself to sticking with a faulty plan; change your course of action and move on.
6. *Failing to set up effective communications.* It is the responsibility of the IC to ensure that radio communications are in place and operating properly. Very few problems at a scene create more headaches and dangers than poor communications. When people cannot effectively talk to each other and goals are not understood by everyone, team members cannot work as safely as they need to. Poor communications can slow a response, so make sure that the key players can talk with each other and that the flow of information is unencumbered.
7. *Failing to accept blame and to learn from one's mistakes.* This is a simple concept: A mistake does not have to be yours for you to learn from it. You are the IC—you are the sole set of shoulders upon which success and failure rests. This is a great luxury that most people never take advantage of. Do not pass up the opportunity when it presents itself. Look at the success and failure of previous incidents, and figure out why they turned out the way they did. If someone made a bad decision, learn from that decision and avoid making it when you are in the same situation. When you do make a mistake—and you will, because we all do—admit it as your own, learn from it, and be determined not to make it again. Experience is a good teacher if you are astute enough to pay attention to it.

■ Note
Take the time to know what you are facing before committing people or resources to action.

■ Note
Poor communications can slow a response.

■ Note
Experience is a good teacher if you are astute enough to pay attention to it.

EVENT REVIEW

An IC or a Site Supervisor should conduct an event review once a response or a project is complete. This action is essential for bringing closure to the operation and providing an opportunity to review what took place, whether those actions were appropriate to the circumstances, whether all the required equipment was in place, and what can be done to improve the operation the next time. OSHA mandates that event reviews take place after emergency response operations.

A properly conducted event review does more than simply provide insight into what went right and what went wrong during an operation; it allows everyone involved to understand why things were done in a particular manner. The review is not used to assign blame or unduly criticize the performance of any one team member, although obvious deficiencies must be addressed. Rather, the incident response was a team effort, so the review should be conducted as a team. It is crucial, however, to be honest and open during review sessions. If a performance truly was below expectation, it should be noted as such. As long as all comments are positive, constructive, and offered as a means to improve performance, honesty is the best policy. Event reviews also serve as a basis for fixing problems that may lead to further incidents or for correcting items to enhance future operations.

SUMMARY

- The Incident Commander's (IC's) job is to bring order and direction to the chaos of an emergency and to ensure the safety of everyone working at the incident or operation.
- The overall scene safety is the responsibility of the IC or the Site Supervisor.
- OSHA mandates the use of the ICS to manage the response to emergencies involving hazardous materials.
- The IC or the Site Supervisor must set incident/operational objectives and priorities on every incident.
- An Incident Action Plan must be developed for every incident. It can be formal or informal but needs to be understood by all the key players in an incident.
- It is the IC's responsibility to attempt to bring rapid control and closure to an event.
- Incident strategy is the overall theme of or "big-picture" thinking for the event. It is the large-scale goal setting for the incident and should be determined by the IC.
- Tactics are the individual actions or procedures used by the team to solve the problem.
- In order to remain effective, the IC or the site supervisor must delegate tasks and supervision when necessary.
- *Span of control* refers to how many persons a supervisor can effectively manage during an emergency.
- A 5:1 ratio of workers to supervisors is the recommended span of control for emergency response and other operations posing significant hazards.
- The Incident Command System (ICS) should be implemented at all emergency incidents, regardless of size.
- The ICS is a modular system characterized by common terminology, unity of command, and span of control.
- The General Staff Sections are: Finance, Operations, Logistics, and Planning.
- The Finance Section deals with accounting, payroll, and documentation.
- The Operations Section deals with getting the job done. The Operations Chief may supervise many different disciplines, such as fire response, medical activities, and hazardous materials actions.
- The Logistics Section deals with procuring services, equipment, and personnel.
- The Planning Section deals with directing the incident on a long-term basis and provides technical support.

- The IC or the Site Supervisor presides over the General Staff, which consists of the scene Safety Officer, the Liaison Officer, and the Information Officer.
- On smaller incidents, the IC or the Site Supervisor can wear many hats and fill many of the ICS functions simultaneously.
- A scene Safety Officer should be in place for all incidents. The IC or the Site Supervisor may fill that role during small incidents.
- Each incident management structure must include some form of communication that is understood by all responders.
- NIMS is mandated for use when incidents involve multiple jurisdictions and require federal or state response. It is similar in form to ICS, uses similar principles and language, and is used nationwide.
- *Unified Command* is when multiple jurisdictions fill a single position within the ICS.

REVIEW QUESTIONS

1. Describe the differences between the ICS and NIMS.
2. What are the four major components of the ICS?
3. Describe *span of control* and identify the optimum level.
4. Define the two parts of unity of command.
5. What is the major purpose and value of a Unified Command?
6. Name the three positions that along with the IC make up the Command Staff. For each position, identify two key responsibilities.
7. Describe the four major components of the General Staff and provide a brief description of each area of responsibility.
8. List five "must-do's" for the IC at an emergency response.
9. List five of the seven common mistakes made by ICs.

ACTIVITY

Your facility has been notified of a strange, chemical-like odor coming from a chemical storage area (this can include cleaning chemicals). As you approach the area you look through the glass door and see whitish vapors coming from a four-pack of containers on the floor. You see a corrosive label on the nearby box and on the side of one of the bottles. People are working in a room connected to this area. The person who reported the spill is not around, and someone states that he may be injured.

1. Size up the situation by considering all the facts, probabilities/possibilities, your own situation, and resources, and then create a plan of action to deal with the incident. List the actions you will take in order of importance, including all applicable PPE, and justify your decisions. Consider the appropriate response based on your agency's ability to handle this incident.

   Make sure to address the following points:

   a. How do you isolate the area?
   b. What will you do if there are more jobs than available responders?
   c. Where might you get information to help you identify which chemicals are involved?
   d. Is this spill likely to migrate? If is it, will the problem get bigger if you don't do something quickly?
   e. How would you contain the spill? What materials would you use?
   f. How would you decontaminate someone who becomes contaminated?
   g. Where does the contaminated equipment go?

PUTTING IT ALL TOGETHER: RESPONSE AND SITE OPERATIONS

Learning Objectives

Upon completion of this chapter, you should be able to:

- Identify the types of operations requiring compliance with the Hazardous Waste Operations and Emergency Response (HAZWOPER) regulation.
- Apply the various components of the HAZWOPER regulations to cleanup operations and emergency response activities.
- List tactics that can be used in site operations and emergency response to control spills.
- Describe the process of dealing with response to releases of unknown hazardous materials.

CASE STUDY

Just prior to leaving for the station where I served as the Deputy Fire Chief, I was alerted to a response to a hazardous materials incident on the major interstate freeway that ran through my jurisdiction. The response included a standard first alarm assignment with a Battalion Chief, two engine companies, and a truck company. The initial dispatch stated that a tank vehicle was leaking on the shoulder of the freeway and that the driver had left the vehicle and was standing alongside the roadway. First-arriving fire companies found that the tank vehicle was carrying ammonium nitrate, a strong oxidizer used as a fertilizer. The leak was small and coming from an area on the bottom of the tank, but the cause of the leak could not be determined.

It took several hours to stabilize the incident and transfer the leaking material into a second tank vehicle. During this time, given the highly reactive nature of the material involved, the interstate was shut down. A Unified Command system was set up with members of the California Highway Patrol and the local fire department. Representatives from the chemical manufacturer, chemical engineers, and representatives from the county environmental health unit also were involved in planning the response and mitigation activities. Despite major disruptions to traffic in the area, no other problems occurred, and the incident was resolved without injury to either civilians or response personnel.

During the post-incident review, it was pointed out that the reason for the operation's success was the deliberate effort to develop a plan prior to implementing action. If actions had taken place before a plan was set up, injuries and additional material release might have occurred. By all accounts, the handling of this emergency response was "textbook."

INTRODUCTION

All too often, the information we learn in the classroom seems esoteric and as if it does not apply to what we do in the field every day. After all, most operations involving hazardous substances are carried out safely and without incident. However, when things are not handled appropriately, the potential for harm can increase dramatically. For example, in the case study incident with the leaking tank vehicle, prompt action may not have been required. A full knowledge of the hazards and operations required, coupled with a strong command and control system using the Incident Command System (ICS), allowed the incident to be resolved without any problems except delayed motorists.

In this book, we have explored the range of regulations pertaining to hazardous materials handling and response, identified several levels of HAZWOPER certification, and noted the Occupational Safety and Health Administration (OSHA) requirements for those working in the field of hazardous materials response and site operations. Legally carrying out most operations involving hazardous substances requires an understanding of those requirements.

Then, we explored the physical and health concerns of the various materials, often interjecting numerous definitions of terms that may seem not to mean a lot at the time that we discussed them. However, without a thorough understanding of the properties, characteristics, and hazards of the materials that we work with, the potential for serious harm can occur.

Next, we discussed the concepts of controlling hazards including a discussion of the range of personal protective equipment (PPE) and respirators that can help us work safely in the field. This included discussing systems for controlling response activities, decontamination procedures, monitoring practices, and other hazardous conditions. Such systems are necessary for effectively maintaining safety in our operations, and without these systems, personnel exposure to materials is likely.

Safety
Applying the concepts learned in this book is vital to working safely in any capacity with a hazardous material.

This final chapter focuses on applying the concepts we have discussed so far. We will apply the information in this book to two actual scenarios to better understand its usefulness in site cleanup operations and emergency response scenarios. These incidents were real: one was similar to the case study described above, and the other is an exercise that tests the ability of responders in the field to develop a plan to work safely in a common cleanup scenario. As you review this information, you will see the importance of what you have learned and will appreciate your ability to use what you have learned to make yourself and others safer.

SITE SAFETY PLAN—CLEANUP SCENARIO

Hazardous material spills are cleaned up all the time. Spills usually are caused by inadvertent actions that result in the release of small amounts of a material. Most of these so-called incidental spills are handled appropriately by the workers who spilled the material. Examples of incidental spills are a chemist who spills a small amount of a reagent or a solvent on a laboratory bench and cleans it up with absorbent materials or the accidental overfilling of a vehicle during refueling operations, which is handled by spreading loose absorbents onto the small spill.

Note
Most spills of most materials are not emergency response operations and may not be subject to the application of the HAZWOPER regulation.

However, some spills are much larger, carry a higher potential for harm, and require a more complex approach to cleanup. These spills may be the result of a major transportation accident in which a tank vehicle tips over and dumps its load into the environment. In these cases, after the emergency phase of the response is handled, the cleanup phase is implemented to remove the contaminants from the area and restore the area to its normal state. Other spills may be the result of leaking underground storage tanks that are removed from the ground, leaving an area of what is referred to as "dirty dirt." Still others occur when abandoned containers of illegally dumped hazardous substances are found. Regardless of a spill's cause, spill cleanup activities are a major part of the HAZWOPER regulations and requirements.

> **■ Note**
> Spills of hazardous materials that pose an unreasonable threat to the environment or to the public are covered by HAZWOPER.

The spill used as the basis for our first exercise involved abandoned drums. The drums were discovered near a new housing development. Figure 12-1 shows a map of the area and identifies the proximity of the leaking drums to the surrounding area. The drums were slowly leaking material into the soil on which they sat. The leaks were small drips and did not constitute an immediate emergency response threat. In fact, the local fire department was called initially and determined that the spill should be cleaned up by a local hazardous waste cleanup contractor.

Now, in our hypothetical exercise scenario, you are in charge of this cleanup effort.

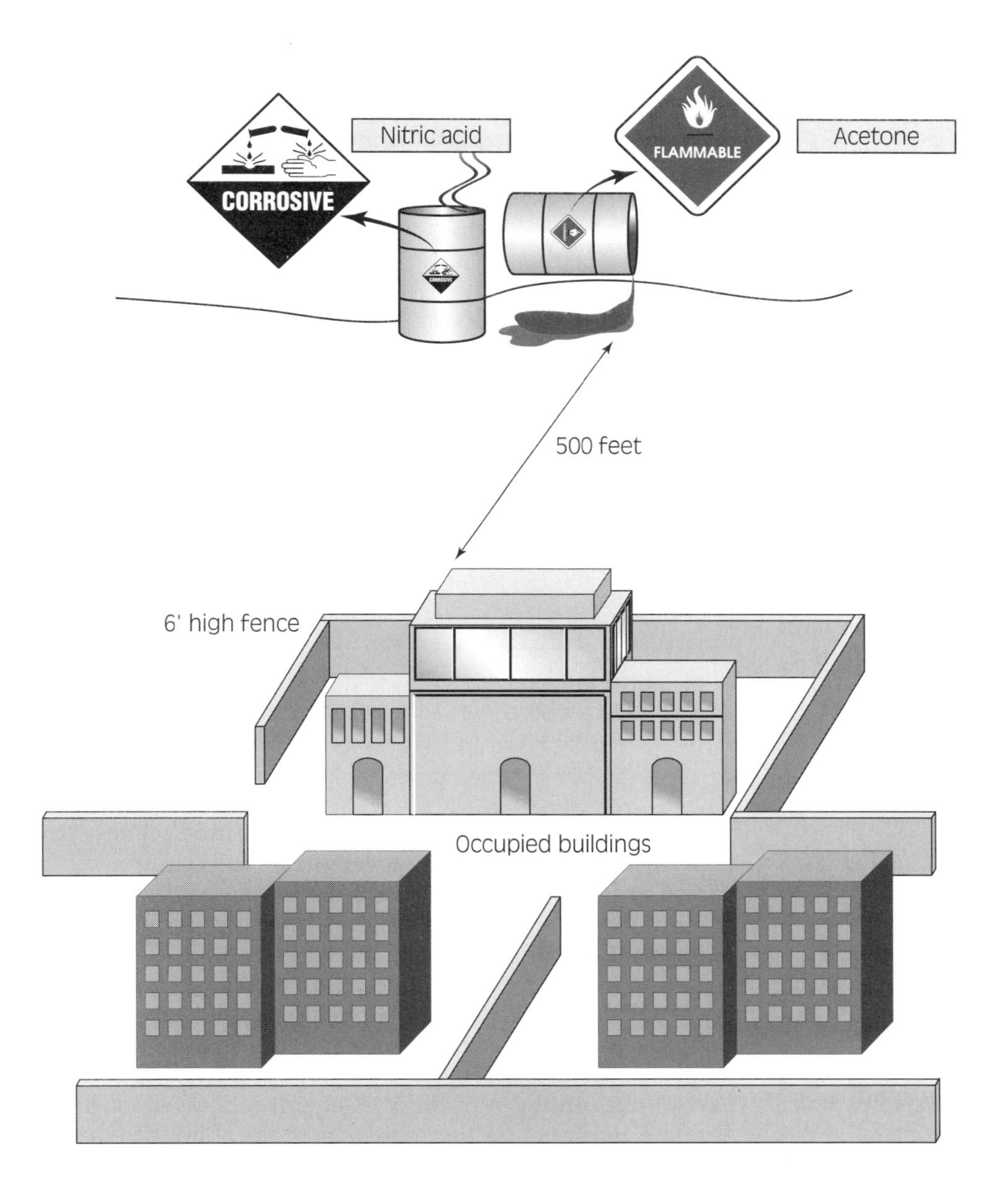

Figure 12-1
A map showing an overview of the spill cleanup site.

The containers involved in the spill are two 55-gal. drums. One is a black metal drum, and the second is made of a blue plastic. The drums have Department of Transportation (DOT) labels affixed to their sides. These labels are visible from a safe distance and indicate that the black drum, which is on its side, contains a flammable liquid. There is a small drip-size leak noted from the bottom of that drum. The blue plastic drum is upright and bears a corrosive label. A small amount of vapor is seen intermittently emitting from the area around the bung of that drum. The drums are approximately 3 ft. apart, and the area where they are located is slightly sloped, as shown in Figure 12-1. The weather is warm and sunny. The temperature currently is 75°F, but is expected to reach more than 100°F by late afternoon. Your job is to evaluate the situation, apply the information from our HAZWOPER study to these operations, and develop a plan to clean up the material.

There often are multiple ways to effectively and safely manage a cleanup operation, similar to emergency response operations. Plans to clean up a release such as the ones described in the preceding paragraph can take on a number of facets. Some organizations may take a conservative approach to such operations and develop detailed and time-consuming programs. Others may choose to be more aggressive in their operations while still safely handling the situation. The approach presented in our discussion is simply one way of applying the information you have acquired from this book. Keep in mind that many other methods can be used. The key to developing any cleanup program is to ensure that the concepts and safeguards presented in this text are considered and factored into the final operations.

To begin this cleanup activity, we might first ask ourselves whether the HAZWOPER regulation even applies in this situation. As we know, many cleanup operations do not fall under the scope of this regulation and are safely handled by those with training other than that mandated by HAZWOPER. As we learned in Chapter 1, incidental spills such as the one in our scenario are not emergency responses and are not "uncontrolled hazardous waste sites." In this case, we conclude that the spill, if allowed to continue, would pose a hazard to people in the area and the surrounding environment. So our program must take into account the requirements of the HAZWOPER regulation and develop a site specific Health and Safety Plan (HASP) for the operations. As we move through our scenario, we will see that the issues that we discuss would be included in the HASP.

■ Note

An important step in any cleanup operation is to determine whether the activities fall within the scope of the HAZWOPER regulation.

Given that the site is a cleanup operation subject to HAZWOPER, we then may want to determine the level of training required for the cleanup operations. Given that the cleanup operations are not emergency responses or operations conducted at a regulated treatment, storage, and disposal facility (TSDF), we know that training must comply with section (e) of the regulation. This requires that those involved be trained to either the General or Occasional Waste Worker level and that a Site Supervisor be present. Given the high potential for exposure to levels above the permissible exposure limit (PEL) for the substances involved, our selection of training for those participating would likely be that they complete the 40-hour initial training for General Site Workers, and that they be briefed on the site hazards and the requirements of the HASP and supervised by the Site

Supervisor during the cleanup operations. If the cleanup operations were going to take multiple days, the certification of the workers would not be complete until they had the required three days of supervised field experience, as was discussed in Chapter 1.

Safety
HAZWOPER regulations must be followed for major cleanup operations, TSDF operations, and emergency responses.

Note
Cleanup workers must be supervised during their initial days working at a hazardous waste cleanup operation.

Note
Developing a HASP is required for cleanup activities covered by HAZWOPER.

A likely second step in this cleanup operation is to evaluate the hazards involved. Recall from Chapter 2 that we need to conduct a hazard and risk assessment to determine the potential for exposure and a program to control it. This involves properly identifying the substances as described in Chapter 5. In this case, the DOT labels and the types of containers may lead us to conclude that the materials probably are those listed on the label. However, personnel should always take into account that materials may be placed into inappropriate containers. Never assume that you are dealing with a given substance until final identification and disposal of a material are made.

Safety
Conducting a hazard and risk assessment is essential to providing for the safety of everyone working on a project.

As part of the hazard and risk assessment, the requirements for protective systems are established. Because engineering controls are not feasible for managing the operations and controlling the hazards, using administrative procedures and PPE is required. A Hot Zone or Exclusionary Zone is established, and personnel are required to remain outside this area of potential harm. Decisions regarding evacuation of the surrounding areas also would take place. The HASP would identify the specific operations that were needed, and rules would be developed to further control the exposure. Personnel would always work from an upwind and upgrade position, never allowing themselves to work directly over the spilled material, and they would try to maintain a safe distance when working in proximity to it. Further, a second team of properly trained personnel may be identified to stand by in the Cold Zone and provide assistance should something unplanned occur.

Safety
Working from an upwind and uphill area provides additional protection for the entry team.

As specific actions are identified, the required PPE and tools can be selected. Selecting the level and the type of PPE is one of personal choice in many cases. Some organizations may choose to be conservative and require high levels of protection, such as the use of Level A, fully encapsulating suits, for most or all hazardous situations. While this provides a high level of protection, remember that, as we have discussed in previous chapters, such protection comes at a cost. The weather in our scenario already places personnel at risk from heat stress, and given the forecast of even higher temperatures, it may be more appropriate to select a lower level of protection or conduct the cleanup operations after dark when the weather is cooler. For the purposes of our exercise, we elect to use Level B nonencapsulating suits that have a broad range of chemical compatibility.

! Safety
Using higher levels of protection can burden personnel with hazards from heat stress.

■ Note
Scheduling nighttime activities can help prevent heat stress.

The tools in our exercise scenario will include air monitoring devices. As we learned in previous chapters, chemically protective clothing is not a barrier against and provides no protection from fire. The use of a combustible gas monitor probably will provide adequate information about the presence or lack of flammable or combustible vapors in the area. As the entry team enters the Hot Zone, they should begin monitoring and evaluating the atmosphere at several levels. As they approach the drums, they also should look for other indications that could help determine the exact material contained in the drums. Many drums bear a label that provides the product's exact name, but this label may not have been visible on initial observation.

Once it is determined that the drums can be safely approached, stop the material leaking from the overturned drum by rolling the drum into a position where the leak is above the level of the liquid. Rolling a drum has several benefits. It is a simple but effective technique for controlling leaks without exposing the drum to additional stress or strain that could cause it to fail. Many hazardous materials teams have been surprised when the drum or container they are attempting to set upright is heavier than they thought or fails catastrophically in the process. Rolling a container to get the leak area above the liquid level prevents these problems and limits the physical strain on workers who are already burdened with layers of protective clothing and equipment. Also, rolling a drum may expose previously unseen labels and provide additional information.

! Safety
Drum rolling is a way to prevent drum failure when controlling releases.

After the first drum has been rolled, the next obvious step is to seal the leak in the upright drum containing corrosive material. This often involves simply tightening the bung using the appropriate bung wrench. The wrench must be made of a non-sparking material given the flammability concerns. If sealing the bung does not seal the vapor release, using an inert putty material is another method of stopping the leak. However, the team assigned to stop the vapor release should be instructed to confirm that the material is a corrosive substance by first inserting wetted pH paper into the vapors using an extension pole. Depending on the amount and corrosivity of the vapors, the team may elect to simply let the vapors continue to release and not to build up inside the drum, which could be weakened from the effects of exposure to the environment and sunlight. Regardless of whether the drum is initially sealed by the entry team, it will be necessary for the drum's contents to be removed and placed into an appropriate new container for later disposal. This can be done using a combination of manually operated pumps that withdraw the material from its original container and place it into the new container.

Safety
Using an extension pole allows sampling from a safer distance.

overpacking
the process of placing a smaller drum inside a larger one for control or disposal

V-roll technique
a method of sliding a leaking drum into a larger salvage drum laid alongside it in a V-position

After the leak in the metal drum is stopped or controlled, that drum may be **overpacked** into a larger salvage drum so that it can be safely uprighted and later removed from a site. The HASP may indicate that the method for this process may be a **V-roll technique**, as seen in Figure 12-2. That figure shows personnel placing a salvage drum alongside the leaking drum so that the two drums create a "V" shape. Using the technique, the drum is not uprighted but simply allowed to slide into the larger drum. Once the leaking drum is rolled into the salvage drum, the lid can be placed onto the salvage drum and the drum uprighted. Using this method limits the potential for failure of leaking containers and controls any spillage that might occur in the process. Keep in mind that appropriate labels must be placed onto the outer drum to identify the contents and their hazards. At some point in the operation, a sample of the leaking material may be taken either for field analysis using the haz-cat method or for later analysis at a laboratory prior to final disposal.

Figure 12-2
An example of a V-roll technique being used at an incident. (Courtesy of the San Ramon Valley Protection District.)

> **■ Note**
> Samples of materials involved in an operation may need to be evaluated before proper disposal occurs.

Our HASP also needs to identify the type and amount of decontamination required by this operation. The decontamination program for our scenario will include an area where two pools are set up, a decontamination team of four, and some mild detergents to help remove debris from the entry personnel.

> **Safety**
> Decontamination activities must be in place prior to site entry for both emergency and nonemergency hazardous materials operations.

So, with our plan to clean up the spilled materials in place, the operations are conducted and the drums are quickly overpacked or the contents transferred to another container. Once this is done, the soil and surrounding environment is monitored to determine the extent of the contamination. Any contaminated soil, dirty dirt, is placed into appropriate containers and taken away as hazardous waste with the other materials. All containers are properly labeled, and the project is completed in just a few hours and without incident.

> **Safety**
> Never put a hazardous material into another container without labeling that outer container.

EMERGENCY RESPONSE INVOLVING AN UNKNOWN SUBSTANCE

The second of our scenarios involved an emergency response to a hazardous materials release. In this real-life case, the materials were in five metal drums that had fallen from a truck. The vehicle had left the scene, making it impossible to identify the materials by using the shipping documents. One of the drums may have had a small leak, as was evident from some liquid that seemed to be in the area under the drum. No other visible indications of releases were present, and no container labels or markings could be seen from outside the Hot Zone, which had been established by the local law enforcement prior to the arrival of the HazMat team.

Essentially, this incident required a response operation to an unknown material. This exercise scenario is challenging because in such cases, all the information required to select the proper protective equipment and determine the techniques required to manage the incident is not known. Such incidents occur on a regular basis and present a huge challenge to even the most experienced teams. Responding to these types of incidents causes considerable debate among the so-called experts because there is really no one best method to handling all of them. The techniques chosen will vary based on your team and its specific protocols and standard procedures. But again, applying the concepts you have learned in this book makes such incidents manageable.

> **■ Note**
> Most hazardous materials response activities can be done correctly using more than one tactic or technique.

When an incident such as this occurs, several key factors determine how to manage it. The first of these factors is knowledge, because knowledge is power. Simply put, the more we know, the less we have to guess and the more control we have. A good analogy is putting together a jigsaw puzzle without the picture of what the completed puzzle looks like. The first few pieces that you look at may not make a lot of sense to you. Soon you find two pieces that fit together and before you know it, the picture starts to take shape. As you continue, more pieces fit together and more of what is really there becomes clear. Only when all the pieces come together does the full picture really reveals itself. Responding to an unknown hazardous materials release is the same process. During emergency response, putting the puzzle together begins with applying the pieces of information presented in this book.

> **! Safety**
> When dealing with a hazardous materials emergency, knowledge is power and promotes safety.

> **■ Note**
> Information often comes in the form of pieces, and these pieces must be assembled much like a jigsaw puzzle.

As with the spill-cleanup scenario, the first step in responding to this second exercise may already be in place, and it may not require any brain activity at the time of the response. That step is determining whether the response falls under the scope of the HAZWOPER regulation. As we know, responses to incidental spills that do not endanger others outside the area of the release are not subject to the rules in the regulation. In cases of serious spills of unknown materials, however, it is clear that we must comply with the HAZWOPER section (q) requirements because they relate to emergency response operations. And this is actually good news because that section requires that we have trained personnel who have specific responsibilities for and constraints on what they can and cannot do. Further, section (q) of the regulation requires that a trained Incident Commander (IC) oversee the operations and that specific procedures be enacted in the response program. Recall from the discussion of the role of the IC that certain steps must be taken when an emergency response operation is initiated. Following those steps, we can apply our course material to effectively and safely manage this response operation exercise.

> **■ Note**
> Section (q) of the HAZWOPER regulation must be followed for all emergency responses to hazardous materials releases.

Step 1: Use the ICS and appoint an IC. Using the ICS is required when an emergency response to a hazardous materials release occurs. As we know, the system promotes a safe operation by tracking personnel, limiting freelancing, ensuring a manageable chain of command, providing one plan of attack, and ensuring consistency in the communication systems used at the site. In the case of a response to an off-site incident, such as the one in our scenario, the incident may fall under the jurisdiction of one or more public safety agencies, such as police or fire departments. Often, using a Unified Command structure that includes representatives of both is a more appropriate system because response operations to complex incidents can require multiple agencies to secure the scene and control the release. For our scenario, a Unified Command structure may be set up initially.

With the Unified Command established, as more units arrive at the scene, various sections are set up. Recall that the ICS is modular, so we can choose to implement only those modules that are required. Given the complexity of this situation, various members of the command staff may initially be appointed. Additionally, a division to secure the site may be initially set up, and law enforcement personnel may assist with this. As other units arrive, they are directed to report to a Staging Area that can be set up near the incident site in a location from which they can be dispatched.

Keep in mind that OSHA mandates that personnel who serve in the role of the IC must be certified to do so. Certification is accomplished by meeting the initial and annual training requirements identified in section (q)(6) of the HAZWOPER regulation. In a Unified Command system, all personnel filling the role of the IC should meet the training and certification requirements.

> **■ Note**
> Operations involving multiple jurisdictions lend themselves to the use of a Unified Command.

Step 2: Identify the hazards and implement controls. A common sense approach to any hazardous materials emergency is to get as many pieces of the puzzle together as you can and start to assemble them before taking action. Remember that knowledge is power and with that power comes your ability to be in control of the circumstances rather than a victim of them. Getting useful information quickly is important to identifying the hazardous conditions that may be present. In some cases, such as in our scenario, the drums may contain nonhazardous materials. But we cannot know that without information. And in our scenario, we do not have a lot of information about the hazards that might be present. In a fixed-facility incident where the inventory and types of materials are known, getting this information is far easier. For example, someone at the site may understand the materials' properties and know where and how the materials are stored. In a transportation incident such as in our scenario, however, we must rely on other information indicators, although it is always helpful to determine whether witnesses can provide information about the vehicle involved and what they observed.

> **■ Note**
> Gathering information at a transportation emergency is generally more challenging than at a fixed facility incident.

Indications about what might be in the drums come primarily from the containers themselves and the area where they are located. Try to determine which type of drum is involved and what its construction material is. Find out whether any facilities in the area commonly use these types of drums. Are there labels, signs, or other markings on the drums that indicate which material it contains? Are the containers leaking and, if so, is there any visible indication of the reaction of the material with air or the environment where the release occurs? Essentially, you must become a detective, quickly assembling the pieces of the puzzle.

Safety
Become a HazMat detective when dealing with unknown materials. Ask a lot of questions and get as much information as possible about what the material might be or might not be.

If the exact material or its hazard class cannot be determined prior to site entry, it will be necessary to enter the site to gain further information. In the next step, we will discuss techniques for the selection of PPE for this incident, but as part of the hazard assessment, one of the early operations may involve obtaining samples of the air around the drums and even some of the material itself. The entry team can test the material with standard pH paper to determine whether the substance is corrosive. Additional test strips can be used to quickly evaluate other properties or components of the material. A sample may need to be obtained so that a haz-cat process can be performed on it or so that it can be sent to a laboratory for analysis using more sophisticated testing equipment than is available in the field. Regardless, until you know which material is present, all actions must be tempered to prevent unnecessary exposure because our selection of PPE for response personnel is largely a guess until more information becomes available. Remember that there is no chemical-proof PPE and that the resistance of a suit to materials is dependent on the suit and the material involved. As we know, some materials can degrade PPE and immediately expose the person wearing the PPE, so limit the entry approach until samples are obtained and more is known about the materials you are dealing with. As further information is obtained, additional controls can be implemented, so initial actions should be geared toward identifying what is present and its hazards to personnel and the environment.

Safety
Initial efforts during emergency releases should be directed at determining which material is present. Until that is done, response efforts must be conservative.

Step 3: Select the appropriate level of protection and PPE for personnel involved in the incident. This step is significantly complicated in situations involving unknown materials. In our exercise, we know nothing about the material in the drums or about the hazards it poses. The substance might not even present a hazard, but assuming this without confirmation could result in serious consequences, including the exposure or death of the entry team members. Because of the limitations of PPE, responding to unknown materials must be done slowly and methodically and with the goal of obtaining information about a material while staying as far away as possible from it. Some technical teams may have access to a mechanical robot that can be remotely controlled from a safe area to enter the release site and obtain a sample of the material. Such machines are not commonly available, however. In most cases such as ours, therefore, the entry team will have to evaluate the site conditions as they enter the area and note any changes that occur. Wetted-down pH strips attached to the front of the entry team suits, the use of multi-gas monitors, and other equipment should be taken into the scene during the initial entry to obtain more information as the incident progresses.

> **Safety**
> Monitoring and sampling equipment are required on initial entry into an area where unknown materials are present.

When you are dealing with an unknown material, there are no hard and fast rules for selecting PPE suits. Many hazardous material emergency response organizations have identified the limitations of their protective equipment and know which pieces of equipment provide protection against the broadest range of materials. These organizations may know, for example, that one type of glove or a particular suit protects against a whole range of solvents, corrosives, and other common materials, so those gloves or that suit may be selected.

When PPE is selected on the basis of an educated guess, it is critical that other protective measures be taken. The entry team should monitor each other at each step of the entry and exploration process. Their approach to the release site should be carefully and slowly measured, and they should stay upwind and upgrade if possible. Indications of what is present may be so subtle that they can be missed by a team that is not paying attention to everything around them.

> **Safety**
> When you are dealing with an unknown substance, select a PPE suit with the broadest range of resistance to the types of materials most likely to be present.

The minimum level of protection selected for any response to an unknown material release should be Level B. Some organizations require the use of Level A for such operations, although this may be overkill during outdoor incidents in which the concentration of materials is limited. Remember that the higher level of protection come with increased problems from limited communication and loss of awareness of the surrounding area. As we have said, Level B protection may actually be safer in some cases, especially during outdoor incidents involving only minimal leaking.

> **Safety**
> Level B is the lowest level of protection for working with unknown materials.

> **Note**
> Level A protection is the highest level of protection, but may not always be the safest level.

Step 4: Limit personnel exposure by minimizing the number of personnel involved in the site entry and operations. This step may seem like common sense action, but it can create considerable issues. When confronted with an unknown situation, the people in charge may feel an urgent need to get as much information as quickly as possible to manage the emergency. They may therefore elect to send multiple teams into the Hot Zone to get that information. One of the issues associated with fire department response to a hazardous materials release is that the time factors of a hazardous materials operation and a fire operation differ. In most fire situations, seconds count because fires grow exponentially with each passing minute. Fire personnel often are action oriented, and given that they are well protected from the hazards associated with fire operations, they can quickly and safely respond to such incidents. However, incidents involving a hazardous material cannot and should not be responded to that quickly, and this is even truer when the material is unknown. Remember, our response efforts need to be like playing a game of chess or assembling a jigsaw puzzle—they need to be slow, methodical, and always forward thinking.

The minimum number for an entry team is two.

Safety

Do not expose more personnel to the hazard than necessary, especially if that hazard is unknown.

The number of personnel entering a site can be only so limited, however, because HAZWOPER requires that we do not commit an individual to entering a hazardous area alone. As we know, personnel working in the Hot Zone must be part of a team of two or more individuals. Typically, teams of two entry personnel are used, although teams with three or more people may be required for more complex operations.

Last, ensure that personnel entering the Hot Zone to perform *offensive actions* are appropriately trained and certified to the correct level. In our exercise, the IC must allow only responders qualified as either Hazardous Materials Technicians or Specialists to conduct these actions.

Safety

Personnel trained to the Technician or Specialist levels are authorized to enter the Hot Zone and conduct offensive actions.

Step 5: Assign backup personnel to further support entry team operations. Before entry is made into the Hot Zone, the IC must designate a backup team to assist in the operations or rescue of entry team personnel should this be necessary. The backup team should be located in the Cold Zone and be outfitted with levels of protection similar to those worn by the entry team. There is some debate as to how many people should make up backup teams, but generally a team of no fewer than two persons should be assigned to this function. Given that these personnel could be used for Hot Zone entry, training and certification to the Technician or Specialist level is required, just as it is for the entry team.

Backup teams should contain no fewer than two persons.

When setting up the system for managing our exercise scenario, the IC should make contingency plans for dealing with possible injuries. We know that OSHA requires that we have "qualified basic life support personnel . . . with medical equipment and transportation capability" standing by at scenes such as the one in our exercise. If such equipment and staff are not present, the IC should immediately request the dispatch of an ambulance or similar unit. That unit may include a paramedic team and should be on site and available to provide appropriate care and transportation. In many cases, however, even in the event of injury, the basic life support units may not be used. The incident management team should provide paramedics with information on the materials that may be present as soon as possible. This provides paramedics with time to research the potential symptoms of exposure and conditions that might be aggravated by exposure to a given material. If a paramedic team is on site, they may also establish contact with the base hospital to begin a dialogue with the staff regarding possible exposures and treatment. Again, the more that is done in advance, the better the results that likely will occur.

Note
As with other operations, it is important that the medical response personnel be kept informed on the progress of identifying the substances present.

Step 6: Assign a qualified Safety Officer and provide him with the authority needed to ensure safety at the scene. As we learned in Chapter 11, an ICS position that often is filled in emergency response situations involving hazardous materials is that of Safety Officer. As we know, this person has broad authority and is responsible for helping assess the hazards presented by the emergency situation and the operations being conducted. The Safety Officer in our scenario may be assigned initially to work in the command post, which should be established near the incident. As the incident operations begin, the Safety Officer may roam the area and monitor the activities, ensuring that the actions being conducted are done in a safe manner. Should Safety Officers note any problems or issues, they have the full authority of the IC to immediately terminate the activities and regroup the teams until a plan is developed to manage the new challenges.

In our scenario, which involves unknown materials, it is vital that the Safety Officer be informed of all new information that could aid in the material identification process. The Safety Officer also may assist in timing the entry team as they begin using the air in their self-contained breathing apparatus (SCBA) units.

Safety
When SCBA equipment is used, someone must be responsible to track the time that personnel are using the breathing air.

Step 7: Ensure that decontamination programs are in place (Figure 12-3). Because emergency activities present a high risk of potential exposure to personnel, an early step in the incident management process is to establish and implement a decontamination program prior to Hot Zone entry. When a situation involves unknown materials, decontamination must be more thorough than it might otherwise require. In our case, a two-pool decontamination system is chosen and four personnel will staff the various stations. Various detergents will be available for use by the decontamination team should they be needed.

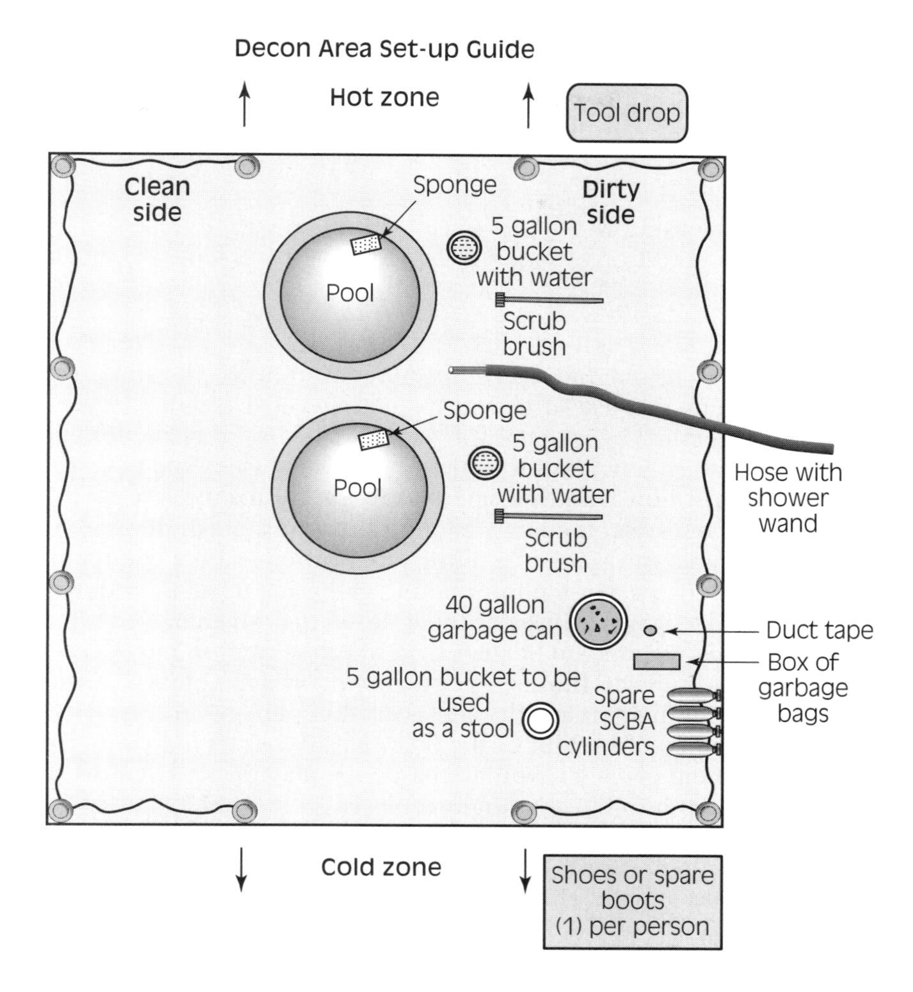

Figure 12-3
A decontamination area set up for emergency response personnel.

> **Safety**
> Decontamination activities must be more thorough when an unknown substance is being dealt with.

Specific Tactics and Operations

With the required elements of the response system in place, the IC may elect to begin breaking up the organizational structure to maintain an effective span of control. Initially, a Safety Officer, an Information Officer, and a Liaison should be assigned. Recall that the people in these positions make up the Command Staff and work directly for the IC without the need for supervision. These key positions allow the IC to focus on the task at hand, which in the case of an unknown material can be daunting.

As the response to an incident evolves, an IC has several options for breaking up the ICS to ensure appropriate supervision and organization. Because our incident is limited to five drums that are not believed to be leaking at this point, the IC may elect to appoint a HazMat group with a leader to oversee the entry and decontamination teams, a security group with a law enforcement team leader, and a medical team led by one of the paramedics. Using this system, the IC (or Unified Command ICs) has only three people reporting directly to him. This is well within the span of control for emergency operations and limits the number of additional personnel who are needed for the activities. If the incident becomes more complex, the ICS can expand and include members of the General Staff who can assist in the major areas of operations, planning, and logistics.

Note
Because of its flexibility, the ICS allows management organizations to grow as an incident grows in complexity.

For our incident, the PPE selection for the unknown materials should include the use of an encapsulating Level B suit that has the broadest range of chemical resistance and gloves with similar characteristics. Some organizations require that several layers of gloves be used when dealing with unknown materials; however, this can lead to a significant decrease in the wearer's dexterity. In our case, the entry team will try to limit their approach to the area with the primary mission of gaining information as they conduct their initial entry.

As part of the information gathering mission in our scenario, the entry team will always approach the area taking advantage of wind and terrain. Because most vapors are heavier than air, they tend to move downgrade and be affected by wind conditions. The entry team will maintain a safe distance from the drums but will approach the area and attempt to get information on the condition of the drums, the extent of the leak that was observed, and any reactions that may be taking place, as well as get any data available from the monitoring instruments taken into the area. In our mission, the team has taken in a multi-gas meter, pH paper, a small scoop on a long handle that can be used to pick up a sample of the spilled material, a container for the sample, and test strips to indicate the material's general characteristics. Test strips are made by a variety of manufacturers and can be used like pH paper strips. In this case, some strips will indicate the presence of specific materials such as chlorine, fluorine, organic materials (such as solvents), and other common substances. It is important to remember that the initial entry is done for the purpose of gathering information. There is no hurry to stop the spill until it is confirmed that this can be done safely.

Safety
Information gathering is a priority when you are responding to the release of an unknown material. Until this is done, other actions should be limited.

Depending on the information gathered, the response efforts will be directed at ensuring the safety of the responders while containing the spill. If testing or haz-cating provides a more definitive analysis of the material, the entry team can then stop the release from the drum using a variety of options, including rolling the drum, overpacking, tightening the bung, or placing inert plugs into the damaged area. Plugs of a variety of materials, including wood, plastics, and a range of rubber substances, can be inserted into small holes on drums to provide a fairly secure seal. Once the drum is sealed and the leak stopped, the incident can be declared stable and a cleanup contractor can be called to the scene to help properly dispose of the materials. The entry team, other involved personnel, and their equipment will be appropriately decontaminated, and everyone can return to their stations. At some later point, a debriefing of the response efforts will be conducted to ensure that the event was appropriately managed and that all necessary steps were taken to ensure safety and an effective response effort.

> **■ Note**
> Debriefing personnel involved in an incident provides vital information to help future response operations.

CONCLUSION

Responding to hazardous materials releases during emergencies and at site operations where hazardous materials and wastes are present can be complicated and present a range of challenges for even the most veteran of personnel. However, as we have seen throughout this book, the basic concepts of handling such incidents are well within most people's ability to comprehend. By applying these concepts we can safely and effectively conduct the operations that are required to control the release situation. As we have seen, one does not need to be a chemist or a chemical engineer to work in this industry. And while advanced degrees may be of value in some areas, most workers in the fields of hazardous materials operations and emergency response become qualified on the basis of a focused course of study such as the one contained in this text.

However, the industry is evolving, and as technology continues to change, the equipment that we use continues to evolve. As this occurs, the price of that equipment is reduced, making it is available to more people who may use it on a regular basis.

It is incumbent on everyone in any course of study in this field to look at that study as a journey, one that is never ending and that requires close attention and ongoing learning. While the concepts discussed in this text serve as the foundation of that learning, their application will change as technology allows us to learn more quickly. Regardless of whether your job is responding to emergencies involving hazardous materials or working in the hazardous waste industry in some capacity, it is vital that you never let your guard down—always presume that the materials that you work with can hurt you even if you have not been hurt by them previously. Most importantly, continue learning.

> **■ Note**
> Hazardous materials personnel must continue to expand their knowledge and continue to learn because the industry continues to evolve.

SUMMARY

- Most spills of hazardous materials are not emergency response operations and may not be subject to the application of the HAZWOPER regulation. However, spills of hazardous materials that pose an unreasonable threat to the environment or to the public are covered by HAZWOPER.
- The key to developing any cleanup plan is ensuring that the concepts and safeguards presented in this book are considered and factored into the final operations.
- An important step in any cleanup operation is determining whether the activities fall within the scope of the HAZWOPER regulation. The HAZWOPER regulation includes:
  - ☐ Training levels
  - ☐ Supervising the work

- ☐ Developing a HASP
- ☐ A hazard and risk assessment
- ☐ Identifying the PPE needed
- ☐ Monitoring and sampling
- ☐ Setting up the decontamination process

- When confronting a spill from a drum that has fallen, use proper drum rolling techniques, such as the V-roll technique, to upright the drum and seal it in an overpack drum.
- Emergency response to a spill or release of a hazardous material also falls within the scope of the HAZWOPER regulation. Steps involved in HAZWOPER emergency response include:
  - ☐ Using the ICS
  - ☐ Identifying the hazards and use of controls
  - ☐ Selecting PPE
  - ☐ Identifying the entry team
  - ☐ Ensuring that Hot Zone personnel perform offensive actions to their level of training
  - ☐ Assigning personnel as needed to ensure that the span of control is not exceeded and that all activities are properly supervised
- Ensure the safety of the site by assigning a Safety Officer for all complex operations.
- Limit actions taken by personnel until monitoring, sampling, and information gathering determine the types of materials present and their hazards.
- A debriefing of personnel involved in the incident provides vital information to improve future response operations.

REVIEW QUESTIONS

1. Explain how to stop a leak in a drum and contain it in a larger salvage or overpack drum.
2. You are at the site of a vehicle accident where a single truck carrying drums in an open trailer has rolled over onto its side, allowing drums to roll out of the trailer onto the road. The driver is alive, but unconscious. There is an odor similar to rotten eggs in the air. You do not see any markings on the drums. What steps would you take to identify what is in the drums?
3. In working with unknown materials, what is the lowest level of protection you would select to use? Explain why this level is most often the appropriate level used. Identify circumstances where higher levels of protection would be indicated.
4. Discuss why a Safety Officer is needed during an incident involving a hazardous material release. Cite at least two examples to support your conclusion.

ACTIVITY

1. *Unknown Spill.* You have been called to investigate an odor emitting from the cargo area of a large delivery truck. The truck is parked at the rear of the building by the loading ramp. The driver tells you that he started feeling dizzy as he was unloading some packages. The truck has a red placard with the number "3" on the bottom, which you can see from a distance. Other people have also stated that they can smell something like

"paint thinner" in the air. The driver is unable to help you with any further identification of the contents of the truck.

You are the IC for this incident. You are responsible for developing an Incident Action Plan for your response that should include the following components:

a. What are your initial actions when arriving at the scene?
b. How would you determine which exact materials are involved? What if that information was not immediately available? What other actions may be required to determine which materials are involved?
c. What level of protection would be required for response personnel?
d. Diagram your ICS for the incident. Include all major groups and positions.

2. *Earthquake.* There has been a major earthquake in the region where you work. The magnitude was such that a laboratory in the area has sustained extensive damage. Materials inside the laboratory have been thrown about in a large room. Experiments that were being conducted were also interrupted, and the materials involved in the experiments are among the spilled materials.

Your firm has been called to the site to clean up the mess left by this natural disaster. Assume that no ongoing reactions or other conditions that would constitute an emergency response are present. As the site Safety Officer, develop a short version of a HASP for this cleanup. Make sure to include the following in your planning:

a. Because power is out to the area, emergency power is required. What issues should be addressed to provide adequate lighting for the project?
b. What level of protection is required for initial entry into the site? Which activities would be conducted first, second, third, etc.?
c. What type of decontamination operations would be required for this operation?
d. How would you identify the materials involved in this operation?

HAZARDOUS WASTE OPERATIONS AND EMERGENCY RESPONSE

29 CFR, Part 1910.120

(a) Scope, application, and definitions.

(1) Scope. This section covers the following operations, unless the employer can demonstrate that the operation does not involve employee exposure or the reasonable possibility for employee exposure to safety or health hazards:

(i) Clean-up operations required by a governmental body, whether Federal, state, local or other involving hazardous substances that are conducted at uncontrolled hazardous waste sites (including, but not limited to, the EPA's National Priority Site List (NPL), state priority site lists, sites recommended for the EPA NPL, and initial investigations of government identified sites which are conducted before the presence or absence of hazardous substances has been ascertained);

(ii) Corrective actions involving clean-up operations at sites covered by the Resource Conservation and Recovery Act of 1976 (RCRA) as amended (42 U.S.C. 6901 et seq);

(iii) Voluntary clean-up operations at sites recognized by Federal, state, local or other governmental bodies as uncontrolled hazardous waste sites;

(iv) Operations involving hazardous waste that are conducted at treatment, storage, disposal (TSD) facilities regulated by 40 CFR Parts 264 and 265 pursuant to RCRA; or by agencies under agreement with U.S.E.P.A. to implement RCRA regulations; and

(v) Emergency response operations for releases of, or substantial threats of releases of, hazardous substances without regard to the location of the hazard.

(2) Application.

(i) All requirements of Part 1910 and Part 1926 of Title 29 of the Code of Federal Regulations apply pursuant to their terms to hazardous waste and emergency response operations whether covered by this section or not.

If there is a conflict or overlap, the provision more protective of employee safety and health shall apply without regard to 29 CFR 1910.5(c)(1).

(ii) Hazardous substance clean-up operations within the scope of paragraphs (a)(1)(i) through (a)(1)(iii) of this section must comply with all paragraphs of this section except paragraphs (p) and (q).

(iii) Operations within the scope of paragraph (a)(1)(iv) of this section must comply only with the requirements of paragraph (p) of this section.

Notes and Exceptions:

(A) All provisions of paragraph (p) of this section cover any treatment, storage or disposal (TSD) operation regulated by 40 CFR parts 264 and 265 or by state law authorized under RCRA, and required to have a permit or interim status from EPA pursuant to 40 CFR 270.1 or from a state agency pursuant to RCRA.

(B) Employers who are not required to have a permit or interim status because they are conditionally exempt small quantity generators under 40 CFR 261.5 or are generators who qualify under 40 CFR 262.34 for exemptions from regulation under 40 CFR parts 264, 265 and 270 ("excepted employers") are not covered by paragraphs (p)(1) through (p)(7) of this section. Excepted employers who are required by the EPA or state agency to have their employees engage in emergency response or who direct their employees to engage in emergency response are covered by paragraph (p)(8) of this section, and cannot be exempted by (p)(8)(i) of this section.

(C) If an area is used primarily for treatment, storage or disposal, any emergency response operations in that area shall comply with paragraph (p)(8) of this section. In other areas not used primarily for treatment, storage, or disposal, any emergency response operations shall comply with paragraph (q) of this section. Compliance with the requirements of paragraph (q) of this section shall be deemed to be in compliance with the requirements of paragraph (p)(8) of this section.

(iv) Emergency response operations for releases of, or substantial threats of releases of, hazardous substances which are not covered by paragraphs (a)(1)(i) through (a)(1)(iv) of this section must only comply with the requirements of paragraph (q) of this section.

(3) Definitions—

Buddy system means a system of organizing employees into work groups in such a manner that each employee of the work group is designated to be observed by at least one other employee in the work group. The purpose of the buddy system is to provide rapid assistance to employees in the event of an emergency.

Clean-up operation means an operation where hazardous substances are removed, contained, incinerated, neutralized, stabilized, cleared-up, or in any other manner processed or handled with the ultimate goal of making the site safer for people or the environment.

Decontamination means the removal of hazardous substances from employees and their equipment to the extent necessary to preclude the occurrence of foreseeable adverse health effects.

Emergency response or responding to emergencies means a response effort by employees from outside the immediate release area or by other designated responders (i.e., mutual aid groups, local fire departments, etc.) to an occurrence which results, or is likely to result, in an uncontrolled release of a hazardous substance. Responses to incidental releases of hazardous substances where the substance can be absorbed, neutralized, or otherwise controlled at the time of release by employees in the immediate release area, or by maintenance personnel are not considered to be emergency responses within the scope of this standard. Responses to releases of hazardous substances where there is no potential safety or health hazard (i.e., fire, explosion, or chemical exposure) are not considered to be emergency responses.

Facility means (A) any building, structure, installation, equipment, pipe or pipeline (including any pipe into a sewer or publicly owned treatment works), well, pit, pond, lagoon, impoundment, ditch, storage container, motor vehicle, rolling stock, or aircraft, or (B) any site or area where a hazardous substance has been deposited, stored, disposed of, or placed, or otherwise come to be located; but does not include any consumer product in consumer use or any water-borne vessel.

Hazardous materials response (HAZMAT) team means an organized group of employees, designated by the employer, who are expected to perform work to handle and control actual or potential leaks or spills of hazardous substances requiring possible close approach to the substance. The team members perform responses to releases or potential releases of hazardous substances for the purpose of control or stabilization of the incident. A HAZMAT team is not a fire brigade nor is a typical fire brigade a HAZMAT team. A HAZMAT team, however, may be a separate component of a fire brigade or fire department.

Hazardous substance means any substance designated or listed under (A) through (D) of this definition, exposure to which results or may result in adverse effects on the health or safety of employees:

[A] Any substance defined under section 101(14) of CERCLA;

[B] Any biologic agent and other disease causing agent which after release into the environment and upon exposure, ingestion, inhalation, or assimilation into any person, either directly from the environment or indirectly by ingestion through food chains, will or may reasonably be anticipated to cause death, disease, behavioral abnormalities, cancer, genetic mutation, physiological malfunctions (including malfunctions in reproduction) or physical deformations in such persons or their offspring.

[C] Any substance listed by the U.S. Department of Transportation as hazardous materials under 49 CFR 172.101 and appendices; and

[D] Hazardous waste as herein defined.

Hazardous waste means—

[A] A waste or combination of wastes as defined in 40 CFR 261.3, or

[B] Those substances defined as hazardous wastes in 49 CFR 171.8.

Hazardous waste operation means any operation conducted within the scope of this standard.

Hazardous waste site or Site means any facility or location within the scope of this standard at which hazardous waste operations take place.

Health hazard means a chemical, mixture of chemicals or a pathogen for which there is statistically significant evidence based on at least one study conducted in accordance with established scientific principles that acute or chronic health effects may occur in exposed employees. The term "health hazard" includes chemicals which are carcinogens, toxic or highly toxic agents, reproductive toxins, irritants, corrosives, sensitizers, hepatotoxins, nephrotoxins, neurotoxins, agents which act on the hematopoietic system, and agents which damage the lungs, skin, eyes, or mucous membranes. It also includes stress due to temperature extremes. Further definition of the terms used above can be found in Appendix A to 29 CFR 1910.1200.

IDLH or Immediately dangerous to life or health means an atmospheric concentration of any toxic, corrosive or asphyxiant substance that poses an immediate threat to life or would interfere with an individual's ability to escape from a dangerous atmosphere.

Oxygen deficiency means that concentration of oxygen by volume below which atmosphere supplying respiratory protection must be provided. It exists in atmospheres where the percentage of oxygen by volume is less than 19.5 percent oxygen.

Permissible exposure limit means the exposure, inhalation or dermal permissible exposure limit specified in 29 CFR Part 1910, Subparts G and Z.

Published exposure level means the exposure limits published in "NIOSH Recommendations for Occupational Health Standards" dated 1986, which is incorporated by reference as specified in § 1910.6, or if none is specified, the exposure limits published in the standards specified by the American Conference of Governmental Industrial Hygienists in their publication "Threshold Limit Values and Biological Exposure Indices for 1987–88" dated 1987, which is incorporated by reference as specified in § 1910.6.

Post emergency response means that portion of an emergency response performed after the immediate threat of a release has been stabilized or eliminated and clean-up of the site has begun. If post emergency response is performed by an employer's own employees who were part of the initial emergency response, it is considered to be part of the initial response and not post emergency response. However, if a group of an employer's own employees, separate from the group providing initial response, performs the clean-up operation, then the separate group of employees would be considered to be performing post-emergency response and subject to paragraph (q)(11) of this section.

Qualified person means a person with specific training, knowledge and experience in the area for which the person has the responsibility and the authority to control.

Site safety and health supervisor (or official) means the individual located on a hazardous waste site who is responsible to the employer and has the authority and knowledge necessary to implement the site safety and health plan and verify compliance with applicable safety and health requirements.

Small quantity generator means a generator of hazardous wastes who in any calendar month generates no more than 1,000 kilograms (2,205) pounds of hazardous waste in that month.

Uncontrolled hazardous waste site means an area identified as an uncontrolled hazardous waste site by a governmental body, whether Federal, state, local or other where an accumulation of hazardous substances creates a threat to the health and safety of individuals or the environment or both. Some sites are found on public lands such as those created by former municipal, county or state landfills where illegal or poorly managed waste disposal has taken place. Other sites are found on private property, often belonging to generators or former generators of hazardous substance wastes. Examples of such sites include, but are not limited to, surface impoundments, landfills, dumps, and tank or drum farms. Normal operations at TSD sites are not covered by this definition.

(b) Safety and health program.

NOTE TO (b): Safety and health programs developed and implemented to meet other federal, state, or local regulations are considered acceptable in meeting this requirement if they cover or are modified to cover the topics required in this paragraph. An additional or separate safety and health program is not required by this paragraph.

(1) General.

(i) Employers shall develop and implement a written safety and health program for their employees involved in hazardous waste operations. The program shall be designed to identify, evaluate, and control safety and health hazards, and provide for emergency response for hazardous waste operations.

(ii) The written safety and health program shall incorporate the following:

(A) An organizational structure;

(B) A comprehensive workplan;

(C) A site-specific safety and health plan which need not repeat the employer's standard operating procedures required in paragraph (b)(1)(ii)(F) of this section;

(D) The safety and health training program;

(E) The medical surveillance program;

(F) The employer's standard operating procedures for safety and health; and

(G) Any necessary interface between general program and site specific activities.

(iii) Site excavation. Site excavations created during initial site preparation or during hazardous waste operations shall be shored or sloped as appropriate to prevent accidental collapse in accordance with Subpart P of 29 CFR Part 1926.

(iv) Contractors and sub-contractors. An employer who retains contractor or sub-contractor services for work in hazardous waste operations shall inform those contractors, sub-contractors, or their representatives of the site emergency response procedures and any potential fire, explosion, health, safety or other hazards of the hazardous waste operation that have been identified by the employer's information program.

(v) Program availability. The written safety and health program shall be made available to any contractor or subcontractor or their representative who will be involved with the hazardous waste operation; to employees; to employee designated representatives; to OSHA personnel, and to personnel of other Federal, state, or local agencies with regulatory authority over the site.

(2) Organizational structure part of the site program.

(i) The organizational structure part of the program shall establish the specific chain of command and specify the overall responsibilities of supervisors and employees. It shall include, at a minimum, the following elements:

(A) A general supervisor who has the responsibility and authority to direct all hazardous waste operations.

(B) A site safety and health supervisor who has the responsibility and authority to develop and implement the site safety and health plan and verify compliance.

(C) All other personnel needed for hazardous waste site operations and emergency response and their general functions and responsibilities.

(D) The lines of authority, responsibility, and communication.

(ii) The organizational structure shall be reviewed and updated as necessary to reflect the current status of waste site operations.

(3) Comprehensive workplan part of the site program. The comprehensive workplan part of the program shall address the tasks and objectives of the site operations and the logistics and resources required to reach those tasks and objectives.

(i) The comprehensive workplan shall address anticipated clean-up activities as well as normal operating procedures which need not repeat the employer's procedures available elsewhere.

(ii) The comprehensive workplan shall define work tasks and objectives and identify the methods for accomplishing those tasks and objectives.

(iii) The comprehensive workplan shall establish personnel requirements for implementing the plan.

(iv) The comprehensive workplan shall provide for the implementation of the training required in paragraph (e) of this section.

(v) The comprehensive workplan shall provide for the implementation of the required informational programs required in paragraph (i) of this section.

(vi) The comprehensive workplan shall provide for the implementation of the medical surveillance program described in paragraph (f) if this section.

(4) Site-specific safety and health plan part of the program.—

(i) General. The site safety and health plan, which must be kept on site, shall address the safety and health hazards of each phase of site operation and include the requirements and procedures for employee protection.

(ii) Elements. The site safety and health plan, as a minimum, shall address the following:

(A) A safety and health risk or hazard analysis for each site task and operation found in the workplan.

(B) Employee training assignments to assure compliance with paragraph (e) of this section.

(C) Personal protective equipment to be used by employees for each of the site tasks and operations being conducted as required by the personal protective equipment program in paragraph (g)(5) of this section.

(D) Medical surveillance requirements in accordance with the program in paragraph (f) of this section.

(E) Frequency and types of air monitoring, personnel monitoring, and environmental sampling techniques and instrumentation to be used, including methods of maintenance and calibration of monitoring and sampling equipment to be used.

(F) Site control measures in accordance with the site control program required in paragraph (d) of this section.

(G) Decontamination procedures in accordance with paragraph (k) of this section.

(H) An emergency response plan meeting the requirements of paragraph (l) of this section for safe and effective responses to emergencies, including the necessary PPE and other equipment.

(I) Confined space entry procedures.

(J) A spill containment program meeting the requirements of paragraph (j) of this section.

(iii) Pre-entry briefing. The site specific safety and health plan shall provide for pre-entry briefings to be held prior to initiating any site activity, and at such other times as necessary to ensure that employees are apprised of the site safety and health plan and that this plan is being followed. The information and data obtained from site characterization and analysis work required in paragraph (c) of this section shall be used to prepare and update the site safety and health plan.

(iv) Effectiveness of site safety and health plan. Inspections shall be conducted by the site safety and health supervisor or, in the absence of that individual, another individual who is knowledgeable in occupational safety and health, acting on behalf of the employer as necessary to determine the effectiveness of the site safety and health plan. Any deficiencies in the effectiveness of the site safety and health plan shall be corrected by the employer.

(c) Site characterization and analysis

(1) General. Hazardous waste sites shall be evaluated in accordance with this paragraph to identify specific site hazards and to determine the appropriate safety and health control procedures needed to protect employees from the identified hazards.

(2) Preliminary evaluation. A preliminary evaluation of a site's characteristics shall be performed prior to site entry by a qualified person in order to aid in the selection of appropriate employee protection methods prior to site entry. Immediately after initial site entry, a more detailed evaluation of the site's specific characteristics shall be performed by a qualified person in order to further identify existing site hazards and to further aid in the selection of the appropriate engineering controls and personal protective equipment for the tasks to be performed.

(3) Hazard identification. All suspected conditions that may pose inhalation or skin absorption hazards that are immediately dangerous to life or health (IDLH) or other conditions that may cause death or serious harm shall be identified during the preliminary survey and evaluated during the detailed survey. Examples of such hazards include, but are not limited to, confined space entry, potentially explosive or flammable situations, visible vapor clouds, or areas where biological indicators such as dead animals or vegetation are located.

(4) Required information. The following information to the extent available shall be obtained by the employer prior to allowing employees to enter a site:

(i) Location and approximate size of the site.

(ii) Description of the response activity and/or the job task to be performed.

(iii) Duration of the planned employee activity.

(iv) Site topography and accessibility by air and roads.

(v) Safety and health hazards expected at the site.

(vi) Pathways for hazardous substance dispersion.

(vii) Present status and capabilities of emergency response teams that would provide assistance to on-site employees at the time of an emergency.

(viii) Hazardous substances and health hazards involved or expected at the site and their chemical and physical properties.

(5) Personal protective equipment. Personal protective equipment (PPE) shall be provided and used during initial site entry in accordance with the following requirements:

(i) Based upon the results of the preliminary site evaluation, an ensemble of PPE shall be selected and used during initial site entry which will provide protection to a level of exposure below permissible exposure limits and published exposure levels for known or suspected hazardous substances and health hazards and which will provide protection against other known and suspected hazards identified during the preliminary site evaluation. If there is no permissible exposure limit or published exposure level, the employer may use other published studies and information as a guide to appropriate personal protective equipment.

(ii) If positive-pressure self-contained breathing apparatus is not used as part of the entry ensemble, and if respiratory protection is warranted by the potential hazards identified during the preliminary site evaluation, an escape self-contained breathing apparatus of at least five minutes' duration shall be carried by employees during initial site entry.

(iii) If the preliminary site evaluation does not produce sufficient information to identify the hazards or suspected hazards of the site an ensemble providing equivalent to Level B PPE shall be provided as minimum protection, and direct reading instruments shall be used as appropriate for identifying IDLH conditions. (See Appendix B for guidelines on Level B protective equipment.)

(iv) Once the hazards of the site have been identified, the appropriate PPE shall be selected and used in accordance with paragraph (g) of this section.

(6) Monitoring. The following monitoring shall be conducted during initial site entry when the site evaluation produces information which shows the potential for ionizing radiation or IDLH conditions, or when the site information is not sufficient reasonably to eliminate these possible conditions:

(i) Monitoring with direct reading instruments for hazardous levels of ionizing radiation.

(ii) Monitoring the air with appropriate direct reading test equipment for (i.e., combustible gas meters, detector tubes) for IDLH and other conditions that may cause death or serious harm (combustible or explosive atmospheres, oxygen deficiency, toxic substances.)

(iii) Visually observing for signs of actual or potential IDLH or other dangerous conditions.

(iv) An ongoing air monitoring program in accordance with paragraph (h) of this section shall be implemented after site characterization has determined the site is safe for the start-up of operations.

(7) Risk identification. Once the presence and concentrations of specific hazardous substances and health hazards have been established, the risks associated with these substances shall be identified. Employees who will be working on the site shall be informed of any risks that have been identified. In situations covered by the Hazard Communication Standard, 29 CFR 1910.1200, training required by that standard need not be duplicated.

NOTE TO PARAGRAPH (c)(7).—Risks to consider include, but are not limited to:

[a] Exposures exceeding the permissible exposure limits and published exposure levels.

[b] IDLH Concentrations.

[c] Potential Skin Absorption and Irritation Sources.

[d] Potential Eye Irritation Sources.

[e] Explosion Sensitivity and Flammability Ranges.

[f] Oxygen deficiency.

(8) Employee notification. Any information concerning the chemical, physical, and toxicologic properties of each substance known or expected to be present on site that is available to the employer and relevant to the duties an employee is expected to perform shall be made available to the affected employees prior to the commencement of their work activities. The employer may utilize information developed for the hazard communication standard for this purpose.

(d) Site control.

(1) General. Appropriate site control procedures shall be implemented to control employee exposure to hazardous substances before clean-up work begins.

(2) Site control program. A site control program for protecting employees which is part of the employer's site safety and health program required in paragraph (b) of this section shall be developed during the planning stages of a hazardous waste clean-up operation and modified as necessary as new information becomes available.

(3) Elements of the site control program. The site control program shall, as a minimum, include: A site map; site work zones; the use of a "buddy system"; site communications including alerting means for emergencies; the standard

operating procedures or safe work practices; and, identification of the nearest medical assistance. Where these requirements are covered elsewhere they need not be repeated.

(e) Training.

(1) General.

(i) All employees working on site (such as but not limited to equipment operators, general laborers and others) exposed to hazardous substances, health hazards, or safety hazards and their supervisors and management responsible for the site shall receive training meeting the requirements of this paragraph before they are permitted to engage in hazardous waste operations that could expose them to hazardous substances, safety, or health hazards, and they shall receive review training as specified in this paragraph.

(ii) Employees shall not be permitted to participate in or supervise field activities until they have been trained to a level required by their job function and responsibility.

(2) Elements to be covered. The training shall thoroughly cover the following:

(i) Names of personnel and alternates responsible for site safety and health;

(ii) Safety, health and other hazards present on the site;

(iii) Use of personal protective equipment;

(iv) Work practices by which the employee can minimize risks from hazards;

(v) Safe use of engineering controls and equipment on the site;

(vi) Medical surveillance requirements including recognition of symptoms and signs which might indicate over exposure to hazards; and

(vii) The contents of paragraphs (G) through (J) of the site safety and health plan set forth in paragraph (b)(4)(ii) of this section.

(3) Initial training.

(i) General site workers (such as equipment operators, general laborers and supervisory personnel) engaged in hazardous substance removal or other activities which expose or potentially expose workers to hazardous substances and health hazards shall receive a minimum of 40 hours of instruction off the site, and a minimum of three days actual field experience under the direct supervision of a trained experienced supervisor.

(ii) Workers on site only occasionally for a specific limited task (such as, but not limited to, ground water monitoring, land surveying, or geophysical surveying) and who are unlikely to be exposed over permissible exposure limits and published exposure limits shall receive a minimum of 24 hours of instruction off the site, and the minimum of one day actual field experience under the direct supervision of a trained, experienced supervisor.

(iii) Workers regularly on site who work in areas which have been monitored and fully characterized indicating that exposures are under permissible exposure limits and published exposure limits where respirators are not necessary, and the characterization indicates that there are no health hazards or the possibility of an emergency developing, shall receive a minimum of 24 hours of instruction off the site, and the minimum of one day actual field experience under the direct supervision of a trained, experienced supervisor.

(iv) Workers with 24 hours of training who are covered by paragraphs (e)(3)(ii) and (e)(3)(iii) of this section, and who become general site workers or who are required to wear respirators, shall have the additional 16 hours and two days of training necessary to total the training specified in paragraph (e)(3)(i).

(4) Management and supervisor training. On-site management and supervisors directly responsible for, or who supervise employees engaged in, hazardous waste operations shall receive 40 hours initial training, and three days of supervised field experience (the training may be reduced to 24 hours and one day if the only area of their responsibility is employees covered by paragraphs (e)(3)(ii) and (e)(3)(iii)) and at least eight additional hours of specialized training at the time of job assignment on such topics as, but not limited to, the employer's safety and health program and the associated employee training program, personal protective equipment program, spill containment program, and health hazard monitoring procedure and techniques.

(5) Qualifications for trainers. Trainers shall be qualified to instruct employees about the subject matter that is being presented in training. Such trainers shall have satisfactorily completed a training program for teaching the subjects they are expected to teach, or they shall have the academic credentials and instructional experience necessary for teaching the subjects. Instructors shall demonstrate competent instructional skills and knowledge of the applicable subject matter.

(6) Training certification. Employees and supervisors that have received and successfully completed the training and field experience specified in paragraphs (e)(1) through (e)(4) of this section shall be certified by their instructor or the head instructor and trained supervisor as having completed the necessary training. A written certificate shall be given to each person so certified. Any person who has not been so certified or who does not meet the requirements of paragraph (e)(9) of this section shall be prohibited from engaging in hazardous waste operations.

(7) Emergency response. Employees who are engaged in responding to hazardous emergency situations at hazardous waste clean-up sites that may expose them to hazardous substances shall be trained in how to respond to such expected emergencies.

(8) Refresher training. Employees specified in paragraph (e)(1) of this section, and managers and supervisors specified in paragraph (e)(4) of this section, shall receive eight hours of refresher training annually on the items specified in paragraph (e)(2) and/or (e)(4) of this section, any critique of incidents that have occurred in the past year that can serve as training examples of related work, and other relevant topics.

(9) Equivalent training. Employers who can show by documentation or certification that an employee's work experience and/or training has resulted in training equivalent to that training required in paragraphs (e)(1) through (e)(4) of this section shall not be required to provide the initial training requirements of those paragraphs to such employees and shall provide a copy of the certification or documentation to the employee upon request. However, certified employees or employees with equivalent training new to a site shall receive appropriate, site specific training before site entry and have appropriate supervised field experience at the new site. Equivalent training includes any academic training or the training that existing employees might have already received from actual hazardous waste site experience.

(f) Medical surveillance

(1) General. Employees engaged in operations specified in paragraphs (a)(1)(i) through (a)(1)(iv) of this section and not covered by (a)(2)(iii) exceptions and employers of employees specified in paragraph (q)(9) shall institute a medical surveillance program in accordance with this paragraph.

(2) Employees covered. The medical surveillance program shall be instituted by the employer for the following employees:

(i) All employees who are or may be exposed to hazardous substances or health hazards at or above the established permissible exposure limit, above the published exposure levels for these substances, without regard to the use of respirators, for 30 days or more a year;

(ii) All employees who wear a respirator for 30 days or more a year or as required by 1910.134;

(iii) All employees who are injured, become ill or develop signs or symptoms due to possible overexposure involving hazardous substances or health hazards from an emergency response or hazardous waste operation; and

(iv) Members of HAZMAT teams.

(3) Frequency of medical examinations and consultations. Medical examinations and consultations shall be made available by the employer to each employee covered under paragraph (f)(2) of this section on the following schedules:

(i) For employees covered under paragraphs (f)(2)(i), (f)(2)(ii), and (f)(2)(iv);

(A) Prior to assignment;

(B) At least once every twelve months for each employee covered unless the attending physician believes a longer interval (not greater than biennially) is appropriate;

(C) At termination of employment or reassignment to an area where the employee would not be covered if the employee has not had an examination within the last six months.

(D) As soon as possible upon notification by an employee that the employee has developed signs or symptoms indicating possible overexposure to hazardous substances or health hazards, or that the employee has been injured or exposed above the permissible exposure limits or published exposure levels in an emergency situation;

(E) At more frequent times, if the examining physician determines that an increased frequency of examination is medically necessary.

(ii) For employees covered under paragraph (f)(2)(iii) and for all employees including of employers covered by paragraph (a)(1)(iv) who may have been injured, received a health impairment, developed signs or symptoms which may have resulted from exposure to hazardous substances resulting from an emergency incident, or exposed during an emergency incident to hazardous substances at concentrations above the permissible exposure limits or the published exposure levels without the necessary personal protective equipment being used:

(A) As soon as possible following the emergency incident or development of signs or symptoms;

(B) At additional times, if the examining physician determines that follow-up examinations or consultations are medically necessary.

(4) Content of medical examinations and consultations.

(i) Medical examinations required by paragraph (f)(3) of this section shall include a medical and work history (or updated history if one is in the employee's file) with special emphasis on symptoms related to the handling of hazardous substances and health hazards, and to fitness for duty including the ability to wear any required PPE under conditions (i.e., temperature extremes) that may be expected at the work site.

(ii) The content of medical examinations or consultations made available to employees pursuant to paragraph (f) shall be determined by the attending physician. The guidelines in the Occupational Safety and Health Guidance Manual for Hazardous Waste Site Activities (See Appendix D, reference #10) should be consulted.

(5) Examination by a physician and costs. All medical examinations and procedures shall be performed by or under the supervision of a licensed physician, preferably one knowledgeable in occupational medicine, and shall be provided

without cost to the employee, without loss of pay, and at a reasonable time and place.

(6) Information provided to the physician. The employer shall provide one copy of this standard and its appendices to the attending physician and in addition the following for each employee:

(i) A description of the employee's duties as they relate to the employee's exposures.

(ii) The employee's exposure levels or anticipated exposure levels.

(iii) A description of any personal protective equipment used or to be used.

(iv) Information from previous medical examinations of the employee which is not readily available to the examining physician.

(v) Information required by §1910.134.

(7) Physician's written opinion.

(i) The employer shall obtain and furnish the employee with a copy of a written opinion from the examining physician containing the following:

(A) The physician's opinion as to whether the employee has any detected medical conditions which would place the employee at increased risk of material impairment of the employee's health from work in hazardous waste operations or emergency response, or from respirator use.

(B) The physician's recommended limitations upon the employees assigned work.

(C) The results of the medical examination and tests if requested by the employee.

(D) A statement that the employee has been informed by the physician of the results of the medical examination and any medical conditions which require further examination or treatment.

(ii) The written opinion obtained by the employer shall not reveal specific findings or diagnoses unrelated to occupational exposure.

(8) Recordkeeping.

(i) An accurate record of the medical surveillance required by paragraph (f) of this section shall be retained. This record shall be retained for the period specified and meet the criteria of 29 CFR 1910.1020.

(ii) The record required in paragraph (f)(8)(i) of this section shall include at least the following information:

(A) The name and social security number of the employee;

(B) Physicians' written opinions, recommended limitations and results of examinations and tests;

(C) Any employee medical complaints related to exposure to hazardous substances;

(D) A copy of the information provided to the examining physician by the employer, with the exception of the standard and its appendices.

(g) Engineering controls, work practices, and personal protective equipment for employee protection. Engineering controls, work practices, personal protective equipment, or a combination of these shall be implemented in accordance with this paragraph to protect employees from exposure to hazardous substances and safety and health hazards.

(1) Engineering controls, work practices and PPE for substances regulated in Subparts G and Z.

(i) Engineering controls and work practices shall be instituted to reduce and maintain employee exposure to or below the permissible exposure limits for substances regulated by 29 CFR Part 1910, to the extent required by Subpart Z, except to the extent that such controls and practices are not feasible.

NOTE TO PARAGRAPH (g)(1)(i): Engineering controls which may be feasible include the use of pressurized cabs or control booths on equipment, and/or the use of remotely operated material handling equipment. Work practices which may be feasible are removing all non-essential employees from potential exposure during opening of drums, wetting down dusty operations and locating employees upwind of possible hazards.

(ii) Whenever engineering controls and work practices are not feasible, or not required, any reasonable combination of engineering controls, work practices and PPE shall be used to reduce and maintain to or below the permissible exposure limits or dose limits for substances regulated by 29 CFR Part 1910, Subpart Z.

(iii) The employer shall not implement a schedule of employee rotation as a means of compliance with permissible exposure limits or dose limits except when there is no other feasible way of complying with the airborne or dermal dose limits for ionizing radiation.

(iv) The provisions of 29 CFR, subpart G, shall be followed.

(2) Engineering controls, work practices, and PPE for substances not regulated in Subparts G and Z. An appropriate combination of engineering controls, work practices, and personal protective equipment shall be used to reduce and maintain employee exposure to or below published exposure levels for hazardous substances and health hazards not regulated by 29 CFR Part 1910, Subparts G and Z. The employer may use the published literature and MSDS as a guide in making the employer's determination as to what level of protection the employer believes is appropriate for hazardous substances and health hazards for which there is no permissible exposure limit or published exposure limit.

(3) Personal protective equipment selection.

(i) Personal protective equipment (PPE) shall be selected and used which will protect employees from the hazards and potential hazards they are likely to encounter as identified during the site characterization and analysis.

(ii) Personal protective equipment selection shall be based on an evaluation of the performance characteristics of the PPE relative to the requirements and limitations of the site, the task-specific conditions and duration, and the hazards and potential hazards identified at the site.

(iii) Positive pressure self-contained breathing apparatus, or positive pressure air-line respirators equipped with an escape air supply shall be used when chemical exposure levels present will create a substantial possibility of immediate death, immediate serious illness or injury, or impair the ability to escape.

(iv) Totally-encapsulating chemical protective suits (protection equivalent to Level A protection as recommended in Appendix B) shall be used in conditions where skin absorption of a hazardous substance may result in a substantial possibility of immediate death, immediate serious illness or injury, or impair the ability to escape.

(v) The level of protection provided by PPE selection shall be increased when additional information or site conditions show that increased protection is necessary to reduce employee exposures below permissible exposure limits and published exposure levels for hazardous substances and health hazards. (See Appendix B for guidance on selecting PPE ensembles.)

NOTE TO PARAGRAPH (g)(3): The level of employee protection provided may be decreased when additional information or site conditions show that decreased protection will not result in hazardous exposures to employees.

(vi) Personal protective equipment shall be selected and used to meet the requirements of 29 CFR Part 1910, Subpart I, and additional requirements specified in this section.

(4) Totally-encapsulating chemical protective suits.

(i) Totally-encapsulating suits shall protect employees from the particular hazards which are identified during site characterization and analysis.

(ii) Totally-encapsulating suits shall be capable of maintaining positive air pressure. (See Appendix A for a test method which may be used to evaluate this requirement.)

(iii) Totally-encapsulating suits shall be capable of preventing inward test gas leakage of more than 0.5 percent. (See Appendix A for a test method which may be used to evaluate this requirement.)

(5) Personal protective equipment (PPE) program. A personal protective equipment program, which is part of the employer's safety and health program required in paragraph (b) of this section or required in paragraph (p)(1) of this section and which is also a part of the site-specific safety and health plan shall be established. The PPE program shall address the elements listed below. When elements, such as donning and doffing procedures, are provided by the manufacturer of a piece of equipment and are attached to the plan, they need not be rewritten into the plan as long as they adequately address the procedure or element.

(i) PPE selection based upon site hazards,
(ii) PPE use and limitations of the equipment,
(iii) Work mission duration,
(iv) PPE maintenance and storage,
(v) PPE decontamination and disposal,
(vi) PPE training and proper fitting,
(vii) PPE donning and doffing procedures,
(viii) PPE inspection procedures prior to, during, and after use,
(ix) Evaluation of the effectiveness of the PPE program, and
(x) Limitations during temperature extremes, heat stress, and other appropriate medical considerations.

(h) Monitoring.

(1) General.

(i) Monitoring shall be performed in accordance with this paragraph where there may be a question of employee exposure to hazardous concentrations of hazardous substances in order to assure proper selection of engineering controls, work practices and personal protective equipment so that employees are not exposed to levels which exceed permissible exposure limits, or published exposure levels if there are no permissible exposure limits, for hazardous substances.

(ii) Air monitoring shall be used to identify and quantify airborne levels of hazardous substances and safety and health hazards in order to determine the appropriate level of employee protection needed on site.

(2) Initial entry. Upon initial entry, representative air monitoring shall be conducted to identify any IDLH condition, exposure over permissible exposure limits or published exposure levels, exposure over a radioactive material's dose limits or other dangerous condition such as the presence of flammable atmospheres, oxygen-deficient environments.

(3) Periodic monitoring. Periodic monitoring shall be conducted when the possibility of an IDLH condition or flammable atmosphere has developed or when there is indication that exposures may have risen over permissible

exposure limits or published exposure levels since prior monitoring. Situations where it shall be considered whether the possibility that exposures have risen are as follows:

(i) When work begins on a different portion of the site.

(ii) When contaminants other than those previously identified are being handled.

(iii) When a different type of operation is initiated (e.g., drum opening as opposed to exploratory well drilling).

(iv) When employees are handling leaking drums or containers or working in areas with obvious liquid contamination (e.g., a spill or lagoon).

(4) Monitoring of high-risk employees. After the actual clean-up phase of any hazardous waste operation commences; for example, when soil, surface water or containers are moved or disturbed; the employer shall monitor those employees likely to have the highest exposures to those hazardous substances and health hazards likely to be present above permissible exposure limits or published exposure levels by using personal sampling frequently enough to characterize employee exposures. The employer may utilize a representative sampling approach by documenting that the employees and chemicals chosen for monitoring are based on the criteria stated in the first sentence of this paragraph. If the employees likely to have the highest exposure are over permissible exposure limits or published exposure limits, then monitoring shall continue to determine all employees likely to be above those limits. The employer may utilize a representative sampling approach by documenting that the employees and chemicals chosen for monitoring are based on the criteria stated above.

NOTE TO PARAGRAPH (h): It is not required to monitor employees engaged in site characterization operations covered by paragraph (c) of this section.

(i) Informational programs. Employers shall develop and implement a program which is part of the employer's safety and health program required in paragraph (b) of this section to inform employees, contractors, and subcontractors (or their representative) actually engaged in hazardous waste operations of the nature, level and degree of exposure likely as a result of participation in such hazardous waste operations. Employees, contractors and subcontractors working outside of the operations part of a site are not covered by this standard.

(j) Handling drums and containers

(1) General.

(i) Hazardous substances and contaminated, liquids and other residues shall be handled, transported, labeled, and disposed of in accordance with this paragraph.

(ii) Drums and containers used during the clean-up shall meet the appropriate DOT, OSHA, and EPA regulations for the wastes that they contain.

(iii) When practical, drums and containers shall be inspected and their integrity shall be assured prior to being moved. Drums or containers that cannot be inspected before being moved because of storage conditions (i.e., buried beneath the earth, stacked behind other drums, stacked several tiers high in a pile, etc.) shall be moved to an accessible location and inspected prior to further handling.

(iv) Unlabeled drums and containers shall be considered to contain hazardous substances and handled accordingly until the contents are positively identified and labeled.

(v) Site operations shall be organized to minimize the amount of drum or container movement.

(vi) Prior to movement of drums or containers, all employees exposed to the transfer operation shall be warned of the potential hazards associated with the contents of the drums or containers.

(vii) U.S. Department of Transportation specified salvage drums or containers and suitable quantities of proper absorbent shall be kept available and used in areas where spills, leaks, or ruptures may occur.

(viii) Where major spills may occur, a spill containment program, which is part of the employer's safety and health program required in paragraph (b) of this section, shall be implemented to contain and isolate the entire volume of the hazardous substance being transferred.

(ix) Drums and containers that cannot be moved without rupture, leakage, or spillage shall be emptied into a sound container using a device classified for the material being transferred.

(x) A ground-penetrating system or other type of detection system or device shall be used to estimate the location and depth of buried drums or containers.

(xi) Soil or covering material shall be removed with caution to prevent drum or container rupture.

(xii) Fire extinguishing equipment meeting the requirements of 29 CFR Part 1910, Subpart L, shall be on hand and ready for use to control incipient fires.

(2) Opening drums and containers. The following procedures shall be followed in areas where drums or containers are being opened:

(i) Where an airline respirator system is used, connections to the source of air supply shall be protected from contamination and the entire system shall be protected from physical damage.

(ii) Employees not actually involved in opening drums or containers shall be kept a safe distance from the drums or containers being opened.

(iii) If employees must work near or adjacent to drums or containers being opened, a suitable shield that does not interfere with the work operation shall be placed between the employee and the drums or containers being opened to protect the employee in case of accidental explosion.

(iv) Controls for drum or container opening equipment, monitoring equipment, and fire suppression equipment shall be located behind the explosion-resistant barrier.

(v) When there is a reasonable possibility of flammable atmospheres being present, material handling equipment and hand tools shall be of the type to prevent sources of ignition.

(vi) Drums and containers shall be opened in such a manner that excess interior pressure will be safely relieved. If pressure cannot be relieved from a remote location, appropriate shielding shall be placed between the employee and the drums or containers to reduce the risk of employee injury.

(vii) Employees shall not stand upon or work from drums or containers.

(3) Material handling equipment. Material handling equipment used to transfer drums and containers shall be selected, positioned and operated to minimize sources of ignition related to the equipment from igniting vapors released from ruptured drums or containers.

(4) Radioactive wastes. Drums and containers containing radioactive wastes shall not be handled until such time as their hazard to employees is properly assessed.

(5) Shock sensitive wastes. As a minimum, the following special precautions shall be taken when drums and containers containing or suspected of containing shock-sensitive wastes are handled:

(i) All non-essential employees shall be evacuated from the area of transfer.

(ii) Material handling equipment shall be provided with explosive containment devices or protective shields to protect equipment operators from exploding containers.

(iii) An employee alarm system capable of being perceived above surrounding light and noise conditions shall be used to signal the commencement and completion of explosive waste handling activities.

(iv) Continuous communications (i.e., portable radios, hand signals, telephones, as appropriate) shall be maintained between the employee-in-charge of the immediate handling area and both the site safety and health supervisor and the command post until such time as the handling operation is completed. Communication equipment or methods that could cause shock sensitive materials to explode shall not be used.

(v) Drums and containers under pressure, as evidenced by bulging or swelling, shall not be moved until such time as the cause for excess pressure is determined and appropriate containment procedures have been implemented to protect employees from explosive relief of the drum.

(vi) Drums and containers containing packaged laboratory wastes shall be considered to contain shock-sensitive or explosive materials until they have been characterized.

Caution: Shipping of shock sensitive wastes may be prohibited under U.S. Department of Transportation regulations. Employers and their shippers should refer to 49 CFR 173.21 and 173.50.

(6) Laboratory waste packs. In addition to the requirements of paragraph (j)(5) of this section, the following precautions shall be taken, as a minimum, in handling laboratory waste packs (lab packs):

(i) Lab packs shall be opened only when necessary and then only by an individual knowledgeable in the inspection, classification, and segregation of the containers within the pack according to the hazards of the wastes.

(ii) If crystalline material is noted on any container, the contents shall be handled as a shock-sensitive waste until the contents are identified.

(7) Sampling of drum and container contents. Sampling of containers and drums shall be done in accordance with a sampling procedure which is part of the site safety and health plan developed for and available to employees and others at the specific worksite.

(8) Shipping and transport.

(i) Drums and containers shall be identified and classified prior to packaging for shipment.

(ii) Drum or container staging areas shall be kept to the minimum number necessary to safely identify and classify materials and prepare them for transport.

(iii) Staging areas shall be provided with adequate access and egress routes.

(iv) Bulking of hazardous wastes shall be permitted only after a thorough characterization of the materials has been completed.

(9) Tank and vault procedures.

(i) Tanks and vaults containing hazardous substances shall be handled in a manner similar to that for drums and containers, taking into consideration the size of the tank or vault.

(ii) Appropriate tank or vault entry procedures as described in the employer's safety and health plan shall be followed whenever employees must enter a tank or vault.

(k) Decontamination—

(1) General. Procedures for all phases of decontamination shall be developed and implemented in accordance with this paragraph.

(2) Decontamination procedures.

(i) A decontamination procedure shall be developed, communicated to employees and implemented before any employees or equipment may enter areas on site where potential for exposure to hazardous substances exists.

(ii) Standard operating procedures shall be developed to minimize employee contact with hazardous substances or with equipment that has contacted hazardous substances.

(iii) All employees leaving a contaminated area shall be appropriately decontaminated; all contaminated clothing and equipment leaving a contaminated area shall be appropriately disposed of or decontaminated.

(iv) Decontamination procedures shall be monitored by the site safety and health supervisor to determine their effectiveness. When such procedures are found to be ineffective, appropriate steps shall be taken to correct any deficiencies.

(3) Location. Decontamination shall be performed in geographical areas that will minimize the exposure of uncontaminated employees or equipment to contaminated employees or equipment.

(4) Equipment and solvents. All equipment and solvents used for decontamination shall be decontaminated or disposed of properly.

(5) Personal protective clothing and equipment.

(i) Protective clothing and equipment shall be decontaminated, cleaned, laundered, maintained or replaced as needed to maintain their effectiveness.

(ii) Employees whose non-impermeable clothing becomes wetted with hazardous substances shall immediately remove that clothing and proceed to shower. The clothing shall be disposed of or decontaminated before it is removed from the work zone.

(6) Unauthorized employees. Unauthorized employees shall not remove protective clothing or equipment from change rooms.

(7) Commercial laundries or cleaning establishments. Commercial laundries or cleaning establishments that decontaminate protective clothing or equipment shall be informed of the potentially harmful effects of exposures to hazardous substances.

(8) Showers and change rooms. Where the decontamination procedure indicates a need for regular showers and change rooms outside of a contaminated area, they shall be provided and meet the requirements of 29 CFR 1910.141. If temperature conditions prevent the effective use of water, then other effective means for cleansing shall be provided and used.

(l) Emergency response by employees at uncontrolled hazardous waste sites—

(1) Emergency response plan.

(i) An emergency response plan shall be developed and implemented by all employers within the scope of paragraphs (a)(1)(i) through (ii) of this section to handle anticipated emergencies prior to the commencement of hazardous waste operations. The plan shall be in writing and available for inspection and copying by employees, their representatives, OSHA personnel and other governmental agencies with relevant responsibilities.

(ii) Employers who will evacuate their employees from the danger area when an emergency occurs, and who do not permit any of their employees to assist in handling the emergency, are exempt from the requirements of this paragraph if they provide an emergency action plan complying with 29 CFR 1910.38.

(2) Elements of an emergency response plan. The employer shall develop an emergency response plan for emergencies which shall address, as a minimum, the following:

(i) Pre-emergency planning.

(ii) Personnel roles, lines of authority, training, and communication.

(iii) Emergency recognition and prevention.
(iv) Safe distances and places of refuge.
(v) Site security and control.
(vi) Evacuation routes and procedures.
(vii) Decontamination procedures which are not covered by the site safety and health plan.
(viii) Emergency medical treatment and first aid.
(ix) Emergency alerting and response procedures.
(x) Critique of response and follow-up.
(xi) PPE and emergency equipment.

(3) Procedures for handling emergency incidents.

(i) In addition to the elements for the emergency response plan required in paragraph (l)(2) of this section, the following elements shall be included for emergency response plans:

(A) Site topography, layout, and prevailing weather conditions.
(B) Procedures for reporting incidents to local, state, and federal governmental agencies.

(ii) The emergency response plan shall be a separate section of the Site Safety and Health Plan.
(iii) The emergency response plan shall be compatible and integrated with the disaster, fire and/or emergency response plans of local, state, and federal agencies.
(iv) The emergency response plan shall be rehearsed regularly as part of the overall training program for site operations.
(v) The site emergency response plan shall be reviewed periodically and, as necessary, be amended to keep it current with new or changing site conditions or information.
(vi) An employee alarm system shall be installed in accordance with 29 CFR 1910.165 to notify employees of an emergency situation, to stop work activities if necessary, to lower background noise in order to speed communication, and to begin emergency procedures.
(vii) Based upon the information available at time of the emergency, the employer shall evaluate the incident and the site response capabilities and proceed with the appropriate steps to implement the site emergency response plan.

(m) Illumination. Areas accessible to employees shall be lighted to not less than the minimum illumination intensities listed in the following Table H-120.1 while any work is in progress:

Table H-120.1 *Minimum illumination intensities in foot-candles*

FOOT-CANDLES	AREA OR OPERATIONS
5	General site areas.
3	Excavation and waste areas, accessways, active storage areas, loading platforms, refueling, and field maintenance areas.
5	Indoors: warehouses, corridors, hallways, and exitways.
5	Tunnels, shafts, and general underground work areas. (Exception: minimum of 10 foot-candles is required at tunnel and shaft heading during drilling, mucking, and scaling. Mine Safety and Health Administration approved cap lights shall be acceptable for use in the tunnel heading.)
10	General shops (e.g., mechanical and electrical equipment rooms, active storerooms, barracks or living quarters, locker or dressing rooms, dining areas, and indoor toilets and workrooms).
30	First aid stations, infirmaries, and offices.

(n) Sanitation at temporary workplaces—

(1) Potable water.

(i) An adequate supply of potable water shall be provided on the site.

(ii) Portable containers used to dispense drinking water shall be capable of being tightly closed, and equipped with a tap. Water shall not be dipped from containers.

(iii) Any container used to distribute drinking water shall be clearly marked as to the nature of its contents and not used for any other purpose.

(iv) Where single service cups (to be used but once) are supplied, both a sanitary container for the unused cups and a receptacle for disposing of the used cups shall be provided.

(2) Nonpotable water.

(i) Outlets for nonpotable water, such as water for firefighting purposes, shall be identified to indicate clearly that the water is unsafe and is not to be used for drinking, washing, or cooking purposes.

(ii) There shall be no cross-connection, open or potential, between a system furnishing potable water and a system furnishing nonpotable water.

(3) Toilet facilities.

(i) Toilets shall be provided for employees according to Table H-120.2.

(ii) Under temporary field conditions, provisions shall be made to assure not less than one toilet facility is available.

(iii) Hazardous waste sites, not provided with a sanitary sewer, shall be provided with the following toilet facilities unless prohibited by local codes:

(A) Chemical toilets;
(B) Recirculating toilets;
(C) Combustion toilets; or
(D) Flush toilets.

(iv) The requirements of this paragraph for sanitation facilities shall not apply to mobile crews having transportation readily available to nearby toilet facilities.

(v) Doors entering toilet facilities shall be provided with entrance locks controlled from inside the facility.

(4) Food handling. All food service facilities and operations for employees shall meet the applicable laws, ordinances, and regulations of the jurisdictions in which they are located.

(5) Temporary sleeping quarters. When temporary sleeping quarters are provided, they shall be heated, ventilated, and lighted.

(6) Washing facilities. The employer shall provide adequate washing facilities for employees engaged in operations where hazardous substances may be harmful to employees. Such facilities shall be in near proximity to the worksite; in areas where exposures are below permissible exposure limits and which are under the controls of the employer; and shall be so equipped as to enable employees to remove hazardous substances from themselves.

Table H-120.2 *Toilet facilities*

Number of employees	Minimum number of facilities
20 or fewer	One.
More than 20, fewer than 200	One toilet seat and 1 urinal per 40 employees.
More than 200	One toilet seat and 1 urinal per 50 employees.

(7) Showers and change rooms. When hazardous waste clean-up or removal operations commence on a site and the duration of the work will require six months or greater time to complete, the employer shall provide showers and change rooms for all employees exposed to hazardous substances and health hazards involved in hazardous waste clean-up or removal operations.

(i) Showers shall be provided and shall meet the requirements of 29 CFR 1910.141(d)(3).

(ii) Change rooms shall be provided and shall meet the requirements of 29 CFR 1910.141(e). Change rooms shall consist of two separate change areas separated by the shower area required in paragraph (n)(7)(i) of this section. One change area, with an exit leading off the worksite, shall provide employees with an area where they can put on, remove and store work clothing and personal protective equipment.

(iii) Showers and change rooms shall be located in areas where exposures are below the permissible exposure limits and published exposure levels. If this cannot be accomplished, then a ventilation system shall be provided that will supply air that is below the permissible exposure limits and published exposure levels.

(iv) Employers shall assure that employees shower at the end of their work shift and when leaving the hazardous waste site.

(o) New technology programs.

(1) The employer shall develop and implement procedures for the introduction of effective new technologies and equipment developed for the improved protection of employees working with hazardous waste clean-up operations, and the same shall be implemented as part of the site safety and health program to assure that employee protection is being maintained.

(2) New technologies, equipment or control measures available to the industry, such as the use of foams, absorbents, adsorbents, neutralizers, or other means to suppress the level of air contaminants while excavating the site or for spill control, shall be evaluated by employers or their representatives. Such an evaluation shall be done to determine the effectiveness of the new methods, materials, or equipment before implementing their use on a large scale for enhancing employee protection. Information and data from manufacturers or suppliers may be used as part of the employer's evaluation effort. Such evaluations shall be made available to OSHA upon request.

(p) Certain Operations Conducted Under the Resource Conservation and Recovery Act of 1976 (RCRA). Employers conducting operations at treatment, storage and disposal (TSD) facilities specified in paragraph (a)(1)(iv) of this section shall provide and implement the programs specified in this paragraph. See the "Notes and Exceptions" to paragraph (a)(2)(iii) of this section for employers not covered.

(1) Safety and health program. The employer shall develop and implement a written safety and health program for employees involved in hazardous waste operations that shall be available for inspection by employees, their representatives and OSHA personnel. The program shall be designed to identify, evaluate and control safety and health hazards in their facilities for the purpose of employee protection, to provide for emergency response meeting the requirements of paragraph (p)(8) of this section and to address as appropriate site analysis, engineering controls, maximum exposure limits, hazardous waste handling procedures and uses of new technologies.

(2) Hazard communication program. The employer shall implement a hazard communication program meeting the requirements of 29 CFR 1910.1200 as part of the employer's safety and program.

NOTE TO §1910.120—The exemption for hazardous waste provided in 1910.1200 is applicable to this section.

(3) Medical surveillance program. The employer shall develop and implement a medical surveillance program meeting the requirements of paragraph (f) of this section.

(4) Decontamination program. The employer shall develop and implement a decontamination procedure meeting the requirements of paragraph (k) of this section.

(5) New technology program. The employer shall develop and implement procedures meeting the requirements of paragraph (o) of this section for introducing new and innovative equipment into the workplace.

(6) Material handling program. Where employees will be handling drums or containers, the employer shall develop and implement procedures meeting the requirements of paragraphs (j)(1)(ii) through (viii) and (xi) of this section, as well as (j)(3) and (j)(8) of this section prior to starting such work.

(7) Training program—

(i) New employees. The employer shall develop and implement a training program which is part of the employer's safety and health program, for employees exposed to health hazards or hazardous substances at TSD operations to enable the employees to perform their assigned duties and functions in a safe and healthful manner so as not to endanger themselves or other employees. The initial training shall be for 24 hours and refresher training shall be for eight hours annually. Employees who have received the initial training required by this paragraph shall be given a written certificate attesting that they have successfully completed the necessary training.

(ii) Current employees. Employers who can show by an employee's previous work experience and/or training that the employee has had training equivalent to the initial training required by this paragraph, shall be considered as meeting the initial training requirements of this paragraph as to that employee. Equivalent training includes the training that existing employees might have already received from actual site work experience. Current employees shall receive eight hours of refresher training annually.

(iii) Trainers. Trainers who teach initial training shall have satisfactorily completed a training course for teaching the subjects they are expected to teach or they shall have the academic credentials and instruction experience necessary to demonstrate a good command of the subject matter of the courses and competent instructional skills.

(8) Emergency response program—

(i) Emergency response plan. An emergency response plan shall be developed and implemented by all employers. Such plans need not duplicate any of the subjects fully addressed in the employer's contingency planning required by permits, such as those issued by the U.S. Environmental Protection Agency, provided that the contingency plan is made part of the emergency response plan. The emergency response plan shall be a written portion of the employer's safety and health program required in paragraph (p)(1) of this section. Employers who will evacuate their employees from the worksite location when an emergency occurs and who do not permit any of their employees to assist in handling the emergency are exempt from the requirements of paragraph (p)(8) if they provide an emergency action plan complying with 29 CFR 1910.38.

(ii) Elements of an emergency response plan. The employer shall develop an emergency response plan for emergencies which shall address, as a minimum, the following areas to the extent that they are not addressed in any specific program required in this paragraph:

(A) Pre-emergency planning and coordination with outside parties.
(B) Personnel roles, lines of authority, training, and communication.
(C) Emergency recognition and prevention.
(D) Safe distances and places of refuge.
(E) Site security and control.
(F) Evacuation routes and procedures.
(G) Decontamination procedures.
(H) Emergency medical treatment and first aid.
(I) Emergency alerting and response procedures.
(J) Critique of response and follow-up.
(K) PPE and emergency equipment.

(iii) Training.

(A) Training for emergency response employees shall be completed before they are called upon to perform in real emergencies. Such training shall include the elements of the emergency response plan, standard operating procedures the employer has established for the job, the personal protective equipment to be worn and procedures for handling emergency incidents.

Exception #1: an employer need not train all employees to the degree specified if the employer divides the work force in a manner such that a sufficient number of employees who have responsibility to control emergencies have the training specified, and all other employees, who may first respond to an emergency incident, have sufficient awareness training to recognize that an emergency response situation exists and that they are instructed in that case to summon the fully trained employees and not attempt control activities for which they are not trained.

Exception #2: An employer need not train all employees to the degree specified if arrangements have been made in advance for an outside fully-trained emergency response team to respond in a reasonable period and all employees, who may come to the incident first, have sufficient awareness training to recognize that an emergency response situation exists and they have been instructed to call the designated outside fully-trained emergency response team for assistance.

(B) Employee members of TSD facility emergency response organizations shall be trained to a level of competence in the recognition of health and safety hazards to protect themselves and other employees. This would include training in the methods used to minimize the risk from safety and health hazards; in the safe use of control equipment; in the selection and use of appropriate personal protective equipment; in the safe operating procedures to be used at the incident scene; in the techniques of coordination with other employees to minimize risks; in the appropriate response to over exposure from health hazards or injury to themselves and other employees; and in the recognition of subsequent symptoms which may result from over exposures.

(C) The employer shall certify that each covered employee has attended and successfully completed the training required in paragraph (p)(8)(iii) of this section, or shall certify the employee's competency for certification of training shall be recorded and maintained by the employer.

(iv) Procedures for handling emergency incidents.

(A) In addition to the elements for the emergency response plan required in paragraph (p)(8)(ii) of this section, the following elements shall be included for emergency response plans to the extent that they do not repeat any information already contained in the emergency response plan:

(1) Site topography, layout, and prevailing weather conditions.

(2) Procedures for reporting incidents to local, state, and federal governmental agencies.

(B) The emergency response plan shall be compatible and integrated with the disaster, fire and/or emergency response plans of local, state, and federal agencies.

(C) The emergency response plan shall be rehearsed regularly as part of the overall training program for site operations.

(D) The site emergency response plan shall be reviewed periodically and, as necessary, be amended to keep it current with new or changing site conditions or information.

(E) An employee alarm system shall be installed in accordance with 29 CFR 1910.165 to notify employees of an emergency situation, to stop work activities if necessary, to lower background noise in order to speed communication; and to begin emergency procedures.

(F) Based upon the information available at time of the emergency, the employer shall evaluate the incident and the site response capabilities and proceed with the appropriate steps to implement the site emergency response plan.

(q) Emergency response program to hazardous substance releases. This paragraph covers employers whose employees are engaged in emergency response no matter where it occurs except that it does not cover employees engaged in operations specified in paragraphs (a)(1)(i) through (a)(1)(iv) of this section. Those emergency response organizations who have developed and implemented programs equivalent to this paragraph for handling releases of hazardous substances pursuant to section 303 of the Superfund Amendments and Reauthorization Act of 1986 (Emergency Planning and Community Right-to-Know Act of 1986, 42 U.S.C. 11003) shall be deemed to have met the requirements of this paragraph.

(1) Emergency response plan. An emergency response plan shall be developed and implemented to handle anticipated emergencies prior to the commencement of emergency response operations. The plan shall be in writing and available for inspection and copying by employees, their representatives and OSHA personnel. Employers who will evacuate their employees from the danger area when an emergency occurs, and who do not permit any of their employees to assist in handling the emergency, are exempt from the requirements of this paragraph if they provide an emergency action plan in accordance with 29 CFR 1910.38.

(2) Elements of an emergency response plan. The employer shall develop an emergency response plan for emergencies which shall address, as a minimum, the following areas to the extent that they are not addressed in any specific program required in this paragraph:

(i) Pre-emergency planning and coordination with outside parties.

(ii) Personnel roles, lines of authority, training, and communication.

(iii) Emergency recognition and prevention.

(iv) Safe distances and places of refuge.

(v) Site security and control.

(vi) Evacuation routes and procedures.

(vii) Decontamination.

(viii) Emergency medical treatment and first aid.

(ix) Emergency alerting and response procedures.

(x) Critique of response and follow-up.

(xi) PPE and emergency equipment.

(xii) Emergency response organizations may use the local emergency response plan or the state emergency response plan or both, as part of their emergency response plan to avoid duplication. Those items of the emergency response plan that are being properly addressed by the SARA Title III plans may be substituted into their emergency plan or otherwise kept together for the employer and employee's use.

(3) Procedures for handling emergency response.

(i) The senior emergency response official responding to an emergency shall become the individual in charge of a site-specific Incident Command System (ICS). All emergency responders and their communications shall be coordinated and controlled through the individual in charge of the ICS assisted by the senior official present for each employer.

NOTE TO PARAGRAPH (q)(3)(i).—The "senior official" at an emergency response is the most senior official on the site who has the responsibility for controlling the operations at the site. Initially it is the senior officer on the first-due piece of responding emergency apparatus to arrive on the incident scene. As more senior officers arrive (i.e., battalion chief, fire chief, state law enforcement official, site coordinator, etc.) the position is passed up the line of authority which has been previously established.

(ii) The individual in charge of the ICS shall identify, to the extent possible, all hazardous substances or conditions present and shall address as appropriate site analysis, use of engineering controls, maximum exposure limits, hazardous substance handling procedures, and use of any new technologies.

(iii) Based on the hazardous substances and/or conditions present, the individual in charge of the ICS shall implement appropriate emergency operations, and assure that the personal protective equipment worn is appropriate for the hazards to be encountered. However, personal protective equipment shall meet, at a minimum, the criteria contained in 29 CFR 1910.156(e) when worn while performing fire fighting operations beyond the incipient stage for any incident.

(iv) Employees engaged in emergency response and exposed to hazardous substances presenting an inhalation hazard or potential inhalation hazard shall wear positive pressure self-contained breathing apparatus while engaged in emergency response, until such time that the individual in charge of the ICS determines through the use of air monitoring that a decreased level of respiratory protection will not result in hazardous exposures to employees.

(v) The individual in charge of the ICS shall limit the number of emergency response personnel at the emergency site, in those areas of potential or actual exposure to incident or site hazards, to those who are actively performing emergency operations. However, operations in hazardous areas shall be performed using the buddy system in groups of two or more.

(vi) Back-up personnel shall be standing by with equipment ready to provide assistance or rescue. Qualified basic life support personnel, as a minimum, shall also be standing by with medical equipment and transportation capability.

(vii) The individual in charge of the ICS shall designate a safety officer, who is knowledgeable in the operations being implemented at the emergency response site, with specific responsibility to identify and evaluate hazards and to provide direction with respect to the safety of operations for the emergency at hand.

(viii) When activities are judged by the safety officer to be an IDLH and/or to involve an imminent danger condition, the safety officer shall have the authority to alter, suspend, or terminate those activities. The safety official shall immediately inform the individual in charge of the ICS of any actions needed to be taken to correct these hazards at the emergency scene.

(ix) After emergency operations have terminated, the individual in charge of the ICS shall implement appropriate decontamination procedures.

(x) When deemed necessary for meeting the tasks at hand, approved self-contained compressed air breathing apparatus may be used with approved cylinders from other approved self-contained compressed air breathing apparatus provided that such cylinders are of the same capacity and pressure rating. All compressed air cylinders used with self-contained breathing apparatus shall meet U.S. Department of Transportation and National Institute for Occupational Safety and Health criteria.

(4) Skilled support personnel. Personnel, not necessarily an employer's own employees, who are skilled in the operation of certain equipment, such as mechanized earth moving or digging equipment or crane and hoisting equipment, and who are needed temporarily to perform immediate emergency support work that cannot reasonably be performed in a timely fashion by an employer's own employees, and who will be or may be exposed to the hazards at an emergency response scene, are not required to meet the training required in this paragraph for the employer's regular employees. However, these personnel shall be given an initial briefing at the site prior to their participation in any emergency response. The initial briefing shall include instruction in the wearing of appropriate personal protective equipment, what chemical hazards are involved, and what duties are to be performed. All other appropriate safety and health precautions provided to the employer's own employees shall be used to assure the safety and health of these personnel.

(5) Specialist employees. Employees who, in the course of their regular job duties, work with and are trained in the hazards of specific hazardous substances, and who will be called upon to provide technical advice or assistance at a hazardous substance release incident to the individual in charge, shall receive training or demonstrate competency in the area of their specialization annually.

(6) Training. Training shall be based on the duties and function to be performed by each responder of an emergency response organization. The skill and knowledge levels required for all new responders, those hired after the effective date of this standard, shall be conveyed to them through training before they are permitted to take part in actual emergency operations on an incident. Employees who participate, or are expected to participate, in emergency response, shall be given training in accordance with the following paragraphs:

(i) First responder awareness level. First responders at the awareness level are individuals who are likely to witness or discover a hazardous substance release and who have been trained to initiate an emergency response sequence by notifying the proper authorities of the release. They would take no further action beyond notifying the authorities of the release. First responders at the awareness level shall have sufficient training or have had sufficient experience to objectively demonstrate competency in the following areas:

(A) An understanding of what hazardous substances are, and the risks associated with them in an incident.

(B) An understanding of the potential outcomes associated with an emergency created when hazardous substances are present.

(C) The ability to recognize the presence of hazardous substances in an emergency.

(D) The ability to identify the hazardous substances, if possible.

(E) An understanding of the role of the first responder awareness individual in the employer's emergency response plan including site security and control and the U.S. Department of Transportation's Emergency Response Guidebook.

(F) The ability to realize the need for additional resources, and to make appropriate notifications to the communication center.

(ii) First responder operations level. First responders at the operations level are individuals who respond to releases or potential releases of hazardous substances as part of the initial response to the site for the purpose of protecting nearby persons, property, or the environment from the effects of the release. They are trained to respond in a defensive fashion without actually trying to stop the release. Their function is to contain the release from a safe distance, keep it from spreading, and prevent exposures. First responders at the operational level shall have received at least eight hours of training or have had sufficient experience to objectively demonstrate competency in the following areas in addition to those listed for the awareness level and the employer shall so certify:

(A) Knowledge of the basic hazard and risk assessment techniques.

(B) Know how to select and use proper personal protective equipment provided to the first responder operational level.

(C) An understanding of basic hazardous materials terms.

(D) Know how to perform basic control, containment and/or confinement operations within the capabilities of the resources and personal protective equipment available with their unit.

(E) Know how to implement basic decontamination procedures.

(F) An understanding of the relevant standard operating procedures and termination procedures.

(iii) Hazardous materials technician. Hazardous materials technicians are individuals who respond to releases or potential releases for the purpose of stopping the release. They assume a more aggressive role than a first responder at the operations level in that they will approach the point of release in order to plug, patch or otherwise stop the release of a hazardous substance. Hazardous materials technicians shall have received at least 24 hours of training equal to the first responder operations level and in addition have competency in the following areas and the employer shall so certify:

(A) Know how to implement the employer's emergency response plan.

(B) Know the classification, identification and verification of known and unknown materials by using field survey instruments and equipment.

(C) Be able to function within an assigned role in the Incident Command System.

(D) Know how to select and use proper specialized chemical personal protective equipment provided to the hazardous materials technician.

(E) Understand hazard and risk assessment techniques.

(F) Be able to perform advance control, containment, and/or confinement operations within the capabilities of the resources and personal protective equipment available with the unit.

(G) Understand and implement decontamination procedures.

(H) Understand termination procedures.

(I) Understand basic chemical and toxicological terminology and behavior.

(iv) Hazardous materials specialist. Hazardous materials specialists are individuals who respond with and provide support to hazardous materials technicians. Their duties parallel those of the hazardous materials technician, however, those duties require a more directed or specific knowledge of the various substances they may be called upon to contain. The hazardous materials specialist would also act as the site liaison with Federal, state, local and other government authorities in regards to site activities. Hazardous materials specialists shall have received at least 24 hours of training equal to the technician level and in addition have competency in the following areas and the employer shall so certify:

(A) Know how to implement the local emergency response plan.

(B) Understand classification, identification and verification of known and unknown materials by using advanced survey instruments and equipment.

(C) Know the state emergency response plan.

(D) Be able to select and use proper specialized chemical personal protective equipment provided to the hazardous materials specialist.

(E) Understand in-depth hazard and risk techniques.

(F) Be able to perform specialized control, containment, and/or confinement operations within the capabilities of the resources and personal protective equipment available.

(G) Be able to determine and implement decontamination procedures.

(H) Have the ability to develop a site safety and control plan.

(I) Understand chemical, radiological and toxicological terminology and behavior.

(v) On scene incident commander. Incident commanders, who will assume control of the incident scene beyond the first responder awareness level, shall receive at least 24 hours of training equal to the first responder operations level and in addition have competency in the following areas and the employer shall so certify:

(A) Know and be able to implement the employer's incident command system.

(B) Know how to implement the employer's emergency response plan.

(C) Know and understand the hazards and risks associated with employees working in chemical protective clothing.

(D) Know how to implement the local emergency response plan.

(E) Know of the state emergency response plan and of the Federal Regional Response Team.

(F) Know and understand the importance of decontamination procedures.

(7) Trainers. Trainers who teach any of the above training subjects shall have satisfactorily completed a training course for teaching the subjects they are expected to teach, such as the courses offered by the U.S. National Fire Academy, or they shall have the training and/or academic credentials and instructional experience necessary to demonstrate competent instructional skills and a good command of the subject matter of the courses they are to teach.

(8) Refresher training.

(i) Those employees who are trained in accordance with paragraph (q)(6) of this section shall receive annual refresher training of sufficient content and duration to maintain their competencies, or shall demonstrate competency in those areas at least yearly.

(ii) A statement shall be made of the training or competency, and if a statement of competency is made, the employer shall keep a record of the methodology used to demonstrate competency.

(9) Medical surveillance and consultation.

(i) Members of an organized and designated HAZMAT team and hazardous materials specialist shall receive a baseline physical examination and be provided with medical surveillance as required in paragraph (f) of this section.

(ii) Any emergency response employees who exhibit signs or symptoms which may have resulted from exposure to hazardous substances during the course of an emergency incident either immediately or subsequently, shall be provided with medical consultation as required in paragraph (f)(3)(ii) of this section.

(10) Chemical protective clothing. Chemical protective clothing and equipment to be used by organized and designated HAZMAT team members, or to be used by hazardous materials specialists, shall meet the requirements of paragraphs (g)(3) through (5) of this section.

(11) Post-emergency response operations. Upon completion of the emergency response, if it is determined that it is necessary to remove hazardous substances, health hazards and materials contaminated with them (such as contaminated soil or other elements of the natural environment) from the site of the incident, the employer conducting the clean-up shall comply with one of the following:

(i) Meet all the requirements of paragraphs (b) through (o) of this section; or

(ii) Where the clean-up is done on plant property using plant or workplace employees, such employees shall have completed the training requirements of the following: 29 CFR 1910.38, 1910.134, 1910.1200, and other appropriate safety and health training made necessary by the tasks they are expected to perform such as personal protective equipment and decontamination procedures.

DIRECTORY OF STATES WITH APPROVED OCCUPATIONAL SAFETY AND HEALTH PLANS (as of October 24, 2007)

Alaska
Department of Labor and Workforce Development
P.O. Box 111149
1111 W. 8th Street, Room 304
Juneau, Alaska 99811-1149
(907) 465-4855

Arizona
Industrial Commission of Arizona
800 W. Washington
Phoenix, Arizona 85007-2922
(602) 542-5795

California
Department of Industrial Relations
1515 Clay Street, Suite 1901
Oakland, CA 94612
510-286-7000

Connecticut
Department of Labor
200 Folly Brook Boulevard
Wethersfield, Connecticut 06109
(860) 263-6000

Hawaii
Department of Labor and Industrial Relations
830 Punchbowl Street
Honolulu, Hawaii 96813
(808) 586-9100

Indiana
Department of Labor
State Office Building
402 West Washington Street, Room W195
Indianapolis, Indiana 46204-2751
(317) 232-2655

Iowa
Division of Labor Services
1000 E. Grand Avenue
Des Moines, Iowa 50319-0209
(515) 281-5387

Kentucky
Department of Labor
1047 U.S. Highway 127 South, Suite 4
Frankfort, Kentucky 40601
(502) 564-3070

Maryland
Division of Labor and Industry
Department of Labor, Licensing and Regulation
1100 North Eutaw Street, Room 611
Baltimore, Maryland 21201-2206
(410) 767-2189

Michigan
Department of Labor and Economic Growth
Occupational Safety and Health Administration
P.O. Box 30004
Lansing, Michigan 48909
(517) 323-1820

Minnesota
Department of Labor and Industry
443 Lafayette Road
St. Paul, Minnesota 55155
(651) 284-5005

Nevada
Division of Industrial Relations
400 West King Street, Suite 400
Carson City, Nevada 89703
(775) 684-7260

New Jersey
Department of Labor and Workforce Development
Office of Public Employees' Occupational Safety & Health (PEOSH)
1 John Fitch Plaza
P.O. Box 386
Trenton, New Jersey 08625-0386
(609) 292-2975

New Mexico
Environment Department
1190 St. Francis Dr., Suite N4050
Santa Fe, New Mexico 87505
(800) 219-6157

New York
Department of Labor
Public Employee Safety and Health Bureau (PESH)
State Office Campus, Room 522
Albany, New York 12240
(518) 457-3518

North Carolina
Department of Labor
1101 Mail Service Center
Raleigh, North Carolina 27699-1101
(919) 807-2796

Oregon
Occupational Safety and Health Division
350 Winter Street, Room 430
P.O. Box 14480
Salem, Oregon 97309-0405
(503) 378-3272

Puerto Rico
Department of Labor
Prudencio Rivera Martínez Building, 20th Floor
505 Muñoz Rivera Avenue
Hato Rey, Puerto Rico 00918
(787) 754-2119

South Carolina
Department of Labor, Licensing, and Regulation
Synergy Business Park, Kingstree Building
110 Centerview Drive
P.O. Box 11329
Columbia, South Carolina 29211
Adrienne R. Youmans, Director (803) 896-7688

Tennessee
Department of Labor and Workforce Development
Division of Occupational Safety & Health (TOSHA)
710 James Robertson Parkway, 3rd Floor
Nashville, Tennessee 37243-0659
(615) 741-2793

Utah
Labor Commission
Occupational Safety & Health
160 East 300 South, 3rd Floor
P.O. Box 146650
Salt Lake City, Utah 84114-6650
(801) 530-6901

Vermont
Department of Labor
5 Green Mountain Drive
P.O. Box 488
Montpelier, Vermont 05601-0488
(802) 828-4000

Virgin Islands
Department of Labor
Division of Occupational Safety and Health
3012 Golden Rock, VITRACO Mall
Christiansted, St. Croix, Virgin Islands 00840
(340) 772-1315

Virginia
Department of Labor and Industry
Powers-Taylor Building
13 South 13th Street
Richmond, Virginia 23219
(804) 371-2327

Washington
Department of Labor and Industries
PO Box 44000
Olympia, Washington 98504-4001
(360) 902-4200

Wyoming
Department of Employment
Workers' Safety and Compensation Division
1510 East Pershing Boulevard—West Wing
Cheyenne, Wyoming 82002
Gary W. Child, Administrator (307) 777-7441

OVERALL EVALUATIONS OF CARCINOGENICITY TO HUMANS AS EVALUATED IN IARC MONOGRAPHS: LIST OF ALL AGENTS EVALUATED TO DATE (as of August 13, 2007)

This list contains all hazards evaluated to date, according to the type of hazard posed and to the type of exposure. Where appropriate, chemical abstract numbers (CAS) are given [in square brackets].

GROUP 1: CARCINOGENIC TO HUMANS

Agents and Groups of Agents

4-Aminobiphenyl [92-67-1]

Arsenic [7440-38-2] and arsenic compounds

Asbestos [1332-21-4]

Azathioprine [446-86-6]

Benzene [71-43-2]

Benzidine [92-87-5]

Benzo[a]pyrene [50-32-8]

Beryllium [7440-41-7] and beryllium compounds

N,N-Bis(2-chloroethyl)-2-naphthylamine (Chlornaphazine) [494-03-1]

Bis(chloromethyl)ether [542-88-1] and chloromethyl methyl ether [107-30-2] (technical-grade)

1,3-Butadiene [106-99-0]

1,4-Butanediol dimethanesulfonate (Busulphan; Myleran) [55-98-1]

Cadmium [7440-43-9] and cadmium compounds

Chlorambucil [305-03-3]

1-(2-Chloroethyl)-3-(4-methylcyclohexyl)-1-nitrosourea (Methyl-CCNU; Semustine) [13909-09-6]

Chromium[VI]

Ciclosporin [79217-60-0]

Cyclophosphamide [50-18-0] [6055-19-2]

Diethylstilboestrol [56-53-1]

Epstein-Barr virus

Erionite [66733-21-9]

Estrogen-progestogen menopausal therapy (combined)

Estrogen-progestogen oral contraceptives (combined)

Estrogens, nonsteroidal

Estrogens, steroidal

Estrogen therapy, postmenopausal

Ethanol [64-17-5] in alcoholic beverages

Ethylene oxide [75-21-8]

Etoposide [33419-42-0] in combination with cisplatin and bleomycin

Formaldehyde [50-00-0]

Gallium arsenide [1303-00-0]

Helicobacter pylori (infection with)

Hepatitis B virus (chronic infection with)

Hepatitis C virus (chronic infection with)

Human immunodeficiency virus type 1 (infection with)

Human papillomavirus types 16, 18, 31, 33, 35, 39, 45, 51, 52, 56, 58, 59, and 66

Human T-cell lymphotropic virus type I

Melphalan [148-82-3]

8-Methoxypsoralen (Methoxsalen) [298-81-7] plus ultraviolet A radiation

MOPP and other combined chemotherapy including alkylating agents

Mustard gas (Sulfur mustard) [505-60-2]

2-Naphthylamine [91-59-8]

Neutrons

Nickel compounds

N′-Nitrosonornicotine (NNN) [16543-55-8] and 4-(N-Nitrosomethylamino)-1-(3-pyridyl)-1-butanone (NNK) [64091-91-4]

Opisthorchis viverrini (infection with)

Oral contraceptives, sequential

Phosphorus-32, as phosphate

Plutonium-239 and its decay products (may contain plutonium-240 and other isotopes), as aerosols

Radioiodines, short-lived isotopes, including iodine-131, from atomic reactor accidents and nuclear weapons detonation (exposure during childhood)

Radionuclides, α-particle-emitting, internally deposited

Radionuclides, β-particle-emitting, internally deposited

Radium-224 and its decay products

Radium-226 and its decay products

Radium-228 and its decay products

Radon-222 [10043-92-2] and its decay products

Schistosoma haematobium (infection with)

Silica [14808-60-7], crystalline (inhaled in the form of quartz or cristobalite from occupational sources)

Solar radiation

Talc containing asbestiform fibres

Tamoxifen [10540-29-1]

2,3,7,8-Tetrachlorodibenzo-para-dioxin [1746-01-6]

Thiotepa [52-24-4]

Thorium-232 and its decay products, administered intravenously as a colloidal dispersion of thorium-232 dioxide

Treosulfan [299-75-2]

Vinyl chloride [75-01-4]

X- and Gamma (γ)-Radiation Mixtures

Mixtures

Aflatoxins (naturally occurring mixtures of) [1402-68-2]

Alcoholic beverages

Areca nut

Betel quid with tobacco

Betel quid without tobacco

Coal-tar pitches [65996-93-2]

Coal-tars [8007-45-2]

Herbal remedies containing plant species of the genus Aristolochia

Household combustion of coal, indoor emissions from

Mineral oils, untreated and mildly treated

Phenacetin, analgesic mixtures containing

Salted fish (Chinese-style)

Shale-oils [68308-34-9]

Soots

Tobacco, smokeless

Wood dust

Exposure Circumstances

Aluminium production

Arsenic in drinking water

Auramine, manufacture of

Boot and shoe manufacture and repair

Chimney sweeping

Coal gasification

Coal-tar distillation

Coke production

Furniture and cabinet making

Haematite mining (underground) with exposure to radon

Involuntary smoking (exposure to secondhand or "environmental" tobacco smoke)

Iron and steel founding

Isopropyl alcohol manufacture (strong-acid process)

Magenta, manufacture of

Painter (occupational exposure as a)

Paving and roofing with coal-tar pitch

Rubber industry

Strong-inorganic-acid mists containing sulfuric acid (occupational exposure to)

Tobacco smoking and tobacco smoke

CLASSIFICATION AND LABELLING SUMMARY TABLES

CLASSIFICATION AND LABELLING SUMMARY TABLES

A2.1 **Explosives** (see Chapter 2.1 for details)

Hazard category	Criteria	Hazard communication elements	
Division 1.1	According to the results of the test in Part I of the *Manual of Tests and Criteria, UN Recommendations on the Transport of Dangerous Goods.*	Symbol	
		Signal word	Danger
		Hazard statement	Explosive; mass explosion hazard
Division 1.2	According to the results of the test in Part I of the *Manual of Tests and Criteria, UN Recommendations on the Transport of Dangerous Goods.*	Symbol	
		Signal word	Danger
		Hazard statement	Explosive; severe projection hazard
Division 1.3	According to the results of the test in Part I of the *Manual of Tests and Criteria, UN Recommendations on the Transport of Dangerous Goods.*	Symbol	
		Signal word	Danger
		Hazard statement	Explosive; fire, blast or projection hazard
Division 1.4	According to the results of the test in Part I of the *Manual of Tests and Criteria, UN Recommendations on the Transport of Dangerous Goods.*	Symbol	**1.4**
		Signal word	Warning
		Hazard statement	Fire or projection hazard
Division 1.5	According to the results of the test in Part I of the *Manual of Tests and Criteria , UN Recommendations on the Transport of Dangerous Goods.*	Symbol	**1.5**
		Signal word	Warning
		Hazard statement	May explode in fire
Division 1.6	According to the results of the test in Part I of the *Manual of Tests and Criteria, UN Recommendations on the Transport of Dangerous Goods.*	Symbol	**1.6**
		Signal word	No signal word
		Hazard statement	No hazard statement

A2.2 Flammable gases (See Chapter 2.2 for details)

Hazard category	Criteria	Hazard communication elements	
1	Gases and gas mixtures, which at 20 °C and a standard pressure of 101.3 kPa: (a) are ignitable when in a mixture of 13% or less by volume in air; or (b) have a flammable range with air of at least 12 percentage points regardless of the lower flammable limit.	Symbol	
		Signal word	Danger
		Hazard statement	Extremely flammable gas
2	Gases or gas mixtures, other than those of category 1, which, at 20 °C and a standard pressure of 101.3 kPa, have a flammable range while mixed in air.	Symbol	No symbol used
		Signal word	Warning
		Hazard statement	Flammable gas

A2.3 Flammable aerosols (See Chapter 2.3 for details)

Hazard category	Criteria	Hazard communication elements	
1	On the basis of its components, of its chemical heat of combustion and, if applicable, of the results of the foam test, for foam aerosols, and of the ignition distance test and enclosed space test, for spray aerosols (see decision logic in 2.3.4.1 of Chapter 2.3).	Symbol	
		Signal word	Danger
		Hazard statement	Extremely flammable aerosol
2	On the basis of its components, of its chemical heat of combustion and, if applicable, of the results of the foam test, for foam aerosols, and of the ignition distance test and enclosed space test, for spray aerosols (see decision logic in 2.3.4.1 of Chapter 2.3).	Symbol	
		Signal word	Warning
		Hazard statement	Flammable aerosol

A2.4 Oxidizing gases (See Chapter 2.4 for details)

Hazard category	Criteria	Hazard communication elements	
1	Any gas which may, generally by providing oxygen, cause or contribute to the combustion of other material more than air does.	Symbol	
		Signal word	Danger
		Hazard statement	May cause or intensify fire; oxidizer

A2.5 Gases under pressure (See Chapter 2.5 for details)

Hazard category	Criteria	Hazard communication elements	
Compressed gas	A gas, which when packaged under pressure is entirely gaseous at -50 °C; including all gases with a critical temperature ≤ -50 °C.	Symbol	
		Signal word	Warning
		Hazard statement	Contains gas under pressure; may explode if heated
Liquefied gas	A gas which when packaged under pressure, is partially liquid at temperatures above -50 °C. A distinction is made between: *i)* *High pressure liquefied gas*: a gas with a critical temperature between –50 °C and +65 °C; and *ii)* *Low pressure liquefied gas*: a gas with a critical temperature above +65 °C.	Symbol	
		Signal word	Warning
		Hazard statement	Contains gas under pressure; may explode if heated
Refrigerated liquefied gas	A gas which when packaged is made partially liquid because of its low temperature.	Symbol	
		Signal word	Warning
		Hazard statement	Contains refrigerated gas; may cause cryogenic burns or injury
Dissolved gas	A gas which when packaged under pressure is dissolved in a liquid phase solvent.	Symbol	
		Signal Word	Warning
		Hazard statement	Contains gas under pressure; may explode if heated

A2.6 Flammable liquids (See Chapter 2.6 for details)

Hazard category	Criteria	Hazard communication elements	
1	Flash point < 23 °C and initial boiling point ≤ 35 °C.	Symbol	
		Signal word	Danger
		Hazard statement	Extremely flammable liquid and vapour
2	Flash point < 23 °C and initial boiling point >35 °C.	Symbol	
		Signal word	Danger
		Hazard statement	Highly flammable liquid and vapour
3	Flash point ≥ 23 °C and ≤ 60 °C.	Symbol	
		Signal word	Warning
		Hazard statement	Flammable liquid and vapour
4	Flash point > 60 °C and ≤ 93 °C.	Symbol	No symbol used
		Signal word	Warning
		Hazard statement	Combustible liquid

A.2.7 Flammable solids (See Chapter 2.7 for details)

<table>
<tr><th>Hazard category</th><th>Criteria</th><th colspan="2">Hazard communication elements</th></tr>
<tr><td rowspan="3">1</td><td rowspan="3">Burning rate test:
Substances other than metal powders:
- wetted zone does not stop fire and
- burning time < 45 seconds or
burning rate > 2.2 mm/s
Metal powders:
- burning time ≤ 5 minutes.</td><td>Symbol</td><td></td></tr>
<tr><td>Signal word</td><td>Danger</td></tr>
<tr><td>Hazard statement</td><td>Flammable solid</td></tr>
<tr><td rowspan="3">2</td><td rowspan="3">Burning rate test:
Substances other than metal powders:
- wetted zone stops the fire for at least 4 minutes and
- burning time < 45 seconds or
burning rate > 2.2 mm/second
Metal powders :
- burning time > 5 minutes
and ≤ 10 minutes.</td><td>Symbol</td><td></td></tr>
<tr><td>Signal word</td><td>Warning</td></tr>
<tr><td>Hazard statement</td><td>Flammable solid</td></tr>
</table>

A2.8 Self-reactive substances (See Chapter 2.8 for details)

Hazard category	Criteria	Hazard communication elements	
Type A	According to the results of tests in the *UN Recommendations on the Transport of Dangerous Goods, Manual of Tests and Criteria,* Part II and the application of the decision logic under 2.8.4.1 of Chapter 2.8.	Symbol	
		Signal word	Danger
		Hazard statement	Heating may cause an explosion
Type B	According to the results of tests in the *UN Recommendations on the Transport of Dangerous Goods, Manual of Tests and Criteria,* Part II and the application of the decision logic under para. 2.8.4.1 of Chapter 2.8.	Symbol	 
		Signal word	Danger
		Hazard statement	Heating may cause a fire or explosion
Type C and D	According to the results of tests in the *UN Recommendations on the Transport of Dangerous Goods, Manual of Tests and Criteria,* Part II and the application of the decision logic under para. 2.8.4.1 of Chapter 2.8.	Symbol	
		Signal word	Danger
		Hazard statement	Heating may cause a fire
Type E and F	According to the results of tests in the *UN Recommendations on the Transport of Dangerous Goods, Manual of Tests and Criteria,* Part II and the application of the decision logic under para. 2.8.4.1 of Chapter 2.8.	Symbol	
		Signal word	Warning
		Hazard statement	Heating may cause a fire
Type G	According to the results of tests in the *UN Recommendations on the Transport of Dangerous Goods, Manual of Tests and Criteria,* Part II and the application of the decision logic under para. 2.8.4.1 of Chapter 2.8.	Signal word	There are no label elements allocated to this hazard category
		Symbol	
		Hazard statement	

A2.9 Pyrophoric liquids (See Chapter 2.9 for details)

Hazard category	Criteria	Hazard communication elements	
1	The liquid ignites within 5 min when added to an inert carrier and exposed to air, or it ignites or chars a filter paper on contact with air within 5 min.	Symbol	
		Signal word	Danger
		Hazard statement	Catches fire spontaneously if exposed to air

A2.10 Pyrophoric solids (See Chapter 2.10 for details)

Hazard category	Criteria	Hazard communication elements	
1	The solid ignites within 5 minutes of coming into contact with air.	Symbol	
		Signal word	Danger
		Hazard statement	Catches fire spontaneously if exposed to air

A2.11 Self-heating substances (See Chapter 2.11 for details)

Hazard category	Criteria	Hazard communication elements	
1	A positive result is obtained in a test using a 25 mm sample cube at 140 °C.	Symbol	[flame symbol]
		Signal word	Danger
		Hazard statement	Self-heating; may catch fire
2	(a) A positive result is obtained in a test using a 100 mm sample cube at 140 °C and a negative result is obtained in a test using a 25 mm cube sample at 140 °C <u>and</u> the substance is to be packed in packages with a volume of more than 3 m^3; or (b) A positive result is obtained in a test using a 100 mm sample cube at 140 °C and a negative result is obtained in a test using a 25 mm cube sample at 140 °C, a positive result is obtained in a test using a 100 mm cube sample at 120 °C <u>and</u> the substance is to be packed in packages with a volume of more than 450 litres; or (c) A positive result is obtained in a test using a 100 mm sample cube at 140 °C and a negative result is obtained in a test using a 25 mm cube sample at 140 °C <u>and</u> a positive result is obtained in a test using a 100 mm cube sample at 100 °C.	Symbol	[flame symbol]
		Signal word	Warning
		Hazard statement	Self-heating in large quantities; may catch fire

A2.12 Substances, which on contact with water, emit flammable gases (See Chapter 2.12 for details)

Hazard category	Criteria	Hazard communication elements	
1	Any substance which reacts vigorously with water at ambient temperatures and demonstrates generally a tendency for the gas produced to ignite spontaneously, or which reacts readily with water at ambient temperatures such that the rate of evolution of flammable gas is equal to or greater than 10 litres per kilogram of substance over any one minute.	Symbol	
		Signal word	Danger
		Hazard statement	In contact with water releases flammable gases which may ignite spontaneously
2	Any substance which reacts readily with water at ambient temperatures such that the maximum rate of evolution of flammable gas is equal to or greater than 20 litres per kilogram of substance per hour, and which does not meet the criteria for category 1.	Symbol	
		Signal word	Danger
		Hazard statement	In contact with water releases flammable gases
3	Any substance which reacts slowly with water at ambient temperatures such that the maximum rate of evolution of flammable gas is equal to or greater than 1 litre per kilogram of substance per hour, and which does not meet the criteria for categories 1 and 2.	Symbol	
		Signal word	Warning
		Hazard statement	In contact with water releases flammable gases

A2.13 Oxidizing liquids (See Chapter 2.13 for details)

Hazard category	Criteria	Hazard communication elements	
1	Any substance which, in the 1:1 mixture, by mass, of substance and cellulose tested, spontaneously ignites; or the mean pressure rise time of a 1:1 mixture, by mass, of substance and cellulose is less than that of a 1:1 mixture, by mass, of 50% perchloric acid and cellulose.	Symbol	
		Signal word	Danger
		Hazard statement	May cause fire or explosion; strong oxidizer
2	Any substance which, in the 1:1 mixture, by mass, of substance and cellulose tested, exhibits a mean pressure rise time less than or equal to the mean pressure rise time of a 1:1 mixture, by mass, of 40% aqueous sodium chlorate solution and cellulose; and the criteria for category 1 are not met.	Symbol	
		Signal word	Danger
		Hazard statement	May intensify fire; oxidizer
3	Any substance which, in the 1:1 mixture, by mass, of substance and cellulose tested, exhibits a mean pressure rise time less than or equal to the mean pressure rise time of a 1:1 mixture, by mass, of 65% aqueous nitric acid and cellulose; and the criteria for categories 1 and 2 are not met.	Symbol	
		Signal word	Warning
		Hazard statement	May intensify fire; oxidizer

A2.14 Oxidizing solids (See Chapter 2.14 for details)

Hazard category	Criteria	Hazard communication elements	
1	Any substance which, in the 4:1 or 1:1 sample-to-cellulose ratio (by mass) tested, exhibits a mean burning time less than the mean burning time of a 3:2 mixture, by mass, of potassium bromate and cellulose.	Symbol	
		Signal word	Danger
		Hazard statement	May cause fire or explosion; strong oxidizer
2	Any substance which, in the 4:1 or 1:1 sample-to-cellulose ratio (by mass) tested, exhibits a mean burning time equal to or less than the mean burning time of a 2:3 mixture (by mass) of potassium bromate and cellulose and the criteria for category 1 are not met.	Symbol	
		Signal word	Danger
		Hazard statement	May intensify fire; oxidizer
3	Any substance which, in the 4:1 or 1:1 sample-to-cellulose ratio (by mass) tested, exhibits a mean burning time equal to or less than the mean burning time of a 3:7 mixture (by mass) of potassium bromate and cellulose and the criteria for categories 1 and 2 are not met.	Symbol	
		Signal word	Warning
		Hazard statement	May intensify fire; oxidizer

A2.15 **Organic peroxides** (See Chapter 2.15 for details)

Hazard category	Criteria	Hazard communication elements	
Type A	According to the results of test series A to H in the *UN Recommendations on the Transport of Dangerous Goods, Manual of Tests and Criteria,* Part II and the application of the decision logic under 2.15.4.1 of Chapter 2.15.	Symbol	
		Signal word	Danger
		Hazard statement	Heating may cause an explosion
Type B	According to the results of test series A to H in the *UN Recommendations on the Transport of Dangerous Goods, Manual of Tests and Criteria,* Part II and the application of the decision logic under 2.15.4.1 of Chapter 2.15.	Symbol	 
		Signal word	Danger
		Hazard statement	Heating may cause a fire or explosion
Type C and D	According to the results of test series A to H in the *UN Recommendations on the Transport of Dangerous Goods, Manual of Tests and Criteria,* Part II and the application of the decision logic under 2.15.4.1 of Chapter 2.15.	Symbol	
		Signal word	Danger
		Hazard statement	Heating may cause a fire
Type E and F	According to the results of test series A to H in the *UN Recommendations on the Transport of Dangerous Good, Manual of Tests and Criteria,* Part II and the application of the decision logic under 2.15.4.1 of Chapter 2.15.	Symbol	
		Signal word	Warning
		Hazard statement	Heating may cause a fire
Type G	According to the results of test series A to H in the *UN Recommendations on the Transport of Dangerous Goods, Manual of Tests and Criteria,* Part II and the application of the decision logic under 2.15.4.1 of Chapter 2.15.	Signal word	There are no label elements allocated to this hazard category
		Symbol	
		Hazard statement	

A2.16 Corrosive to metals (See Chapter 2.16 for details)

Hazard category	Criteria	Hazard communication elements	
1	Corrosion rate on steel or aluminium surfaces exceeding 6.25 mm per year at a test temperature of 55 °C.	Symbol	
		Signal word	Warning
		Hazard statement	May be corrosive to metals

A2.17 Acute toxicity (See Chapter 3.1 for details)

Hazard category	Criteria	Hazard communication elements	
1	$LD_{50} \leq$ 5 mg/kg bodyweight (oral) $LD_{50} \leq$ 50 mg/kg bodyweight (skin/dermal) $LC_{50} \leq$ 100 ppm (gas) $LC_{50} \leq$ 0.5 (mg/l) (vapour) $LC_{50} \leq$ 0.05 (mg/l) (dust,mist)	Symbol	
		Signal word	Danger
		Hazard statement	Fatal if swallowed. (oral) Fatal in contact with skin (dermal) Fatal if inhaled (gas, vapour, dust, mist)
2	LD_{50} between 5 and less than 50 mg/kg bodyweight (oral) LD50 between 50 and less than 200 mg/kg bodyweight (skin/dermal) LC50 between 100 and less than 500 ppm (gas) LC50 between 0.5 and less than2.0 (mg/l) (vapour) LC50 between 0.05 and less than 0.5 (mg/l) (dust, mist)	Symbol	
		Signal word	Danger
		Hazard Statement	Fatal if swallowed. (oral) Fatal in contact with skin (dermal) Fatal if inhaled (gas, vapour, dust, mist)
3	LD_{50} between 50 and less than 300 mg/kg bodyweight (oral) LD_{50} between 200 and less than 1000 mg/kg bodyweight (skin/dermal) LC_{50} between 500 and less than 2500 ppm (gas) LC_{50} between 2.0 and less than 10.0 (mg/l) (vapour) LC_{50} between 0.5 and less than 1.0 (mg/l) (dust, mist)	Symbol	
		Signal word	Danger
		Hazard statement	Toxic if swallowed. (oral) Toxic in contact with skin (dermal) Toxic if inhaled (gas, vapour, dust, mist)

Continued on next page

Hazard category (cont'd)	**Criteria**	**Hazard communication elements**	
4	LD_{50} between 300 and less than 2000 mg/kg bodyweight (oral) LD_{50} between 1000 and less than 2000 mg/kg bodyweight (skin/dermal) LC_{50} between 2500 and less than 5000 ppm (gas) LC50 between 10.0 and less than 20.0 (mg/l) (vapour) LC50 between 1.0 and less than 5.0 (mg/l) (dust, mist)	Symbol	!
		Signal word	Warning
		Hazard statement	Harmful if swallowed. (oral) Harmful in contact with skin (dermal) Harmful if inhaled (gas, vapour, dust, mist)
5	LD_{50} between 2000 and 5000 (oral or skin/dermal) For gases, vapours, dusts, mists, LC_{50} in the equivalent range of the oral and dermal LD_{50} (i.e., between 2000 and 5000 mg/kg bodyweight). See also the additional criteria • Indication of significant effect in humans • Any mortality at Category 4 • Significant clinical signs at Category 4 • Indication from other studies.	Symbol	No symbol
		Signal word	Warning
		Hazard statement	May be harmful if swallowed (oral) May be harmful in contact with skin (dermal) May be harmful if inhaled (gas, vapour, dust, mist)

A2.18 Skin corrosion/irritation (See Chapter 3.2 for details)

<table>
<tr><th>Hazard category</th><th>Criteria</th><th colspan="2">Hazard communication elements</th></tr>
<tr><td rowspan="3">1

Corrosive Including sub-categories A, B, and C; see Chapter 3.2, Table 3.2.1</td><td rowspan="3">1. For Substances and Tested Mixtures:
• Human experience showing irreversible damage to the skin;
• Structure/activity or structure property relationship to a substance or mixture already classified as corrosive;
• pH extremes of ≤ 2 and ≥ 11.5 including acid/alkali reserve capacity;
• Positive results in a valid and accepted in vitro skin corrosion test; or
• Animal experience or test data that indicate that the substance/mixture causes irreversible damage to the skin following exposure of up to 4 hours (See Table 3.2.1).
2. If data for a mixture are not available, use bridging principles in 3.2.3.2.
3. If bridging principles do not apply,
(a) For mixtures where substances can be added: Classify as corrosive if the sum of the concentrations of corrosive substances in the mixture is $\geq 5\%$ (for substances with additivity); or
(b) For mixtures where substances cannot be added: $\geq 1\%$. See 3.2.3.3.4.</td><td>Symbol</td><td></td></tr>
<tr><td>Signal word</td><td>Danger</td></tr>
<tr><td>Hazard statement</td><td>Causes severe skin burns and eye damage</td></tr>
<tr><td colspan="4" align="right">Continued on next page</td></tr>
</table>

<table>
<tr><th>Hazard category (cont'd)</th><th>Criteria</th><th colspan="2">Hazard communication elements</th></tr>
<tr><td rowspan="3">2

Irritant

(applies to all authorities)</td><td rowspan="3">1. For Substances and Tested Mixtures
• Human experience or data showing reversible damage to the skin following exposure of up to 4 hours;
• Structure/activity or structure property relationship to a substance or mixture already classified as an irritant;
• Positive results in a valid and accepted in vitro skin irritation test; or
• Animal experience or test data that indicate that the substance/mixture causes reversible damage to the skin following exposure of up to 4 hours , mean value of $\geq$ 2.3 $<$ 4.0 for erythema/eschar or for oedema , or inflammation that persists to the end of the observation period, in 2 of 3 tested animals (Table3.2.2).
2. If data for a mixture are not available, use bridging principles in 3.2.3.2.
3. If bridging principles do not apply, classify as an irritant if:
(a) For mixtures where substances can be added: the sum of concentrations of corrosive substances in the mixture is $\geq$ 1% but $\leq$ 5%; the sum of the concentrations of irritant substances is $\geq$ 10%; or the sum of (10 x the concentrations of corrosive ingredients) + (the concentrations of irritant ingredients) is $\geq$ 10%; or
(b) For mixtures where substances cannot be added: $\geq$ 3%. (See 3.2.3.3.4).</td><td>Symbol</td><td>!</td></tr>
<tr><td>Signal word</td><td>Warning</td></tr>
<tr><td>Hazard statement</td><td>Causes skin irritation</td></tr>
</table>

Continued on next page

<table>
<tr><th>Hazard category (cont'd)</th><th>Criteria</th><th colspan="2">Hazard communication elements</th></tr>
<tr><td rowspan="3">3

Mild Irritant

(applies to some authorities)</td><td rowspan="3">1. For Substances and Tested Mixtures
• Animal experience or test data that indicates that the substance/mixture causes reversible damage to the skin following exposure of up to 4 hours, mean value of $\geq$ 1.5 < 2.3 for erythema/eschar in 2 of 3 tested animals (See Table 3.2.2).
2. If data for a mixture are not available and the bridging principles in 3.2.3.2.
3. If bridging principles do not apply, classify as mild irritant if:
• For mixtures where substances can be added the sum of the concentrations of irritant substances in the mixture is $\geq$ 1% but $\leq$ 10%;
• For mixtures where substances cannot be added: the sum of the concentrations of mild irritant substances is $\geq$ 10%;
• the sum of (10 x the concentrations of corrosive substances) + (the concentrations of irritant substances) is $\geq$ 1% but $\leq$ 10%; or
• the sum of (10 x the concentrations of corrosive substances) + (the concentrations of irritant substances) + (the concentrations of mild irritant substances) is $\geq$ 10%.</td><td>Symbol</td><td>None</td></tr>
<tr><td>Signal word</td><td>Warning</td></tr>
<tr><td>Hazard statement</td><td>Causes mild skin irritation</td></tr>
</table>

A2.19 Serious eye damage / eye irritation (See Chapter 3.3 for details)

Hazard category	Criteria	Hazard communication elements	
1 **Irreversible Effects**	1. *For Substances and Tested Mixtures* • Classification as corrosive to skin; • Human experience or data showing damage to the eye which is not fully reversible within 21 days; • Structure/activity or structure property relationship to a substance or mixture already classified as corrosive; • pH extremes of < 2 and > 11.5 including buffering capacity; • Positive results in a valid and accepted *in vitro* test to assess serious damage to eyes; or • Animal experience or test data that the substance or mixture produces either (1) in at least one animal, effects on the cornea, iris or conjunctiva that are not expected to reverse or have not reversed; or (2) in at least 2 of 3 tested animals a positive response of corneal opacity ≥ 3 and/or iritis >1.5 (see Table3.3.1). 2. *If data for a mixture are not available*, use bridging principles in 3.3.3.2. 3. *If bridging principles do not apply*, (a) For mixtures where substances can be added: Classify as Category 1 if the sum of the concentrations of substances classified as corrosive to the skin and/or eye Category 1 substances in the mixture is $\geq 3\%$; or (b) For mixtures where substances cannot be added:≥ 1 (see 3.3.3.3.4).	Symbol	
		Signal word	Danger
		Hazard statement	Causes serious eye damage

Continued on next page

<table>
<tr><th>Hazard category (cont'd)</th><th>Criteria</th><th colspan="2">Hazard communication elements</th></tr>
<tr><td rowspan="3">2A

Irritant</td><td rowspan="3">1. Substances and tested mixtures
• Classification as severe skin irritant;
• Human experience or data showing production of changes in the eye which are fully reversible within 21 days;
• Structure/activity or structure property relationship to a substance or mixture already classified as an eye irritant;
• Positive results in a valid and accepted in vitro eye irritation test; or
• Animal experience or test data that indicate that the substance/mixture produces a positive response in at least 2 of 3 tested animals of : corneal opacity ≥ 1, iritis ≥ 1, or conjunctival edema (chemosis) ≥ 2 (Table 3.3.2).
2. If data for a mixture are not available, use bridging principles in 3.3.3.2.
3. If bridging does not apply, classify as an irritant (2A) if:
(a) For mixtures where substances can be added: the sum of the concentrations of skin and/or eye Category 1 substances in the mixture is $\geq$ 1% but $\leq$ 3%; the sum of the concentrations of eye irritant substances is $\geq$ 10%; or the sum of (10 x the concentrations of skin and/or eye category 1 substances) + (the concentrations of eye irritants) is $\geq$ 10%;
(b) For mixtures where substances cannot be added: the sum of the concentrations of eye irritant ingredients is $\geq$ 3% (see 3.3.3.3.4).</td><td>Symbol</td><td>!</td></tr>
<tr><td>Signal word</td><td>Warning</td></tr>
<tr><td>Hazard statement</td><td>Causes serious eye irritation</td></tr>
<tr><td rowspan="3">2B

Mild Irritant</td><td rowspan="3">1. For Substances and tested mixtures
• Human experience or data showing production of mild eye irritation;
• Animal experience or test data that indicate that the lesions are fully reversible within 7 days (see Table 3.3.2).
2. If data for a mixture are not available, use bridging principles in 3.3.3.2.
3. If bridging does not apply, classify as an irritant (2B) if:
(a)For mixtures where substances can be added: the sum of the concentrations of skin and/or eye Category 1 substances in the mixture is $\geq$ 1% but $\leq$ 3%; the sum of the concentrations of eye irritant substances is $\geq$ 10%; or the sum of (10 x the concentrations of skin and/or eye category 1 substances) + (the concentrations of eye irritants) is $\geq$ 10%;
(b) For mixtures where substances cannot be added: the sum of the concentrations of eye irritant ingredients is $\geq$ 3% (see 3.3.3.3.4).</td><td>Symbol</td><td>No symbol</td></tr>
<tr><td>Signal word</td><td>Warning</td></tr>
<tr><td>Hazard statement</td><td>Causes eye irritation</td></tr>
</table>

A2.20 **Respiratory sensitizer** (See Chapter 3.4 for details)

Hazard category	Criteria	Hazard communication element	
1	1. *For Substances and Tested Mixture* If there is human evidence that the individual substance induces specific respiratory hypersensitivity, and/or Where there are positive results from an appropriate animal test. 2. *If these mixture meets the criteria* set forth in the "Bridging Principles" through one of the following: (a) Dilution; (b) Batching; (c) Substantially Similar Mixture. 3. *If bridging principles do not apply*, classify if any individual respiratory sensitizer in the mixture has a concentration of: ≥ 1.0% Solid/Liquid ≥ 0.2% Gas.	Symbol	
		Signal word	Danger
		Hazard statement	May cause allergic or asthmatic symptoms or breathing difficulties if inhaled

A2.21 Skin sensitizer (See Chapter 3.4 for details)

<table>
<tr><th>Hazard category</th><th>Criteria</th><th colspan="2">Hazard communication element</th></tr>
<tr><td rowspan="3">1</td><td rowspan="3">1. For Substances and tested mixture
If there is evidence in humans that the individual substance can induce sensitization by skin contact in a substantial number of persons, or
Where there are positive results from an appropriate animal test.
2. If the mixture meets the criteria set forth in the “Bridging Principles” through one of the following:
(a) Dilution;
(b) Batching;
(c) Substantially similar mixture.
4. If bridging principles do not apply, classify if any individual skin sensitizer in the mixture has a concentration of:
≥ 1.0% Solid/Liquid/Gas.</td><td>Symbol</td><td>!</td></tr>
<tr><td>Signal word</td><td>Warning</td></tr>
<tr><td>Hazard Statement</td><td>May cause allergic skin reaction</td></tr>
</table>

A2.22 Mutagenicity (See Chapter 3.5 for details)

Hazard Category	Criteria for classification	Hazard communication elements	
1 (Both 1A and 1B)	Known to induce heritable mutations or regarded as if it induces heritable mutations in the germ cells of humans (see criteria in 3.5.2) or mixtures containing ≥0.1 % of such a substance.	Symbol	
		Signal word	Danger
		Hazard statement	May cause genetic defects (state route of exposure if it is conclusively proven that no other routes of exposure cause the hazard)
2	Causes concern for man owing to the possibility that it may induce heritable mutations in the germ cells of humans (see criteria in 3.5.2) or mixtures containing ≥1.0 % of such a substance.	Symbol	
		Signal word	Warning
		Hazard Statement	Suspected of causing genetic defects (state route of exposure if it is conclusively proven that no other routes of exposure cause the hazard)

A2.23 **Carcinogenicity** (See Chapter 3.6 for details)

Hazard category	Criteria	Hazard communication elements	
1 (both 1A and 1B)	Known or Presumed Human Carcinogen including mixtures containing ≥ 0.1% of such a substance.	Symbol	
		Signal word	Danger
		Hazard statement	May cause cancer (state route of exposure if it is conclusively proven that no other routes of exposure cause the hazard.
2	Suspected human carcinogen Including mixtures containing more than ≥ 0.1 or ≥1.0 % of such a substance (see Notes 1 and 2 in Table 3.6.1 of Chapter 3.6).	Symbol	
		Signal word	Warning
		Hazard statement	Suspected of causing cancer (state route of exposure if it is conclusively proven that no other routes of exposure cause the hazard*)

* *Some authorities will choose to label according to this provision, others may not.*

A2.24 (a) Toxic to reproduction (See Chapter 3.7 for details)

<table>
<tr><th>Hazard category</th><th>Criteria</th><th colspan="2">Hazard communication elements</th></tr>
<tr><td rowspan="3">1
(Both 1A and 1B)</td><td rowspan="3">Known or presumed human reproductive toxicants (see criteria in 3.7.2.2.1 to 3.7.2.6.0 of Chapter 3.7)
or mixtures containing ≥ 0.1% or ≥0.3 % of such a substance
(see notes 1 and 2 of Table 3.7.1, Chapter 3.7).</td><td>Symbol</td><td></td></tr>
<tr><td>Signal word</td><td>Danger</td></tr>
<tr><td>Hazard statement</td><td>May damage fertility or the unborn child (state specific effect if known) (state route of exposure if it is conclusively proven that no other routes of exposure cause the hazard)</td></tr>
<tr><td rowspan="3">2</td><td rowspan="3">Suspected human reproductive toxicants (see criteria in 3.7.2.2.1 to 3.7.2.6.0 of Chapter 3.7)
or mixtures containing ≥ 0.1% or ≥3.0 % of such a substance
(see Notes 3 and 4 of Table 3.7.1, Chapter 3.7).</td><td>Symbol</td><td></td></tr>
<tr><td>Signal word</td><td>Warning</td></tr>
<tr><td>Hazard statement</td><td>Suspected of damaging fertility or the unborn child (state specific effect if known) (state route of exposure if it is conclusively proven that no other routes of exposure cause the hazard)</td></tr>
<tr><td colspan="4" align="right">Continued on next page</td></tr>
</table>

A2.24 (b) **Effects on or via lactation** (See Chapter 3.7)

Hazard category (cont'd)	Criteria	Hazard communication elements	
Special category	Substances which cause concern for the health of breastfed children (see criteria in 3.7.2.2.1 to 3.7.2.6.0 and 3.7.3.4 of Chapter 3.7).	Symbol	No symbol
		Signal word	No signal word
		Hazard Statement	May cause harm to breast-fed children

A2.25 Target organ systemic toxicity following single exposure (See Chapter 3.8 for details)

Hazard category	Criteria	Hazard communication elements	
1	Reliable evidence on the substance or mixture (including bridging) of an adverse effect on specific organ/systems or systemic toxicity in humans or animals. May use guidance values in Table 3.8.1, Category 1 criteria as part of weight of evidence evaluation. May be named for specific organ/system. Mixture that lacks sufficient data, but contains Category 1 ingredient at a concentration of ≥ 1.0 to ≤ 10.0% for some authorities; and ≥10.0% for all authorities.	Symbol	
		Signal word	Danger
		Hazard statement	Causes damage to organs (state all organs affected, if known) through prolonged or repeated exposure (state route of exposure if it is conclusively proven that no other routes of exposure cause the hazard)
2	Evidence on the substance or mixture (including bridging) of an adverse effect on specific organ/systems or systemic toxicity from animal studies or humans considering weight of evidence and guidance values in Table 3.8.1, Category 2 criteria. May be named for specific organ/system affected. Mixture that lacks sufficient data, but contains Category 1 ingredient: ≥ 1 but ≤10% for some authorities; and /or contains Category 2 ingredient: ≥ 1 to ≤10% for some authorities; and ≥10% for all authorities.	Symbol	
		Signal word	Warning
		Hazard statement	May causes lamage to organs (state all organs affected, if known) through prolonged or repeated exposure (state oute of exposure if it is conclusively proven that no other routes of exposure cause the hazard)

A2.26 Target organ systemic toxicity following repeat exposure (See Chapter 3.9 for details)

<table>
<tr><th>Hazard category</th><th>Criteria</th><th colspan="2">Hazard communication elements</th></tr>
<tr><td rowspan="3">1</td><td rowspan="3">Reliable evidence on the substance or mixture (including bridging) of an adverse effect on specific organ/systems or systemic toxicity in humans or animals. May use guidance values in Table 3.91 as part of weight of evidence evaluation. May be named for specific organ/system.
Mixture that lacks sufficient data, but contains Category 1 ingredient: ≥ 1 to ≤ 10% for some authorities; and ≥10% for all authorities.</td><td>Symbol</td><td></td></tr>
<tr><td>Signal word</td><td>Danger</td></tr>
<tr><td>Hazard statement</td><td>Causes damage to organs (state all organs affected, if known) through prolonged or repeated exposure (state route of exposure if it is conclusively proven that no other routes of exposure cause the hazard)</td></tr>
<tr><td rowspan="3">2</td><td rowspan="3">Evidence on the substance or mixture (including bridging) of an adverse effect on specific organ/systems or systemic toxicity from animal studies or humans considering weight of evidence and guidance values in Table 3.9.2 criteria. May be named for specific organ/system.
Mixture that lacks sufficient data, but contains Category 1 ingredient: ≥ 1.0 but ≤10% for some authorities (see Note 3 of Table 3.9.3)
and /or
contains Category 2 ingredient: ≥ 1.0 or ≥10%.</td><td>Symbol</td><td></td></tr>
<tr><td>Signal word</td><td>Warning</td></tr>
<tr><td>Hazard statement</td><td>May cause damage to organs (state all organs affected, if known) through prolonged or repeated exposure (state route of exposure if it is conclusively proven that no other routes of exposure cause the hazard)</td></tr>
</table>

A2.27 Acute hazards to the aquatic environment (See Chapter 3.10 for details)

Hazard category	Criteria	Hazard communication elements	
1	1. *For Substances and Tested Mixtures:* • $L(E)C_{50} \leq 1mg/L$ where $L(E)C_{50}$ is either fish 96hr LC_{50}, crustacea 48hr EC LC_{50} or aquatic plant 72 or 96hr ErC_{50}. 2. *If data for a mixture are not available,* use bridging principles (see 3.10.3.4). 3. *If bridging principles do not apply,* (a) For mixtures with classified ingredients: The summation method (see 3.10.3.10.3.5.5) reveals: • [Concentration of Acute 1] × M > 25% where M is a multiplying factor (see 3.10.3.5.5.5). (b) For mixtures with tested ingredients: The additivity formula (see 3.10.3.5.2 and 3.10.3.5.3) reveals: • $L(E)C_{50} \leq 1mg/L$. (c) For mixtures with both classified and tested ingredients: The combined additivity formula and summation method (see paragraphs 3.10.3.5.2 to 3.10.3.5.5.3) reveal: • Concentration of Acute 1 × M > 25%. 4. *For mixtures with no usable information for one or more relevant ingredients,* classify using the available information and add the statement: "x percent of the mixture consists of component(s) of unknown hazards to the aquatic environment".	Symbol	[environment symbol]
		Signal word	Warning
		Hazard statement	Very toxic to aquatic life

Continued on next page

<table>
<tr><th>Hazard category (cont'd)</th><th>Criteria</th><th colspan="2">Hazard communication elements</th></tr>
<tr><td rowspan="3">2</td><td rowspan="3">1. For Substances and Tested Mixtures:
• $1mg/L < L(E)C_{50} \leq 10mg/L$
where $L(E)C_{50}$ is either fish 96hr LC_{50}, crustacea 48hr EC LC_{50} or aquatic plant 72 or 96hr ErC_{50}.

2. If data for a mixture are not available, use bridging principles (see 3.10.3.4).

3. If bridging principles do not apply,
(a) For mixtures with classified ingredients: The <u>summation</u> method (see 3.10.3.5.5.1 to 3.10.3.5.5.3) reveals:
• [Concentration of Acute 1] × M × 10 + (Concentration of Acute 2] > 25%
where M is a multiplying factor (see 3.10.3.5.5.5).

(b) For mixtures with tested ingredients: The <u>additivity</u> formula (see 3.10.3.5.2-3.10.3.5.3) reveals:
• $1mg/L < L(E)C_{50} \leq 10mg/L$.

(c) For mixtures with both classified and tested ingredients: The combined <u>additivity</u> formula and <u>summation</u> method (see 3.10.3.5.2-3.10.3.5.5.3) reveal:
• [Concentration of Acute 1] × M x 10 + [Concentration of Acute 2] > 25%.

4. For mixtures with no usable information for one or more relevant ingredients, classify using the available information and add the statement: "x percent of the mixture consists of component(s) of unknown hazards to the aquatic environment".</td><td>Symbol</td><td>No symbol used</td></tr>
<tr><td>Signal word</td><td>No signal word</td></tr>
<tr><td>Hazard statement</td><td>Toxic to aquatic life</td></tr>
<tr><td colspan="4">Continued on next page</td></tr>
</table>

<table>
<tr><th>Hazard category (cont'd)</th><th>Criteria</th><th colspan="2">Hazard communication elements</th></tr>
<tr><td rowspan="3">3</td><td rowspan="3">1. For Substances and Tested Mixtures:
• $10mg/L < L(E)C_{50} \leq 100mg/L$
where $L(E)C_{50}$ is either fish 96hr LC_{50}, crustacea 48hr EC LC_{50} or aquatic plant 72 or 96hr ErC_{50}.

2. If data for a mixture are not available, use bridging principles (see 3.10.3.4).

3. If bridging principles do not apply,

(d) For mixtures with classified ingredients:
The <u>summation</u> method (see 3.10.3.5.5.1 to 3.10.3.5.5.3) reveals:
• [Concentration of Acute 1] × M x 100
+ [Concentration of Acute 2] × 10
+ [Concentration of Acute 3] > 25%
where M is a multiplying factor (see 3.10.3.5.5.5).

(e) For mixtures with tested ingredients:
The additivity formula (see 3.10.3.5.2-3.10.3.5.3) reveals:
• $10mg/L < L(E)C_{50} \leq 100mg/L$.

(f) For mixtures with both classified and tested ingredients:
The combined <u>additivity</u> formula and <u>summation</u> method (see 3.10.3.5.2 to 3.10.3.5.5.3) reveal:
• [Concentration of Acute 1] × M x 100
+ [Concentration of Acute 2] × 10
+ [Concentration of Acute 3] > 25%.

4. For mixtures with no usable information for one or more relevant ingredients, classify using the available information and add the statement:
"x percent of the mixture consists of component(s) of unknown hazards to the aquatic environment".</td><td>Symbol</td><td>No symbol used</td></tr>
<tr><td>Signal word</td><td>No signal word</td></tr>
<tr><td>Hazard statement</td><td>Harmful to aquatic life</td></tr>
</table>

A2.28 Chronic hazards to the aquatic environment (See Chapter 3.10 for details)

Hazard category	Criteria	Hazard communication elements	
1	1. *For Substances:* • $L(E)C_{50} \leq 1mg/L$; and • Lack the potential to rapidly biodegrade and/or have the potential to bioaccumulate (BCF≥ 500 or if absent log Kow ≥ 4) where $L(E)C_{50}$ is either fish 96hr LC_{50}, crustacea 48hr EC LC_{50} or aquatic plant 72 or 96hr ErC_{50}. 2. *For Mixtures,* use bridging principles (see 3.10.3.4). 3. *If bridging principles do not apply,* • [Concentration of Chronic 1] x M > 25% where M is a multiplying factor (see 3.10.3.5.5.5). 4. *For mixtures with no usable information for one or more relevant ingredients,* classify using the available information and add the statement: "x percent of the mixture consists of component(s) of unknown hazards to the aquatic environment".	Symbol	
		Signal word	Warning
		Hazard statement	Very toxic to aquatic life with long lasting effects
2	1. *For Substances:* • 1 mg/L < $L(E)C_{50} \leq 10$ mg/L; and • Lack the potential to rapidly biodegrade and/or have the potential to bioaccumulate (BCF≥ 500 or if absent log Kow ≥ 4); unless • Chronic NOECs > 1mg/L. 2. *For Mixtures,* use bridging (see 3.10.3.4). 3. *If bridging principles do not apply,* • [Concentration of Chronic 1] x M x 10 + [Concentration of Chronic 2] > 25% where M is a multiplying factor (see 3.10.3.5.5.5). 4. *For mixtures with no usable information for one or more relevant ingredients,* classify using the available information and add the statement: "x percent of the mixture consists of component(s) of unknown hazards to the aquatic environment".	Symbol	
		Signal word	No signal word
		Hazard statement	Toxic to aquatic life with long lasting effects

Continued on next page

<table>
<tr><th>Hazard category (Cont'd)</th><th>Criteria</th><th colspan="2">Hazard communication elements</th></tr>
<tr><td rowspan="3">3</td><td rowspan="3">1. For Substances:
• $10\ mg/L < L(E)C_{50} \le 100\ mg/L$; and
• Lack the potential to rapidly biodegrade and/or have the potential to bioaccumulate (BCF$\ge$ 500 or if absent log Kow ≥ 4); unless
• Chronic NOECs > 1mg/L.

2. For Mixtures, use bridging principles (see 3.10.3.4).

3. If bridging principles do not apply,
• [Concentration of Chronic 1] × M × 100 + [Concentration of Chronic 2] × 10 + [Concentration of Chronic 3] $> 25\%$
where M is a multiplying factor (see 3.10.3.5.5.5).

4. For mixtures with no usable information for one or more relevant ingredients, classify using the available information and add the statement: "x percent of the mixture consists of component(s) of unknown hazards to the aquatic environment".</td><td>Symbol</td><td>No symbol used</td></tr>
<tr><td>Signal word</td><td>No signal word</td></tr>
<tr><td>Hazard statement</td><td>Harmful to aquatic life with long lasting effects</td></tr>
<tr><td rowspan="3">4</td><td rowspan="3">1. For Substances:
• poorly soluble and no acute toxicity is observed up the water solubility
• Lack the potential to rapidly biodegrade and/or have the potential to bioaccumulate (BCF$\ge$ 500 or if absent log Kow ≥ 4); unless
• Chronic NOECs > 1mg/L.

2. For Mixtures, use bridging principles (see 3.10.3.4).

3. If bridging principles do not apply,
• Sum of concentrations of components classified as Chronic 1, 2, 3 or 4 $> 25\%$.

4. For mixtures with no usable information for one or more relevant ingredients, classify using the available information and add the statement: "x percent of the mixture consists of component(s) of unknown hazards to the aquatic environment".</td><td>Symbol</td><td>No symbol used</td></tr>
<tr><td>Signal word</td><td>No signal word</td></tr>
<tr><td>Hazard statement</td><td>May cause long lasting harmful effects to aquatic life</td></tr>
</table>

GLOSSARY

The following glossary explains terms commonly used to describe chemical properties and hazards. Some terms are used extensively on chemical labels and materials safety data sheets (MSDSs). In most cases, the terms are explained from a work site standpoint based on health and safety. These definitions may differ from the more technical ones used in a chemistry or physics course, but do serve the purpose of providing working information necessary for field use.

Absorbents Any material that is capable of "soaking up" a spilled liquid. Commonly known as "3M pads" or "spill socks." Absorbents come in many shapes and sizes. For spills on a flat surface such as a concrete pad, the best type is the sausage type absorbent boom that can be placed around the spill. Absorbents are also useful for stopping liquids from entering a storm drain or sewer. Be aware that all absorbents that are used and absorb hazardous materials must be disposed of, as would the hazardous material itself.

Absorption In *respiratory protection,* the processes by which the material inside a cartridge traps the airborne contaminant by taking it into the medium. In *toxicology,* the passage of toxic materials through some body surface into body fluids and tissue. Generally speaking, this refers to passage through the skin, eyes, or mucous membranes. In *decontamination,* it is the process where the contaminant is picked up by a neutral material.

Acceptable entry conditions The conditions that must exist in a Permit-Required Confined Space to allow entry and to ensure that employees involved in the operation can safely enter into and work within the space.

ACGIH (American Council of Governmental Industrial Hygienists) Founded in 1938, this is an organization comprised of persons employed by official government units responsible for programs of industrial hygiene, education, and research. It was founded for the purpose of determining standards of exposure to toxic and otherwise harmful materials in workroom air. The standards are revised annually. See TLV (Threshold Limit Value).

Acid An inorganic or organic compound that (1) reacts with metals to yield hydrogen, (2) reacts with a base to form a salt, (3) dissociates in water to yield hydrogen ions, (4) has a pH of less than 7.0, and (5) neutralizes bases or alkalis. All acids contain hydrogen and turn litmus paper red. They are corrosive to human tissue and should be handled with care. Examples include sulfuric and hydrochloric acid.

Action Level The exposure level (concentration in air) at which OSHA regulations to protect employees take effect (29 CFR 1910.1001–1047); for example, workplace air analysis, employee training, medical monitoring, and recordkeeping. Exposure at or above action level is termed occupational exposure. Exposure below this level can also be harmful. This level is generally half the PEL.

Acute exposure Exposure of short duration, usually to relatively high concentrations or high amounts of the material.

Acute health effect Adverse effect on a human or animal which has severe symptoms developing rapidly and coming quickly to a crisis. Acute effects are often immediate and can be severe or life threatening. Also see Chronic effect.

Acute toxicity The ability of a substance to do damage (generally systemic) as a result of a one-time exposure from a single dose of or exposure to a material. This exposure is generally brief in duration.

Adhesion A union of two surfaces that are normally separate.

Adiabatic heat The technical definition is descriptive of a system in which no net heat loss or gain is allowed.

Administrative controls Include work rules, schedules, standard procedures, and other methods to limit the potential exposure of the worker.

Adsorption The processes by which the material inside a cartridge traps the airborne contaminant by attaching to the contaminant.

Aerosol A fine aerial suspension of particles sufficiently small in size to confer some degree of stability from sedimentation (e.g., smoke or fog).

Air Bill A shipping paper, prepared from a Bill of Lading, that accompanies each piece in an air shipment.

Air-line respirator A respirator that is connected to a compressed breathing air source by a hose of small inside diameter. The air is delivered continuously or intermittently in a sufficient volume to meet the wearer's breathing requirements from a tank or compressor located in a remote location.

Air monitoring The process of evaluating the air in a given work environment. This process can be accomplished using a variety of instrumentation depending on the anticipated hazards. In relation to confined spaces, this is a program that must be carried out by the Confined Space Attendant during confined space entry. Air monitoring includes the use of a Combustible Gas Indicator (see CGI) and a colorimetric tube system (see Colorimetric tubes) to determine if the confined space is oxygen deficient, has a combustible gas, or a known hazardous gas such as hydrogen chloride or ammonia. Air monitoring must be conducted before entry is made, any time conditions change (temperature, humidity, hot work), and on a regular basis during the time entry personnel are in the confined space.

Air-purifying respirator A respirator that uses chemicals to remove specific gases and vapors from the air or that uses a mechanical filter to remove particulate matter. An air-purifying respirator must be used only when there is sufficient oxygen to sustain life and the air contaminant level is below the concentration limits of the device. This type of respirator is effective for concentrations of substances that are generally no more than 10 times the threshold limit value (TLV) of the contaminant but never more than the IDLH value, and if the contaminant has warning properties (odor or irritation) below the TLV.

Air-reactive materials Substances that ignite when exposed to air at normal temperatures. Also called pyrophoric.

Alkali Any chemical substance that forms soluble soaps with fatty acids. Alkalis are also referred to as bases. They may cause severe burns to the skin. Alkalis turn litmus paper blue and have pH values from 7.1 to 14.0. Sodium hydroxide, sodium bicarbonate and ammonium hydroxide are all examples of alkalis.

Allergic reaction An abnormal physiological response to chemical or physical stimuli by a sensitive person.

Alpha particle A small, charged particle emitted from the nucleus of an unstable atom. The small particle is essentially a helium nucleus consisting of two neutrons and two protons. This particle is of high energy and is thrown off by many radioactive elements.

Alpha radiation Emission of a charged particle from the nucleus of an unstable atom.

Alveoli The plural of alveolus. Microscopic air sacs in the lungs, which are surrounded by a blood supply, where oxygen is delivered to the blood, and where hazardous wastes are exchanged.

Ambient temperature The temperature of the surrounding area.

Anesthetic A chemical that causes a total or partial loss of sensation. Exposure to anesthetic can cause impaired judgment, dizziness, drowsiness, headache, unconsciousness, and even death. Examples include alcohol, paint remover, and degreasers.

ANFO Stands for Ammonium Nitrate and Fuel Oil and is an example of a common blasting agent.

Anhydrous material A chemical compound containing no free water. Anhydrous ammonia, for example, would be virtually pure ammonia with only trace amounts of water.

Anion A negatively charged ion. This is created when an atom gains an electron in its orbit. The addition of a negatively charged electron results in an imbalance between the protons and electrons. Because there are more electrons in this case, it shifts the electrical charge to the negatively charged state.

ANSI (American National Standards Institute) ANSI is a privately funded, voluntary membership organization that identifies industrial and public needs for national consensus standards and coordinates development of such standards.

Antidote A remedy to relieve, prevent, or counteract the effects of a poison.

APR See Air-purifying respirator.

Aquatic toxicity The adverse effects to marine life that result from exposure to a toxic substance.

Asbestos A fibrous form of silicate minerals that has many uses in society. Although found naturally in California, its prime use is in fireproof fabrics, brake linings, gaskets, roofing compositions, electrical and mechanical insulation, paint filler, chemical filters, reinforcing agent in rubber and plastics, and tile flooring. Its prime hazard is being a carcinogen (see Carcinogen). It is highly toxic by inhalation of dust particles.

Asphyxiant A vapor or gas that can cause unconsciousness or death by suffocation (lack of oxygen). Most simple asphyxiants are harmful to the body only when they become so concentrated that they reduce oxygen in the air (normally about 21%) to dangerous levels (17% or lower). Asphyxiation is one of the principal potential hazards of working in confined and enclosed spaces.

Asphyxiate The ability to suffocate through the exclusion of oxygen.

ASTM (American Society for Testing and Materials) ASTM is the world's largest source of voluntary consensus standards for materials, products, systems, and services. ASTM is a resource for sampling and testing methods, health and safety aspects of materials, safe performance guidelines, and effects of physical and biological agents and chemicals.

Asymptomatic Showing no symptoms.

atm Atmosphere, a unit of pressure equal to 760 mmHg (mercury) at sea level.

Atomic number The number of protons in the nucleus of a particular atom. This number determines the element's position on the periodic table. See also Proton.

Atomic symbol A one- or two-letter designation for a given element. These symbols are found on the periodic table and pertain to only one element. Symbols are also used in chemical formulas to identify the components of a given substance. LiF, for example, is called lithium fluoride and is made up of the elements lithium and fluorine.

Atomic weight The total weight of any atom is the sum of all its subatomic parts—the protons, neutrons, and electrons.

Atomizing Creating small mist droplets through the introduction of pressure.

Atoms Small particles that make up all matter; they are composed of protons, neutrons, and electrons. They can be stable or unstable based on the relationship between particles.

Authorized attendant See Confined Space Attendant.

Authorized entrant See Confined Space Entrant.

Autoignition temperature The lowest temperature to which a material will ignite spontaneously or burn. This is sometimes referred to as ignition temperature. Almost all materials have an autoignition temperature because this is the level to which most materials must be heated in order to begin to burn.

Backdraft explosion An explosion caused when air is suddenly introduced into a confined area that contains high levels of heat and combustible gases.

Base See Alkali.

Benign The term expressing that the disease (usually tumor) is not recurrent, not progressing, and/or not malignant.

Beta particle A by-product of radioactive decay from an unstable nucleus. The process changes a neutron into a proton and subsequently emits an electron from its orbit. The released electron can take the form of a positron (positive charge) or a negatron (negative charge). This type of decay involves a change in the atomic number but no change in the mass of the atom. Beta particles are of high velocity and in some cases exceed 98% of the speed of light.

Beta radiation Particulate radiation that is much smaller than alpha particles, moves faster, and has the ability to travel greater distances from the nucleus.

Bill of Lading A standard shipping document carried in trucks that ship hazardous goods.

Biodegradable Capable of being broken down into innocuous products by the action of living things. If the material is biodegradable, it is generally less hazardous to the environment.

Blanking or blinding The absolute closure of a pipe, line, or duct by the fastening of a solid plate (such as a spectacle blind or a skillet blind) that completely covers the bore and that is capable of withstanding the maximum pressure of the pipe, line, or duct with no leakage beyond the plate.

Blasting agents The lowest explosion group, and often used to initiate other explosive charges.

BLEVE (Boiling Liquid Expanding Vapor Explosion) A major failure of a closed liquid container into two or more pieces. It is usually caused when the temperature of the liquid is well above its boiling point at normal atmospheric pressure.

Boiling point The temperature at which a liquid changes to a vapor state at a given pressure. The boiling point is usually expressed in degrees Fahrenheit at sea level pressure (760 mmHg, or 1 atmosphere). For mixtures, the initial boiling point or the boiling range may be given. Some examples of flammable materials with low boiling points include:

Propane	–44°F
Anhydrous ammonia	–28°F
Butane	31°F
Gasoline	>100°F
Ethylene glycol	387°F

BOM, or BuMines Bureau of Mines, US Department of Interior.

Bonding The interconnecting of two objects by means of a clamp and bare wire. Its purpose is to equalize the electrical potential between the objects to prevent a static discharge when transferring a flammable liquid from one container to another. The conductive path is provided by clamps which make contact with the charged object and a low resistance flexible cable which allows the charge to equalize. See Grounding.

Bottle watch A person who monitors the outside air supply when an air-line system is in use.

Brisance An expression of the shattering effect of a particular explosive material.

Bulk shipment A shipment of specific amounts of hazardous materials above the limits found in 49 CFR, Part 171.8.

C Centigrade, a unit of temperature that has a scale ranging from 0 to 100, with 0 being the temperature at which water freezes and 100 the temperature at which water boils.

"C" or Ceiling The maximum allowable human exposure limit, usually for an airborne substance, that is not to be exceeded even momentarily. Also see PEL and TLV.

Cancer, carcinoma An abnormal multiplication of cells that tends to infiltrate other tissues and metastasize (spread). Each cancer is believed to originate from a single "transformed" cell that grows (splits) at a fast, abnormally regulated pace, no matter where it occurs in the body. Cancer is the second most common cause of death in the United States. Most cancers are caused by our lifestyle—that is smoking and diet.

Carcinogen A material that either causes cancer in humans or, because it causes cancer in animals, is considered capable of causing cancer in humans. Findings are based on the feeding of large quantities of a material to test animals or by the application of concentrated solutions to the animals' skin. A material is considered a carcinogen if: (1) the International Agency for Research on Cancer (IARC) has evaluated and found it a carcinogen or potential carcinogen, (2) the National Toxicology Program's (NTP's) Annual Report on Carcinogens lists it as a carcinogen or potential carcinogen, (3) OSHA regulates it as a carcinogen, or (4) one positive study has been published. Following is a listing of one breakdown of carcinogens. It is the system used by the ACGIH in their classification of carcinogens.

A1—Confirmed Human Carcinogen: The agent is carcinogenic to humans based on the weight of evidence from epidemiological studies of or convincing clinical evidence in exposed humans.

A2—Suspected Human Carcinogen: The agent is carcinogenic in experimental animals at dose levels, by route(s) of administrations, at site(s), of histologic type(s), or by mechanism(s) that are considered relevant to worker exposure. Available epidemiological studies are conflicting or insufficient to confirm an increased risk of cancer in exposed humans.

A3—Animal Carcinogen: The agent is carcinogenic in experimental animals at relatively high dose, by route(s) of administration, at site(s), of histologic types(s), or by mechanisms(s) that are not considered relevant to worker exposure. Available epidemiological studies do not confirm an increased risk of cancer in exposed humans. Available evidence suggests that the agent is not likely to cause cancer in humans except under uncommon or unlikely routes or levels of exposure.

A4—Not Classifiable as a Human Carcinogen: There are inadequate data on which to classify the agent in terms of its carcinogenicity in humans and/or animals.

A5—Not Suspected as a Human Carcinogen: The agent is not suspected to be a human carcinogen on the basis of properly conducted epidemiological studies in humans. These studies have sufficiently long follow-up, reliable exposure histories, sufficiently high dose, and adequate statistical power to conclude that exposure to the agent does not convey a significant risk of cancer to humans. Evidence suggesting a lack of carcinogenicity in experimental animals will be considered if it is supported by other relevant data.

Carcinogenic Capable of causing cancer.

Carcinogenicity The ability to cause or contribute to cancer.

CAS Number (CAS Registration Number) An assigned number used to identify a chemical. CAS stands for Chemical Abstracts Service, an organization that indexes information published in Chemical Abstracts by the American Chemical Society and that provides index guides by which information about particular substances may be located in the abstracts. The CAS number is concise, unique means of material identification.

Catalyst A substance that modifies (slows, or more often quickens) a chemical reaction without being consumed in the reaction.

Cation A positively charged ion. This is created when an atom loses an electron from its orbit. The reduction of a negatively charged electron results in an imbalance between the protons and electrons. Because there are more protons in this case, it shifts the electrical charge to the positively charged state.

Caustic See Alkali.

cc Cubic centimeter is a volume measurement in the metric system, which is equal in capacity to one milliliter (ml). One quart is about 946 cubic centimeters.

Central nervous system (CNS) The term that generally refers to the brain and spinal cord. These organs supervise and coordinate the activity of the entire nervous system. Sensory impulses are transmitted into the central nervous system and motor impulses are transmitted out. The remainder of the nervous system that is not part of the CNS is generally classified as the peripheral nervous system.

CERCLA (Comprehensive Environmental Response, Compensation, and Liability Act of 1980) The Act requires that the Coast Guard provide a national response capability that can be used in the event of a hazardous substance release. The Act also provides for a fund (the Superfund) to be used for the cleanup of abandoned hazardous waste disposal sites.

CFR (Code of Federal Regulations) These are the codes that codify the various federal regulations. The most important example for workplace safety is book number 29 (29 CFR) that contains the OSHA regulations. Other examples include 40 CFR that contains EPA regulations and 49 CFR that contains Department of Transportation regulations.

CGI (Combustible Gas Indicator) A mechanical device that can detect any gas or vapor with a defined flash point (see Flash point) and Lower Explosive Limit (see LEL), if the concentration is high enough. This includes both flammable and combustible materials. Most CGIs also have the capability to monitor oxygen levels in the atmosphere. This is especially important for monitoring areas that have the potential to be oxygen deficient (below 19.5% oxygen) or oxygen enriched (above 23% oxygen).

Chain of custody Documentation of the who, where, how, and what of a sample; includes appropriately labeling the samples, identifying who picked them up, where they were acquired, how they were preserved, and who received them.

Chemical Any element, chemical compound, or mixture of elements and/or compounds where chemical(s) are distributed.

Chemical cartridge respirator See Air-purifying respirator.

Chemical family A group of single elements or compounds with a common general name. Examples: acetone, methyl ethyl ketone (MEK), and methyl isobutyl ketone (MIBK) are of the ketone family; acrolein, furfural, and acetaldehyde are of the aldehyde family.

Chemical Hygiene Plan A written program developed and implemented by the employer setting forth procedures, equipment, personal protective equipment, and work practices that (1) are capable of protecting employees from the health hazards presented by hazardous chemicals used in the particular workplace and (2) meet the requirements of the laboratory standard.

Chemical name The name given to a chemical in the nomenclature system developed by the International Union of Pure and Applied Chemistry (IUPAC) or the Chemical Abstracts Service (CAS).

Chemical pneumonia A condition caused when hazardous materials are drawn into the lungs, causing fluid to build.

Chemical structure The arrangement within the molecule of atoms and their chemical bonds.

Chemistry The branch of study concerning the composition of chemical substances and their effects and interactions with one another. This includes the study of atoms and their behavior, the makeup of chemical compounds, the reactions that occur between these compounds and the energies resulting from those reactions. Chemistry is not an isolated discipline, however, and it includes theories from thermodynamics, physics, and biology.

CHEMTREC (Chemical Transportation Emergency Center) CHEMTREC is a national center established by the Chemical Manufacturers Association (CMA) to relay pertinent emergency information concerning specific chemicals on requests from individuals and response agencies. CHEMTREC has a 24 hour toll free telephone number (800-424-9300) to help agencies who respond to chemical transportation emergencies.

Chiming a drum Describes moving a drum by rolling it on the bottom ring.

Chronic exposure Long-term contact with a substance or repeated exposures to the material. Such exposures often result in long-term health effects such as the development of cancers or organ damage.

Chronic health effect Adverse health effects that result from exposures to the same material multiple times and generally at a lower level or dose.

Chronic toxicity Harmful systemic effects produced by long-term, low-level exposure to chemicals. Typically, this period of time is considered to be more than three months.

Class A explosive Detonating or otherwise of maximum explosion hazard. The nine types of Class A explosives are defined in Sec. 173.53 of 49 CFR.

Class A Flammable Materials These are fires involving solid organic materials including wood, cloth, paper, and many plastics.

Class B Explosive In general, the type of explosive that functions by rapid combustion rather than detonation and include some explosive devices such as special fireworks, flash powders, etc. Flammable hazard (Sec. 173.88 of 49 CFR).

Class C Explosive Certain types of materials that are manufactured articles containing Class A or Class B Explosives, or both, as components but in restricted quantities, and certain types of fireworks. These present a minimum hazard.

Class IA Flammable Liquids A class of flammable liquids with a flash point below 73°F and a boiling point below 100°F.

Class IB Flammable Liquids A class of flammable liquids with a flash point below 73°F and a boiling point at or above 100°F.

Class IC Flammable Liquids A class of flammable liquids with a flash point at or above 73°F and below 100°F.

Class II Liquids A class of combustible liquids with a flash point at or above 100°F and below 140°F.

Class III Liquids A class of flammable liquids with a flash point above 140°F.

Clean Air Act Federal law enacted to regulate and reduce air pollution and administered by the Environmental Protection Agency (EPA).

Clean Water Act (CWA) Federal law enacted to regulate and reduce water pollution. The CWA is administered by the EPA.

Clinical toxicology Designates within the realm of medical science an area of professional emphasis concerned with diseases caused by, or uniquely associated with, toxic substances. This area of toxicology generally deals with drug overdoses and its treatment.

Closed-circuit system Reuses the exhaled air by recycling it through filters inside the backpack where more oxygen is added to it.

CO (carbon monoxide) CO is a colorless, odorless, flammable, and very toxic gas produced by the incomplete combustion of carbon. It is a by-product of many chemical processes.

CO_2 (carbon dioxide) CO_2 is a heavy, colorless gas that is produced by the combustion and decomposition of organic substances and as a by-product of many chemical processes. CO_2 will not burn and is relatively nontoxic (although high concentrations, especially in confined spaces, can create hazardous oxygen-deficient environments).

COC Cleveland Open Cup is a flash point test method.

Cold Zone See Zones.

Colorimetric detector tubes These are sealed glass tubes, that, when the tips are broken off and an air sample is drawn through with a bellows device, will turn chemicals in the tube a different color when exposed to a specific air contaminant. There are several different brands available, with the most common being the Draeger system.

Combustible A term used by NFPA, DOT, and others to classify on the basis of flash points certain liquids that burn. NFPA generally defines "combustible liquids" as having a flash point of 100°F (38.7°C) or higher. In 1992, DOT modified its definition to a liquid with a flash point greater than 140°F. Also see Flammable. Another use for the term is with nonliquid substances such as wood and paper. In this case, materials which are capable of burning are often referred to as combustible or as "ordinary combustibles."

Combustible Liquid NFPA classifies this as any liquid having a flash point at or above 100°F (37.8°C). Combustible Liquids are also referred to as either Class II or Class III liquids depending on their flash point. Note that DOT has a definition that is different—see Combustible above.

Command Post The single place where management operations are conducted.

Compatibility chart Document supplied by the manufacturer of the CPC that provides data on degradation and permeation.

Compound A compound can be defined as the chemical combination of two or more elements that results in the creation of unique properties and a definite, identifiable composition of the substance.

Compressed gas (1) A gas or mixture of gases having, in a container, an absolute pressure exceeding 40 psi at 70°F (21.1°C); or (2) a gas or mixture of gases having, in a container, an absolute pressure exceeding 104 psi at 130°F (54.4°C) regardless of the pressure at 70°F (21.1°C); or (3) a liquid having a vapor pressure exceeding 40 psi at 100°F (37.8°C) as determined by ASTM D-323-72.

Concentration The relative amount of a substance when combined or mixed with other substances. Examples: 2 ppm hydrogen sulfide in air, or a 50% caustic solution.

Confined space Any area as defined by OSHA that has all of the following characteristics: (1) is large enough so that a person can bodily enter, (2) is not designed for continuous occupancy, and (3) has limited ingress and egress. To be classified as a "permit-required" confined space, the area must also possess, or have the potential to possess, one of the following: (1) an oxygen-deficient atmosphere, (2) a potentially hazardous atmosphere (toxicity), (3) a flammable atmosphere, (4) the risk of entrapment due to converging walls or sloping floors, (5) risk of engulfment, or (6) any other recognized serious health or safety hazard. Examples of confined spaces include storage tanks, pits, vaults, or chambers.

Confined Space Attendant Whenever an entry is made into a confined space, an attendant is required. Entry can be as simple as sticking ones head into an opening or as complex as entering an underground storage tank. The attendant is responsible for ensuring entry personnel have the correct respiratory protection and chemical protective clothing. The attendant will also conduct monitoring when required, ensure communications are in place, and ensure the entry team acts in a safe and competent manner. At no time is the attendant allowed to leave the confined space area.

Confined Space Entrant The title of the person who enters a space and performs work.

Confined Space Permit This form is filled out by a competent authority such as a qualified Confined Space Attendant, Industrial Hygienist, Safety Professional, or Marine Chemist. The form will state what the space was tested for, report that it is safe for entry, mentions what specific (if any) equipment is required, and is signed.

Consist A rail shipping paper containing a list of cars in the train by order. Those containing hazardous materials are indicated. Some railroads include information on emergency operations for the hazardous materials on the train with the consist.

Consumer commodity Means a material that is packaged or distributed in a form intended and suitable for sale through retail sales agencies or instrumentalities for consumption by individuals for purposes of personal care or household use. This term also includes drugs and medicines (see ORM-D).

Contact/absorption The process by which materials enter the body by passing into or through the skin, eyes, or mucous membranes.

Container Any bag, barrel, bottle, box, can, cylinder, drum, reaction vessel, storage tank, or the like that contains a hazardous chemical. For the purposes of an MSDS or the Hazard Communication Standard, pipes or piping systems are not considered to be containers.

Container profiles A form of identification that matches the types of containers with the types of materials commonly transported in them.

Contamination Reduction Zone See Zones.

Controls These are methods required by OSHA to eliminate the hazards or reduce the risk when working with hazardous materials or hazardous operations. They consist of, in this order:

Engineering controls: The first level of controlling hazards in the workplace and include the use of built-in safety systems such as guardrails, ventilation systems, and machine-guarding devices. Engineering controls largely eliminate the hazard and do not generally require employee action because they are most often automatic.

Administrative controls: These types of controls reduce the exposure by alerting the employee to the hazard or controlling the amount of exposure through rules and practices. These types of controls can include work rules, schedules, standard procedures, signage, barrier tape, and other methods to limit the potential exposure of the worker.

PPE: Personal protective equipment that is required to be worn by the employee during the time of potential exposure to the hazard.

Corrosive A material that causes visible destruction of, or irreversible alterations in, living tissue by chemical action at the site of contact. For example, a chemical is considered to be corrosive if, when tested on the intact skin of albino rabbits by the method described by the US Department of Transportation in Appendix A to 49 CFR, Part 173, it destroys or changes irreversibly the structure of the tissue at the side of contact following an exposure period of 4 hours.

CPSC Consumer Product Safety Commission has responsibility for regulating hazardous materials when they appear in consumer goods. For CPSC purposes, hazards are defined in the Hazardous Substances Act and the Poison Prevention Packaging Act 1970.

Cryogenic Gases that are cooled to a very low temperature, usually below −150°F (−101°C), to change to a liquid. Also called refrigerated liquids. Some common cryogens include liquid oxygen, nitrogen, and helium.

Cryogens Liquified gases with a boiling point below −150°F.

Cutaneous toxicity See Dermal toxicity.

Cutting, welding, and burning Any action involving the use of oxyacetylene or heliarc welding on metal that will create sparks, heat, or fumes that may collect in one space or effect the atmosphere in a nearby space. Welding also applies to using "Weld-on" type glues in affixing PVC piping and joints. All of the actions may change the hazards and conditions, especially in a confined space.

Dangerous Cargo Manifest A cargo manifest listing the hazardous materials on board a ship and their location.

Dangerous Goods (Canada) Any product, substance, or organism included by its nature or by the regulation in any of the classes listed in the schedule (UN 9 Classes of Hazardous Materials).

DASHO Designated Agency Safety and Health Official is the executive official of a federal department or agency who is responsible for safety and occupational health matters within a federal agency and who is so designated or appointed by the head of the agency.

Decomposition Breakdown of a material or substance (by heat, chemical reaction, electrolysis, decay, or other processes) into parts or elements or simpler compounds.

Decon The shortened term for *decontamination*.

Decontamination corridor The area in the Warm Zone where decon operations are conducted.

Decontamination The systematic process of removing hazardous materials from personnel and their equipment. It is necessary to reduce the potential for exposure to personnel and to minimize the spread of contamination from the site. Decontamination is sometimes abbreviated to the term decon. The term can also be expanded to include area cleanup (decon).

Defensive Actions conducted from outside the hazardous areas and are limited to protecting nearby areas from the effects of a release of a hazardous substance.

Deflagration A rapid combustion of a material occurring in the explosive mass at subsonic speeds. The event is usually caused by contact with a flame source but may also be caused by mechanical heat or friction.

Degradation A form of decontamination that involves altering the chemical hazards present. *Also,* the rapid failure of the CPC caused by the hazardous substance dissolving the CPC fabric.

Delayed effects Effects from chemical exposure that are not seen for up to several hours after the initial event or exposure.

Demand systems Supply air into the mask as the user inhales and creates a negative pressure inside the mask.

Density The mass (weight) per unit volume of a substance. For example, lead is much more dense than aluminum.

Dermal Relating to the skin.

Dermal toxicity Adverse effects resulting from skin exposure to a substance. The term was ordinarily used to denote effects in experimental animals.

Descriptive toxicology The branch of toxicology concerned directly with toxicity testing.

Designated area An area that may be used for work with select carcinogens, reproductive toxins, and substances with a high degree of acute toxicity. A designated area may be the entire laboratory, or a device such as a laboratory hood.

Detonation An extremely rapid decomposition of an explosive material. This decomposition propagates throughout the explosive agent at supersonic speeds and is accompanied by pressure and temperature waves. This detonation could be initiated by mechanical friction, impact, and/or heat.

DHHS US Department of Health and Human Services (replaced US Department of Health, Education, and Welfare). NIOSH and the Public Health Service are part of DHHS.

Diatomic gas A gaseous element requiring two atoms to achieve stability. Examples of diatomic gases are oxygen and nitrogen.

Diffuse The ability of a gas to move around and occupy the area of release.

Diffusion The movement of a material through the fabric of the CPC from higher to lower concentrations.

Dike A barrier constructed to control or confine hazardous substances and prevent them from entering sewers, ditches, streams, or other flowing waters.

Dilution A term used in decontamination as the process of using water to flush the contaminants from personnel, the environment, or equipment.

Diurnal effects Daily weather changes such as barometric pressure, wind direction, temperature, fog, or rain.

Doffing Taking a respirator off.

DOL The US Department of Labor. OSHA and MSHA are part of DOL.

Donning Putting a respirator on.

Dose The amount of a given material or chemical that enters the body of an exposed organism in a given period of time. The time can be as short as a few seconds (to inject a substance) or as long as a lifetime (in the case of chronic exposures).

- Internal dose refers to the amount of a chemical that is absorbed by the body.
- Biologically effective dose refers to the amount that interacts with a particular target tissue or organ.

Dose = Concentration × Time × Exchange Rate

Where the "exchange rate" refers to the breathing rate (e.g., m^3/h) or ingestion rate (liters or grams/day).

Note that if the internal dose is to be calculated, then another factor must be included, the fraction of the chemical which is taken up. This fraction is the retention factor, which ranges from 0 to 1.0.

DOT US Department of Transportation regulates transportation of chemicals and other substances.

Dust Solid particles in the air created through operations such as grinding, sanding, demolition, or drilling.

Dynamite An industrial high explosive that is moderately sensitive to shock and heat. The main ingredients are nitroglycerin or sensitized ammonium nitrate. This material is then distributed in diatomaceous earth or a mass of hydrated silica.

Edema An abnormal accumulation of clear watery fluid in the tissues.

Electromagnetic radiation The propagation of waves of energy of varying electric and magnetic fields through space. The waves move at the speed of light from matter in the form of photons or energy packets. The strength of the waves depends on their individual frequency. See also Gamma-radiation.

Electron A subatomic particle having a negative electric charge. The electron has mass, which is approximately $\frac{1}{1837}$th of a proton. Electrons surround the nucleus of atoms, and are equal in number to the protons for the given element they are orbiting. Electrons that have been separated from an atomic orbit are said to be *free electrons*. Electrons are the subatomic particle responsible for bonding and are pivotal in the formation of compounds.

Element One of the 109 recognized substances that comprise all matter at the atomic level. Elements are the building blocks for the compounds formed by chemical reactions. A listing of the elements can be found on the periodic table. Some examples include sodium, oxygen, neon, and carbon.

Emergency Any occurrence such as, but not limited to, equipment failure, rupture of containers, or failure of controls which results in an uncontrolled release of a hazardous chemical into the workplace. Emergencies generally involve exposure of personnel outside the immediate area to a hazard.

Emergency Response Plan A plan that is developed in advance of an emergency situation that identifies the actions to be taken by all employees at the site in the event of an emergency.

Emergency Response Planning Guideline (ERPG) These are values established to assist emergency response planning when a release involves exposure of the public to the substance. There are three levels developed to describe the effects of the exposure at each level.

Employee An individual employed in a workplace who may be exposed to hazardous chemicals in the course of his or her assignments.

Encapsulating suit Level B suit that encapsulates the breathing-air system under the CPC.

End-of-Service-Life Indicator (ESLI) The system that warns the user that the respirator cartridge is reaching the end of its useful life and needs to be changed.

Endothermic A description of a process that ultimately absorbs heat and requires large amounts of energy for initiation and maintenance.

Engineering controls The first level of controlling hazards in the workplace; includes the use of built-in safety systems such as guardrails, ventilation systems, and machine-guarding devices.

Engulfment The surrounding and effective capture of a person by a liquid or finely divided solid substance that can be aspirated to cause death by filling or plugging the respiratory system or that can exert enough force on the body to cause death by strangulation, constriction, or crushing.

Entry Supervisor The person responsible for ensuring that the confined space operation is conducted safely; the person responsible often completes the Confined Space Entry Permit.

Entry Team The personnel who enter an area where contamination is present; they can be part of a cleanup group or an emergency response team.

Entry The action by which a person passes through an opening into a Permit-Required Confined Space. Entry includes ensuing work activities in that space and is considered to have occurred as soon as any part of the entrant's body breaks the plane of an opening into the space.

Environmental toxicity Information obtained as a result of conducting environmental testing designed to study the effects on aquatic and plant life.

Environmental toxicology The branch of toxicology dedicated to developing an understanding of "chemicals in the environment and their effect on man and other organisms." Environmental media may include air, groundwater, surface water, and soil. Impact may be from fish to philosopher. This area of toxicology is (or should be) the most significant for the development of risk assessments.

EPA (Environmental Protection Agency) Established in 1970, the federal EPA is required to ensure the safe manufacture, use and transportation of hazardous chemicals. The State of California has also established California EPA, which follows the same guidelines on the state level.

Epidemiology Science concerned with the study of disease in a general population. Determination of the incidence (rate of occurrence) and distribution of a particular disease (as by age, sex, or occupation) which may provide information about the cause of the disease.

Epithelium The covering of internal and external surfaces of the body.

Escape pack A small cylinder worn on the belt of the user of an air-line system to allow the user to disconnect from the air line in the event of an emergency.

Etiologic agent A material or substance that is capable of causing disease. The material could contain a viable microorganism, or its toxin which causes or may cause human disease. A biohazard or biologic agent.

Evaporation rate The rate at which a material will vaporize (evaporate) when compared to the known rate of vaporization of a standard material. The evaporation rate can be useful in evaluating the health and fire hazards of a material. The designated standard material is usually normal butyl acetate (NBUAC or n-BuAc), with a vaporization rate designated as 1.0. Vaporization rates of other solvents or materials are then classified as:

Fast evaporating if greater than 3.0. Examples: methyl ethyl ketone = 3.8, acetone = 5.6, hexane = 8.3.

Medium evaporating if ranging from 0.8 to 3.0. Examples: 190 proof (95%) ethyl alcohol = 1.4, VM&P naphtha = 1.4, MIBK = 1.6

Slow evaporating if less than 0.9. Examples: xylene = 0.6, isobutyl alcohol = 0.6, normal butyl alcohol = 0.4, water = 0.3, mineral spirits = 0.1.

Evaporation The process of a liquid entering the air as it dries out; this is caused by a number of factors, including heat and the liquid.

Exclusionary Zone See Zones.

Exothermic An expression of a reaction or process that evolves energy in the form of heat. The process of neutralization evolves heat—this is considered to be an exothermic reaction.

Expansion ratio The amount of gas released from a given volume of liquid with a liquified gas.

Explosive material A chemical that causes a sudden, almost instantaneous release of pressure, gas, and heat when subjected to sudden shock, pressure, or high temperature. Explosives are broken down into various classifications by the Department of Transportation.

Exposure An event in which a pollutant (chemical toxicant in the present context) has contact with an organism, such as a person, over a certain interval of time. Exposure is thus the product of concentration and time (e.g., $mg/m^3 \times h$).

Exposure limits The levels of exposure established by studies that determine the safe levels to enable workers to maintain a margin of safety when functioning in contaminated atmospheres.

Extremely hazardous substance Chemicals determined by the Environmental Protection Agency (EPA) to be extremely hazardous to a community during an emergency spill or release as a result of their toxicities and physical/chemical properties.

Eye protection Recommended safety glasses, chemical splash goggles, face shields, etc., to be utilized when handling a hazardous material.

F Fahrenheit is a scale for measuring temperature. On the Fahrenheit scale, water boils at 212°F and freezes at 32°F.

FACOSH Federal Advisory Council for Occupational Safety and Health is a joint management–labor council that advises the Secretary of Labor on matters relating to the occupational safety and health of federal employees.

FDA US Food and Drug Administration.

Fetus The developing young in the uterus from the seventh week of gestation until birth.

FFSHCs Field Federal Safety and Health Councils are organized throughout the country to improve federal safety and health programs at the field level and within a geographic location.

Field calibration The process of quickly checking the monitoring equipment prior to use in the field.

FIFRA Federal Insecticide, Fungicide, and Rodenticide Act regulates poisons, such as chemical pesticides, sold to the public and requires labels that carry health hazard warnings to protect users. It is administered by EPA.

Fission The splitting of an atom's nucleus that results in the release of considerable amounts of energy.

Fit factor This refers to the number that defines the ratio between the levels of particulates inside the mask versus those measured outside the mask.

Flammable A chemical that includes one of the following categories:

Flammable Aerosol: An aerosol that, when tested by the method described in 16 CFR 1500.45, yields a flame projection exceeding 18 inches at full valve opening or a flashback (a flame extending back to the valve) at any degree of valve opening.

Flammable Gas: (1) A gas that, at ambient temperature and pressure, forms a flammable mixture with air at a concentration of 13% by volume or less or (2) a gas that, at ambient temperature and pressure, forms a range of flammable mixtures with air wider than 12% by volume, regardless of the lower limit.

Flammable Liquid: Any liquid having a flash point below 100°F (37.8°C), except any mixture having components with flash points of 100°F (37.8°C) or higher, the total of which make up 99% or more of the total volume of mixture.

Class IA: A class of flammable liquids with a flash point below 73°F and a boiling point below 100°F.

Class IB: A class of flammable liquids with a flash point below 73°F and a boiling point at or above 100°F.

Class IC: A class of flammable liquids with a flash point at or above 73°F and below 100°F.

Flammable range The concentration of gas or vapor in air that will burn if ignited. It is expressed as a percentage that defines the range between a Lower Explosive Limit (LEL) and an upper explosive limit (UEL). A mixture below the LEL is too "lean" to burn; a mixture above the UEL is too "rich" to burn.

Flammable Solid A solid, other than a blasting agent or explosive as defined in 24 CFR 1910.109 (A), that is liable to cause fire though friction, absorption of moisture, spontaneous chemical change, or retained heat from manufacturing or processing, or which can be ignited readily and when ignited burns so vigorously and persistently as to create a serious hazard. A substance is a flammable solid if, when tested by the method described in 16 CFR 1500.44, it ignites and burns with a self-sustained flame at a rate greater than one-tenth of an inch per second along its major axis.

Flash fire Fire that involves the ignition of flammable vapors accumulated in a given area.

Flash point The minimum temperature at which a liquid gives off a vapor in sufficient concentration to ignite when tested by the following methods:

1. Tagliabue Closed Tester (see American National Standard Method of Test for Flash Point by Tag Closed Tester, Z11.24 1979 [ASTM D5-79]) for liquids with a viscosity of less than 45 Saybolt Universal Seconds (SUS) at 100°F (37.8°C) that do not have a tendency to form a surface film under test.

2. Pensky-Martens Closed Tester (see American National Standard Method of Test for Flash Point by Pensky-Martens Closed Tester, Z11.771979 [ASTM D9-79]) for liquids with a viscosity equal to or greater than 45 SUS at 100°F (37.8°C), or that contain suspended solids, or that have a tendency to form a surface film under test.

3. Setaflash Closed Tester (see American National Standard Method of Test for Flash Point by Setaflash Closed Tester [ASTM D 3278-78]).

Organic peroxides, which undergo auto-accelerating thermal decomposition, are excluded from any of the flash point determination methods specified above. For practical purposes, this temperature denotes the point at which vapors will be emitted from a liquid and will be ignitable in the presence of a flame or ignition source. It is probably the single most important fire term used in our study.

Flash protection A form of protection that involves a flash suit.

Forbidden The hazardous material that must not be offered or accepted for transportation.

Forensic toxicology The branch of toxicology concerned with medico-legal aspects of the harmful effects of chemicals on man and animals.

Formula The scientific expression of the chemical composition of a material (e.g., water is H_2O, sulfuric acid is H_2SO_4, sulfur dioxide is SO_2).

FRA First Responder Awareness level. The first level of training under the HAZWOPER standard for persons who would likely come across a hazardous materials release and whose job it would be to report the spill to the correct agency.

Freelancing The self-initiated actions taken by unassigned personnel during an operation without direction from the IC or the person in charge.

FRO First Responder Operations level. The second level of training under the HAZWOPER standards. This level is designed for those personnel whose role it is to respond to a hazardous materials release and initiate defensive control operations.

Fully encapsulating suit Complete coverall of the body inside the suit; no body area is ever exposed.

Fume A solid condensation particle of extremely small diameter commonly generated from molten metal as metal fume.

g/kg Grams per kilogram is an expression of dose used in oral and dermal toxicology testing to denote grams of a substance dosed per kilogram of animal body weight. Also see kg (kilogram).

Gamma-radiation This form of radiation is essentially electromagnetic radiation. The wavelength of γ-rays is shorter than X-rays and are emitted as photons. Photons are massless and are considered to be pure energy. See also Electromagnetic radiation.

Gases Materials that form one of the three states of matter. Nonparticulate forms of matter that move freely when released and can occupy the entire area of release.

General exhaust A system for exhausting air containing contaminants from a general work area. Also see Local exhaust.

General Site Worker An employee who works at a cleanup site regulated by the HAZWOPER regulation and whose work exposes or potentially exposes him to high levels of hazardous substances.

Genetic Pertaining to or carried by genes. Hereditary.

Gestation The development of the fetus from conception to birth.

Grounding The procedure used to carry an electrical charge to ground through a conductive path. A typical ground may be connected directly to a conductive water pipe or to a grounding bus and ground rod. See Bonding.

Half-life It is the time required for the decay process to reduce the energy production to one-half of its original value. This means that half the atoms are present; therefore, half the radioactive energy is occurring. This time period varies from isotope to isotope. Half-lives can range from millionths of a second to more than a million years.

Hand protection Specific type of gloves or other hand protection required to prevent the harmful exposure to hazardous materials.

HASP (Health and Safety Plan) A plan that is written for a specific cleanup activity required under the HAZWOPER regulation. It is broad in its scope and very detailed, covering the specific operations that must be conducted.

HazMat Team Those employees who are trained to respond to an emergency response involving hazardous substances and who will assume an aggressive role in stopping the release.

Hazard and risk assessment The process of determining the hazards of a particular material and interjecting action to be taken.

Hazard assessment The process that identifies the potential harm that a material presents.

Hazard categorizing (haz-cat) Field analysis that can be performed by appropriate trained personnel working at an emergency response site.

Hazard Communication The title of the OSHA regulation that is sometimes known as the Employee Right-to-Know regulation. It requires employers to make information available to employees on all of the hazardous substances in the workplace to which they may be exposed.

Hazard warning Words, pictures, symbols, or combination thereof presented on a label or other appropriate form to inform of the presence of various materials.

Hazardous atmosphere An atmosphere that may expose employees to the risk of death, incapacitation, impairment of ability to self-rescue (e.g., escape unaided from a permit space), injury, or acute illness from one or more of the following causes:

1. Flammable gas, vapor, or mist in excess of 10% of its Lower Flammable Limit (LFL)

2. Airborne combustible dust at a concentration that meets or exceeds its LFL

NOTE: This concentration may be approximated as a condition in which the dust obscures vision at a distance of 5 feet (1.52 m) or less.

3. Atmospheric oxygen concentration below 19.5% or above 23.5%

4. Atmospheric concentration of any substance for which a dose or a permissible exposure limit is published in Subpart G, Occupational Health and Environmental Control, or in Subpart Z, Toxic and Hazardous Substances, of this part and which could result in employee exposure in excess of its dose or permissible limit;

NOTE: An atmospheric concentration of any substance that is not capable of causing death, incapacitation, impairment of ability to self-rescue, injury, or acute illness due to its health effects is not covered by this provision.

5. Any other atmospheric condition that is immediately dangerous to life or health.

NOTE: For air contaminants for which OSHA has not determined a dose or permissible exposure limits, other sources of information, such as Material Safety Data Sheets that comply with the Hazard Communication standard 1910.1200 of this part, published information, and internal documents can provide guidance in establishing acceptable atmospheric conditions.

Hazardous chemical Any chemical whose presence or use is a physical hazard or a health hazard. For purposes of the OSHA Laboratory Standard, a chemical for which there is significant evidence, based on at least one study conducted in accordance with established scientific principles, that acute or chronic health effects may occur in exposed employees. The term "health hazard" includes chemicals which are carcinogens, toxic or highly toxic agents, reproductive toxins, irritants, corrosives, sensitizers, hepatotoxins, nephrotoxins, neurotoxins, agents which are on the hematopoietic system, and agents that damage the lungs, skin, eyes, or mucous membranes.

Hazardous Materials Specialist An individual trained to the fourth level of HAZWOPER emergency response training. His actions are similar to that of the Hazardous Materials Technician, but may also include training on the handling of specific materials and interaction with outside agencies.

Hazardous Materials Technician An individual trained to the third level of HAZWOPER emergency response training and whose job function involves an aggressive/offensive response to a release of a hazardous material. Personnel at this level are trained to select and use appropriate chemical protective equipment that will allow them to approach a release for the purpose of stopping the release.

Hazardous substance Any substance designated under the Clean Water Act and the Comprehensive Environmental Response, Compensation and Liability Act (CERCLA) as posing a threat to waterways and the environment when released (US Environmental Protection Agency).

Hazardous wastes Discarded materials regulated by the Environmental Protection Agency because of public health and safety concerns. Regulatory authority is granted under the Resource Conservation and Recovery Act (US Environmental Protection Agency).

HAZWOPER The term used to describe the Hazardous Waste Operations and Emergency Response regulation.

Heat stress Conditions such as heat cramps, heat exhaustion, and heat stroke caused by excess heat in the body.

Heavy metals Metals that are poisonous.

Hemoglobin The part of the red blood cell that transports oxygen to the body's cells.

Hepatotoxin A substance that causes injury to the liver.

Heroic dose An expression of exposure levels used in animal testing. Heroic doses are referred to as the maximum tolerated doses of the test chemical. Essentially, it is the highest nonlethal dose the animal can tolerate during the testing process.

Heterogeneous A substance that does not display a uniform makeup. The composition of heterogeneous substances or mixtures differs from sample point to sample point.

Highly Toxic Defined by OSHA as a chemical falling within any of the following categories:

1. A material with a median lethal dose (LD50) of 50 milligrams or less per kilogram of body weight when administered orally to albino rats weighing between 200 and 300 grams each.

2. A material with a median lethal dose (LD50) of 200 milligrams or less per kilogram of body weight when administered by continuous contact for 24 hours (or less if death occurs within 24 hours) with the bare skin of albino rabbits weighing between 2 and 3 kilograms each.

3. A material that has a median lethal concentration (LC50) in air of 200 parts per million by volume or less of gas or vapor, or 2 milligrams per liter or less of mist, fume, or dust, when administered by continuous inhalation for 1 hour (or less if death occurs within 1 hour) to albino rats weighing between 200 and 300 grams each.

High-order explosives Materials capable of detonation that can cause significant damage.

Homogeneous A substance that displays a uniform composition throughout.

Hot Work activities Activities that generate heat or open flame, such as welding and cutting torch operations.

Hot Work Permit The employer's written authorization to perform operations (e.g., riveting, welding, cutting, burning, and heating) capable of providing a source of ignition.

Hot Zone See Zones.

Hydrolysis A chemical decomposition of a substance by water. The products of the decomposition results in the formation of two or more new substances.

Hydrophobic filters Those that swell when they contact liquids.

Hygroscopic The ability of a substance to absorb moisture from the air.

Hypergolic Substances that spontaneously ignite on contact with another. Many hypergolic materials are used as rocket fuels.

Hyperthermia A condition where excess heat builds inside the body.

Hypocalcemia A condition characterized by abnormally low levels of calcium in the blood.

IARC International Agency for Research on Cancer. One of the leading agencies that lists and identifies carcinogens and suspected carcinogens.

IDLH (Immediately Dangerous to Life or Health) The term is usually expressed in parts per million and reflects the atmospheric level of any toxic, corrosive, or asphyxiant that poses a danger to life or would cause significant and irreversible health effects or would impair the ability of an individual to escape the area. In many cases, this value is based on an exposure of 30 minutes.

Ignitability The ability of a material to be set on fire.

Ignition temperature The minimum temperature to which a fuel in air must be heated in order to start self-sustained combustion independent of the heating source.

Immediate health effects Generally occur within minutes following the exposure.

Impervious A material that does not allow another substance to pass through or penetrate it.

Incident Action Plan An emergency response plan that is used to identify the specific hazards and operations that must be conducted in the event of an emergency response to a hazardous substance. It is different than the Site Safety Plan in that it is less detailed and deals only with the emergency phase of the incident.

Incident Command System (ICS) A management system that is used for managing emergency response operations.

Incident Commander The person responsible for all operations at a hazardous materials emergency.

Incompatible Materials that could cause dangerous reactions by direct contact with one another.

Inerting The displacement of the atmosphere in an area such as a Permit-Required Confined Space by a noncombustible gas (such as nitrogen) to such an extent that the resulting atmosphere is noncombustible.

NOTE: This procedure often produces an IDLH oxygen-deficient atmosphere.

Infectious substances Materials capable of producing disease.

Ingestion Taking in by the mouth.

Inhalation Breathing in of a substance in the form of a gas, vapor, fume, mist, or dust. The route of entry that includes exposure via the respiratory system.

Inhibitor A chemical added to another substance to prevent an unwanted chemical change.

Injection Materials getting through the skin through openings or with pressure.

Insoluble Incapable of being dissolved in a liquid.

Intrinsically safe Safe to use in flammable atmospheres and does not create sparks or ignition sources.

Ion An atom or molecule that has acquired a positive or negative charge by gaining or losing an electron. The atom or molecule is no longer considered neutral in this state.

Ionization potential (IP) A unique measure of the energy required to displace the electrons in a compound by using an ultraviolet light.

Ionization The formation of ions. This ion formation occurs when a neutral molecule of an inorganic solid, liquid, or gas undergoes a chemical change. These highly energetic, short-wavelength rays are capable of causing mutations in cell nuclei and DNA. These changes in the cell structures of the body may cause cancer or other long-term disease processes.

Ionizing radiation Radiation that is capable of causing the ionization of solids, liquids, or gases either directly or indirectly. This process can be caused by alpha, beta, or gamma radiation.

Irritant A chemical that is not corrosive but that causes a reversible inflammatory effect on living tissue by chemical action at the site of contact. A chemical is a skin irritant if, when tested on the intact skin of albino rabbits by the methods of 16 CFR 1500.41 for 4 hours exposure or by other appropriate techniques, it results in an empirical score of 5 or more. A chemical is an eye irritant if so determined under the procedure listed in 16 CFR 1500.42 or other appropriate techniques.

Irritating An irritating material, as defined by DOT, is a liquid or solid substance which, upon contact with fire or when exposed to air, gives off dangerous or intensely irritating fumes (not including poisonous materials). See Poison (Class A) and Poison (Class B).

Isolation The process by which an area such as a Permit-Required Confined Space is removed from service and completely protected against the release of energy and material into the area by such means as: blanking or blinding, or removing sections of lines, pipes, or ducts; a double block and bleed system; lockout or tagout of all sources of energy; or blocking or disconnecting all mechanical linkages.

Isotonic solution Solution that contains similar amounts of material commonly found in the body.

Isotope One of two or more types of atoms of an element. These atoms have the same atomic number but a differing number of neutrons. Uranium-238 as compared to uranium-235 is an example of an isotope. U-238 has 92 protons and 146 neutrons. U-235 has 92 protons and 143 neutrons. See Radioactive isotope.

kg Kilogram is a metric unit of weight, about 2.2 US pounds. Also see g/kg, ng, and mg.

Kinetic energy The energy used to make items move.

L Liter is a metric unit of capacity. A US quart is about $\frac{9}{10}$th of a liter.

Lab pack A process of placing smaller containers such as glass jars inside an open-head drum for proper shipment or disposal.

Label Notice attached to a container, bearing information concerning its contents.

Laboratory A facility where the "laboratory use" of hazardous chemicals occurs. It is a workplace where relatively small quantities of hazardous chemicals are used on a nonproduction basis.

Laboratory-type hood A device located in a laboratory, enclosed on five sides with a moveable sash or fixed partial enclosure on the remaining side, constructed and maintained to draw air from the laboratory and to prevent or minimize the escape of air contaminants into the laboratory. A hood allows a worker to conduct chemical manipulations in the enclosure without inserting any portion of the body other than hands and arms. Walk-in hoods with adjustable sashes meet the above definition, provided the sashes are adjusted during use so that the airflow and the exhaust of air contaminants are not compromised and employees do not work inside the enclosure during the release of airborne hazardous chemicals.

Land breezes/shore breezes Those that occur when a land mass cools faster than an adjacent body of water and creates an offshore or onshore breeze.

LC Lethal concentration is the concentration of a substance being tested that is expected to kill. It is a measurement of the material in the air and is expressed in either ppm or ppb.

LC50 The concentration of a material in air that is expected to kill 50% of a group of test animals with a single exposure (usually 1 to 4 hours). The LC50 is expressed as parts of material per million parts of air, by volume (ppm) for gases and vapors, or as micrograms of material per liter of air (mcg/l) or milligrams of material per cubic meter of air (mg/m^3) for dusts and mists, as well as for gases and vapors.

LChi The airborne concentration of a chemical that will be fatal to 100% of a test population.

LClo Lethal concentration low; lowest concentration of a gas or vapor as measured in the air which is capable of killing a specified species over a specified time.

LD Lethal dose is the quantity of a substance being tested that will kill. It is a measurement of the actual dose taken in via ingestion, injection, or dermal exposure and is expressed in mg/kg.

LD50 A single dose of a material expected to kill 50% of a group of test animals. The LD50 dose is usually expressed as milligrams or grams of material per kilogram of animal body weight (mg/kg or g/kg). The material may be administered by mouth or applied to the skin.

LDhi The concentration of a chemical by dermal contact or absorption that will be fatal to 100% of a test population.

LDlo Lethal dose low: lowest administered dose of a material capable of killing a specified test species.

LEL, or LFL Lower Explosive Limit, or Lower Flammable Limit, of a vapor or gas. The lowest concentration (lowest percentage of the substance in air) that will produce a flash of fire when an ignition source (heat, arc, or flame) is present. At concentrations lower than the LEL, the mixture is too "lean" to burn. Also see UEL.

Lethality The term used to describe the ability of the material to kill the exposed subject.

Level A protection Protection by means of a fully encapsulating, gastight suit.

Level B protection PPE with the inclusion of a SAR.

Level C protection PPE with an air-purifying respirator.

Level D protection The lowest level of protection, which includes PPE without a respirator.

Levels of protection Groupings of CPC and respirators to create a system of protection.

Limited quantity Means the maximum amount of hazardous material; as specified in those sections applicable to the particular hazard class for which there are specific exceptions from the requirements of this subchapter. See Secs. 173.118, 173.118(a), 173.153, 173.244, 173.306, 173.345, and 173.364 of 49 CFR.

Liquified gases Gases that become liquids through a combination of temperature and pressure.

Local effects Effects of chemical exposure that occur at the site where the material contacts the body.

Local exhaust A system for capturing and exhausting contaminants from the air at the point where the contaminants are produced (welding, grinding, sanding, or other processes or operations). Also see General exhaust.

Lockout/tagout (LOTO) programs A series of steps to control energy in the workplace.

Long-term health effects Effects that do not show until months or years following exposure.

Lower Explosive Limit (LEL) The lowest concentration of a vapor in air that will form an ignitable mixture.

Lower Flammable Limit (LFL) and Upper Flammable Limit (UFL) Terms sometimes used in place of the more common LEL and UEL.

Low-order explosives Materials in which reactions create pressure waves that move more slowly than those created by a detonation.

m Meter is a unit of length in the metric system. One meter is about 39 inches.

m^3 Cubic meter is a metric measure of volume, approximately 35.3 cubic feet or 1.3 cubic yards.

Malignant Tending to become progressively worse and to result in death.

Malleable Substances exhibiting the properties of flexibility, the ability to bend or be hammered into thin sheets. Most metallic elements are considered to be malleable.

Matter Anything that has mass and occupies space. Matter is generally accepted to found in three basic forms—solid, liquid, and gas. Most of the subatomic particles that make up matter are invisible to the naked eye.

Mechanical exhaust A powered device, such as a motor-driven fan or air steam venturi tube, for exhausting contaminants from a workplace, vessel, or enclosure.

Mechanical filter respirator A respirator used to protect against airborne particulate matter like dusts, mists, metal fume, and smoke. Mechanical filter respirators do not provide protection against gases, vapors, or oxygen-deficient atmospheres.

Melting point The temperature at which a solid substance changes to a liquid state.

Metabolism Physical and chemical processes taking place among the ions, atoms, and molecules of the body. Metabolism is an important element in the study of how the toxic materials that we take into our bodies are processed by it.

mg Milligram is a metric unit of weight which is one-thousandth of a gram.

mg/kg Milligrams of substance per kilogram of body weight is an expression of toxicological dose.

mg/m^3 Milligrams per cubic meter is a unit for expressing concentrations of dusts, gases, or mists in air.

Micron Micrometer is a unit of length equal to one-millionth of a meter. A micron is approximately $\frac{1}{23,000}$th of an inch.

Miscibility See Solubility.

Mist Suspended liquid droplets generated by condensation from the gaseous to the liquid state, or by breaking up a liquid into a dispersed state, such as splashing, foaming, or atomizing. Mist is formed when a finely divided liquid is suspended in air.

Mitigation An operational term used in the industry to describe the process of bringing a chemical release under control. The terms used to describe the tactics and strategies used to handle an emergency or cleanup operation.

Mixture Any combination of two or more chemicals if the combination is not, in whole or part, the result of a chemical reaction.

ml Milliliter is a metric unit of capacity, equal in volume to one cubic centimeter (cc), or approximately one-sixteenth of a cubic inch. One-thousandth of a liter.

mmHg Millimeters (mm) of mercury (Hg) is a unit of measurement for gas pressure.

Molecular weight Weight (mass) of a molecule based on the sum of the atomic weights of the atoms that make up the molecule.

mppcf Million particles per cubic foot is a unit for expressing concentration of particles of a substance suspended in air. Exposure limits for mineral dusts (silica, graphite, Portland cement, nuisance dusts, and others), formerly expressed as mppcf, are now more commonly expressed in mg/m.

mrem Unit of absorbed dose of radiation taken into body tissues, representing one-thousandth of a rem.

MSDS Material Safety Data Sheet. The written information on a specific chemical compound that expresses such items as physical hazards, signs and symptoms of exposure, toxicology information, and other pertinent data.

MSHA Mine Safety and Health Administration, US Department of Labor.

Mutagen A substance or agent capable of altering the genetic material in a living cell.

MW See Molecular weight.

Nausea Tendency to vomit, feeling of sickness at the stomach.

NCI National Cancer Institute is that part of the National Institutes of Health which studies cancer causes and prevention as well as diagnosis, treatment, and rehabilitation of cancer patients.

Near miss An incident that could have resulted in an exposure or accident.

Neonatal The first four weeks after birth.

Nephrotoxin A substance that causes injury to the kidneys.

Neurotoxin A material that affects the nerve cells and may produce emotional or behavioral abnormalities.

Neutralization A neutralization reaction occurs when a mutual reaction occurs between an acid and a base. The products of the reaction include a salt compound, water, and heat. This reaction can occur between organic and nonorganic materials.

Neutron An elementary subatomic particle existing in the nucleus having a mass of 1.009 amu (atomic mass units). The neutron has no electrical charge and exists in the nucleus of all atoms except hydrogen. This atom is composed of one proton and one electron. To find the amount of neutrons in an atom, subtract the atomic number from the atomic weight. The remainder will be the number of neutrons.

NFPA National Fire Protection Association is an international membership organization which promotes/improves fire protection and prevention and establishes safeguards against loss of life and property by fire. Best known on the industrial scene for the National Fire Codes, 13 volumes of codes, standards, recommended practices, and manuals developed (and regularly updated) by NFPA technical committees. Among these are NFPA 704, the code for showing hazards of materials as they might be encountered under fire or related emergency conditions, using the familiar diamond-shaped label or placard with appropriate numbers or symbols, and NFPA 471 and 472 that cover practices for hazardous materials incidents, and procedures for responding to hazardous materials incidents.

ng Nanogram, one-billionth of a gram.

NIOSH (National Institute for Occupational Safety and Health) A government agency under the Department of Health and Human Services that is responsible for investigating the toxicity of workroom environments and all other matters relating to safe industrial practice. NIOSH publishes "The Pocket Guide to Chemical Hazards," which is an excellent resource on health hazards relating to hazardous materials. Other activities that NIOSH is involved in include testing and certifying respiratory protective devices and air sampling detector tubes.

Nonliquified gases A gas other than a gas in solution that under the charging pressure is entirely gaseous at 70°F (21°C).

Nonencapsulating suit Covers the body only and leaves the breathing-air system outside the suit.

Non-entry rescue An operation in which the attendant uses the retrieval equipment worn by entrants to pull them out of a confined space.

Nonflammable Gas Any compressed gas other than a flammable compressed gas.

Nonflammable Not easily ignited, or if ignited, not burning rapidly.

Non-Permit Confined Space A confined space that does not contain or, with respect to atmospheric hazards, have the potential to contain any hazard capable of causing death or serious physical harm.

Non-sparking tools Tools made from beryllium–copper or aluminum–bronze greatly reduce the possibility of igniting dusts, gases, or flammable vapors. Although these tools may emit some sparks when striking metal, the sparks have a low heat content and are not likely to ignite most flammable liquids.

NOx Oxides of nitrogen, which are undesirable air pollutants. NOx emissions are regulated by the EPA under the Clean Air Act.

NPIRS National Pesticide Information Retrieval System is an automated database operated by Purdue University containing information on EPA registered pesticides, including reference file MSDSs.

NRC National Response Center is a notification center that must be called when significant oil or chemical spills or other environment related accidents occur. The toll free telephone number is 1-800-424-8802.

NTP National Toxicology Program. The NTP publishes an "Annual Report on Carcinogens."

Occasional Site Worker Those cleanup workers at a HAZWOPER-regulated site whose exposure to hazardous materials is below the established PEL for the material.

Odor A description of the smell of the substance.

Odor threshold The lowest concentration of a substance's vapor, in air, that can be smelled.

OFAP Office of Federal Agency Programs is the organizational unit of OSHA, which provides federal agencies with guidance to develop and implement occupational safety and health programs for federal employees.

Offensive actions The actions allowed to be performed by Hazardous Materials Technicians. These include entering the hazardous areas with appropriate levels of protection, rescuing exposed personnel, and stopping the release of the hazardous substances.

Olfactory Relating to the sense of smell.

On-Scene Incident Commander The fifth level of emergency responder under the HAZWOPER regulation. These individuals direct the activities of all emergency response personnel at the scene.

Open-circuit unit Allows exhaled air to leave the mask and not be reused.

Oral toxicity Adverse effects resulting from taking a substance into the body by mouth. Ordinarily used to demote effects in experimental animals.

Oral Used in or taken into the body through the mouth.

Organic materials Materials that contain hydrogen and carbon.

Organic peroxide An organic compound that contains the bivalent –O–O– structure and may be considered a structural derivative of hydrogen peroxide where one or both of the hydrogen atoms have been replaced by an organic radical.

Organogenesis The development of tissues into different organs in embryonic development. This period begins approximately 18 days into the pregnancy in humans.

ORMs A class of materials used by the Department of Transportation which does not meet the definition of a hazardous material but poses some risk when transported in commerce. The materials are broken into classifications called ORM-A, ORM-B, ORM-C, ORM-D, and ORM-E. The use of these terms is reduced with the new DOT regulations.

ORM-A: A material which has an anesthetic, irritating, noxious, toxic, or other similar property and which can cause extreme annoyance or discomfort to passengers and crew in the event of leakage during transportation [49 CFR, Sec. 173.500(b)(1)].

ORM-B: A material (including a solid when wet with water) capable of causing significant damage to a transport vehicle from leakage during transportation. Materials meeting one or both of the following criteria are ORM-B materials: (1) A liquid substance that has a corrosion rate exceeding 0.250 inch per year (IPY) on aluminum (nonclad 7075-T6) at a test temperature of 130°F. An acceptable test is described in NACE Standard TM-01-60. (2) Specifically designated by name in Sec. 172.101 [(Sec. 173.500(b)(2)) of 49 CFR].

ORM-C: A material that has other inherent characteristics not described as an ORM-A or ORM-B but which make it unsuitable for shipment, unless properly identified and prepared for transportation. Each ORM-C material is specifically named in Sec. 172.101 [(Sec. 173.500(b)(4)) of 49 CFR].

ORM-D: A material such as a consumer commodity which, though otherwise subject to the regulations of this subchapter, presents a limited hazard during transportation due to its form, quantity, and packaging. They must be materials for which exceptions are provided in Sec. 172.101. A shipping description applicable to each ORM-D material or category of ORM-D materials is found in Sec. 172.101 [(Sec. 173.500(b)(4)) of 49 CFR].

ORM-E: A material that is not included in any other hazard class, but is subject to the requirements of this subchapter. Materials in this class include hazardous wastes and hazardous substances as defined in Sec. 171.8 of 49 CFR.

Orthostatic hypotension The low blood pressure that is found when you stand up.

OSHA (Occupational Safety and Health Administration) On both the federal and state levels, OSHA is responsible for establishing and enforcing standards for exposure of workers to harmful materials in industrial atmospheres and other matter affecting the health and well-being of industrial workers. Federal OSHA is part of the US Department of Labor.

Overpack drum Used to place one drum inside another.

Overpacking The process of placing one smaller drum inside a larger one for control or disposal.

Overpressure The pressure created over and above the normal or ambient pressure.

Oxidation A chemical reaction that brings about an oxidation reaction. An oxidizing agent may (1) provide the oxygen to the substance being oxidized (in which case the agent has to be oxygen or contain oxygen) or (2) it may receive electrons being transferred from the substance undergoing oxidation. Chlorine is a good oxidizing agent for electron-transfer purposes, even though it contains no oxygen.

Oxidizer A chemical other than a blasting agent or explosive that initiates or promotes combustion in other materials, causing fire either by itself or through the release of oxygen or other gases.

Oxygen-deficient atmosphere An atmosphere containing less than 19.5% oxygen by volume.

Oxygen-enriched atmosphere An atmosphere containing more than 23.5% oxygen by volume.

Particulate matter A solid or liquid matter that is dispersed in a gas, or insoluble solid matter dispersed in a liquid. The prime hazard of particulate matter is inhalation along with the possibility of the matter lodging in the lung tissue. Asbestos fibers are especially dangerous when captured by lung tissues.

Pathogen A disease-causing bacterium or virus capable of producing disease.

PCB An abbreviation for polychlorinated biphenyl. This chemical compound was commonly used as a cooling agent in electrical transformers. It is an aromatic hydrocarbon compound consisting of two benzene nuclei combined with two chlorine atoms. This compound is highly toxic.

PEL The Permissible Exposure Limit represents the 8-hour Time Weighted Average exposure limit for the material and is set by OSHA.

PEL (Permissible Exposure Limit) Permissible exposure limit is an exposure limit established by OSHA's regulatory authority. It is generally the 8-hour Time Weighted Average (TWA) limit, but also could be expressed as the maximum concentration exposure limit. The PEL is usually established for airborne hazards.

Penetrations Openings in the PPE.

Periodic table An arrangement of the elements in such a form as to emphasize the similarities of their physical and chemical properties.

Permanent gas A gas that cannot be liquified by pressure alone.

Permeation rate The speed at which hazardous materials seep into CPC.

Permeation The gradual migration of a material through the fabric by diffusion.

Permissible Exposure Limit (PEL) The level of exposure established by OSHA that an employee can be exposed to on an average basis over an eight-hour work period; exposure above this level would require the use of some types of respiratory protection.

Permit-Required Confined Space A confined space that contains potential safety hazards.

Personal protective equipment (PPE) Items worn by the individual to provide protection from a hazard in the workplace; examples include chemical resistive gloves, safety glasses, or some type of respirator.

Pesticides Materials that kill various forms of pests.

pH scale The scale relating the hydrogen ion concentration to that of a given standard solution. A pH of 7 is neutral. Numbers increasing from 7 to 14 indicate greater alkalinity. Numbers decreasing from 7 to 0 indicate greater acidity.

Physical effects The non-health effects that hazardous materials can produce. Examples include the ability to catch fire or create explosive conditions.

Physical hazard Describes a chemical for which there is scientifically valid evidence that it is a combustible liquid, a compressed gas, explosive, flammable, an organic peroxide, an oxidizer, pyrophoric, unstable (reactive) or water reactive.

Placards $10\frac{3}{4}$ inch square diamond markers required on transporting vehicles, like trucks, rail cars, or freight containers 640 cubic feet or larger.

PMCC Pensky-Martens Closed Cup. See Flash point.

Pneumoconiosis A condition of the lung in which there is permanent deposition of particulate matter and tissue reaction to its presence. It may range from relatively harmless forms of iron oxide deposition to destructive forms of silicosis.

Poison A material falling into one of the following two categories.

Class A: A DOT term for extremely dangerous poisons, poisonous gases or liquids that, in very small amounts, either as gas or as vapor of the liquid, mixed with air, are dangerous to life. Examples: phosgene, cyanogen, hydrocyanic acid, and nitrogen peroxide.

Class B: A DOT term for liquid, solid, paste, or semisolid substances, other than Class A poisons or irritating materials, that are known (or presumed on the basis of animal tests) to be so toxic to humans that they are hazardous to health during transportation.

Polar The description of a molecule where the positive and negative charges are permanently separated. This differs from nonpolar substances in which the electrical charges may coincide. Polar molecules ionize in water and conduct an electrical current. Water is the most common polar substance. Nonpolar substances include gasoline, diesel fuel, and most hydrocarbons. If a substance is polar, it will mix with water.

Polymerization A chemical reaction in which one or more small molecules combine to form larger molecules. A hazardous polymerization is such a reaction that takes place at a rate that releases large amounts of energy. If hazardous polymerization can occur with a given material, the MSDS usually will list conditions that could start the reaction and, because the material usually contains polymerization inhibitors, the length of time during which the inhibitor will be effective.

Potential energy Stored energy or energy that has not been released and still exists in a system.

Powered air-purifying respirator (PAPR) A type of APR with a small motor that draws air through the filters and provides a positive pressure into the mask or hood.

ppb (parts per billion) Parts per billion is the concentration of a gas or vapor in air, parts (by volume) of the gas or vapor in a billion parts of air. Usually used to express extremely low concentrations of unusually toxic gases or vapors, also the concentration of a particular substance in a liquid or solid.

PPE (personal protective equipment) The correct clothing and respiratory equipment that is needed to perform a job involving hazardous materials and to protect the worker. PPE includes proper boots, gloves, splash protective clothing, gas protective clothing, Tyvek suits, eye protection, hearing protection, air-purifying respirators (see APR), and air-supplying respirators (see SCBA). It is important that all PPE be used properly and when required.

ppm (parts per million) Parts per million is the concentration of a gas or vapor in air, parts (by volume) of the gas or vapor in air, parts (by volume) of the gas or vapor in a million parts of air; also the concentration of a particular substance in a liquid or solid.

Pressure See psi.

Pressure-demand systems Supply air into the mask as the user inhales and monitor for leakage. If the system detects a leak, it will provide additional air and create positive pressure inside the mask.

Prohibited condition Any condition in a Permit-Required Confined Space that is not allowed by the permit during the period when entry is authorized.

Protection factor The number assigned by OSHA to illustrate the level of protection for a type of respiratory protection equipment.

Proton A basic subatomic particle existing in the nucleus of all atoms. A proton has mass and an atomic weight of 1 amu (atomic mass unit). The number of protons is also expressed as the atomic number for a given element. Carbon has an atomic number of 6, which means it has six protons in the nucleus.

Proximity suits Specialized high-temperature clothing worn when personnel need to enter areas of extremely high temperatures.

psi (pounds per square inch) Pounds per square inch (for MSDS purposes) is the pressure a material exerts on the walls of a confining vessel or enclosure. For technical accuracy, pressure must be expressed as psig (pounds per square inch gauge) or psia (pounds per square inch absolute). Absolute pressure is gauge pressure plus sea level atmospheric pressure, or psig plus approximately 14.7 pounds per square inch. Also see mmHg.

Pulmonary Relating to, or associated with, the lungs.

Pulmonary edema Fluid in the lungs.

PVC (polyvinyl chloride) A synthetic thermoplastic polymer that is used widely in society. Due to its resistance to most acids, fats, and oils, it is used for piping or for specific types of PPE.

Pyrophoric materials A chemical that ignites spontaneously in air at a temperature of 130°F (54.4°C) or below.

Radiation Energy that is released as waves or particles.

Radiation Safety Officer This person is in charge of the program that handles radiation-related issues at fixed facilities. This person develops training programs, safe handling procedures, and emergency response information.

Radiation sickness A group of symptoms associated with a radiation exposure. These symptoms run from mild to extreme and include nausea, vomiting, malaise, etc.

Radioactive isotope Also referred to as radioisotope. A radioactive isotope of any element. These isotopes can be naturally occurring or artificially created by bombarding an atom with neutrons. See Isotope.

Radioactive materials Materials that spontaneously emit ionizing radiation from the nucleus of an atom. See Ionizing radiation.

RCRA Resource Conservation and Recovery Act is environmental legislation aimed at controlling the generation, treatment, storage, transportation, and disposal of hazardous wastes. It is administered by EPA.

Reaction A chemical transformation or change. The interaction of two or more substances to form new substances.

Reactive See Unstable.

Reactivity Chemical reaction with the release of energy. Undesirable effects, such as pressure buildup, temperature increase, formation of noxious, toxic, or corrosive by-products, may occur because of the reactivity of a substance to heating, burning, direct contact with other materials, or other conditions in use or in storage.

Reducing agent In a reduction reaction (which always occurs simultaneously with an oxidation reaction), the reducing agent is the chemical or substance which (1) combines with oxygen or (2) loses electrons in the reactions. See Oxidation.

Relative humidity the ratio of the actual water vapor pressure to the saturation of vapor pressure.

Reproductive toxin A substance that affects either male or female reproductive systems and may impair the ability to have children; the term includes chromosomal damage (mutagenesis) and effects on fetuses (teratogenesis).

Respiratory protection Devices that protect the wearer's respiratory system from exposure to airborne contaminants by inhalation. Respiratory protection is used when a worker must work in an area where he/she might be exposed to concentrations in excess of the allowable exposure limit.

Respiratory system The breathing system that includes the lungs and the air passages (trachea, larynx, mouth, and nose) to the air outside the body, plus the associated nervous and circulatory supply.

Retention factor The time it takes for the body to rid itself of a hazardous material.

Retrieval system The equipment (including a retrieval line, chest or full-body harness, wristlets if appropriate, and a lifting device or anchor) used for non-entry rescue of persons from permit spaces.

Risk assessment The process used to determine how the hazards will potentially harm those involved in the operation.

Routes of Entry The means by which materials may gain access to the body; for example, inhalation, ingestion, injection, and absorption.

Saponification The process where a base dissolves fat and turns the fat into water-soluble materials.

SAR See Supplied-air respirator.

SARA Superfund Amendments and Reauthorization Act of 1986. SARA was the actual forerunner to the HAZWOPER regulation in that it mandated the Occupational Safety and Health Administration (OSHA) to develop a worker protection regulation for those who work with hazardous wastes.

Secondary exposure Results from the material contacting another source that is then taken into the body; an example would be eating food with hands contaminated by a substance.

Secondary hazard Ability of gases to create an oxygen-deficient atmosphere.

Segregation of hazards Involves storing materials apart so that they cannot react with one another.

Select carcinogen A chemical that meets one of the following criteria:

1. It is regulated by OSHA as a Carcinogen.
2. It is listed under the category "known to be carcinogens" in the National Toxicology Program's latest Annual Report on Carcinogens.
3. It is listed under Group 1 ("carcinogenic to humans") by the International Agency for Research on Cancer Monographs.
4. It is listed in Group 2A or 2B by IARC or under the category "reasonably anticipated to be carcinogens" by NTP, and causes statistically significant tumor incidence in experimental animals after inhalation exposure of 6–7 hours per day, 5 days per week, for a significant portion of a lifetime to dosages of less than 10 mg/m^3; or after repeated skin application of less than 300 mg/kg of body weight per week; or after oral dosages of less than 50 mg/kg of body weight per day.

Self-contained breathing apparatus (SCBA) A respiratory protection device that consists of a supply or a means of usable air, oxygen, or oxygen-generating material carried by the wearer. If an air-purifying respirator cannot be used due to one of the conditions listed under APR, then a worker must protect himself with an SCBA. Sometimes known as Scott Packs (a brand name), SCBAs and air-line respirators deliver air to the user from a tank (SCBA) or an air line. Although they negate all of the problems of an APR, they still have problems of their own. Air-line systems are bulky and more difficult to work with due to the attached air line. SCBAs are heavy, require more training, are more expensive, and reduce the mobility of the worker. Both, however, have the advantage of having positive pressure in the facemask that greatly reduces the risk of exposure if there is a leak in the mask by pushing the material out instead of sucking it in as with an APR.

Self-contained units Breathing apparatus where the user carries all components required for the system.

Self-rescue An operation in which entrants leave a confined space on their own power when an emergency occurs.

Sensitizer A chemical that causes a substantial proportion of exposed people or animals to develop an allergic reaction in normal tissue after repeated exposure to the chemical.

SETA Setaflash Closed Tester. See Flash point.

Shipping papers A shipping order, Bill of Lading, manifest, waybill, or other shipping document issued by the carrier.

Silicosis A disease of the lungs caused by the inhalation of silica dust.

Size-up A mental evaluation of the problem that takes into account the issues you will face.

Skin absorption Ability of some hazardous chemicals to pass directly through the skin and enter the bloodstream.

Skin toxicity See Dermal toxicity.

SLUD A mnemonic denoting the effects of exposure to certain types of pesticides (mostly Organophosphates and Carbamates). It stands for salivation, lacrimation, urination, and defecation.

Solubility in water A term expressing the percentage of a material (by weight) that will dissolve in water at ambient temperature. Solubility information can be useful in determining spill cleanup methods and re-extinguishing agents and methods for a material.

Solvation The process that occurs when an acid mixes with water. Hydrogen ions are released, which then bond with some of the water molecules to become H_3O.

Solvent A substance, usually a liquid, in which other substances are dissolved. The most common solvent is water.

SOx Oxides of sulfur.

Span of control The number of people that one supervisor can effectively supervise at any one time.

Spanned To have an approximate calibration for a certain number of chemical compounds.

Specific chemical identity The chemical name, Chemical Abstracts Service (CAS) Registry Number, or any precise chemical designation of a substance.

Specific gravity The weight of a material compared to the weight of an equal volume of water is an expression of the density (or heaviness) of a material. Insoluble materials with specific gravity of less than 1.0 will float on water. Insoluble materials with specific gravity greater than 1.0 will sink in water. Most (but not all) flammable liquids have specific gravity less than 1.0 and, if not soluble, will float on water, an important consideration for fire suppression.

Spontaneously combustible A material that ignites as a result of retained heat from processing, will oxidize to generate heat and ignite, or absorbs moisture to generate heat and ignite.

Stability The ability of a material to remain unchanged. For MSDS purposes, a material is stable if it remains in the same form under expected and reasonable conditions or storage or use. Conditions which may cause instability (dangerous change) are stated; for example, temperatures above 150°F; shock from dropping.

Stable atom An atom that is not in the process of radioactive decay or the formation of an ion. Stable atoms have equal numbers of protons and electrons without an unusual imbalance of protons to neutrons.

State Plan State A state or US territory that has its own OSHA programs.

States of matter The three forms in which all materials exist; they include gas, liquid, and solid.

STEL A term denoting one of the occupational exposure limits for workers. It stands for Short Term Exposure Limit and was developed by the ACGIH. The term represents the maximum exposure limit for workers based on a 15 minute exposure. This level would presumably be harmful if it were exceeded for more than a 15 minute exposure.

Stratification The process by which gases and vapors settle into layers within an area based on their weight.

Strong acid An acid with a pH of 2.5 or less.

Strong base A base with a pH of 12.5 or greater.

Subcutaneous Beneath the layers of skin.

Sublimation The ability of a material to pass from a solid state to a gas state without becoming a liquid. Mothballs are examples of materials that sublimate.

Subsidiary label A second DOT label placed onto the package of specific types of hazardous materials to identify a second hazard of the material.

Supplied-air respirator (SAR) Air-line respirator or self-contained breathing apparatus. This type of respirator is different than the air-purifying respirator in that the air that is breathed by the individual does not come from the atmosphere in the area where work is performed. Air that is breathed with a SAR system comes either from an air bottle carried by the individual or from an air-line system that either uses bottles or compressors that supply air from an outside clean-air source.

Support Zone See Zones.

Surface area The amount of area exposed from a given system. For example, the surface area of the skin of the human body is approximately 25 square feet. Surface area also greatly influences vapor production in spilled liquids. The larger the surface area, the greater the vapor production.

Synonym Another name or names by which a material is also known. Methyl alcohol, for example, is also known as methanol or wood alcohol.

Systemic effects Occur in a body system as a result of the material entering the body and getting into the blood stream.

Systemic poison A poison which spreads throughout the body, affecting all or some of the body systems and organs. Its adverse effect is not localized in one spot or area. Carbon monoxide has systemic effects upon exposure.

Systemic toxicity Adverse effects caused by a substance which affects the body in a general rather than local manner.

Tactics Methods or procedures used to deploy various tactical units (resources) to achieve objectives.

Target organ An organ adversely affected by the hazardous materials that enter the body.

Target organ effects The following is a target organ categorization of effects that may occur, including examples of signs and symptoms and chemicals which have been found to cause such effects. These examples are presented to illustrate the range and diversity of effects and hazards found in the workplace and the broad scope employers must consider in this area, but are not intended to be all-inclusive.

Hepatotoxins: Substances that produce liver damage. Signs and symptoms—jaundice, and liver enlargement. Chemicals—carbon tetrachloride and nitrosamines.

Nephrotoxins: Substances that produce kidney damage. Signs and symptoms—edema and proteinuria. Chemicals—halogenated hydrocarbons and uranium.

Neurotoxins: Substances that produce their primary toxic effects on the nervous system. Signs and symptoms—narcosis, behavioral changes, and decrease in motor functions. Chemicals—mercury and carbon disulfide.

Agents that act on blood hematopoietic system: Chemicals that decrease hemoglobin function and deprive the body tissues of oxygen. Signs and symptoms—cyanosis and loss of consciousness. Chemicals—carbon monoxide and cyanides.

Agents which damage the lung: Chemicals which irritate or damage the pulmonary tissue. Signs and symptoms—cough, tightness in chest, shortness of breath. Chemicals—silica and asbestos.

Reproductive toxins: Chemicals which affect the reproductive capabilities, including chromosomal damage (mutations) and effects on fetuses (teratogenesis). Signs and symptoms—birth defects and sterility. Chemicals—lead and DBCP.

Cutaneous hazards: Chemicals which affect the dermal layer of the body. Signs and symptoms—defatting of the skin, rashes, and irritation. Chemicals—ketones and chlorinated compounds.

Eye hazards: Chemicals which affect the eye or visual capacity. Signs and symptoms—conjunctivitis and corneal damage. Chemicals—organic solvents and acids.

Target organ toxin: A toxic substance that attacks a specific organ of the body. For example, exposure to carbon tetrachloride can cause liver damage.

TCC Tag (Tagliabue) Closed Cup. See Flash point.

TCL Toxic concentration low, the lowest concentration of a gas or vapor capable of producing a defined toxic effect in a specified test species over a specified time.

TDL Toxic dose low, lowest administered dose of a material capable of producing a defined toxic effect in a specified test species.

Teratogen A substance or agent that can cause malformations in the fetus of a pregnant female exposed to it.

Testing The process by which the hazards that may confront personnel such as entrants of a confined space are identified and evaluated. Testing includes specifying the tests that are to be performed in the permit space.

Threshold The dividing line between effect and no effect levels of exposure.

Tight fitting Refers to the type of respirator that is strapped to the face of the user and a seal is formed between the mask and the skin.

Title 29 or 29 CFR The section of regulation where Federal OSHA Safety Regulations are found.

TLV (Threshold Limit Value) A set of standards established by the American Conference of Governmental Industrial Hygienists for concentrations of airborne substances in workroom air. They are Time Weighted Averages based on conditions that it is believed workers may be repeatedly exposed to day after day without adverse effects. The TLV values are revised annually and provide the basis for the safety regulations of OSHA.

TLV/C (Threshold Limit Value/Ceiling) Another term that denotes an occupational exposure limit. This term represents the maximum level or amount of a substance that a worker can be exposed to at any given time. It is usually given in either ppm, ppb, or mg/m^3.

TLV/TWA (Threshold Limit Value/Time Weighed Average) The TLV/TWA is a term used to denote one of the occupational exposure limits for types of hazardous materials. The term was developed by the ACGIH, which denotes the average amount of a material that the average worker can be exposed to in the course of a typical 8 hour day, 5 day work week. At or below this level, it is presumed that a worker would suffer no ill effects from exposure to a given substance. It is usually given in either ppm, ppb, or mg/m^3 of air and is for airborne substances.

Tool drop The designated area just inside the Hot Zone and adjacent to the decon corridor where equipment and tools are kept pending later use, disposal, or decontamination.

Toxic material A substance as defined by OSHA as falling within any of the following categories:

1. A substance that has median lethal dose (LD50) of 50 milligrams or more per kilogram of body weight but not more than 500 milligrams per kilogram of body weight when administered orally to albino rats weighing between 200 and 300 grams each.

2. A substance with a median lethal dose (LD50) of more than 200 milligrams or less per kilogram of body weight but not more than 1000 milligrams per kilogram of body weight when administered by continuous contact (dermal) for 24 hours (or less if death occurs within 24 hours) with the bare skin of albino rabbits weighing between 2 and 3 kilograms each.

3. A chemical that has a median lethal concentration (LC50) in air of more than 200 parts per million but not more than 2000 parts per million by volume or vapor, or 2 milligrams per liter but not more than 20 milligrams per liter of mist, fume, or dust, when administered by continuous inhalation for 1 hour (or less if death occurs within 1 hour) to albino rats weighing between 200 and 300 grams each.

Toxicologist A person trained to examine the nature of these adverse effects and to assess the probability of their occurrence.

Toxicology The study of the adverse effects of chemicals and other hazardous substances on living organisms.

TSCA Toxic Substances Control Act is federal environmental legislation (administered by EPA) that regulates the manufacture, handling, and use of materials classified as "toxic substances."

TSDF (Treatment, Storage, and Disposal Facility) Regulated sites where hazardous wastes are taken for final disposal or treatment.

TWA (Time Weighted Average) TWA exposure is the airborne concentration of a material to which a person is exposed, averaged over the total exposure time, generally the total workday (8–12 hours). Also see TLV/TWA.

UEL, or UFL Upper explosive limit, or upper flammable limit, of a vapor or gas; the highest concentration (highest percentage of the substance in air) that will produce a flash of fire when an ignition source (heat, arc, or flame) is present. At higher concentrations, the mixture is too "rich" to burn. Also see LEL.

Unified Command This is the system of command used in the ICS where multiple agencies or jurisdictions will assign a person to share key responsibilities within the ICS, including that of IC.

Uniform Hazardous Waste Manifest A standard shipping document required by the EPA for shipments of hazardous wastes.

Universal precautions Actions that are required to be taken to prevent the potential for infection.

Unstable atom Atom in which the number of protons and neutrons do not equal the number of electrons, and as a result emits radiation.

Unstable reactive A chemical that, in the pure state or as produced or transported, will vigorously polymerize, decompose, condense, or become self-reactive under conditions of shock, pressure, or temperature.

Unstable Tending toward decomposition or other unwanted chemical change during normal handling or storage.

USDA US Department of Agriculture.

Vapor The gaseous form of a solid or liquid substance as it evaporates.

Vapor density The weight of a vapor or gas compared to the weight of an equal volume of air is an expression of the density of the vapor or gas. Materials lighter than air have vapor densities less than 1.0. (Propane, hydrogen sulfide, butane, chlorine, and sulfur dioxide, e.g., have vapor densities greater than 1.0.) All vapors and gases will mix with air, but the lighter materials will tend to rise and dissipate (unless confined). Heavier vapors and gases are likely to concentrate in low places, along or under floors, in sumps, sewers and manholes, in trenches and ditches, where they may create fire or health hazards.

Vapor pressure The pressure exerted by a saturated vapor above its own liquid in a closed container. When quality control tests are performed on products, the test temperature is usually 100°F, and the vapor pressure is expressed as pounds per square inch (psig or psia), but vapor pressures reported on MSDSs are in millimeters of mercury (mmHg) at 68°F (20°C), unless stated otherwise.

Three facts are important to remember:

1. Vapor pressure of a substance at 100°F will always be higher than the vapor pressure of the substance at 68°F (20°C).
2. Vapor pressures reported on MSDSs in mmHg are usually very low pressures; 760 mmHg is equivalent to 14.7 pounds per square inch.
3. The lower the boiling point of a substance, the higher its vapor pressure.

Vascular areas Areas that have a high concentration of blood vessels.

Ventilation See General exhaust, Local exhaust, and Mechanical exhaust.

Viscosity The ability or the resistance of a liquid material to flow. The higher the viscosity, the less that the material will flow. Maple syrup is a substance with a high viscosity. Water is less viscous and would tend to flow easier. Viscosity will assist you in determining if a liquid has a high risk of spreading out and potentially finding its way to a storm drain or sewer before adequate containment or cleanup procedures can be initiated.

Volatility The ability of a material to rapidly change from a liquid state to a gaseous state.

V-roll technique The method to slide a leaking drum into a larger salvage drum laid in a V position near each other.

Warm Zone See Zones.

Warning properties Properties such as odor, taste, or irritation that alert us to the presence of a material.

Waste Site Supervisor A trained and certified individual under the HAZWOPER regulation whose job includes oversight of other certified HAZWOPER waste site cleanup personnel.

Water reactive Means any solid substance (including sludges and pastes) that, by interaction with water, is likely to become spontaneously flammable or to give off flammable or toxic gases in dangerous quantities.

Waybill The shipping paper used by the railroads indicating origin, destination, route, and product. There is a waybill for each car, and this waybill is generally carried by the conductor.

Zeroing Bringing a monitoring device to a baseline or zero readout on gauges or digital readouts.

Zones Although zoning is usually applied to hazardous materials spills and isolation, setting up work zones such as Hot or Exclusionary Zone assists any task involving hazardous materials. By isolating the work area and letting other personnel know of the hazards, an employer can reduce the possibility of other workers being injured or exposed. The following work zones are areas designated for specific operations to occur:

Hot Zone: The area where the potential for hazards exists. This zone is sometimes referred to as the Exclusionary Zone or the Red Zone.

Cold Zone: The area where no contamination or hazards are present. This zone may also be referred to as the Support or Green Zone.

Warm Zone: The area just outside the Hot Zone where decontamination operations take place. The decontamination corridor can be found in the Warm Zone. This zone is sometimes referred to as the Contamination Reduction Zone or the Yellow Zone.

ACRONYMS

ACGIH	American Conference of Governmental Industrial Hygienists
AFFF	Aqueous Film-Forming Foams
ANFO	ammonium nitrate and fuel oil
ANSI	American National Standards Institute
APR	air-purifying respirator
ASPECT	Airborne Spectral Imagery of Environmental Contaminants Technology
BLEVE	boiling liquid expanding vapor explosion
CERCLA	Comprehensive Environmental Response, Compensation and Liability Act
CFR	Code of Federal Regulations
CGI	combustible gas indicator
CHEMTREC	Chemical Transportation Emergency Center
CO	carbon monoxide
CPC	chemical protective clothing
CTA	critical task analysis
DOL	Department of Labor
DOT	Department of Transportation
EMS	Emergency Medical Services
EOC	emergency operations center
EPA	Environmental Protection Agency
ERG	Emergency Response Guidebook
ERPG	Emergency Response Planning Guideline
ESLI	End-of-Service-Life Indicator
FID	flame ionization detector
FRA	First Responder Awareness
FRO	First Responder Operations
GC	gas chromatograph
GHS	Globally Harmonized System for the Classification and Labeling of Chemicals
HASP	Health and Safety Plan
HAZWOPER	Hazardous Waste Operations and Emergency Response
HEPA	Highly Efficient Particulate Air
HIV	Human Immunodeficiency Virus
HMIS	Hazardous Materials Identification System
HSPD	Homeland Security Presidential Directive
IARC	International Agency for Research on Cancer
IC	Incident Commander
ICS	Incident Command System
IDLH	Immediately Dangerous to Life or Health
IP	ionization potential
IPA	isopropyl alcohol
IV	intravenous
JHA	job hazard analysis
JSA	job safety analysis
LEL	Lower Explosive Limit
LEPCs	Local Emergency Planning Committees
LFL	Lower Flammable Limit
LOTO	lockout/tagout
MEKP	methyl ethyl ketone peroxide
MSDS	material safety data sheet
NCI	National Cancer Institute
NCP	National Contingency Plan
NFPA	National Fire Protection Association
NIC	NIMS Integration Center
NIMS	National Incident Management System
NIOSH	National Institute of Occupational Safety and Health
NPL	National Priority List
NRC	National Response Center

NTP	National Toxicology Program
OSHA	Occupational Safety and Health Administration
PAPR	powered air-purifying respirator
PBI	polybenzimidazole
PEL	Permissible Exposure Limit
PID	photoionization detector
PPE	personal protective equipment
PRCS	Permit-Required Confined Space
PSI	Pounds per square inch
RCRA	Resource Conservation and Recovery Act
SAR	supplied-air respirator
SARA	Superfund Amendment and Reauthorization Act
SCBA	self-contained breathing apparatus
SEMS	Standardized Emergency Management System
TB	tuberculosis
TLV	threshold limit value
TNT	trinitrotoluene
TSCA	Toxic Substances Control Act
TSD	treatment, storage, and disposal
TSDF	treatment, storage, and disposal facility
TWA	Time Weighted Average
UEL	Upper Explosive Limit
UFL	Upper Flammable Limit
UV	ultraviolet

INDEX

D

I

R

S